Carnivore Ecology and Conservation

Techniques in Ecology and Conservation Series

Series Editor: William J. Sutherland

Bird Ecology and Conservation: A Handbook of Techniques
William J. Sutherland, Ian Newton, and Rhys E. Green

Conservation Education and Outreach Techniques
Susan K. Jacobson, Mallory D. McDuff, and Martha C. Monroe

Forest Ecology and Conservation: A Handbook of Techniques
Adrian C. Newton

Habitat Management for Conservation: A Handbook of Techniques
Malcolm Ausden

Conservation and Sustainable Use: A Handbook of Techniques
E.J. Milner-Gulland and J. Marcus Rowcliffe

Invasive Species Management: A Handbook of Principles and Techniques
Mick N. Clout and Peter A. Williams

Amphibian Ecology and Conservation: A Handbook of Techniques
C. Kenneth Dodd, Jr.

Insect Conservation: A Handbook of Approaches and Methods
Michael J. Samways, Melodie A. McGeoch, and Tim R. New

Remote Sensing for Ecology and Conservation: A Handbook of Techniques
Ned Horning, Julie A. Robinson, Eleanor J. Sterling, Woody Turner, and Sacha Spector

Marine Mammal Ecology and Conservation: A Handbook of Techniques
Ian L. Boyd, W. Don Bowen, and Sara J. Iverson

Carnivore Ecology and Conservation: A Handbook of Techniques
Luigi Boitani and Roger A. Powell

Carnivore Ecology and Conservation

A Handbook of Techniques

Edited by
Luigi Boitani
and
Roger A. Powell

OXFORD
UNIVERSITY PRESS

Great Clarendon Street, Oxford, OX2 6DP,
United Kingdom

Oxford University Press is a department of the University of Oxford.
It furthers the University's objective of excellence in research, scholarship,
and education by publishing worldwide. Oxford is a registered trade mark of
Oxford University Press in the UK and in certain other countries

First published 2012

Published in the United States of America by Oxford University Press
198 Madison Avenue, New York, NY 10016, United States of America

British Library Cataloguing in Publication Data
Data available

Library of Congress Control Number: 2011938082

ISBN 978–0–19–955853–7

Foreword

Animals that must hunt and kill for at least part of their living are inherently interesting to many people. Perhaps that is because humans evolved to make our living that way as well, and carnivores often compete with us to this very day. Wolves, bears, lions, tigers, leopards, lynx, mink, weasels, and foxes, and a wide variety of their relatives, have long grabbed the human imagination. In any case, carnivores comprise a very significant contingent of the world's wildlife, and many books have been written about them.

This book is distinct from its predecessors primarily through its emphasis on techniques for dealing with carnivores: how to sample them, capture them for study, handle them, monitor them, and even how to help minimize their competition with us. It is a very helpful book that fills an important niche and comes at the right time.

In many parts of the world carnivores are persecuted, while in other parts they are being restored. Thus societies remain interested in carnivores for one reason or another, and science serves society's interest through numerous carnivore studies. The authors of this book's chapters have conducted a significant proportion of those studies for many years, and the editors for even longer.

Both editors are well qualified to produce this book, having studied and worked with carnivores and their conservation for decades. I had the great opportunity of partnering with Luigi Boitani in 1974, early in his career, when we spent a month in Italy's Abruzzo Mountains live-trapping, radio-collaring, and tracking wolves. I had presented my paper "Current Techniques in the Study of Elusive Wilderness Carnivores" at the Eleventh International Congress of Game Biologists in Stockholm in September 1973. It covered my experiences live-trapping and radio-tracking wolves, fishers, martens, and lynxes as well as a literature review of current techniques used to study other carnivores. I like to think of that paper as a germ that helped spawn the present book. Luigi attended the Stockholm meeting, sought to apply my techniques with wolves in Italy, and asked me to join him there to get started. I eagerly agreed. Little did I realize then that 40 years later, Luigi and Roger Powell would devote a whole book to techniques for studying carnivores.

During the same general period when I met Luigi, I also met Roger Powell. Roger had joined my research team as a summer intern on a wolf–deer project in the Superior National Forest of Minnesota, where we had also been radio-tracking lynx, martens, and fishers on the side. The duties clearly agreed with him, for a few years later he began his own carnivore study, this one involving fishers. That study became his dissertation topic, and I became one of his advisors.

That was all long ago, and the field has advanced greatly and blossomed. Now instead of merely locating an animal via telemetry (a feat in itself years ago), one searches the profuse literature, decides on study objectives, carefully plans the study's design, and chooses from any of the many high-tech radio-collars on the commercial market that will best serve the objectives.

However, dealing with the most appropriate technology to study carnivores is only a small part of carnivore investigations now. The data currently obtainable has opened many new carnivore research vistas, and Boitani and Powell and their collaborators have assembled a set of chapters that nicely address that array. An early chapter on carnivore surveys, for example, is basic, for such surveys are of special importance, both spatially and temporally. In some areas and with some species, just obtaining a general idea of numbers and distribution can be very important. Mapping such distributions plays a major role in these studies, and non-invasive sampling is particularly valuable, especially with endangered or rare species and in inaccessible areas. These subjects are well covered in this book.

In some areas of the world and with certain carnivores, detailed counts are required annually. Sometimes with such counts it is valuable to estimate various demographic parameters, and radio-telemetry often facilitates those estimates. To collar carnivores, it is necessary to capture and handle them, allowing considerable amounts of valuable data to be collected at that time. Once a carnivore is radio-collared, data can be obtained about its movements, activity, home range or territory, and dispersal. Often data about the creature's predation and food habits can also be collected, as well as information about its reproductive behavior. Several chapters of this book deal with these subjects.

A subsidiary type of information, not directly related to a collared carnivore's movements, involves cause-specific mortality, including that from intraspecific strife and diseases. Learning all this basic ecological, physiological, and behavioral information then greatly aids in deriving mitigation measures for minimizing depredation on livestock and other conflicts with humans, as well as facilitating methods of restoring carnivores, monitoring the results, and furthering conservation efforts. Addressing those issues further rounds out this fine compendium.

Thus all in all, this book, edited by Luigi Boitani and Roger Powell, will be of great use not only to carnivore researchers, but also to wildlife biologists throughout the world who deal with carnivores, and it should stand as a milestone in the carnivore-ecology and techniques literature for many years to come.

L. David Mech
US Geological Survey and University of Minnesota, USA

Contents

List of contributors

Cheryl S. Asa Saint Louis Zoo, 1 Government Drive, St. Louis, MO 63110, USA. E-mail: Asa@stlzoo.org

Julie Betsch Wildlife Biology Program, Department of Ecosystem and Conservation Sciences, University of Montana, Missoula, MT 59812 USA. E-mail: juliebetsch@gmail.com

Luigi Boitani Department of Biology and Biotechnologies, Sapienza Università di Roma, Viale dell'Università 32, 00185 Roma, Italy. E-mail: luigi.boitani@uniroma1.it

Urs Breitenmoser Institute of Veterinary Virology, Vetsuisse Faculty, University of Berne, Länggass-Strasse 122, CH-3012 Bern, Switzerland. E-mail: u.breitenmoser@kora.ch

Christine Breitenmoser-Würsten KORA, Thunstrasse 31, CH-3074 Muri, Switzerland. E-mail: ch.breitenmoser@kora.ch

Marc R. L. Cattet Canadian Cooperative Wildlife Health Centre, Western College of Veterinary Medicine, University of Saskatchewan, 52 Campus Drive, Saskatoon, Saskatchewan, S7N 5B4, Canada. E-mail: marc.cattet@usask.ca

Paolo Ciucci Department of Biology and Biotechnologies, Sapienza Università di Roma, Viale dell'Università 32, 00185 Roma, Italy. E-mail: paolo.ciucci@uniroma1.it

Deana L. Clifford California Department of Fish and Game, Wildlife Investigations Lab, 1701 Nimbus Road, Rancho Cordova, CA 95670, USA and Wildlife Health Center, University of California, Davis, CA 95616, USA. E-mail: dlclifford@ucdavis.edu

Hilary S. Cooley US Fish and Wildlife Service, Idaho Fish and Wildlife Office, Boise, ID 83709, USA. E-mail: hilarycooley@gmail.com

David Christianson Department of Ecology, Montana State University, 310 Lewis Hall, Bozeman, Montana, USA. E-mail: davealanchris@yahoo.com

Kerry R. Foresman Division of Biological Sciences, The University of Montana, Missoula, MT 59812, USA. E-mail: foresman@mso.umt.edu

Mark R. Fuller US Geological Survey, Forest and Rangeland Ecosystem Science Center, and Boise State University - Raptor Research Center, 970 Lusk St., Boise, Idaho 83706 USA. E-mail: mark_fuller@usgs.gov

Todd K. Fuller Department of Environmental Conservation, University of Massachusetts, 160 Holdsworth Way, Amherst, MA 01003-9285 USA. E-mail: tkfuller@eco.umass.edu

Mourad W. Gabriel Integral Ecology Research Center, 102 Larson Heights Road, McKinleyville, CA 95519 USA and Veterinary Genetics Laboratory, University of California, Davis, 95616 USA. E-mail: mwgabriel@ucdavis.edu

Jean-Michel Gaillard UMR CNRS 5558 "Biometrie et Biologie Evolutive", Bât. G. Mendel, Université Lyon 1, 43 boulevard du 11 novembre 1918, 69622 Villeurbanne Cedex, France. E-mail: jean-michel.gaillard@univ-lyon1.fr

Eric M. Gese US Department of Agriculture, Wildlife Services, National Wildlife Research Center, Department of Wildland Resources, Utah State University, Logan, UT 84322-5230, USA. E-mail: eric.gese@usu.edu

Duncan Halley Norwegian Institute for Nature Research, Tungasletta 2, NO-7485 Trondheim, Norway. E-mail: Duncan.Halley@nina.no

Mark Hebblewhite Wildlife Biology Program, Department of Ecosystem and Conservation Sciences, University of Montana, Missoula, MT, 59812, USA. E-mail: mark.hebblewhite@cfc.umt.edu

K. Ullas Karanth Wildlife Conservation Society, Centre for Wildlife Studies, 26-2, Aga Abbas Ali Road (Apt:403), Bangalore, Karnataka-560042, India. E-mail: ukaranth@wcs.org

Marcella J. Kelly Department of Fish and Wildlife Conservation, 146 Cheatham Hall, Virginia Tech, Blacksburg, VA 24061-0321. USA. E-mail: makelly2@vt.edu

Frederick F. Knowlton US Department of Agriculture, Wildlife Services, National Wildlife Research Center, Department of Wildland Resources, Utah State University, Logan, UT 84322-5230, USA. E-mail: ffknowlton@msn.com

John D.C. Linnell Norwegian Institute for Nature Research, Tungasletta 2, NO-7485 Trondheim, Norway. E-mail: John.Linnell@nina.no

Annette Mertens Institute of Applied Ecology, Via B. Eustachio 10, 00161 Rome, Italy. E-mail: mertens.annette@gmail.com

Bernardo Mesa Department of Fish and Wildlife Conservation, 318 Cheatham Hall, Virginia Tech, Blacksburg, VA 24061-0321. USA. E-mail: bmesa@vt.edu

L. Scott Mills Wildlife Biology Program, Department of Ecosystem and Conservation Sciences, University of Montana; Missoula, MT, 59812 USA. E-mail: LScott.mills@umontana.edu

Michael S. Mitchell US Geological Survey, Montana Cooperative Wildlife Research Unit, 205 Natural Science Building, University of Montana, Missoula, MT 59812, USA. E-mail: mike.mitchell@umontana.edu

Alessio Mortelliti Department of Biology and Biotechnologies, Sapienza Università di Roma, Viale dell'Università 32, 00185 Roma, Italy. E-mail: alessio.mortelliti@uniroma1.it

James D. Nichols Patuxent Wildlife Research Center, 12100 Beech Forest Dr., Laurel, MD 20708-4017, USA. E-mail: jnichols@usgs.gov

Erlend B. Nilsen Norwegian Institute for Nature Research, Tungasletta 2, NO-7485 Trondheim, Norway. E-mail: Erlend.Nilsen@nina.no

John Odden Norwegian Institute for Nature Research, Tungasletta 2, NO-7485 Trondheim, Norway. E-mail: john.odden@nina.no

Morten Odden Faculty of Forestry and Wildlife Management, Hedmark University College, Evenstad, NO-2480 Koppang, Norway. E-mail: morten.odden@hihm.no

Manuela Panzacchi Norwegian Institute for Nature Research, Tungasletta 2, NO-7485 Trondheim, Norway. E-mail: manuela.panzacchi@nina.no

Kenneth H. Pollock Centre for Fish and Fisheries Research, Murdoch University, Murdoch, WA, Australia. E-mail: K.Pollock@murdoch.edu.au

Roger A. Powell, Department of Biology, North Carolina State University, Raleigh, NC 27695-7617, USA. E-mail: newf@ncsu.edu

Gilbert Proulx Alpha Wildlife Research & Management Ltd., 229 Lilac Terrace, Sherwood Park, Alberta, T8H 1W3 Canada. E-mail: gproulx@alphawildlife.ca

Carlo Rondinini Department of Biology and Biotechnologies, Sapienza Università di Roma, Viale dell'Università 32, 00185 Roma, Italy.
E-mail: carlo.rondinini@uniroma1.it

Michael K. Stoskopf Department of Companion Animal and Special Species Medicine, College of Veterinary Medicine, North Carolina State University, 4700 Hillsborough St., Raleigh, North Carolina, 27606, USA. E-mail: michael_stoskopf@ncsu.edu

Carole Toïgo ONCFS, 5 allée de Bethléem, ZI Mayencin, 38610 Gières, France.
E-mail: c.toigo@oncfs.gouv.fr

Greta M. Wengert Integral Ecology Research Center, 102 Larson Heights Road, McKinleyville, CA 95519 USA and Veterinary Genetics Laboratory, University of California, Davis, CA 95616 USA. E-mail: gmwengert@ucdavis.edu

Claudia Wultsch Department of Fish and Wildlife Conservation, 318 Cheatham Hall, Virginia Tech, Blacksburg, VA 24061-0321. USA. E-mail: wultschc@vt.edu

Barbara Zimmermann Faculty of Forestry and Wildlife Management, Hedmark University College, Evenstad, NO-2480 Koppang, Norway.
E-mail: barbara.zimmermann@hihm.no

1

Introduction: research and conservation of carnivores

Luigi Boitani and Roger A. Powell

This is a book about carnivores but, more so, it is a book about techniques for studying carnivores. The emphasis is on the diverse ways that researchers and managers study carnivores, from documenting presence and absence and counting numbers; to studying individuals and populations remotely or interactively; to understanding movements, habitat, physiology, and disease; to helping populations recover or limiting damage to livestock. The ways one can study carnivores are as diverse as the carnivores themselves.

The diversity of carnivores contributes to the diversity of study techniques. The Carnivora includes some 230+ species, the exact number depending on the species' concept used and systematic techniques used to generate phylogenies. Carnivores live throughout all continents except Antarctica, from sea level to > 5000 m (snow leopards, *Uncia uncia*), and in all habitats, from deserts to rain forests, from tropics to the arctic, and including densely populated urban areas. Although we often associate carnivores with wilderness and remote areas, many have adapted to human-made habitats. Some stone martens (*Martes foina*) live in dense urban centers of some European cities, often resting in attics of houses and under the hoods of cars. Raccoons (*Procyon lotor*) have colonized many North American cities, also sometimes resting in attics but more often resting and denning in hollow trees or, like striped skunks (*Mephitis mephitis*), under houses. In fact, except for those that are strict specialists (e.g. giant pandas, *Ailuropoda melanoleuca*; black-footed ferrets, *Mustela nigripes*), carnivores can be remarkably flexible in their use of human-made habitats, depending on the level of persecution. Black bears (*Ursus americanus*) make winter dens under people's houses, wolves (*Canis lupus*) and black and brown bears (*Ursus arctos*) scavenge in dumps, and tigers (*Panthera tigris*) and polar bears (*Ursus maritimus*) sometimes hunt people in towns and villages. The research used as examples in this book spans the diversity of carnivore habitats from areas with no permanent human inhabitants, through areas with various levels of sparse human occupation, to areas with dense human populations and highly altered habitats.

The diversity of carnivores, however, does not end with habitats. Carnivores span over four orders of magnitude in weight, from female least weasels (*Mustela rixosa* formerly *Mustela nivalis*) weighing less than 50 g to male polar bears reaching 600 kg. They vary similarly in densities, from urban racoons with densities exceeding 100/km^2 to wolverines (*Gulo gulo*) and far northern bears (polar, brown, and black) and wolves with home ranges of 100s to >1000s of km^2. And, while black-footed ferrets are recovering from a population low of 10 individuals and other carnivores are similarly endangered, small Asian mongooses (*Herpestes javanicus*) are invasive on the West Indies and Hawaiian Islands.

Although "carnivore" means meat eater, members of the Carnivora have diets that span the entire spectrum. Some are strict carnivores (many felids and mustelids), many scavenge, have some level of omnivory (canids to most ursids and procyonids), or are insectivorous (some mongooses, canids, and aardwolves, *Proteles cristatus*), and giant pandas are strictly vegetarian. For predatory carnivores, hunting strategies include ambush, stalking, chasing, and hunting in groups. Indeed, many carnivores are highly social and have highly complex social behaviors and capabilities, to the extent that humans domesticated wolves to become dogs (*Canis familiaris*), with which they have since coevolved.

The consequence of this "diverse diversity" of carnivores, and the diversity of human relationships with carnivores, is that developing a book on techniques for studying carnivores has been challenging. The diversity within the group is truly astounding and makes generalizations nigh onto impossible, except, perhaps, for one: in all their diverse personifications, carnivores are iconic. They all have charisma, from tigers, lions (*Panthera leo*), brown and polar bears, to the weasels, whose personalities outsize their bodies. We, the editors of this book, admit to being awestruck by carnivores and deeply moved by them. Where carnivores are endangered, they are flagship animals that capture human attention and, thereby, provide protection for other species and sometimes whole ecological communities. Within this book, our authors present diverse techniques for studying carnivores and present diverse perspectives on the objectives and goals possible for study.

In addition, because of the diversity of carnivores, a book of study and research techniques has the potential to be applicable to many other mammals as well. Carnivores are elusive and require diverse, and often sophisticated, techniques to get information on their ecology and behavior; these techniques can be used with animals from other groups. Nonetheless, no technique, no matter how advanced or sophisticated, is of much value unless a researcher or a manager understands the animals being studied. A researcher's goal should be to predict how carnivores think as they live in their individual environments. Thinking like an animal is the best technique and is the overarching mind frame needed to make the most of any technique. As Mike Mitchell has called it, we should seek to "crawl inside their furbrains" to understand how they work.

Beyond good, tried-and-true techniques and the latest technological advances, this book also emphasizes the conceptual framework needed to plan, to design, and to implement research in ways that optimize the use of good techniques. Many authors in the book refer to the rigorous application of the scientific method, noting that research starts with (1) solid hypotheses based on the biology of the animals, (2) explicit and acceptable assumptions, (3) sound experimental design, and (4) rigorous application of appropriate field and analytical techniques. Rapid advances in technical and analytical capabilities cannot substitute for sound research planning. In fact, advanced capabilities *require* the strongest of scientific frameworks to avoid having the techniques drive the research, which inevitably leads to unproductive research.

Similarly, today's conservation needs call for evidence-based action: explicit evidence showing the need for conservation action and explicit evidence showing the effectiveness of specific techniques.

The study of carnivores has a long history. The early monographs by Murie (1940, 1944), Errington (1943), and Mech (1966) on coyotes (*Canis latrtans*), minks (*Mustela vison*), and wolves, the work of the Craigheads (1956) on predator communities, and then the monographs by Schaller (1967, 1972) and Kruuk (1972b) on tigers, lions, and spotted hyaenas (*Crocuta crocuta*), established a solid foundation for research on carnivores. These early researchers obtained their hard-earned data from long, arduous hours in the field using little of what we would call "modern technology." Their research endures because their data were, and still are, solid. Starting with the advent of telemetry, with research on carnivores in the 1960s (Craighead and Craighead 1971), "modern technology" began making good data easier to collect and opened a diversity of possibilities for research. Indeed, Errington and Murie would have had trouble conceiving of the potential information available using DNA collected from carnivores, often remotely, because their early research was done before DNA was known to carry genetic codes. Since those early studies, the literature on the ecology, behavior, and conservation of carnivores has expanded exponentially, making this handbook of research techniques possible.

The rich methodology now available for the study of carnivores opens many opportunities and challenges not possible only a few years ago. Key challenges in ecology and behavior of carnivores include the following.

1. The basic natural histories are unknown for many species, especially in developing countries. New (and future) techniques in remote sampling offer possibilities for obtaining basic information on the most elusive carnivores in remote locations.
2. We need more studies of known species in new ecological contexts. Ecological, behavioral, and evolutionary theory, and the responses to carnivores in well-studied contexts, can provide solid hypotheses for how, and most importantly why, carnivores should respond in new situations.
3. Carnivore guilds, resources partitioning, niches, competition, intra-guild predation and mutualisms (yes, mutualisms) are only narrowly understood, if at all. Is intra-guild predation a special type of interference competition, as predicted by behavioral theory, or simply an extension of interference not available to non-predatory competitors who lack weapons? Are so few cases of mutualism documented because few exist, because each case must result from learning by individual animals, or because biologists in Western society are programmed to see competition but not mutualisms?
4. The community-wide effects of predation are just beginning to be understood and need further study. The indirect effects of wolves on riparian vegetation and hydrology in Yellowstone National Park sparked well-deserved excitement among biologists and conservationists. Surely such effects are widespread among carnivores.
5. Why and how do animals use habitats, what do habitats provide, and what are their biological functions? Are habitats important to carnivores because they provide direct benefits (den or rest sites, for example), because they affect prey abundance, or because they affect the abilities of carnivores to catch prey? We cannot answer these simple questions for most carnivores. Indeed, we do not understand habitat from the animals' points of view for any but a couple of carnivore species.

Key challenges for conservation are, unfortunately, still many and include the following.

1. Most carnivore species are endangered and many will soon start vanishing. Lions are predicted to go extinct in the wild by 2030. With valiant efforts, black-footed ferrets have been pulled back from extinction (indeed, they were considered extinct in the 1970s) but all wild populations are threatened by presently unsolvable problems of endemic diseases. Some carnivore populations are so poorly known that conservation status cannot be defined (e.g. *Mustela felipei, M. africana, M. nudipes, M. kathiah* just to mention a few within a single, narrow taxon).

2. Ecological functions of carnivores within communities are poorly understood, putting the ecological integrity of communities in danger as carnivore population become low.
3. Coexistence of carnivores with humans, especially large carnivores, depends on developing strategies to deal with livestock depredation, a complex issue that involves the integration of biological as well as social and economic aspects.
4. Similarly, some urban carnivores compete with humans, often for space. Stone martens and raccoons consistently damage the buildings they inhabit, and coyotes expand their hunting ranges into residential areas, often killing pets.
5. Invasive carnivores cause conservation problems for other species, either via predation (e.g. stoat, *Mustela erminea*, predation on native birds in New Zealand, including the iconic brown kiwis, *Apteryx mantelli*); via competition (e.g. American minks outcompeting European minks, *Mustela lutreola*); or via hybridization (e.g. coyotes hybridizing with red wolves, *Canis rufus*, in the only free-living red wolf population, in coastal North Carolina, USA).
6. Feral and free-ranging cats (*Felis catus*) constitute serious invasive-predator problems. Domestic cats prey on endangered species and have caused many species to become endangered (cats have caused more endangerment than any other species except humans); compete with other predators, some endangered; and hybridize with European wildcats (*Felis sylvestris*). In addition, cats are consistently provisioned by humans, intentionally or unintentionally, exacerbating all the problems. To a lesser extent, domestic dogs also cause similar conservation problems.

This book presents the techniques now available to tackle these ecological and conservation challenges. Forty-one authors, chosen for their backgrounds and experience pertinent to the specific needs of each chapter, have contributed to the 17 chapters, presenting information gained from hundreds of cumulative years of research on, and management of, carnivores. We hope that the book becomes a standard resource for researchers, managers, and conservationists who study and manage carnivores. It is also appropriate for graduate students and for graduate reading courses.

The book has been designed to be read from front to back. It is divided informally into four sections: some introductory concepts (Chapters 2 and 3), data collection (Chapters 4–7), data analysis and design (Chapters 8–13), and human–carnivore interactions for conservation and mitigation (Chapter 14–17). Each section builds on the sections that come before it. Nonetheless, the chapters have been written so that readers can choose to read individual chapters. Each chapter cites the other chapters that introduce critical, background concepts, showing readers where to turn to "fill in" information on narrow topics.

Chapters 2 and 3 cover survey design and mapping. These chapters highlight how research and conservation goals and objectives dictate study design, which then dictates the techniques to be used. This point is repeated elsewhere in the book: goals dictate design, which dictates technique, not the other way around. Paolo Ciucci and Alessio Mortelliti have great experience in planning and implementing field experiments under rigorous sampling design, and Carlo Rondinini is a leading author in the field of species distribution models.

Chapter 4 introduces the many noninvasive study methods available to researchers and managers today. In her research on cheetahs (*Acinonyx jubatus*), Marcella Kelly was one of the first people to use computer programming to identify individual mammals from coat patterns; her coauthors complement her experience with the diversity of noninvasive techniques. Often, however, research requires having animals in hand. Thus, Chapter 5 provides thorough information on how to humanely live-trap and kill-trap and handle carnivores. The chapter includes extensive tables on traps, sets, drugs, and handling techniques. Gilbert Proulx has extensive experience with testing traps for humane capture, and he and his coauthors have handled diverse carnivores. Once a carnivore is in hand, one should collect as much data as possible (Chapter 6); doing so may prevent the need to capture other animals (or the same animals) in the future. Kerry Foresman has handled a wide diversity of carnivores and teaches a course on making the most of having an animal in hand. Mark and Todd Fuller, in Chapter 7, introduce the diversity of telemetry equipment now available for use with carnivores and highlight how best to use different types of equipment. Together, they have decades of experience working with the most advanced telemetry techniques.

In Chapter 8, Ken Pollock, Ullas Karanth, and Jim Nichols present state-of-the art approaches to using diverse data to understand demographics of populations. Ken and his coworkers have been driving forces in research on statistical approaches to population data. Chapter 9 starts with another discussion of research design, emphasizing that researchers and managers must understand the concepts pertinent to their goals before they can design research. In this chapter, I (R.A. Powell) define terms and concepts related to movements and home ranges and discuss how different ways of analyzing data are appropriate for different goals. In Chapter 10, Mike Mitchell and Mark Hebblewhite emphasize the importance of understanding what habitat is for carnivores, and that it must be defined functionally, not descriptively, from the animals' perspective. These authors have shown how research on habitat must answer "why" questions. Erlend Nilsen and coworkers, in

Chapter11, present techniques for studying and analyzing carnivores' diets, techniques that go far beyond the dogmatic standard of scat analysis. Nilsen and his coauthors have decades of joint experience in studying predator–prey relationships and quantifying the impact of predation on the dynamics of prey populations. Cheryl Asa (Chapter 12), with tremendous experience studying reproductive endocrinology of carnivores, provides an overview of many techniques available for physiological studies of carnivores. Greta Wengert and her coworkers (Chapter 13) introduce their cutting-edge approaches to investigating mortality and diseases of carnivores, and explain clearly how researchers and managers without background in pathology can still collect samples and data allowing the most up-to-date analyses.

Chapter 14, by John Linnell and coauthors, tackles the difficult concepts of how to deal with carnivores and people, where carnivores kill livestock. Managing large carnivore populations in human-dominated landscapes is not an easy task and Linnell and his coauthors have built their extensive expertise on a diversity of situations on all continents. Michael Stoskopf (Chapter 15), who has chaired the red wolf recovery implementation team, a science committee that guides the research and management of the reintroduced population of red wolves, covers the topic of reintroducing and otherwise restoring extirpated and endangered populations of carnivores. He emphasizes the importance of not only populations and demographics, but also of health and disease. In Chapter 16, Eric Gese and his coauthors present approaches to monitoring, again emphasizing that objectives and goals must precede study design, which then dictates techniques. For years, they have been using techniques ranging from the traditional to the most advanced in monitoring carnivore populations in diverse ecological contexts. Finally, the Breitenmoser team (Chapter 17) provide a tremendous overview of the techniques to assess conservation status and the most appropriate approaches for planning conservation measures. Urs and Christine Breitenmoser are responsible for research and monitoring of the large carnivores in Switzerland and are co-chairs of the IUCN/SSC Cat Specialist Groups. They have first-hand experience in the implications of conservation of carnivores in areas with high human densities. Their chapter is a stimulating and unconventional view of what conservation means when a compromise with human activities is necessary.

We hope you enjoy the book, that you read it and learn and become motivated, and that you turn to it as a resource for years to come.

2

Designing carnivore surveys

Luigi Boitani, Paolo Ciucci, and Alessio Mortelliti

In ecology, a study aiming to collect data over a relatively broad spatial scale and through some sampling scheme is often called a survey. It is generally aimed at defining the status of an ecological element (species, habitat, vegetation type, water quality, etc.) by measuring the values of one or more attributes (distribution, abundance, richness, allelic frequencies, species' composition, etc.) of that element. The title of this chapter, for example, is actually imprecise, as it indicates the ecological element, the carnivores, but does not indicate the attributes to be surveyed. Without appropriate a priori qualification, a survey does not imply any predefined precision, resolution, scale, and reliability of the data to be obtained and it is open to many misuses. Without specification of the variables to be "surveyed" and why, a survey risks being used to look for a posteriori patterns but lacking key elements and attributes.

Attributes can be assessed using a huge variety of quantitative or qualitative field techniques. With the exception of the simple case of surveying a species' presence in an area, the goal of a survey is generally to obtain an estimate or an index of the attribute of interest, not its absolute value, and is often used to indicate a first reconnaissance of a poorly known ecological factor. For example, a survey of a species' abundance will result in an estimate or an index of population size, whereas a census will yield an absolute number of all individuals (Thompson *et al.* 1998; Bibby 2004).

In wildlife ecology, surveys are most frequently intended to define the distributions and abundances of species and their habitats, the primary features that define the status of animal populations. Therefore, a survey is a first descriptive step in the series of increasingly complicated study designs aimed at more advanced ecological questions. Surveys are rarely designed to explain *why* and *how* ecological processes occur. Nonetheless, as surveys are often prompted by specific conservation and management needs, such as establishing protected area or mitigating human–wildlife conflicts, their designs are deeply influenced by their intended purposes.

Whereas a survey is the assessment of the status of an attribute at one time and area, the repetition of the same survey at the same location at more than one time allows inference about change. This repetition is generally called monitoring, but we prefer the conceptual distinction made by Greenwood and Robinson (2006) between surveillance as "....repeatedly surveying something to measure how it changes," and monitoring, which "...entails setting targets" such as repeatedly measuring something against a desired value that is the objective of management. While the conceptual differences between surveillance and monitoring are obvious, the words have been confused in the scientific literature (Yoccoz *et al.* 2001). Surveys, surveillance-monitoring and targeted-monitoring, are all based on sampling a population of interest and, to allow meaningful inference, they all require statistically robust designs and careful planning. Neither surveillance- nor targeted-monitoring is the mere repetition of single surveys, they require a higher level of design to detect specified levels of change (see Chapter 16; Elzinga *et al.* 2001; McComb *et al.* 2010). Some surveys, such as to confirm the presence of a species in a certain area, can collect data opportunistically, but the great majority of surveys (and all monitoring) require data to be collected systematically in space and time, through precise sampling protocols. Haphazard collections do not allow inference and are, often, a waste of precious resources.

This chapter describes the conceptual framework needed to design and to plan a survey of carnivore distribution and occupancy, although much of the same framework applies to surveys of species' abundance and other population states. We use the definition of survey offered by Long and Zielinski (2008: 8) "the attempt to detect a species at one or more sites within the study area, where 'attempt' involves one or more field sampling occasions, through proper methods, procedures and sampling design."

This chapter is not a cookbook of field protocols for surveys, because the diversity of carnivore populations and their habitats precludes generalizations; instead, it focuses on the key planning steps that are crucial for obtaining meaningful data. While field protocols for a variety of species and research objectives have been published elsewhere (e.g. Braun 2005; Long and Zielinski 2008) the discussion of the conceptual framework for a carnivore survey has seldom been presented. We assume that the reader has the knowledge of the basic terms and concepts of elementary statistics, as we use them to discuss the framework within which the protocols for surveying carnivores in different ecological contexts and for all research and management objectives can be developed. Chapters 4 and 5 discuss field techniques to find evidence of carnivores' presence and Chapters 3 and 8 discuss uses of survey data to map species' ranges and to estimate population sizes.

2.1 Challenges of surveying carnivores

For a biologist planning a survey of carnivore distribution and occupancy, the downside is that carnivores, due to their natural histories, introduce extraordinary sampling and logistical challenges for conducting successful and reliable surveys. Carnivores live at low densities, are elusive, often nocturnal, highly mobile, and difficult to observe or catch; individuals have relatively large home-ranges, often of different sizes for males and females, and populations occur over large and often remote areas; territorial species may spread over vast extents, often with clumped distributions in disjoint ranges (e.g. Koen *et al.* 2008); finally, many species leave signs and tracks that are ambiguous and not easily identified (Heinemeyer *et al.* 2008).

Due to carnivores' expectedly low detection rates, substantial survey efforts and especially efficient (often costly) field techniques are required to achieve adequate precision at a proper spatial or temporal scale, whatever the objectives of the survey (e.g. distribution or abundance). In addition, due to large individual home-ranges, especially in territorial species, researchers need to conduct surveys over large enough areas to produce biologically and statistically meaningful results. Often, this requirement adds to what would already be unrealistically high costs and logistical complexity. On the other side, the smaller the geographic extent of the survey or the size of the sampling units, the more likely model assumptions (e.g. closure) will be violated.

In short, carnivores often require survey conditions where "a lot of zeros occur in the data," ultimately affecting the reliability (i.e. bias and precision) of the estimators (McDonald 2004). This fact is why practitioners of carnivore surveys must address the challenges right from the beginning of planning a survey, striving to find the most efficient combination of sampling schemes and effective field techniques. Although complex to define, proper survey design for carnivores should strive to be "cost and time effective and unbiased across the landscape" (Koen *et al.* 2008: 24). Compared to traditionally adopted approaches, which involved live-captures and canonical statistical frameworks (e.g. Otis *et al.* 1978; Seber 1982), the emerging, efficient, noninvasive techniques for studying carnivores, the new robust yet flexible analytical and simulation procedures, and efficient sampling schemes are the tools researchers have at hand today (see Chapters 4 and 8).

2.2 Planning a survey

To plan a successful survey, researchers should carefully determine: (1) the final goals and specific objectives of the survey; (2) the type of data needed; (3) the survey procedures (i.e. sampling strategies and survey methods) expected to provide reliable inferences most efficiently (Yoccoz *et al.* 2001; MacKenzie and Royle 2005). In addition, central to any survey design, is recognizing that the probability of detection of a species is <1 at every survey site, no matter what the survey effort or method adopted. Successful surveys require more rigorous standards than simple recordings of natural track and sign, based on poor and inconsistent survey designs. Researchers must contemplate a proper combination of adequate sampling and efficient field methods to accommodate imperfect detection. Sophisticated noninvasive survey methods, coupled with recent modeling techniques (Long *et al.* 2008a), allow researchers to conduct carnivore surveys over large areas and at multiple scales, obtaining reliable inferences on population states well beyond simple presence/absence data (e.g. Koen *et al.* 2008). Consequently, conducting a survey at the species' or population level is not simple. The planning process is intimidating, requiring the assistance of a statistician and a data analyst, logistics are daunting, and costs can be unfeasibly high.

Inferential survey design depends on its goals and objectives, and it should be efficient. Its key elements include: (1) sampling details (study area, sample unit characteristics, selection criteria); (2) survey protocol (detection method, sampling season, survey duration), and (3) statistical considerations (i.e. precision of estimates). For carnivores, no perfect survey design exists. The optimal survey protocol (the protocol that includes the best compromises to deal with constraints) for a given species and site can be inadequate for the same species elsewhere. Researchers need to assess the adequacy of a survey using variance-based criteria for the population state of interest (abundance, occupancy). Given the complexity of factors interacting at a local scale, adequate survey design must address the specific location and the survey's objectives, the biology and behavior of the species, its distribution and abundance, the extent and characteristics of the geographic region, and the resources and time available. When planning a survey, researchers must consider all these factors carefully and evaluate how they dictate the analytical framework. However complex survey design becomes, researchers must always keep the survey objectives clearly in mind (Long *et al.* 2008a; Royle *et al.* 2008). Objectives come first, they will drive the survey's design and field work.

2.2.1 Fundamentals of survey design: establishing goals and objectives

Carnivore surveys can be designed to meet many different objectives. Most often, surveys estimate occupancy, distribution, or relative abundance (Koen *et al.* 2008; Long and Zielinski 2008). Sometimes, however, researchers wish to make inferences to detailed demographic or ecological objectives. Researchers might want compare attributes of carnivore populations in time and space, assess the effects of development projects on carnivore populations, or evaluate the details of how carnivores respond to management interventions. If researchers repeat well-designed surveys through time (e.g. Pollock 1982; MacKenzie *et al.* 2003), they can assess population characteristics and processes (demographic and genetic structure, natality, survival, and recruitment) and relate them to specific conservation and management goals. By interpreting longitudinal survey data properly following management interventions, biologists can evaluate progress toward a stated conservation or management goal (see Chapter 8).

Different objectives require different sampling designs, different types of data, and different resources. A researcher's ability to make inferences, and how accurate and reliable parameter estimates will be, depends on the target species' behavior, its density and distribution, and the logistical constraints of the survey, and more (McDonald 2004). Carnivore surveys have inherent difficulties of ensuring adequate sampling effort and accounting for imperfect detection properly. Therefore, a researcher must assess carefully, at an early stage of survey planning, both the technical and the statistical feasibilities of meeting chosen objectives, thereby avoiding an inconclusive survey. A researcher must choose realistic yet functional objectives, given the pertinent conservation or management issues. For example, Sargeant *et al.* (2005), in a swift fox (*Vulpes velox*) survey, traded estimates of abundance for larger scale, occupancy-based measures of population status, having realized that estimating population abundance would have required prohibitively high costs and intensive sampling effort.

2.2.2 Fundamentals of survey design: carnivore survey data

Survey objectives targeting a population, a species, or a habitat require different data. For populations, the pertinent data can vary from simple presence/absence, to counts of natural or elicited track and sign, to repeated identification of individuals, or a mixture. Whenever feasible, researchers should target detection of individual animals (i.e. develop capture histories), and apply well-known and operationally efficient models to infer demographic states (Royle *et al.* 2008). A multitude of noninvasive survey techniques can sample individual carnivores over large areas (see Chapter 4; Long *et al.* 2008a). If logistical and financial constraints preclude collecting such data at the required intensity and spatial scale for individual detection, the researcher must determine whether other data can be used to meet the survey objectives. Researchers can use count data, for example, to make inferences about population occupancy state and dynamics, provided the survey can be designed to accommodate imperfect detection. If, however, survey design cannot accommodate imperfect detection in presence/absence or count data, then researchers often target surveys to make inference on relative abundance, which ignores detection bias or assumes it is constant across space and time. Despite the common, historic use of such in surveys, measures of relative abundance are controversial (Thompson *et al.* 1998; Anderson 2001) and found to be unreliable for carnivores (Royle *et al.* 2008).

2.2.3 Fundamentals of survey design: sampling design, methods, and protocols

To meet survey objectives that are feasible, given pertinent conditions and constraints (i.e. target population, study area, logistical and financial constraints), a researcher should try to find the most efficient combination of sampling design and survey methods (the optimal design given the constraints). Sampling design includes where, how, and how much, and how often to sample. Sampling methods must be chosen to detect representative individuals in the target population and include what specifically to measure, specifically where, and specifically how.

To develop sampling design, a researcher must first delineate the boundaries of the survey population and then decide how to divide the space into meaningful sampling units (i.e. the individual units where counts or measurements are actually recorded). Choosing sampling units includes choosing where they will be located within the study area, and determining a representative sample.

Next, the researcher must carefully develop the *survey protocol*, detailing which field techniques, and under which conditions, they should be used to detect individuals within sampling units (i.e. the actual sampling) when, how often, and for how long. Answering these "which" and "how" questions (Yoccoz *et al.* 2001) requires critical, realistic, a priori assessment of available resources (time, funds, trained personnel, equipment, etc.). If financial or logistical constraints preclude using the survey protocol, the researcher must seek more feasible options by re-examining the specifics of the survey protocol, then, if necessary, the specifics of the sampling design, then survey design, and finally up to survey objectives.

Since detectability of carnivores is low, researchers must plan every level of their surveys to avoid producing unreliable results. Efficient but effective sampling design, possibly encompassing large study areas, is required as are efficient, effective, and feasible sampling methods. Sampling methods, in particular, must be chosen to ensure reasonably high detection rates, which will likely translate into detection probabilities and sample sizes adequate for analyses and modeling procedures required to produce results that meet the objectives. Researchers must choose carefully among new noninvasive field techniques (see Chapter 4; Karanth *et al.* 2004b; Long *et al.* 2008a) and among the diverse sampling strategies that have been designed specifically for rare and elusive species (e.g. Manly 2004; McDonald 2004; Smith *et al.* 2004).

For reasons obvious to statisticians, but often less so to field biologists, one must sample carnivores according to statistically sound sampling schemes. Knowing the number, spatial distribution, and independence of sampling units needed to achieve adequate sample sizes and the necessary level of precision is critical if a researcher is to be able to draw inferences from the chosen statistical analyses and modeling. Dealing with low-density carnivores and their elusive behavior often leads researchers to adopt convenient "sampling" designs and methods (Anderson 2001). Such strategies, usually regarded as haphazard, incidental, or opportunistic, take many different forms, which are widely documented in the carnivore literature: interviews with local residents, verified reports of the species' presence, incidentally retrieved carcasses, or, more frequently, data collected according to non-probabilistic sampling designs. Such "data" may appear to provide evidence of a species' presence, but their numerous potential sources of bias preclude their use in inferential surveys (Aubry and Jagger 2006; McKelvey *et al.* 2008). First, they provide presence-only information, at best, without estimates of error, and they offer no insight regarding the presence of the species elsewhere (i.e. where no sampling was conducted). Second, they have an unmeasured but potentially large geographical bias towards areas inhabited by people. Despite their severe limitations, incidental or opportunistic data sometimes provide insight about a species' past or recent distribution in remote areas or insight that can be incorporated into inferential survey (Koen *et al.* 2008).

2.2.4 Fundamentals of survey design: statistically formalizing survey objectives

Sampling design must meet the assumptions of the analytical framework for a researcher to be able to reach any inferences or conclusions about a target population (Long and Zielinski 2008). Consequently, the conceptual framework of the survey design, the sampling design, and the sampling methods all must be related explicitly to a specific analytical method. This requires "a rendering in statistical terms of the why, what and how questions" (Royle *et al.* 2008: 294). One must formalize a priori the relationship (the dependency) between the sample data and the population state (e.g. abundance, occurrence), so that sampling methods produce data that meet the assumptions of the chosen statistical analyses (Royle *et al.* 2008). In addition to biological, technical, and logistical considerations, the statistical framework introduces essential elements that a researcher must include when designing a carnivore survey. Ambiguous or lack of a priori attention to the inferential, analytical framework can make a survey useless.

2.3 Dealing with false absence

A false absence occurs when members of a species are considered absent from a site when some are actually present. False absences are a plague of carnivore surveys; they cause bias in parameter estimates (Gu and Swihart 2004) and increase the risk of spurious results, inaccurate interpretations of results, and wrong conclusions. See Chapter 8 for additional material pertinent to this section.

False absences can occur if the probability of detecting members of a particular species is <1 (MacKenzie *et al.* 2002). This issue can not be overlooked for carnivores because so many are elusive. Even if the target carnivores are abundant at a site, the probability of detecting one (the probability that an animal will leave a track at a scent-station or will trigger a camera) may be low. If the objective of a survey is to obtain an unbiased estimate of the probability of presence in a given area, one must partition the variation in the data into factors affecting detectability (e.g. soil type, weather, trap efficacy, site-specific forest structure) and factors affecting occupancy (e.g. habitat type, prey abundance; MacKenzie *et al.* 2006).

To address false absences, MacKenzie *et al.* (2002) incorporated detection history data into a maximum likelihood estimation model for the estimate of separate occupancy and detection probabilities. Through a logit-link function, researchers can also model detection and occupancy probabilities as functions of site or sampling covariates in an analogous way to an ordinary logistic regression.

In the framework proposed by MacKenzie *et al.* (2002), a *site* is a sampling unit, which could be a camera trap, a scent station (single or cluster), a transect (Linkie *et al.* 2006; Smith *et al.* 2007) or even a 2000-km^2 grid in an expert opinion survey (Karanth *et al.* 2009). In carnivore surveys a "visit" to a site is equivalent to a single sampling occasion and it occurs within a sampling period, such as a period of activation of a camera trap (e.g. a single night, a week, or even a month). A sequence of sampling periods generates a detection history at a site, which can be written as a sequence of 1s and 0s corresponding to the detection or non-detection of the target carnivores. If an animal is detected in the first sampling but not in the subsequent two periods at site i, then site i has the detection history "100" and the detection likelihood:

$$\Psi_i \mathrm{p}_{i1} {}^*(1 - \mathrm{p}_{i2})(1 - \mathrm{p}_{i3}),$$

where Ψ_i is the probability of the species being present in site i, p_{ij} is the probability of detecting the species in the j-th visit.

Within this framework, a site can have the detection history "000"; the site has three sampling periods but no animals were ever detected. In this case the likelihood statement for the site is:

$$\Psi_i(1 - p_{i1})(1 - p_{i2})(1 - p_{i3}) + (1 - \Psi_i),$$

where the term $(1-\Psi_i)$ is the probability that members of the species are absent at the site.

2.3.1 The fast growing family of occupancy models

The basic model (MacKenzie *et al.* 2002) assumes population closure throughout a entire survey (i.e. members of the species were present throughout the survey; Chapter 8). This assumption can be relaxed to include the possibility of extinction and colonization (Chapter 8; MacKenzie *et al.* 2003). Additional parameterizations include (see Chapter 8) 1) multi-state occupancy (MacKenzie *et al.* 2009), where occupancy can be categorized in multiple states, such as by sex, age classes, or an index of abundance (e.g. many, few, none); 2) multiple-species occupancy (MacKenzie *et al.* 2004), where the detection probability or occupancy pattern of 1 species influences the occupancy pattern or detection probability of other species (e.g. a visit by a non-target animal to a trap affects the detection of a target animal; Mortelliti *et al.* 2010); and 3) detection histories gathered using multiple methods (e.g. survey methods included camera trapping and searches for track and sign; Nichols *et al.* 2008).

The family of occupancy models is experiencing dynamic adaptive radiation (see also Chapter 8).

2.3.2 Assumptions of occupancy models: the importance of a priori planning

The possibility of extracting a great deal of information (e.g. detection, colonization, and extinction probabilities) from extensive survey data is extremely tempting. Nevertheless, most carnivores are mobile, wide-ranging animals, with high risk of violating the assumptions needed for data analysis. Therefore, a priori planning cannot be overemphasized. Extensive knowledge of the biology of the target species is needed to design surveys that do not violate assumptions or that can accommodate violations and still meet objectives. Occupancy analyses require multiple visits at some sampling sites.

Population closure within and between sampling periods, but within the same season, is the first important assumption of occupancy models. Violation of this assumption leads to biased estimates of parameter values (Rota *et al.* 2009) and the appropriate strategy to tackle violations of this assumption depend on the characteristics of the target species, the sampling units, the scale at which the research is carried out and, most importantly, the objectives of the research.

If movement in and out of each site is random, the occupancy estimator may not be biased but the probability of presence switches to a probability of a site being *used* by the species. The detection probability parameter now includes an additional confounding factor, which is the availability of the species at the time of sampling (see Chapter 8; Kendall and White 2009). If movement is not random (often the case with carnivores), the estimator will be biased. One approach to dealing with the violation is to pool data. For example, in a detection history "110" for which we suspect that the population was open between the second and third surveys, we could pool data to a history "11" (Kendall and White 2009). Interpretation of the detection probability parameter must be changed accordingly. If the temporal scale of movements of the target species (e.g. daily movements of a wolf pack) is comparable to the sampling interval (e.g. daily snow tracking), then a second approach is to extend the sampling interval (e.g. weekly instead of daily). Again, interpretation of the parameters changes accordingly, e.g. the detection probability value is the probability of detecting tracks of the previous week.

A third approach is to adopt larger sampling units (e.g. the size of the home range of a wolf pack), at the cost of increased total survey expenses. Again, interpretation of the presence and detection probability parameters changes in a scale-dependent fashion.

When estimating population density using closed population capture–recapture methods, movements of peripheral individuals (whose home ranges extend beyond the study area) extend the boundaries of the effective study area. A large literature on estimating densities of small mammals on trapping grids suggests remedies for this problem (e.g. White and Shenk 2001; Boulanger *et al.* 2004). Estimating abundance only, and not density, avoids this problem.

A second explicit assumption is *no false presences in the data*, i.e. a species is never misidentified. Carnivore surveys are particularly prone to this risk, especially for signs of presence. Using only confirmed data (Karanth *et al.* 2009) or investing resources in genetic confirmation are two options to avoid violation of this assumption. Otherwise, false presence can sometimes be handled statistically (Royle and Link 2006).

A third important assumption is no *unmodeled heterogeneity in detection probability*. Allocate time and resources to measure potentially important, biologically meaningful sampling covariates.

A fourth crucial assumption is that *detection histories at each location are independent*. Appropriate spacing of sampling units (e.g. placing camera traps more than twice the radius of individual home-ranges) may reduce the risk of violating this assumption.

To handle multiple, simultaneous visits by target animals to a single visit, each visit may be considered as separate. Individual-specific heterogeneity in the detection probability can be handled with an "individual" covariate (MacKenzie *et al.* 2006). Another approach is to use spatial replication, and survey at subsites could be considered as a single visit to the site. This strategy may reduce the cost of visiting the site and may be implemented by single observers, but it should be done with caution since it may introduce bias. This bias may be removed if sampling locations are chosen with replacement, or the target species is highly mobile over a short period of time (the case for most carnivores) (Kendall and White 2009).

2.3.3 Some practical issues

Many techniques are used to gather detection histories, such as hair snares, track plates, scent-stations, camera traps, and searches for track and signs. Since presence/absence data are usually collected at the species' level, individuals need not be identified. Nevertheless, trap-happy and trap-shy individuals bias estimates of detection probability.

Another issue is that very low detection probabilities (generally <0.3) lead to overestimating occupancy probability (MacKenzie *et al.* 2002). This issue may be tackled by enlarging sampling units, pooling data, or thinking creatively on how to maximize detection probability (e.g. spending more resources for more effective trapping devices).

A final issue is that even the simplest occupancy model requires the estimation of a relatively high number of parameters, more than an equivalent logistic regression analysis. The information-theoretic approach, which is the default in PRESENCE software (http://www.mbr-pwrc.usgs.gov/software/), will prize the most parsimonious models; nevertheless, we will still need a data-rich matrix of detection histories from which to extract a reasonable amount of biologically meaningful information. This complication has two implications: always attempt to keep the ratio of the (number of covariates)/(number of cases) relatively low; we cannot fit occupancy models to matrices with few detections.

2.3.4 Designing an occupancy study

Clearly, designing an occupancy study requires a great deal of a priori work. *Post hoc* application of occupancy models is often unsuccessful. Optimal sampling design is crucial for optimizing use of funds. A key design question is: what is the optimal number of visits per site vs. number of sites to be sampled needed for an occupancy estimate with a >10% precision? To answer this question, two pieces of information are required: estimates of the probability of occupancy, and detection probability. These estimates are usually not available without either a pilot study or published estimates for the same species and similar environment. With no guidance, an educated guess still works better than allocating sampling effort haphazardly. In addition, a general rule of thumb is that, for rare species, survey many sampling units with low intensity but, for common species, survey few sampling units with high intensity (MacKenzie and Royle 2005).

Develop a study-specific cost function that is a simple equation, where the cost of moving to a site or the man-hour cost for technicians are linked to the number of visits (k) and the number of sites (s). Once the cost function is implemented, designing a study either in terms of (a) minimizing the cost for a desired level of precision, or (b) minimizing the variance given a fixed total budget, is possible. Minimize the cost function by finding the optimal values for k and s (easily performed on spreadsheet software such as the Microsoft Excel with add-in Solver ©).

Several sampling designs work within this framework, such as a standard design (all sites are surveyed k times), a removal design (sampling is interrupted once the target species is first detected), or double sampling (repeat surveys are conducted at a subset of sites; MacKenzie and Royle 2005). To be able to generalize results, use a probabilistic sampling scheme.

When it comes to analyzing the data and interpreting results, remember that an information-theoretic approach allows ranking of the relative abilities of each hypothesis (called a model) to predict the data used to test the hypotheses. Nevertheless, when possible, attempt an absolute measurement of model fit (MacKenzie and Bailey 2004; Moore and Swihart 2005).

Regrettably, many published occupancy models do not report parameter estimates (the βs) or their level of precision, even though reports provide (a) important clues on the models' reliability (i.e. large standard errors suggest high uncertainty and low power) and (b) quantitative predictions that other scientists can use to make preliminary inferences about their study areas and, most importantly, to estimate ψ and p to design their own (optimized) occupancy study.

2.4 Key issues for developing a survey design

Careful a priori considerations of the key components of a survey design can go a long way toward achieving results that lead to reliable inferences.

2.4.1 Target population and spatial extent of the survey

Once the objectives of a survey have been formalized, one needs to define clearly the area and the (biological) population over which to conduct the survey. Without a clear definition of the survey area, one cannot plan a quantitative survey, chose a proper sampling design, choose sampling methods, or assess logistics of the survey. In sampling terms, the area chosen is the "sampling frame," and the elements of the population therein represent the statistical population over which inferences have to be drawn. Depending on whether one's objective is to estimate occupancy or abundance, the statistical population will be the complete collection of all sampling sites or the animals therein. When a well-defined population of the target animals exists, physical features of the population's biological boundaries can be used to define the survey area. When no biologically distinct population of the target animals exists, the survey area must be chosen using geographical features or administrative and jurisdictional boundaries. In this case, however, researchers must realize that individual carnivores and their populations rarely will be contained within arbitrary boundaries (e.g. Linnell *et al.* 2008). With particular reference to capture–recapture surveys aimed to estimate population abundance, it is critical that researchers are able to account for the closure assumption. Individuals whose home range extends beyond the edges of the study area, make the effectively surveyed area problematic to be quantified, although several remedies have been suggested to account for this source of bias (e.g. White and Shenk 2001; Boulanger *et al.* 2004; Silver *et al.* 2004; Jackson *et al.* 2006). To match biologically relevant scales with site-specific management needs, researchers could also consider a multiscale approach encompassing site, landscape, and range-wide scales (Koen *et al.* 2008; McComb *et al.* 2010), thereby providing local managers with site-specific inferences while controlling for larger scale population processes. Whether the study population is designated using biological or geographic boundaries, the survey area must be consistent with the conservation and management objectives for the survey.

Financial and logistical constraints often conflict with the desired geographic extent of a survey and with sampling intensity and resolution, so that the feasibility of the intended survey scale and sampling design should be realistically evaluated, based on the accessibility and other characteristics of the study area (i.e. land cover, topography, climate). At fixed costs, the larger the sampling area, the lower the sampling intensity and the resolution of the data. Given the accuracy and precision needed for analyses, if funding limits the survey area to a size too small to meet the survey's objectives, then the researcher should reconsider the objectives.

2.4.2 Attribute to measure

No matter what the specific target (species, population, habitat) and objectives of a survey, one or more attributes must be chosen to be measured. Because attributes inevitably vary in space or time, they are more properly considered variables (McComb *et al.* 2010). Their measurement can be qualitative (presence/absence), semi-quantitative (visual estimation of density, cover, conditions, etc.), or quantitative (number of individuals, number of tracks, weight, etc.; Elzinga *et al.* 2001). An ideal attribute is easy and inexpensive to measure, is informative and sensitive enough to meet survey objectives, and, for carnivores in particular, measuring it has low impact on the target animals. Measurements of attributes constitute the data that a researcher analyzes to make inferences.

Choice of an attribute depends on the life history of the target species, on the distribution and density of the target population, on the terrain and the local vegetative communities, and on the field techniques that can be used realistically in a given survey. Noninvasive survey techniques exist to measure a great variety of attributes appropriate for carnivore surveys (Long *et al.* 2008a). Spontaneous or elicited vocalizations can be used (e.g. wolves, *Canis lupus*; Harrington and Mech 1982), or visual counts of distinctive social groups (e.g. female bears with cubs (*Ursus* spp.); Knight *et al.* 1995; Keating *et al.* 2002). Given the choice, researchers should select attributes of low inherent variability as a low sampling error enhances the efficiency of the sampling design. Field personnel should be able to measure attributes accurately under difficult field conditions.

2.4.3 Sampling design

2.4.3.1 Probabilistic sampling

Because it is clearly unrealistic to measure a given attribute across the entire target population, one must consider the target population or the study area as a collection of sampling units (Cochran 1977; Thompson 2002). The entire collection of sampling units is the "sampling frame," comprising the statistical population over which inferences will be drawn (Scheaffer *et al.* 1996). Sampling units can be individual animals within the target population or spatial units (plots, quadrats, strip transects) within the study area (Thompson *et al.* 1998). Sometimes attributes are measured in all sampling units, or more often a representative number of sampling units can be chosen.

How sampling units are chosen to be measured affects a researcher's ability to make inferences. Choosing a truly representative sample of sampling units requires some form of probability-based sampling, which will allow a researcher to draw inductive inferences about sampling units not visited (McDonald 2004). Probabilistic sampling schemes are well known: simple and stratified random sampling, systematic sampling, Latin square and ranked set sampling, adaptive sampling (e.g. Thompson *et al.* 1998; Krebs 1999; Elzinga *et al.* 2001; Thompson 2002; Williams *et al.* 2002b).

If sampling units are selected according to non-probabilistic criteria (e.g. purposive, haphazard, and convenience sampling; Thompson *et al.* 1998; Krebs 1999; Anderson 2001), such as when transects are selected close to roads because they are accessible, they are not representative of the non-sampled units and their measurements cannot be used to make inferences to the entire population (Anderson 2001). A critical requisite for proper field sampling is the "willingness to look like a fool to people who are not accustomed to thinking in terms of probability sampling" (D. Whitney quoted by McDonald 2004). Although researchers must always strive to apply probabilistic-based sampling schemes, survey areas or situations will stymie the best efforts: in these cases, measurements refer to sampled units only, and researchers should acknowledge the potential bias in the sample data and interpret their survey's results accordingly.

Probabilistic sampling is not required if the aim of the survey is qualitative (i.e. to document the presence of a species in an area). In this case, the opportunistic spread of survey locations across suitable habitats is an efficient sampling choice (Elzinga *et al.* 2001; Long and Zielinski 2008), even though this sampling design does not account for incomplete detectability.

In capture–recapture surveys used to estimate population abundance, the study area is not partitioned into discrete spatial sampling units because the individuals within the population are the elements of an indefinite sampling frame, and all need to be available for sampling during the survey. In these cases, subdividing the study area into grid cells, all of which are sampled (e.g. hair-snagging grids for bears: Woods *et al.* 1999; Kendall *et al.* 2008, 2009), spreads detection effort evenly throughout the target population, maximizing capture probability and minimizing capture heterogeneity.

2.4.3.2 Adaptive cluster sampling

The most canonical sampling designs were developed for moderately abundant to abundant species (Thompson *et al.* 1998) and are not necessarily the most efficient for carnivore populations (but see McDonald 2004). In sampling terms, *efficiency* of a sampling design entails high precision (small variance) with given sampling costs, and it is useful to evaluate alternative sampling designs. Sampling efficiency is primarily affected by how individuals are dispersed across a landscape, which is usually unknown. Nonetheless, approximate prior knowledge or educated guesses on their distribution might suffice for choosing an appropriate sampling scheme (Krebs 1999). Because carnivores occur at low densities, often in clustered distributions and with low probabilities of detection (Thompson 2004), adaptive cluster sampling and its derivates, such as adaptive stratified random, two-phase adaptive stratified sampling, and sequential sampling, are appropriate (Thompson *et al.* 1998; Krebs 1999; Christman 2004; Manly 2004; Smith *et al.* 2004). These designs come in many variations that can be incorporated into two-stage sampling designs (Manly 2004; Smith *et al.* 2004). Adaptive cluster sampling entails visiting an initial, random set of sampling units, followed by a continued search on sampling units adjacent to those where target animals were initially detected. By doing this, the areas occupied by clusters of individuals are disproportionately, but more efficiently, sampled and the disproportionate sampling is accommodated by using unbiased estimators of abundance (Thompson 1990). Adaptive cluster sampling may be logistically difficult because the final sample size (effort) is not known a priori (McDonald 2004). Particularly favorable conditions for adaptive sampling include situations where individuals are dispersed in rare clusters (Smith *et al.* 2004), and where travelling cost between sampling units is high. Adaptive sampling requires an independent estimate of detection probability (Thompson and Seber 1996).

2.4.3.3 Stratification

Carnivores are rarely dispersed at random within their population range. Most often, they aggregate in areas of higher habitat suitability and prey density. As a consequence, randomly locating sampling units across the study area may not be justified. Many sampling units will be located where target individuals are absent, leading to low sampling efficiency, as detection histories with many zeros would inflate the overall sampling variance. A more efficient alternative in these cases is to *stratify* the statistical population by partitioning it into subpopulations, called strata by statisticians. Stratification in carnivore surveys usually entails grouping sampling units into strata according to how likely they are to contain target animals, with more sampling units allocated to high-likelihood strata (Becker *et al.* 1998; Koen *et al.* 2008). Stratification is an efficient, expedient way to spread sampling effort across a large area with fixed sampling costs. To work, it must include sampling of strata with low probability of detecting target animals (McDonald 2004).

Sampling strata are defined on the basis of tentative or previous information on how population density is expected to vary across the survey area. A probabilistic sampling design should be adopted separately within each stratum, and stratum-specific estimates of population parameters are combined to make inferences that apply to the entire population (Cochran 1977; Thompson *et al.* 1998).

Researchers use stratification for various reasons (Krebs 1999), but with the goal of narrow confidence intervals for final estimates. Specifically, by stratifying, researchers (hope to) reduce the variance of the measurements within each sampling stratum with respect to the overall population variance (i.e. the variance obtained without stratification).

Theoretically, identifying strata should be guided by variance-based criteria known across the statistical population. In practice, however, the distribution of variance is rarely known a priori, and auxiliary environmental variables (e.g. habitat types, topography, prey distributions) are often intuitively used, implying an approximate relationship exists between those variables and the expected variances among strata (Thompson *et al.* 1998; Krebs 1999).

Deciding how many sampling units to measure in each stratum is critical. Sampling may be proportional to the areas of the strata (proportional allocation) or, more formally, in proportion to the stratum-specific sampling variance and cost to survey (optimal allocation; Krebs 1999). In practice, once the strata have been delineated, sampling units within each stratum are allocated proportionally to the expected density of target animals, assuming this is proportional to within-stratum population variance. Although practical, such allocation of sampling units is clearly tentative and possibly far from being optimal (Manly 2004). Alternatively, a two-phase approach ensures a more adequate allocation of sampling units (Manly 2004).

2.4.3.4 Size, configuration, and spacing of sampling units

With the exception of "plotless methods" (Thompson *et al.* 1998), sampling units must be defined. At a logistical level, size and shape of sampling units must depend on the behavior and distribution of the target animals, the attribute to be measured, and the size of the study area. Size and shape of sampling units must also depend on the inferential framework of the survey (model assumptions, the precision required) and the objectives. Size of the sampling units affects both the proportion of sampling units that can be sampled and the intensity of the sampling effort (number of sessions, survey length), and both of these factors affect accuracy and precision of results. For carnivores, sampling methods rarely include complete counts within sampling units, mitigating sampling bias due to plot shape (cf. Krebs 1999; Thompson *et al.* 1998).

Unbiased estimators generally require independent measurements among sampling units. Thus, researchers should choose sampling units large enough to limit the chances that a single individual will be detected in more than one. To have reasonable detection and occupancy probabilities within each sampling unit, size of the sampling unit should be chosen to match the scale at which individuals or social groups generally move (e.g. average seasonal or annual home-range size; Kendall and McKelvey 2008; Long and Zielinski 2008). Long and Zielinski (2008) suggested sizing sampling units at least twice the radius of individual home-ranges of carnivores. For small sampling units, individuals are not available to be counted when they are temporarily outside a given sampling unit. For large sample units, sampling effort per unit (search time, transect length, station density) will be relatively small, if it is fixed per sampling unit. A large survey area may require large sample units to keep costs down. For example, to assess wolverine (*Gulo gulo*) occupancy in areas of <100 000 and >100 000 km^2, 100 and 1000 km^2 sampling units have been used (Koen *et al.* 2008). Although autocorrelation in sample data can be accommodated a posteriori (McComb *et al.* 2010), it is best avoided by spacing units apart according to biologically-based criteria (e.g. Long and Zielinski 2008).

2.4.4 Sampling effort

Researchers can measure a survey's overall effort in terms of: (1) the number of sampling units sampled; (2) the number of sampling sessions, or the length of the survey (temporal replication of sampling units is require to account for incomplete detectability); and (3) the density of detection devices or search time within sampling units. Financial and logistical constraints dictate the upper limit for survey effort. Increasing 1, 2, or 3 should improve the reliability of a carnivore survey in terms of increasing sample sizes and, thereby, increasing detection probability and precision.

For any given population and its abundance and distribution, and given the survey objectives, which effort component contributes most to survey efficiency is usually unclear. A clear tradeoff exists among different strategies, especially if detection and occupancy probabilities are taken simultaneously into account (i.e. more sampling units vs. more sampling sessions; Bailey *et al.* 2007). Whereas, in general, researchers should choose short survey lengths for both statistical (e.g. closure assumption) and logistical reasons (Long and Zielinski 2008), for carnivore surveys, too few or too short sampling sessions may fail to detect animals at an adequate rate (Gompper *et al.* 2006). In capture–recapture applications, for example, length and number of sampling occasions should be high enough to achieve statistically adequate sample sizes, corresponding to encounter histories with a limited number of non-detections. To model heterogeneity in capture probability, adequately requires at least 5–8 trapping sessions (Otis *et al.* 1978; Williams *et al.* 2002b) and large sample sizes. Accordingly, more intensive sampling in capture–recapture surveys could be traded for smaller study areas, even though this trade increases the risk of violating the geographic closure assumption.

For surveys where method-specific detection probability (p) is known or can be estimated a priori, the probability of not detecting the animals at sites where they are actually present ($1-p$) can be used to estimate the number of sampling sessions (K) needed to minimize false-negative error rates (i.e. $[1-p]^K$; Field *et al.* 2005; Campell *et al.* 2008; Long and Zielinski 2008). Obviously, detection probability is proportional to both the number of sampling sessions and the number of traps or devices (e.g. hair snares, track plates, camera traps) activated in each sampling unit. To account for device failure, a minimum of two devices should be placed per sampling unit (Long and Zielinski 2008). Changing locations of detection devices within sampling units can increase detection probability and reduce heterogeneity (e.g. Boulanger *et al.* 2002). Increasing search time per sampling occasion or transect length per sampling unit increases sampling effort but also increases detection probability.

Ultimately, the tradeoff between the number of sampling units and sampling occasions depends on the expected abundance and distribution of the animals in the area (MacKenzie *et al.* 2006). For surveys of rare, sparse carnivores, inferences to occupancy states may be achieved more reliably by sampling less intensively a larger portion of the study area with the intention of increasing the number of sampling units with positive detections (Long and Zielinski 2008). Software is available for exploring the tradeoff between more sampling units versus more sampling occasions, while accounting for estimator precision and enhanced detection probability (Program GENPRES; Bailey *et al.* 2007) (http://www.mbr-pwrc.usgs.gov/software/).

2.4.5 Tackling system variability: measures of precision and their meaning

Survey planning should strive for reliable inferences. That means ensuring unbiased parameter estimates with acceptable precision. Unbiased (or sufficiently unbiased) estimates are the product of unbiased estimators, adequate sampling procedures, and correctly applied survey methods. If not accommodated, bias causes erroneous and misleading inferences and seriously jeopardizes the validity of any management implications emerging from a survey. Unlike sampling error, bias cannot be controlled by increasing sample size. On the other hand, precision of the parameter estimates is a function of the sampling error associated with measuring the chosen attribute. Different sources of variability compound the overall variance within a survey. These sources of variability are conveniently grouped into *process variability*, caused by temporal fluctuations and spatial heterogeneity in a given population attribute, and *system variability*, caused by a combination of among-unit variation (if not all sampling units were sampled) and within-unit (or enumeration) variation (detection of animals within sampling units is probably incomplete; Thompson *et al.* 1998). System variability is generally high in carnivore populations due to their clumped distributions and their low abundance and detectability.

As both process and system variability represent biological and statistical realities, point estimates (i.e. not accompanied by any measure of statistical uncertainty) are misleading and should never be reported (Krebs 1999). To quantify precision, several measures of variability (i.e. variance, standard error, confidence intervals, coefficient of variation) can be computed from the sample data (Thompson *et al.* 1998 presented a succinct yet rigorous treatment of how to quantify precision in the context of surveys and monitoring programs). In addition, depending on the sampling stage, variance can be quantified both as sample (and enumeration) variance and as estimator variance (Cochran 1977; Thompson *et al.* 1998; Williams *et al.* 2002b). To make this point clear, we will present both variance formulas for estimating abundance in the simple case of random sampling, assuming counts within selected sampling units are complete (no enumeration variance) and following notation by Thompson *et al.* (1998).

Sample variance ($\hat{S}^2_{N_i}$), a function of both the spread of values in a sample and the sample size, is given by:

$$\hat{S}^2_{N_i} = \frac{\left(\sum_{i=1}^{u}[(N_i)] - \overline{N}\right)^2}{u-1},$$

where u is the sample size, i.e. the number of sampling units randomly selected for the survey among the total number available (U) in the sampling frame; N_i is the value of the counts or measurements within sampling unit i; $\overline{N}$ is the sample mean. *Estimator variance* [$\widehat{Var}(\widehat{N})$], a function of the sample variance, the sample size, and the portion of sampling units sampled, is given by:

$$\widehat{Var}(\widehat{N}) = U^2\left[\left(1-\frac{u}{U}\right)\frac{\hat{S}^2_{N_i}}{u}\right].$$

Cochran (1977), Lancia *et al.* (1994), Thompson *et al.* (1998), Krebs (1999), and Williams *et al.* (2002b) provided formulas for estimator variance incorporating the enumeration subcomponent (i.e. two-stage sampling designs) and for other sampling designs.

The *sample variance* stems from the interaction between the among-unit subcomponent of system variability and the spatial component of process variability, and it is a useful measure of the efficiency of a sampling scheme. Sample variance does not provide the final measure of an estimates' precision; although this fact terrifies many biologists (as the total number of sampling units enters as a quadratic multiplicative factor in the variance formula, see above), it is the *estimator variance*, or its equivalent standard error:

$$[\widehat{SE}(\widehat{N}) = \sqrt{\widehat{Var}(\widehat{N})}]$$

that measures the precision of our inference.

The *confidence interval* (CI) of a parameter estimate, commonly used as an expression of precision, is obtained from the standard error (assuming a normal distribution) of the estimate and reflects the spread of the underlying sampling distribution. The more efficient a survey design (i.e. the narrower the underlying sampling distribution of the estimate), the smaller the width of CI about the estimate. Confidence intervals are based both on the (arbitrarily chosen) confidence level, i.e. $100(1-\alpha)\%$, and the estimator standard error, which reflects the precision of the survey design. In practice, the CI is interpreted as the range of values within which we are $100(1-\alpha)\%$ confident that the true population parameter is included. For instance, at a given level of confidence (e.g. 95%), an abundance estimate of, say, 90–110 river otters is much more precise (and useful) than an estimate of 40–160 otters.

The *coefficient of variation* (CV), obtained by scaling the sample standard deviation by the mean (or the standard error of the parameter estimate by the estimate), is a relative measure of precision. Sample CVs can be used to assess the sample size needed to achieve a desired level of (relative) precision (Krebs 1999), and estimator CVs provide the currency with which to compare the efficiency of alternative sampling designs or different survey protocols. For example, Becker *et al.* (2004) showed that puma (*Puma concolor*), wolf, and wolverine surveys conducted in different localities in North America using transect intercept probability sampling differed greatly in precision, with CVs ranging from 13 to 74%. Through a regression model based on these surveys' details, they estimated a minimum sampling intensity (e.g. km transect/1000 km^2) to obtain reliable abundance estimates (i.e. expected CV of about 10%).

The measures of precision listed above not only allow one to assess how much different components of the sampling error of our survey design affect the reliability of our estimates, but also provide important indications regarding how to increase the efficiency of survey design. Can we reduce enumeration variance by adopting more effective detection techniques or by increasing sampling effort? Or should we modify size, shape, and number of sampling units? Would it be otherwise a better choice to design alternative stratification criteria or sampling procedures (i.e. match more closely the underlying distribution and density of the population)?

For any quantitative survey objective, we need to minimize system variability by reducing one or both of its components. Ideally, the among-unit variability can be reduced by one or more of the following options (Elzinga *et al.* 2001): (1) adopting more efficient sampling schemes, such as adopting adaptive cluster instead of simple random sampling; (2) improving the efficiency of stratification and allocation criteria by, for example, using variance-based criteria; (3) modifying size and shape of sampling units; (4) increasing the sample size. Reducing sampling error is critical not only to improve the precision (i.e. reliability) of our estimates, but also to allow meaningful comparisons of survey results in space or time. Sampling error is, in fact, directly related to the probability of revealing a true difference between two or more population states (i.e. statistical power). Although prospective power analysis pertains more to monitoring programs than to single surveys (Elzinga *et al.* 2001), preliminary estimates of sample variance can be obtained from the literature (e.g. Gibbs 2000) or, more rigorously, from *ad hoc* pilot studies.

For detection probability, as well as for an estimate of absolute density, unbiased parameter estimators usually assume that sample data are statistically independent (but see: Sargeant *et al.* 1998, 2005; Royle *et al.* 2008). Although the effect of violating the independence assumption depends on the specific survey objectives and analytical framework (Long and Zielinski 2008; Royle *et al.* 2008), autocorrelated data generally lead to biased (over- or underestimated) estimates whose true variance is underestimated (i.e. Type I error underscored; Krebs 1999). Controlling for autocorrelation patterns in the data, is therefore, important for the correct interpretation of the observed variance.

2.4.6 Field methods

In carnivore surveys, sampling within sampling units usually takes either of two common configurations (Campbell *et al.* 2008): station-based (e.g. track stations, track plates, live-traps, hair snares, camera traps, audio recordings) or transect-based (ground and aerial transects, scat-detection dogs). Neither guarantees complete detectability, which must be remembered when choosing the size of the sampling units.

Given the expectedly low probability of detection in most carnivore species, a successful carnivore survey essentially rests on the proper choice of the most efficient field technique(s). These vary with the species and the survey conditions (survey objectives, expected distribution and density of the population, extent and characteristics of the study area, resources available). A multitude of survey methods are potentially available to survey the great diversity of carnivore species in their diverse environments (e.g. Zielinski and Kucera 1995; see also Chapters 4, 5, and 8). In particular, diversification of noninvasive field and lab techniques is facilitating an unprecedented upsurge of reliable carnivore surveys (Long *et al.* 2008a). Robust statistical and analytical frameworks for analyzing survey data are well known and under continuous development (e.g. Chapter 8; Williams *et al.*

2002b; Amstrup *et al.* 2005; MacKenzie *et al.* 2006; Royle *et al.* 2008), allowing powerful and reliable inferences from seemingly simple presence/absence data.

Nevertheless, no single method can be universally effective for any species in all situations. Making proper choices requires, on one side, understanding the species' biology and behavior and, on the other side, understanding basic sampling requirements dictated by the statistical formalization of a survey's objectives. For a given survey, choice of the proper field methods must be viewed in the context of the statistical requirements. In case detection probabilities are deemed inadequate by using a single survey method or, if the survey is geared toward multiple objectives or multiple species, researchers should contemplate more than one detection method (Campbell *et al.* 2008). For example, Becker *et al.* (2004) enhanced the performance of probability sampling using telemetred individuals. Similarly, more than one technique can be used to recapture individuals in capture–recapture surveys, and, to increase capture probability and more efficiently model capture heterogeneity, researchers can consolidate individual encounter histories recurring to multiple-data sources (e.g. Boulanger *et al.* 2008). In short, when dealing with carnivores we must master technical details (e.g. know the most efficient track-plate or hair-snare design, the optimal configuration within a sample unit). We must also be familiar enough with the fundamentals of inferential survey statistics to design surveys that allow inference that, in turn, allows us to meet objectives (Royle *et al.* 2008).

3

Mind the map: trips and pitfalls in making and reading maps of carnivore distribution

Carlo Rondinini and Luigi Boitani

A wide range of theoretical and applied analyses in animal ecology, biogeography, and conservation biology involve the production or use of maps of species' distributions. These include studies from individual (home range) to population (regional) level, to continental and global level. The reasons for producing maps of distributions vary from assessing the structural connectivity of landscapes, to predicting the spread of invasive species, to detecting zones of transition among faunal assemblages, to identifying conservation priority sites that maximize the return on investment of conservation money globally. It comes, therefore, as no surprise that the number of species' distribution maps produced at various scales grows.

Species' distributions are dynamic over time. Individuals live in different places at different times and, therefore, in theory, an appropriate distribution model is a probability density function across the study region. But because individual home-ranges shift, contract, expand, local populations go extinct, new sites are colonized, and habitat is converted by humans, the form of the probability density function would slowly but continuously change. Therefore, any map of a species' distribution is necessarily an abstract and simplified representation of a complex reality. Any map is a model, with its specific assumptions, approximations and errors. Because the availability of modeling tools to develop species' distribution maps is continuously increasing, processes from very fine to broad scale are relevant to the interpretation of modern maps.

The true distribution of a species is impossible to map, but it can be approximated by two useful concepts: the extent of occurrence (EOO) and the area of occupancy (AOO) (Gaston 1991). The extent of occurrence identifies the region encompassing all localities where a species has been recorded; the area of occupancy is a subset of the extent of occurrence, which excludes all areas within the extent of occurrence that are not occupied by the species, because they are unsuitable or presently not occupied (Gaston 2003). Depending on the data and method used to make them, species' distribution maps can be closer to an extent of occurrence (geographic range maps) or to an area of occupancy. This is because at increasingly small scales, more and more holes appear in a species' distribution, which are overlooked at broad scales.

Species' distribution maps always contain two types of error, although in variable proportion: they can erroneously indicate that a species is present or absent, which are, respectively, referred to as errors of commission and omission (Fielding and Bell 1997). Maps closer to the extent of occurrence are prone to commission errors because they overestimate the area actually occupied by a species. By definition, the area of occupancy should be free from both commission and omission errors, but in practice areas of occupancy are obtained in one of two ways: either by excluding the portions of the extent of occurrence that are perceived to be unsuitable for the species, therefore reducing commission errors at the expense of a potential increase in omission errors; or by extrapolating from known occurrences, but in this case not all omission errors are removed.

Not all maps are equal or equally useful for all purposes, but distinguishing a useful map from a useless one (for any given purpose) by simply looking at it may be impossible. To distinguish a useful map, one must understand the relevant information, the metadata, that should be attached to maps. For a map of species' distribution this information includes the method used to make the map; the data used; if the map is based on expert knowledge, the expert's name(s); and, if data are point occurrences, how they were collected, their biological significance, their time span, and how they were extrapolated. Understanding the different types of maps is essential to place them in the appropriate context, acknowledge their limitations, and use them appropriately (Rondinini *et al.* 2006b).

3.1 Maps based on expert knowledge

3.1.1 Geographic range maps

Maps based on expert knowledge translate opinion and non-quantitative information into quantities. Two pieces of expert knowledge are usually translated into a map: the limits of the species' geographic distribution, and the species' habitat; the former can be used to draw polygonal geographic ranges. Although these products are often considered "true maps" (as opposed to "models"), they, too, are models. For these maps, the models are the algorithms in the heads of the experts and, therefore, the models are implicit and the data are undocumented. Yet, given the scarcity of hard data on the distribution of many species (carnivores are no exception, e.g. most tropical small cats), this may be the only way to map their distribution. When possible, gathering species' experts in a workshop (e.g. Schipper *et al.* 2008) should be considered to reduce bias due to individual knowledge. Range maps based on expert opinion vary widely in the level of detail across species, range sizes (usually small-range species are better mapped than large-range ones), and geographic regions, reflecting variable survey intensity. These maps are a useful tool for biogeographic analyses and for identifying broad regions of conservation interest, but are unsuitable for species' management and conservation planning because they are too coarse, especially for large-range species.

3.1.2 Deductive habitat suitability models (HSM)

The knowledge of species' habitat can be used to produce deductive HSMs, i.e. models of the habitat potentially used by a species. These maps are particularly useful if intersected with geographic range maps, to identify the areas potentially suitable for (therefore assumed to be potentially used by) a species within its range. These models, which, like all models, need to be evaluated before they can be used for any application, are named deductive as opposed to inductive HSMs, which are based on the extrapolation of the species' habitat from the habitat type recorded at known species' occurrences (Corsi *et al.* 2000). Inductive HSMs are discussed later in this chapter.

Deductive HSMs have been developed especially in North America based on the habitat suitability index (HSI), i.e. an analytical, species-specific function (drawn from data and expert knowledge) relating the amount of a given habitat feature to the level of suitability for a species. By combining values of HSI for different habitat features, an overall suitability index is obtained for each point in the study region. Compared to expert-based range maps, expert-based HSMs are more documented, because the species–habitat association used to produce the model is recorded and the algorithm applied is explicit. Deductive HSMs have been produced for mammals, including carnivores, in Africa (Rondinini *et al.* 2005; Boitani *et al.* 2008), North America (National Biological Information Infrastructure 2010), Central America (Jenkins and Giri 2008), Southeast Asia (Catullo *et al.* 2008). In general, deductive HSMs are useful to model broad taxa on a regional to global scale, due to the lack of a good, unbiased sample of known occurrences for most species. Deductive HSMs are very powerful tools for identifying areas that have been exploited and converted to human-dominated land use and as such are no longer suitable habitat, and in predicting changes in species' distributions associated with changes in land use and climate.

The number and type of variables that can be used to produce deductive HSMs is more restricted than for inductive HSMs, because expert knowledge is usually limited to a few variables, e.g. the vegetation types used by a species, the elevation limits of its distribution, the association with water. For generalist carnivores this may result in deductive HSMs with a high proportion of suitable area, because environmental variables with a less evident association with suitability for these species (e.g. human activities and attitudes) cannot be incorporated in the model. For conservation, however, this weakness of the models positively contributes to reducing the possibility of incurring in Type II error and missing important conservation action.

3.2 Maps based on species' occurrence surveys

Surveys of species' occurrence (Chapter 2) often produce data in the form of point localities with attached information on: the individual(s) surveyed (related to taxonomy, biometry, behavior, abundance, etc.), the site (structure/habitat type, biological community, etc.), and time. This type of data is particularly suitable for analysis in a geographical information system (GIS) to produce maps. Not all types of data are suitable for the production of all types of maps. The correct use of survey data to produce maps depends on the purpose for which the data have been collected, which in turn (should have) guided the sampling strategy for data collection. But because no technical limitation impedes the use of any point data to produce any maps, *survey data can be misused to produce maps of little use for biogeography and conservation (or even worse, misleading maps).*

3.2.1 Types of data

Species' occurrence data are usually collected through:

- radio-telemetry;
- systematic surveys (trapping, sighting, scent-stations, ...);
- occasional observations (possibly gathered opportunistically a posteriori from other sources, e.g. museum specimens).

With radio-telemetry, many repeated data samples are collected on relatively few individuals, which is particularly true for most studies on carnivores. When mapped, radio-telemetry data are usually clumped in few areas, leaving large empty spaces in the areas that were not used by the animals monitored. Clumping may in part reflect avoidance of some habitat types, but it is also due to data autocorrelation (i.e. the position of one data point in space depends on the position of the previous data point collected). While a number of techniques exists to reduce autocorrelation from radio-telemetry data (Johnson *et al.* 2008), these are effective at the level of individual home-ranges, but cannot eliminate the clumping over a large region due to unsampled individuals.

Good radio-telemetry data provide robust information on where individuals were at the time of sampling. This dependability allows robust inferences on individuals (internal anatomy of individual home-ranges, preference for a habitat type over another, and possibly behavior) but sometimes with limited capability to generalize to the whole population over a larger region. The possibility to generalize data depends on the size of home ranges and on the variability of habitat availability in the region in relation to the individual choices in habitat use. For large carnivores, with large home-ranges and low population density, it is more feasible to radio-track a representative sample of the entire population of the study region. When this is the case, the data collected can also be used to map the overall distribution of the population (e.g. Falcucci *et al.* 2009). For small carnivores, with small home ranges and high population densities, the number of individuals surveyed through radio-telemetry is usually a small fraction of the entire population in the study region, and the point data collected are unlikely to be representative of the population distribution. For carnivores with restricted niches, data on the habitat preferences of relatively few individuals can be robustly extrapolated (i.e. models are likely to be positively tested) to map the potential distribution of the population over the study region through HSMs. On the other hand, for adaptable carnivores (e.g. canids, many mustelids) in study regions that are highly variable in habitat types, it is likely that different individuals use different habitat types according to availability. In this case, unsampled individuals may live in unsurveyed habitat types and this reduces the generality of the HSM.

Systematic surveys are usually aimed at the collection of few (if any) repeated data on each of many individuals of a population, which may or may not be individually recognized (recapture). The location of survey data should be defined a priori on the basis of one of a number of possible sampling strategies (see Chapter 2).

Systematic survey data may include information on individuals, e.g. the frequency of occurrence at each sampling site (if individual recognition is possible) and individual behavior (if data were collected through direct observations), but in general they are more suited for making inferences at the population level (distribution, frequency of use of different habitat types). The use that can be made of systematic survey data depends on the accuracy of the absence estimate (see Chapter 2) and the sampling strategy used for data collection. If a probability of absence is estimated, so that "true" absences are detected correctly, data collected on regular grids can be used to draw the geographic range of a population (Gaston 1991) and detect the gaps between subpopulations. Scale (resolution and size of the study region) is fundamental in this respect. Many carnivores are capable of long-distance movements to close even large gaps and to occupy previously unused suitable habitat, therefore the knowledge, at least approximate, of the species' dispersal capability is required. Regular grids and random points are usually not the best strategy to collect data if the aim is to extrapolate the geographic distribution of the population through HSMs, because they do not guarantee even (or proportional) sampling of all habitat types in the study region. Random survey data stratified by habitat type are suitable for the extrapolation of the geographic distribution of the population through HSMs, under the assumption that preference for habitat types is proportional to use (assumption that is not necessarily correct, see Garshelis 2000). When "true" absences are detected correctly, the data are suitable for models based on presence/absence data, otherwise presence-only methods can be used (see below).

Occasional data (e.g. from museum specimens, genetic evidence, camera traps, trustworthy sightings, etc.) collected opportunistically, e.g. for the compilation of atlases of species' distribution, result in records being spatially biased towards places that are easily accessed, taxonomically biased towards species that are relatively conspicuous, and temporally biased, due to irregular recordings over time (Keller and Scallan 1999; Polasky *et al.* 2000; Funk and Richardson 2002). "True" absences cannot be inferred from this type of data, because no information is available on the intensity of sampling in sites where the species has not been recorded. Occasional data collected a posteriori have generally been accumulated across large time spans and this should be accounted for in their analysis. Often, however, it is assumed that species' distributions are static, and that these data represent a snapshot of the distribution taken at the present time. This ignores the dynamic nature of species' distributions due to dispersal or shifts in distribution due to changed land-use or environmental conditions. Occasional data may contain a number of false presences because of positional errors, errors in species' identification, and habitat conversion occurred since the original collection of the data point. These errors are usually difficult to assess, making them much less robust than radio-telemetry and survey data for statistical and biological inference (McKelvey *et al.* 2008).

3.2.2 Biological significance and time relevance

At each point locality surveyed, the outcome can be one of the following (MacKenzie *et al.* 2006; and see Chapter 2):

1. detected (present);
2. not detected:

2.1 estimated absent (with x% probability);
2.2 not found.

With radio-telemetry data, usually only outcome 1 is recorded (although depending on the technique used to locate individuals, their absence could be also estimated with reasonable certainty). On the other hand, with occasional data, even outcome 1 can be uncertain, due to the many potential sources of error in space and time. With survey data, presence can be certain (if taxonomic and positional errors can be excluded), "not detected" takes the form of either 2.1 or 2.2 depending on the design of the survey (Chapter 2). In the latter case, when a species is not found but no probability can be attached to it, the information can be highly uncertain (MacKenzie *et al.* 2002; Wintle *et al.* 2005) and should not be used for any inferences.

What is the meaning of presence data for developing and interpreting maps? Even where an animal is detected, not all locations are equal, although this is difficult to tell after they have been transformed into coordinates on a map. Animals move in the environment for a variety of reasons, including physiological (feeding, resting, hibernating, traveling, migrating, etc.) and social (mating, defending territory, dispersing, etc.). Not all the habitat they use is of high quality (i.e. literally, increases their fitness) and this is especially true for carnivores. Animals may move through low-quality habitat to reach a patch of high-quality habitat, or because different patches of habitat are necessary to fulfill different biological needs (e.g. eating and drinking; Garshelis 2000). Much of the area actually occupied by species may represent sink habitat that is unable to sustain a population without the contribution of immigrating individuals (Tyre *et al.* 2001). For these reasons, a dataset of point localities of an animal population is not necessarily representative of the (best) habitat used. This is especially true with occasional data, mostly sightings and road kills, which are often biased towards easily accessible areas (roads, urban areas) (Reddy and Davalos 2003) or between different habitat types, and can be different across taxonomic groups in the same area (Freitag *et al.* 1998). Furthermore, species are often under-recorded in the core of their range, while outlying occurrences are recorded. For small carnivores in particular, road kills or other dead animal reports are a common source of point data, but they are mostly representative of the distribution of sink areas, not of high-quality habitat. To avoid this potential weakness, maps should be thematic, i.e. should be based on a coherent subset of data (maps of feeding areas, mortality areas, etc.). Finally, in regions with high human pressure on the environment, animals may live in the suboptimal habitat left, not in their optimal habitat.

Therefore, even if the dataset is representative of the habitat used by the population, this may not reflect the ideal species' habitat.

While estimated absence with a high probability may appear to have a straightforward meaning, it has not, especially for carnivores. Time, on multiple scales, is fundamental to interpret absence. Most carnivores are highly mobile for their size and do not use their home ranges evenly throughout the year, therefore seasonal home-ranges can differ widely. For example, male polecats and many other small mustelids may appear transient during the breeding season, covering much more space than used when they maintain (foraging) home ranges. As a result, the distribution of an animal's locations will be clumped during foraging and dispersed during the breeding season. On a larger timescale, dispersal capability and (meta) population dynamics should also be considered. For example, wolves and other large carnivores do not occupy all the suitable habitat in a region at any given time, but are capable of long-distance (several hundreds kilometers) dispersal to establish populations in suitable sites that were not used for many years. Therefore, the interpretation of absence data should be based on the knowledge of the species' biology.

3.2.3 Extrapolating points to map the distribution of a population

Species' occurrence data can be directly represented on maps, or extrapolated using a variety of techniques (Figure 3.1). Five broad types of maps can be developed:

(1) simple point localities;
(2) buffers around point localities;
(3) grid cells containing point localities;
(4) contour lines around point localities (pattern, autocorrelation, geographic models);
(5) HSMs fitted on point localities.

Because population distributions are multiscalar and change over time, any map of the distribution of a population (even type 1 above) is a model, i.e. an abstract representation of reality, and as such it is subject to underlying assumptions on time, scale, and extrapolation method.

The simplest possible map of population distribution, i.e. a map representing the coordinates of localities where individuals were found, is the only map that does not involve any extrapolation. As such, this type of map is highly valuable to develop any other more sophisticated maps, because it displays the original data. It is also the most unrealistic estimate of the distribution: in fact it assumes that no positional errors whatsoever exist in the data, that all individuals in the population have been sampled, and that they don't move. Point locality data seriously

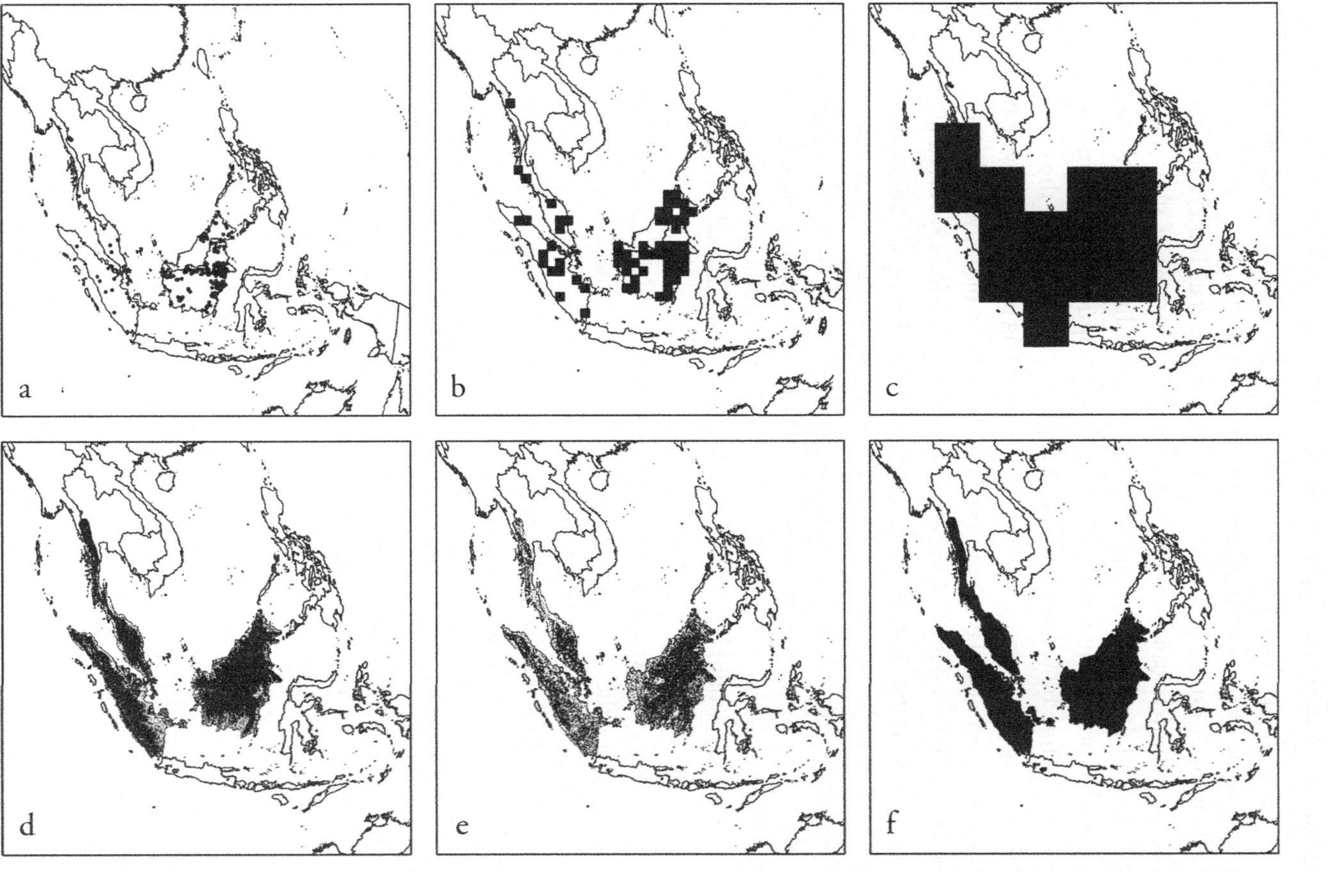

Fig. 3.1 Six distribution maps for the linsang (*Prionodon linsang*) in Southeast Asia: (a) point locations where the species was known to occur (from Catullo *et al.* 2008); (b) 100-km grid cells containing the point locations shown in (a); (c) 500-km grid cells containing the point locations shown in (a); (d) inductive habitat suitability model developed with the software Maxent and based on the point locations in (a), a map of land cover and a map of elevation, clipped to the geographic range shown in (f); (e) deductive (expert-based) habitat suitability model based on the same variables as in (d); (f) known species' geographic range from the IUCN Red List of Threatened Species (www.iucnredlist.org). In (a), (b), (c), and (f), black means estimated presence, white estimated absence. In (d) and (e), black is maximum suitability, white is lowest suitability, with shades of grey indicating intermediate suitability. Each map is very likely to contain false absences and false presences in variable amounts that are correlated to the size of the estimated area of presence. For the two-habitat suitability models, if a threshold is used to define areas of presence and absence, the amount of commission and omission errors depends on it.

underestimate the actual area of occupancy because they are usually sparse and discontinuous, and they inevitably misplace the actual area of occupancy because species' distributions change over time. In order to produce useful maps, point locality data need to be extrapolated using a variety of techniques.

A very simple form of extrapolation is the creation of (circular) buffers around point locations. The buffers can be used to relax the assumptions of the simple point locality map, and the buffer size should be chosen accordingly. Buffers of the size of an estimated error polygon would only relax the assumption of no positional error, and would therefore be too small for most applications. Buffers in the order of magnitude of a home range (individual or group, depending on the social system of the species mapped) would relax assumptions about mobility of individuals over time, and would identify the limits of the area where the individuals surveyed are expected to be found during their normal ranging activity. For some intensively studied populations of large carnivores, whose home ranges are large, the assumption that all individuals (or all groups) in the region have been surveyed can be realistic. In this case, a map of point localities surrounded by home-range wide buffers would be a reasonable estimate of the population distribution in a short time span. Still, this map would be useless to predict long-term seasonal, dispersal, and colonization dynamics.

A very popular form of extrapolation of point locality data is obtained by superimposing a grid to the set of points, and considering as occupied all the cells that contain one or more point data. This extrapolation is routinely applied for the production of distribution atlases. This method has many disadvantages with respect to the buffer extrapolation. Usually it is not an informed extrapolation, because the size of the grid cell is not chosen to match a biologically relevant size. Therefore, there is a tradeoff with smaller grid cells lessening the likelihood of commission errors but increasing the likelihood of omission errors simply by chance. Species' distributions, and patterns of species' richness, are sensitive to scaling (Stoms 1992). Using larger grid cells, or scaling up, reduces the impact of spatial biases, but results in a decreased resolution, which in turn enhances the oasis effect, by which grid cells are assigned the same weighting, regardless of the number of occurrences they contain (Lawes and Piper 1998). In addition, while buffers are centered around point localities, grid cells are not, increasing the likelihood that a whole grid cell is considered occupied due to a point location occurring close to its margin and/or in a small suitable patch within the cell. Finally, grid cells are misleading because the cells cover the study region evenly, but the data represented on the map are only presences. The empty cells can be easily misinterpreted as absences, when in reality they are an irresolvable mix of absences, under sampled and non-sampled cells.

More sophisticated forms of interpolation can be based on the geographic pattern of distribution of the known point localities. They belong to two groups: those aimed at the definition of hard limits of the area where the population occurs (geographic range), and those aimed at the estimation of the density of use of the study region by the population. The methods that aim to identify the geographic ranges implicitly assume a Boolean distribution of the population: present inside, absent outside. This assumption is almost always false because species select their habitat among those available and do not occupy their range entirely (Gaston 1993). These methods include the minimum convex polygon (MCP) surrounding either all or a proportion (e.g. 95%) of the known localities where the population has been recorded, and expert-based polygons. An additional assumption of the MCP is that all outermost locations of the population have been detected, which can be the case in relatively small study regions and for well-known species. Polygons based on expert opinion can overcome this limitation, but at the expense of subjectivity and lack of repeatability. Methods that identify the limits of the population range can be useful for biogeographic analysis, e.g. to identify transition zones between different communities, but are less suitable for fine-scale analysis, e.g. conservation-related analysis aimed at setting protected areas, because they contain many false positives (commission errors). This is especially true for carnivores, whose population densities are usually low to very low and, therefore, whose geographic ranges are largely unoccupied.

Extrapolation and interpolation methods that aim to estimate the density of use of the study region include kernel algorithms (Worton 1989), and regression with geographic coordinates (Lichstein *et al.* 2002). Kernel algorithms perform a nonparametric smoothing that transforms a map of point localities in a continuous map of utilization distribution (UD), a probability density function of the use of the study region by the population. The UD can be represented as a 3D volume, where x and y are the geographic coordinates of the study region, and z is the density of the probability to find an individual of the population. The key parameter that can be tweaked in the analysis is the smoothing distance, often named *h*, which is the geographic distance within which point localities contribute to the same local peak of the UD in the z dimension. Smaller values of *h* make the UD sharper by fitting it more closely to the point locations, while larger values of *h* make it smoother and more loosely fitting on data points. From the UD, contour lines can be drawn to include fixed proportions (e.g. 95, 75, 50%) of the UD, thus defining cores and peripheries of the population range. Geographic models use the geographic coordinates of point locality data on presences (and absences) to extrapolate the density of use of the potential area where the population can be found (Elith and Leathwick 2009). These models are a special case of the HSMs described below, where the only structural information on habitat is geographic location.

Kernel algorithms and geographic models may also be collectively called "structural" models, because they only rely on the spatial structure of point locality data, ignoring habitat information. They rely on the autocorrelation of presence data in space, therefore assume even sampling (and even detectability) across the study region, so that the variation in the density of points of presence depends only on variation in population density. Also, because these methods do not take into consideration any relationship between species and habitats, they implicitly assume that individuals do not show any habitat preference. While this is unlikely to be the case, even for generalist species, such as foxes and raccoons, the distribution of populations is inherently autocorrelated, so that the likelihood of finding an individual in one site depends on the presence of other individuals in the surroundings, as well as on habitat type. For example, a sink habitat may have a high chance of being occupied if it is close to a source population. For this reason, these methods should perform well in predicting species' presence within a small region, even if they do tell nothing about the quality of the site where an individual is found.

3.2.4 Inductive HSM

Unlike the methods outlined previously, the extrapolation of inductive HSMs relies on variables that are external to the distribution of point localities, namely the variables that describe the habitat type. Knowledge of species–habitat relationships is inferred from the habitat where the species has been recorded and the output is a semiquantitative or quantitative probability of species' occurrence or abundance (Corsi *et al.* 2000). Inductive HSMs use spatially-incomplete information (species' occurrences), to generate spatially comprehensive predictions of species' distributions, avoiding many of the problems of scale inherent in the manual construction of range maps and giving information about variation in likelihood of occurrence or abundance. While the extrapolation and interpolation methods described previously tend to smooth the differences between adjacent habitats, inductive HSMs emphasize differences in habitat suitability by providing spatially explicit probabilities or likelihoods of species' occurrence or abundance. To estimate habitat suitability for the species of interest across the study region, inductive HSMs use maps of environmental variables, which may include vegetation type, elevation, water, level of human disturbance, and climate (Guisan and Zimmermann 2000). Based on the values of the environmental variables in the sites where the species was recorded, they extrapolate the potential presence of the species elsewhere.

Methods for the development of inductive HSMs can be divided into those that use only data on species' presence, and those that use data on presence and absence (or pseudo-absence). Modeling techniques based on presence-only data include Mahalanobis distance (Clark *et al.* 1993), ecological niche factor analysis (ENFA) (Hirzel *et al.* 2002), maximum entropy models (Maxent) (Phillips *et al.* 2006). All these methods refer implicitly or explicitly to the concept of ecological niche. Locations where the species have been found are considered to be representative of the ideal habitat conditions (as described by the environmental variables chosen). For each site in the study region, the distance to this ideal condition is measured (in the multidimensional space of the same environmental variables), and defines the level of suitability of the site; this is maximum where the conditions correspond to those of sites where the species occurs, and decreases as the ecological distance increases.

These types of models rely on a double assumption: that the point locality data collected reflect the best habitat conditions for the population in the study region, and that these habitat conditions correspond to the ideal habitat conditions for that population. The first statement may be unrealistic for a variety of reasons, in particular if the sample of locations is biased or if the biological significance of the point data collected has been ignored or misunderstood. Although inductive HSMs have extraordinary value in accommodating varying intensities of sampling, the resulting predicted distributions may still reflect this bias (Kadmon *et al.* 2004) and, when studying carnivores, it should be explicitly considered and resolved. The second assumption is hard to test but its potential consequences should be considered. The ideal habitat of a species is where the individual fitness is maximized (Krebs 2009). Ideal habitat may no longer be available to the species due to human-induced conversion, and individuals may live in remnant suboptimal habitat. Assuming that the best available habitat is the ideal habitat for a species can lead to mistakes; for example, in the quantifications of extinction debts (Tilman *et al.* 2002), reintroductions, assisted migration in response to climate change (Hoegh-Guldberg *et al.* 2008). Techniques based only on presence data do not calculate directly a probability of presence or absence, although for some of them (including ENFA and MAXENT), methods exist to identify a cutoff for interpreting low suitability levels as absences. The relative performance of these models is reviewed in Elith *et al.* (2006). They found that methods relying on presence-only data perform well, as compared to methods based on presence and absence data, and that the novel techniques outperform the most established ones.

Modeling methods based on presence and absence data use a variety of regression techniques (general linear models, generalized linear models, generalized additive models, etc.) to estimate the probability (conditional to the data and method used; Wilson *et al.* 2005) that the species of interest is present across the study region. The different techniques have been discussed thoroughly elsewhere in the literature (Scott *et al.* 2002; Elith and Leathwick 2009), therefore they are not covered in detail here. It is worth remarking here that, in addition to the assumptions of presence-only HSMs, these models assume that absence has been correctly estimated. As discussed in Chapter 2, presences can be easily overlooked, making this assumption difficult to defend. This is particularly true for carnivores, which are usually elusive species and highly mobile over large home-ranges. Sometimes, in the absence of robust absence data, these models are fitted to presence and pseudo-absence data (Elith *et al.* 2006), i.e. point localities where the species was not recorded (and sometimes chosen because they are very different in terms of habitat type from the points where it was recorded). This may be a reasonable practice for species that are well-known habitat specialists, but it would be dangerous to apply to a generalist species (many carnivores), unless the population of interest has been extensively sampled, because habitat choice may be highly variable among different individuals, and unsampled individuals may behave unexpectedly. We strongly caution against the use of modeling techniques that require absence or pseudo-absence data of carnivore.

Further assumptions in common among the techniques for the development of inductive HSMs regard data quality, quantity, and spatial structure. Low-quality data include those that contain errors (positional, taxonomical) and those that are biologically irrelevant (e.g. an individual crossing a low suitability habitat) or misleading (e.g. data biased towards accessible areas, or locations of dead animals). These data may introduce errors or bias in the estimated suitability of a habitat type. Data quantity should be evaluated not only from a statistical, but also from a biological point of view. For most HSMs it has been suggested that 30–50 (sometimes fewer) data points are sufficient to develop a robust model (Elith *et al.* 2006; Phillips *et al.* 2006). While this is statistically correct, it is not necessarily biologically sensible. Considerations of the number of individuals sampled (also in terms of the fraction of the population of interest that they represent), and of the seasonal variability of behavior, are necessary to evaluate the usefulness of the maps produced. An evaluation (at least visual) of the spatial structure of the data is also necessary to assess the extent to which an inductive HSM can be extrapolated to distant places. Many carnivores (e.g. leopards, jaguars, wolverines, foxes) have very large geographic ranges, and use very different types of habitat in different portions of their ranges. Point data from one portion of a range would produce very poor predictions of the suitability of other, distant sites (e.g. point data for the leopard in African forest would be poor predictors of suitability in Asian semi-deserts).

3.2.5 Caveats and limitations of deductive and inductive HSM

The availability of maps of environmental variables represents a limiting factor for the predictive power of HSMs. This is best illustrated with the example of semi-aquatic carnivores, e.g. river otters (*Lutra lutra*, *Lontra canadensis*). The distribution of these carnivores is dependent on the presence of sometimes small water courses, which are often poorly mapped. The choice in this case is between incorporating in the HSM an inaccurate or incomplete layer of water courses, which would introduce false absences due to some water courses not being mapped correctly, or not using the layer and introduce a number of false positives due to the fact that a key limiting factor of otter distribution is missing.

Another limitation of HSMs is related to the resolution at which environmental variables are mapped. Unless the maps are compiled for the purpose of modeling the species of interest, their resolution often does not match the scale at which the species uses the environment. As a result, individuals of low-vagility species might be capable of inhabiting fragments of habitat that are much smaller than the resolution of existing maps, introducing errors of omission. These omission errors might be biased to particular types of land cover (e.g. fragmented habitats might be omitted as good habitat if existing maps are too coarse to identify small fragments), and can in turn result in geographic biases in the predicted distribution data. Habitat that is fragmented due to anthropogenic reasons tends to occur in low-altitude areas that are more readily accessible and therefore the distribution of the species might be incorrectly associated with high-altitude areas.

The majority of species' distribution models are limited to the relationships between species and the environment, and do not take into account historical and biogeographical factors affecting species' distributions. This problem is often circumvented by constraining the output of models to the known EOO of the species (Rondinini *et al.* 2005; Boitani *et al.* 2008) or can be accommodated through the inclusion of spatial variables (linear and exponential terms derived from geographic coordinates; Lichstein *et al.* 2002). Furthermore, the probability of a species' occurrence depends on many factors that are unlikely to be included as explanatory variables in an HSM. Such factors might include prey density, inter-specific competition, population dynamics, individual behavior (Van Horne 1983), and human activities and attitudes that threaten the population. For carnivores in particular, humans are often a primary cause of mortality (legal or illegal killings due to conflicts, accidental mortality). When this is the case, further insights on the potential distribution of the population can be gained by overlaying to the traditional HSM, based on environmental variables, a second model, predicting potential mortality areas on the basis of human-related variables (density of human population, livestock, other potential sources of conflict, roads, etc.). This technique, also known as double-layer modeling (Naves *et al.* 2003; Falcucci *et al.* 2009), may aid the identification of potential sink areas, where habitat suitability and mortality are both high.

The evaluation of the predictive power of HSMs should always be carried out with data independent from those used for model development, to avoid overoptimistic results (Guisan and Zimmermann 2000). Even so, the result depends on some properties of data and models, namely prevalence, sensitivity, and specificity. Data *prevalence* is the proportion of sites where the species was detected, and model prevalence is the proportion of sites where the species is predicted present. *Sensitivity* is the proportion of sites where the species is correctly predicted present, and *specificity* the proportion of sites where it is correctly predicted absent. For carnivores, data prevalence is usually low, because of low population density and low detection probability. On the other hand, model prevalence is often high, because individuals are often found in a variety of habitats, therefore the high proportion of sites is predicted suitable. This has two consequences. One is that the maps of carnivore distribution derived from inductive HSMs will provide an optimistic picture of the population distribution, with many suitable areas. Yet, due to the low density of carnivore populations, many of these suitable areas will not be occupied at any given time (Gaston 1993). Therefore these maps should be interpreted, much more than for other taxa, as *potential* distributions. The second consequence is that, even by chance, sensitivity tends to be high and specificity low. When no reliable absence data are available, only sensitivity can be measured, therefore the HSM will appear to perform well, even if in reality it assigns suitability at random. In such cases, to avoid overestimating the predictive power of the HSM, a comparison with random Monte Carlo simulations can be performed to test whether the sensitivity of models is significantly higher than random (Catullo *et al.* 2008).

4

Noninvasive sampling for carnivores

Marcella J. Kelly, Julie Betsch, Claudia Wultsch, Bernardo Mesa, and L. Scott Mills

Now is an exciting time to study carnivore ecology via noninvasive sampling methods. Technological and methodological advances, and new techniques for data analysis, have contributed to a rapid increase in noninvasive carnivore studies. These studies complement and extend inferences from traditional sampling regarding individuals, populations, and communities. Today, researchers can estimate size and survival rate for a population, estimate historic and current rates of movement across fragmented landscapes, and measure carnivore stress loads without ever catching, handling, or even seeing a single animal. Noninvasive sampling is the gathering of data without capturing, handling, or otherwise physically restraining individual animals. The techniques usually imply that a target animal is not observed during data collection and, presumably, is unaffected by data collection. Although direct animal observations for behavioral studies and for distance sampling may also be considered noninvasive, we do not include these direct observation methods. Noninvasive data-collection methods include sign surveys, diet analyses, camera trapping, DNA extraction, and endocrine (see Chapter 12) or disease monitoring (see Chapter 13) from scats and hair. Perhaps a better name is "less invasive" because we do not really know the impact, for example, of removing scat samples found in a jaguar (*Panthera onca*) habitat for 2 months. Such a study might disrupt marking behavior and unknown impacts could arise in a study site from the presence of a trained scent-detecting dog locating scats. Nonetheless, the term has gained familiarity and become conventional (Long *et al.* 2008a).

Why use noninvasive sampling? The advantages are numerous. Capture and handling are highly stressful and potentially dangerous to both humans and animals, especially with large carnivores. Invasive studies require more permitting, especially with endangered species, and often suffer issues with animal care and use committees. In addition, capture, handling, and subsequent monitoring are usually expensive, logistically difficult, time consuming, and result in small samples sizes, limiting population-level inferences, especially for elusive, low-density, or trap-shy animals. By contrast, noninvasive techniques can produce larger sample sizes, reducing bias, increasing precision, and broadening the scope of potential hypotheses. Noninvasive field sampling is often relatively simple to employ and to standardize, training inexperienced people can be easy and studies can cover large areas. Finally, noninvasive sampling is less likely to induce a trap response in animals, again reducing human-induced bias.

While noninvasive techniques supply new information and hold great promise, we do not suggest that they should replace all traditional capture and handling studies, such as those to obtain information about body condition, to collect blood, or to affix transmitters for studies of movements, home ranges, and habitat selection (Chapters 6, 7, 8, 9, and 10). This chapter echoes and extends the recent book on this topic (Long *et al.* 2008a), and focuses on recent advances.

4.1 Methods of noninvasive sampling

4.1.1 Sign surveys

Naturalists have sampled carnivores noninvasively for decades. Skillful, field-based identification of tracks, scats, kills, bones, and hair have illuminated much of what we know about distribution and habits of carnivores, and have instilled a deep appreciation for natural history. In fact, identification of animal sign can be quite reliable in some instances. For example, Prugh and Ritland (2005) identified coyote (*Canis latrans*) scats by morphology with >90% accuracy in the Alaska Range, despite the presence of three other similarly sized carnivores. In other cases, however, scat identification by morphology alone is prone to error. For example, 18% of scats identified with high confidence by experienced field collectors as marten (*Martes martes*) scats were actually from foxes (*Vulpes vulpes*, Davison *et al.* 2002). Misidentification of carnivore sign in the field occurs more often when the target species is rare (Prugh and Ritland 2005).

As with scat, identification of tracks in snow, dirt, and mud can be useful and at times reliable. However, identification problems can arise due to substrate quality and animal movements (Heinemeyer *et al.* 2008). If concerns about uncertainty can be ameliorated, track surveys can be effective and inexpensive for occurrence and distributional studies. Snow tracking has been used widely in the US (Zielinski and Kucera 1995), Canada, and Scandinavia (Pellikka *et al.* 2005; Hellstedt *et al.* 2006) to monitor populations of diverse carnivores. In open landscapes, snow tracking even can be conducted from helicopters or planes (Heinemeyer and Copeland 1999).

Carnivore presence can also be determined from hair (from scats, kill sites, or hair snags) using macro- or microscopic examination of hair morphology (Raphael 1994; Teerink 2003). Where all sympatric carnivores and other species with similar hair patterns can be catalogued, hair morphology *may* be diagnostic for species' identification (Oli 1993; Gonzalez-Esteban *et al.* 2006). In many cases, unfortunately, no diagnostic visual or microscopic characteristics exist for species' identification, e.g. black versus grizzly bears (*Ursus Americana vs U. arctos*, Woods *et al.* 1999); and hairs from different parts of the body may have different morphology.

Other sign, such as scrapes, tree nests, latrines, and kills, can also be used to survey for specific carnivores. Identifying sign is a terrific natural history skill, but sign surveys by themselves supply limited information, due to species' misidentification and inability to distinguish individuals. Assuming, however, that species' identity from sign surveys is accurate, "occupancy modeling" (Chapters 2, 11, 16; MacKenzie and Nichols 2004; MacKenzie *et al.* 2006) allows researchers to combine detection/non-detection histories with spatial modeling to estimate and to predict species' occurrence across a landscape. By incorporating estimates of detectability from sign surveys directly, this approach corrects the inherent negative bias present in naïve occupancy estimates (MacKenzie *et al.* 2003; Tyre *et al.* 2003).

4.1.2 Genetic sampling

Noninvasive collection of genetic samples is limited only by the creativity and natural history knowledge of the investigator. Carnivore hairs and scats are the two most commonly collected genetic samples. Hairs are often obtained via snags or rub devices (Figure 4.1). To sample bears, researchers have strung barbed wire around bait, and bears leave hair on the wire when approaching the bait (Woods *et al.* 1999; Mowat and Strobeck 2000; Kendall *et al.* 2009). Sampling bears' natural rub trees can detect bears not sampled by barbed wire corrals (Boulanger *et al.* 2008; Stetz *et al.* 2010). After McDaniel *et al.* (2000) published a protocol for a baited hair-collecting pad, using roofing nails for Canada lynx surveys, this collection device was used to sample Eurasian lynxes (Schmidt and Kowalczyk 2006), ocelots (*Leopardis pardalis*, Weaver *et al.* 2005), and felids in the tropics (Castro-Arellano *et al.* 2008). Rub pads and backtracking putative lynx tracks in snow to collect scats and hairs is more efficient than rub pads alone (McKelvey *et al.* 2006). Zielinski *et al.* (2006) used glue tips to collect hair from small forest carnivores.

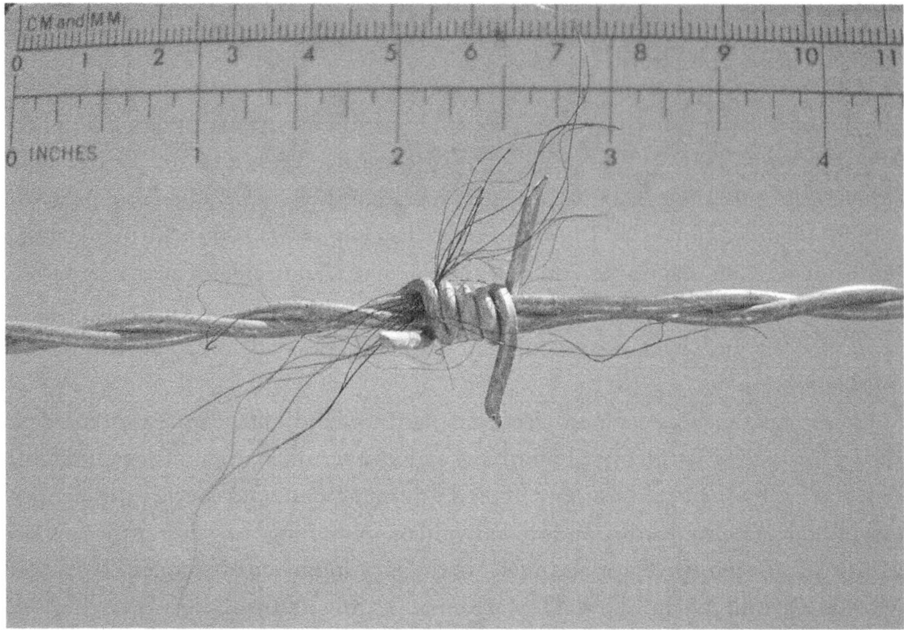

Fig. 4.1 A barbed wire hair snag "capturing" black bear hair for later DNA analysis. As part of a road ecology study, barbed wire was strung along the entirety of an 11-mile stretch of a highway, which was due to be widened. In addition to locating hotspots of road crossing (and identifying areas for potential underpasses), this study examined which sex and individuals were more likely to cross roads and where. Photo courtesy of J. Andrew Trent, Virginia Tech.

Hair-snag devices are usually inexpensive and easy to install but require carnivores to find them (potentially necessitating baits and species-specific attractants) and to rub against them. For some species, the amount of DNA left may be very small (e.g. single hairs or hair fragments). Hairs with follicles provide higher quality DNA extracts than do scats, which have more agents that inhibit and prevent amplification. A single hair, however, usually yields much less DNA than feces. Multiple hairs can usually be pooled to increase DNA yield for species' detection studies, because diagnostic bands for multiple species can be simultaneously visualized. When individual identity is required, however, pooling multiple hairs is risky because it can create false, "new" genotypic individuals (see Alpers *et al.* 2003; Roon *et al.* 2005). Researchers must accept the low DNA yield from single hairs, or perhaps develop a hair snag that allows only one animal to rub it (Beier *et al.* 2005; Bremner-Harrison *et al.* 2006).

Fecal DNA originating from cells sloughed from the intestinal lining and extracted from scat samples can be collected from elusive carnivores, which often deposit scat at prominent sites for intra- and interspecific communication (Gorman and Trowbridge 1989; Barja *et al.* 2005). Typically, scats are collected by walking transects and searching visually. Efficiency can be increased by following animal tracks in the snow, sand, mud, or dust (McKelvey *et al.* 2006; Ulizio *et al.* 2006; Marucco *et al.* 2008).

Researchers also can increase scat-collection rates, even over large, remote areas, by using scent-detecting or scat-detector dogs (*Canis familiaris*; Hurt *et al.* 2000; Wasser *et al.* 2004; Smith *et al.* 2005, 2003; Long *et al.* 2007; MacKay *et al.* 2008). Detector dogs commonly are trained and handled following protocols applied for search-and-rescue dogs (MacKay *et al.* 2008). The dogs must have a strong, object-oriented drive towards a toy or food, which serves as a reward after successful detection. High-performing detection dogs are hard-working, energetic, focused, bold individuals, and are selected independent of breed or sex (Svartberg 2002; Maejima *et al.* 2007; Rooney *et al.* 2007). They require much attention and focused care from trained, professional handlers. A handler needs to learn a dog's behavior and body language to interpret detection alerts correctly under difficult field conditions. Handlers must have knowledge of scent-direction patterns in challenging environments (Shivik 2002; Gazit and Terkel 2003). After a dog alerts its handler (Figure 4.2), the handler investigates the find, being careful about body language so as not to affect the dog's response, and decides if the dog was successful and deserves a reward. Handlers should carry and use target scats during periods of low scat detections to keep dogs motivated and reliable (i.e. prevent false detections to get its toy).

Using scat dogs in a survey design depends on study objectives, habitat, and characteristics of the target species and budget (scat dogs can be expensive). Established trails and roads may be used in some cases, as when DeMatteo *et al.* (2009) surveyed for bush dog scats within 15 m of both sides of trails and roads through thick tropical vegetation. In other cases, opportunistic searches may be made within survey grid cells (Wasser *et al.* 2004; Wultsch 2008; Figure 4.3), or following predefined transect routes (Smith *et al.* 2006; Long *et al.* 2007). If scat dogs follow the trails of individuals of the target species, scats will not be a random or representative sample from the population. This bias is important for some studies. Finally, study design must include tests of scat dogs to document error rate for each individual.

Once collected, samples must be properly stored to inhibit enzymes that degrade DNA. In general, hair samples are easy to store. In dry environments, simply place hairs in individual paper envelopes; in humid climates, dry them quickly and

Fig. 4.2 Example of a sit alert by a scat detection dog, Billy, upon finding a felid scat in Belize, Central America.

completely with silica gel drying agent. For scats, DNA quality can vary depending on collection location, environmental conditions (e.g. UV light, humidity, mold) and the region of the scat sampled (Wultsch 2008). Dry scat samples in silica gel, with at least 5X desiccant per part of sample. Samples can be frozen, but repeated freeze–thaw should be avoided (do not use household freezers with self-defrost). A scraping of a scat to obtain shed epithelial cells can be put into buffer solution in the field, preferably into screw-top tubes to prevent leakage. Liquid storage techniques, such as ethanol (>95%) and DET buffers (at 5–10 parts per part of sample), are excellent for DNA preservation, but ethanol has a tendency to leak.

Other sample types can be stored similarly. For example, DNA collected from saliva with a Q-tip-like swab, either directly from a carnivore or from a prey item bite wound to determine the species and identity of the predator (Sundqvist *et al.* 2008), can be allowed to dry in a paper envelope (avoid plastic bags) or, for longer term storage, placed in Longmire buffer. Urine samples can be collected with swabs or collected directly from snow and kept frozen until DNA extraction (e.g. Hedmark *et al.* 2004; Sastre *et al.* 2009).

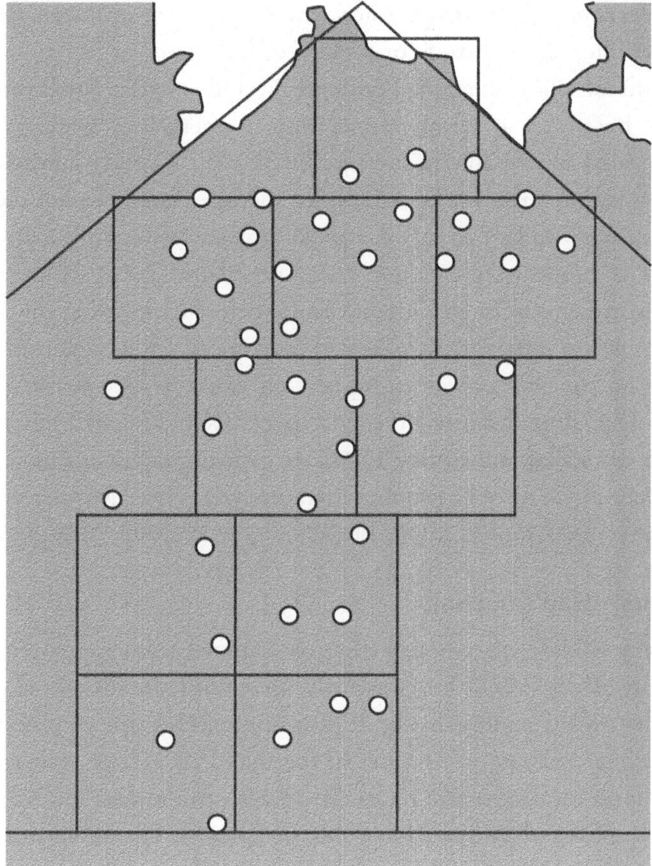

Fig. 4.3 An example of a survey designed specifically for capture–recapture estimates of abundance of felids and to compare two noninvasive survey techniques (remote cameras and molecular scatology) in Belize, Central America. Remote cameras were placed at 1.5–3.0-km intervals for ocelots and jaguars and were operational for 2.5 months. Camera data were collapsed such that every 10 days was one encounter occasion for ~8 encounter occasions for capture–recapture population estimates. A 4 × 4 km grid was superimposed over the camera grid and a scat-dog team searched opportunistically for felid scat for a minimum of 5 km per grid cell. After completing all grid cells in roughly 10 days, the scat-dog team repeated the survey up to five times to create five encounter occasions for mark–recapture. Two camera stations were left out of the analysis due to difficulties in reaching those stations with the scat dog.

Obtaining at least 5–10 known tissue or blood samples—preferably from the same animals for which noninvasive samples were collected—allows optimization of laboratory protocols and determination of genotyping errors (see Section 4.2.1). High-quality samples may include ear tissue punches < or ~5 drops of blood.

Tissue samples can be desiccated in silica, stored in ethanol, or frozen. Blood can be stored dry on filter paper, frozen, or preserved in buffer.

The benefit of silica, ethanol, Longmire or DET buffer preservation is that samples can be stored at room temperature. Nonetheless, freezing samples at –20°C (or colder) is advisable to increase DNA yield. Getting samples to the lab for extraction within a few weeks or months, increases amplification success. Field personnel must be vigilant to guard against cross-contaminating samples during collection. The proper steps include using new latex gloves with each sample, sterilizing instruments with alcohol and flame before each collection, and storing different samples in different, well-labeled containers. Data organization is facilitated through a bar code system using peel-off labels to link physical samples to information (e.g. time, location) on data sheets (Kendall and McKelvey 2008). Methods for extracting and storing DNA are evolving rapidly; readers should see reviews and check the forensic genetics literature (e.g. Oyler-McCance and Leberg 2005; Schwartz and Monfort 2008; Beja-Pereira *et al.* 2009; Morling 2009).

4.1.3 Camera-trap sampling

Photographing wildlife via remotely triggered cameras (camera trapping) emerged in 1877 (Guggisberg 1977) but was little used until the invention of infrared, automatically tripped cameras in the 1980s. Cameras became commercially available, lightweight, and easy to operate. In the mid-1990s, large-scale camera grids were linked with capture–mark–recapture analysis to estimate animal abundance (Karanth 1995; Karanth and Nichols 1998, Chapter 5). The 2000s brought digital camera technology. Widespread, remote camera use has resulted in an increase in carnivore inventories, due to the ability to photograph multiple species at the same site (e.g. Barea-Azcon *et al.* 2007; Datta *et al.* 2008; Tobler *et al.* 2008; Can and Togan 2009; Johnson *et al.* 2009; Pettorelli *et al.* 2009).

Two infrared trigger mechanisms exist in remote camera technology: active and passive infrared systems. An active infrared beam is triggered by an animal breaking an infrared beam that passes from a transmitting unit through the detection zone to a receiving unit. A passive infrared system is triggered by the heat difference between the animal and the environment as the animal moves past a heat and/or motion sensor (Kays and Slauson 2008). While pressure pad and baited string-trip cameras are still useful (King *et al.* 2007), most modern studies use passive infrared systems. Camera flashes at night may cause aversion and may be potentially damaging to the eyes of mammals (Schipper 2007), yet some carnivores appear attracted to a flash, especially large felids (personal experience). Digital camera options now include white flash and infrared flash, but image quality is still low with the latter.

Digital camera durability and reliability are increasing and, most importantly, they do not have the 36-exposure limit of film cameras. Some passive digital systems can transmit images wirelessly to a base station or laptop computer. Additionally, many models can collect short video sequences.

4.1.4 Endocrine/hormone sampling

Hormones affect physiological processes that maintain homeostasis allowing an animal to cope with its environment. The emerging discipline of "conservation physiology" seeks to understand the physiological responses of animals to environments altered by human disturbance (Wikelski and Cooke 2006). Noninvasive endocrine tools are employed to monitor wildlife populations and individuals (Berger *et al.* 1999; Foley *et al.* 2001; Garnier *et al.* 2002; Sands and Creel 2004; Cockrem 2005). Immunoassays can measure the concentration of select hormones and their metabolites from noninvasively collected samples such as scats (Wasser *et al.* 1988, Creel *et al.* 1997, Barja *et al.* 2008), urine (Thompson and Wrangham 2008, Braun *et al.* 2009), saliva (Queyras and Carosi 2004), and hair (Koren *et al.* 2002). Due to metabolic clearance rates and gut transit time, fecal metabolite concentration represents a cumulative concentration over time (Schwarzenberger 1996). The length of time depends on the species and the stressor, and requires background research in controlled conditions.

Two types of steroid hormones are commonly assessed in noninvasive studies of wildlife endocrinology: adrenal and gonadal. Adrenal hormones, including glucocorticoids (GLCs), also known as stress hormones, are commonly measured as an indicator of overall physiological condition of an individual or a population. Gonadal hormones, such as progestagens, estrogens, and androgens, are used to determine puberty, estrous, ovulation, pregnancy, abortion, and sex (Brown and Wildt 1997; Morato *et al.* 2004; Sanson *et al.* 2005; Graham *et al.* 2006; Dehnhard *et al.* 2008; Herrick *et al.* 2010).

In biological samples, hormones can be assessed through both quantitative and qualitative techniques. Immunoassays and spectrometric techniques can detect small concentrations and immunoassay techniques are used widely in wildlife physiology (Chapter 12). While conservation physiology is an exciting, emerging discipline, multiple cautions exist. Background hormone levels, time-lags in endocrine response, impacts of age, sex, social status, and microflora on metabolite levels, can confound assessments of potential stressors. So far, most studies are correlative and do not directly address cause and effect (Millspaugh and Washburn 2004; Chapter 10). Although chronically elevated GLCs induced by a persistent stressor *can* have negative effects on an organism, including behavioral, reproductive, metabolic, immune, and neurological functions, to date GLCs have not been linked to meaningful measures of fitness or population dynamics, and are often linked only indirectly to potential stressors.

4.2 Recent tools and advances in noninvasive sampling

4.2.1 Noninvasive DNA techniques

DNA can be collected by sampling hair, scats, urine, regurgitates, saliva, or nearly any other sloughed piece from an animal. The current deluge of noninvasive genetic sampling for carnivores traces its roots to a single development in the late 1980s: the invention and commercialization of the polymerase chain reaction (PCR). PCR "amplifies" DNA, producing millions of copies of the original template DNA, so that researchers can decipher the genetic makeup of organisms from noninvasively collected samples that may be of poor quality or small quantity.

A DNA marker is a sequence of DNA amplified via PCR. Fragment analyses separate targeted pieces of DNA by size. For species' identification, fragments often are amplified from mtDNA because the high copy number increases the probability of amplification for low-quantity, low-quality samples. Often, the size of the amplified DNA itself is not diagnostic for different species, so the amplified product is broken into species-specific pieces, known as RFLPs (restriction fragment length polymporphisms). Different-sized fragments, diagnostic for each species, are produced depending on whether and how mutations have changed the DNA sequences recognized by the endonuclease (Figure 4.4). RFLPs have been applied to differentiate endangered species, such as the San Joaquin kit foxes from other cooccurring canid species (red fox, grey fox, coyote, domestic dog) (Paxinos *et al.* 1997) and to identify species of felids, ursids, and mustelids (Mills *et al.* 2000a; Riddle *et al.* 2003; Vercillo *et al.* 2004; Colli *et al.* 2005; Bidlack *et al.* 2007; Livia *et al.* 2007).

These fragment approaches are fast and inexpensive but they can be limited by potential variation in mtDNA fragment lengths among individuals within a species. In addition, species sampled must be known a priori, and primers and restriction enzymes must have been identified. Where carnivore species are little known, amplified fragments can be sequenced. A nucleotide sequence must then be compared to known sequences archived in a sequence database (e.g. GenBank). Direct sequencing is expensive (though prices are dropping) and can contaminate the signal of the carnivore with that of its prey, if the sample is scat.

A rapidly developing variant of sequenced mtDNA fragments for species' identification is named "The Barcode of Life Initiative" (Savolainen *et al.* 2005; Ratnasingham and Hebert 2007; www.barcodinglife.org). DNA barcoding

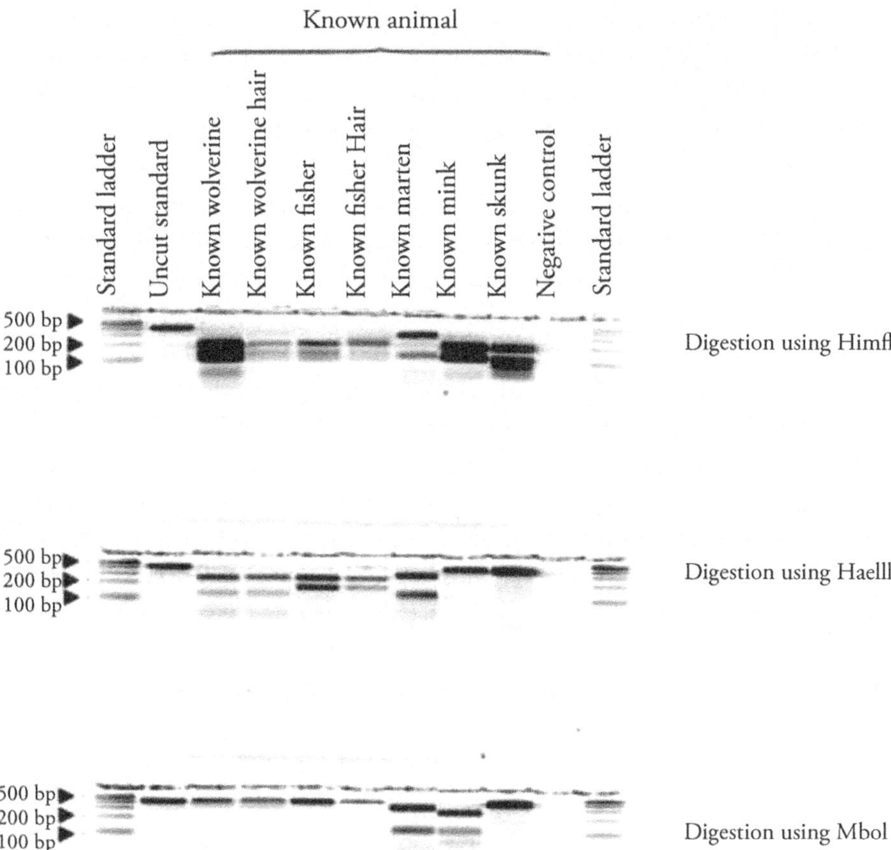

Fig. 4.4 An example of RFLP (restriction fragment length polymporphisms) fragment analysis of mtDNA to distinguish different forest mustelids of the northern USA, using single hairs from noninvasive snags (from Riddle *et al.* 2003). After amplifying the cytochorome *b* region of mtDNA with PCR, the DNA was digested with three different restriction enzymes, creating species-specific fragments that collectively distinguish among different species. The first and last lanes are a molecular ladder that helps to determine size of the bands, and the uncut standard contains a PCR product from a wolverine not subjected to the restriction digests; the negative control is pure water to check for contamination. An example for practice: the first restriction digest (*Hin*fI) distinguishes between marten (with two fragments, of 329 and 113 bp in size) and wolverine (with three fragments of 212, 132, and 98 bp), but wolverine has exactly the same bands as fisher (which appear lighter but are still present). So the next digest (*Hae*III) distinguishes between wolverine (259, 140, and 43 bp) and fisher (259 and 183 bp). Thus multiple restriction enzymes are like multiple morphological characteristics that we might use to tell different species apart.

depends on standardized analyses of a specific DNA region for all species on earth. For animals, the accepted barcode region is a 648-bp region of the mitochondrial Cytochrome c oxidase subunit I (referred to as *cox1* or COI). Barcoding is well-suited to noninvasively collected carnivore samples and has standardized methodology.

Microsatellites, or simple-sequence repeats (SSRs), are short sequences of nuclear DNA repeated between 5 and 100 times, which are widely used for individual-level questions in carnivores. Microsatellite loci typically have high variation within species and are codominant, with alleles displaying Mendelian inheritance. Thus, microsatellites are well-suited to traditional population genetic models and to distinguishing individuals. Sets of highly polymorphic microsatellite loci have been identified for many different carnivore species. Likewise, single nucleotide polymorphisms (SNPs) can be used to address individual-level questions related to genetic variation and population structure, parentage and relatedness, and individual identity (Morin *et al.* 2004; Morin and McCarthy 2007).

The future for carnivore genetic sampling lies in complementing the neutral markers described above with markers that describe, or are linked to, genes of known coding function. These markers, only now being developed for wildlife species, will bring us one step closer to describing fitness attributes directly. These may range from behaviors (e.g. sprint speed), to morphology (e.g. muscle structure), to physiology (e.g. biochemical processing of nutrients).

For carnivore sex determination, a gene present only on the male Y chromosome, such as the SRY gene (testis determining factor), will amplify and be detected in males but not in females. To control for amplification failure, this method usually requires co-amplifying 1 or more microsatellite loci as a control. A second approach amplifies a portion of DNA with alleles of different size residing on both the X and Y chromosomes (Shaw *et al.* 2003). Carnivore sex determination has been applied to fecal samples from sympatric felids in North America (Pilgrim *et al.* 2005) and Asia (Wei *et al.* 2008).

Any time a recorded genotype, or molecular marker, deviates from the actual genotype or marker, a genotyping error has occurred. Although improvements in lab technology and techniques have decreased many forms of genotyping error, it remains an inescapable issue made more prevalent with the low-quantity, low-quality DNA yields of noninvasively collected samples. If unaccounted for, genotyping error could compromise all uses of noninvasively collected DNA samples, including species and individual identity, paternity analysis, occupancy and abundance, gene flow, forensics, and behavior. Some standards for minimizing and measuring genotyping error include following strict protocols, genotyping each specimen multiple times to obtain consensus genotypes (Waits and Paetkau

2005), using statistical metrics to determine levels of genotyping error (McKelvey and Schwartz 2004), and proper use of blind controls (Mills 2002). Once estimated, genotyping error can be incorporated explicitly into parameter estimates (e.g. abundance: Lukacs and Burnham 2005; paternity: Kalinowski *et al.* 2007; Knapp *et al.* 2009).

Genotyping error has been extensively evaluated for abundance estimation, where failing to account for genotyping errors can cause successive captures from the same individual to appear to be from different individuals, biasing the estimates of abundance high (Waits and Leberg 2000). The opposite problem, a low bias, arises when different animals fail to be distinguished due to having too few loci or having too little variation. This phenomenon, termed the "shadow effect" because different animals appear as identical genetic shadows of each other (Mills *et al.* 2000b), decreases as many, highly variable loci become available for most species.

In short, genotyping error is an important consideration in designing and implementing noninvasive genetic studies. It may cost more (e.g. by running each sample multiple times), and it may make the analysis more complicated, but the reward will be more precise and unbiased estimates.

4.2.2 Using noninvasive DNA data

Genetic data can be used to estimate species' distributions. Berry *et al.* (2007) used fecal DNA sampling to provide range information for invasive but cryptic red foxes, a devastating pest, in Tasmania. Canada lynx range distribution on national forest land across the USA was surveyed with mtDNA obtained from hair collected both from baited hair-traps and from backtracking tracks in snow (Mills 2002; McKelvey et al 2006). Nicholson and van Manen (2009) used hair samples to document that site occupancy for black bears decreased after completion of a new highway in North Carolina and that the decrease was not a function of distance from the highway, rather, the highway affected the entire study area.

The ability of noninvasive genetic sampling to identify individuals makes available the entire body of capture–mark–recapture methods for abundance estimation (Chapter 5). Capture–recapture studies use two basic genetic approaches. The first uses multiple discrete "capture" occasions (Otis *et al.* 1978; Huggins 1989). Hair traps, for example, are spread across a landscape in grid-like fashion (Tredick and Vaughan 2009), or opportunistically with regular or clustered spacing (Kendall *et al.* 2009; Robinson *et al.* 2009). Capture periods last several days to weeks, depending on frequency of returning to the traps to collect hair, and data are analyzed with closed capture models (e.g. Program MARK). The second approach uses continuous trapping, wherein individuals can be "captured" multiple times (e.g. hair snared on multiple traps) within a single trapping occasion

(Miller *et al.* 2005; Petit and Valiere 2006). The samples resemble random draws from the population with replacement and can be analyzed with closed capture models in CAPWIRE (Miller *et al.* 2005) or BAYESN (Gazey and Staley 1986; Petit and Valiere 2006).

Noninvasive genetic sampling can extend capture–recapture techniques over time (e.g. years) to obtain multiple estimates of population size. Such data can yield estimates of population growth, survival, and recruitment. In the Alaska range, Prugh *et al.* (2005) used fecal genotyping of 834 scat samples over a 3-year period to estimate coyote survival rates. Interestingly, they found that radio-collared individuals had higher survival rates than uncollared individuals, but that survival did not differ between the sexes. Marucco *et al.* (2008) genotyped 1399 scats from 14 sampling sessions of wolves recolonizing the Western Alps in Italy and France. Using open-population models and AIC model selection they documented that young wolves had lower apparent annual survival than adults, that survival rates were lower in the summer than in the winter, and that population growth over 7 years was positive ($\lambda = 1.04$) but lower than that recorded for other recolonizing wolf populations.

Noninvasive genetic sampling provides researchers with new approaches to use landscape genetics to elucidate conservation challenges. Although the seminal theory for quantifying population structure from genetic data dates back to Sewall Wright (1931), the field of "landscape genetics" has blossomed as a recent integrative discipline to understand how landscape features affect animal movement and local adaptation (Storfer *et al.* 2007; Balkenhol *et al.* 2009; Sork and Waits 2010). F_{ST} (Wright's measure of genetic distance) or coalescent approaches provide relative measures of gene flow, assuming equilibrium between genetic drift increasing divergence and gene flow decreasing it. Because this equilibrium would have been achieved many generations previously, before recent human-caused population fragmentation, these measures may essentially provide a window into historic levels of connectivity. By contrast, current levels of gene flow (interpopulation movement followed by breeding), analogous to immigration-and-reproduction events measured by radio-telemetry, can be estimated by genetic assignment tests. As an example, Proctor *et al.* (2005) used hair traps to survey for grizzly bears on both sides of a major highway just north of the US–Canada border. They used two approaches to test for migrants. First they used area-specific allele frequencies in a likelihood-based assignment test (Paetkau *et al.* 1995). Individual were "assigned" to the area with the highest probability of occurrence. Second, they used a model-based clustering method in program STRUCTURE (Pritchard *et al.* 2000), which clusters individuals into groups through iterative assignments and probabilities of origin. Individuals that are repeatedly assigned to a group other than where they were captured, are considered putative migrants. The results were striking: a surprisingly small amount of migration that was heavily sex-biased occurred (mostly males and only one female), suggesting that demographic connection had been severed across their entire range in southern Canada by the highway and associated settlements.

It is now axiomatic that maintaining genetic variation is important, both to maintain long-term evolutionary potential in a changing environment and to minimize the short-term demographic effects of inbreeding depression (Frankham 2005; Mills 2007; Chapter 9). Noninvasive genetic samples can provide estimates of genetic variation (e.g. heterozygosity and polymorphism) and effective population size (Tallmon *et al.* 2008). More importantly, sampling over time, or in fragmented vs. control sites, can document potential *decreases* in heterozygosity, which can translate into decreases in survival or reproductive rates via inbreeding depression. Inbreeding depression, in turn, can decrease population growth rate and decrease population viability (Mills and Smouse 1994).

Noninvasive genetic sampling has also become important for assessing the taxonomic status of individuals. For example, the primary threat to the persistence of reintroduced, endangered red wolves (*Canis rufus*), is hybridization with coyotes; a microsatellite nDNA test of wild-born pups (Adams and Waits 2007) allows managers to detect and remove hybrids before they can interbreed with the extant red wolves. Additional insights into hybridization can be revealed using mtDNA, whose maternal inheritance indicates the direction of hybridization. Because coyote mtDNA is found in gray wolves but not vice versa, hybridization between coyotes and wolves occurs by way of male wolves mating with female coyotes (Lehman *et al.* 1991).

Of course, the fact that genetic sampling can provide diagnostic identification of both species and individuals has immediate and broad implications for forensics and solving wildlife crimes involving carnivores. Millions and Swanson (2006) hypothesized that bobcats in Michigan were being poached from the Lower Peninsula (LP) but registered by hunters as harvested from the Upper Peninsula (UP), where bag limits were higher. Microsatellites markers and assignment tests documented that some bobcats claimed as harvested in the UP were genetically assigned to the LP. In a more condemning example, Caniglia *et al.* (2009) extracted DNA from wolf canine teeth on a necklace to show that the teeth belonged to six individual Italian wolves (a legally protected species), including a male and a female wolf recently found dead.

Genotyping individuals within a population allows researchers to calculate relatedness and thereby to examine social and mating structure. Gotelli *et al.* (2007) used fecal DNA analysis to determine paternity of cheetah (*Acinonyx jubatus*) cubs and found that adult females were surprisingly promiscuous. Not only did 43% of litters have multiple fathers, females mated with unrelated males within an estrus cycle and mated with different males in subsequent breeding seasons.

4.2.3 Data collection, handling, and analyses with remote cameras

Type of cameras used, placement, and duration of camera-trapping studies depend on the goals of the study (Figure 4.3), land cover, and budget. Deploying large numbers of remote cameras is expensive. Study design, deployment, site selection, equipment management, minimizing theft and wildlife damage, camera expense, reliability, and sensitivity have been reviewed by Swann *et al.* (2004), Kays and Slauson (2008), and Long and Zielinski (2008). Several websites compare camera performance and prices. Regardless of study goals, one must plan for the substantial data-management required in any camera-trapping study. Sifting through photographs (either film or digital) and entering them into a useful database often takes more time than deploying and monitoring cameras in the field. While a study may target only one carnivore species, entering all data on all non-target species, including humans, is important, as this information can become useful for determining potential competitors, distribution of prey, linking trapping rates (photos taken) of the target carnivores to trapping rates of prey, and human use of the study site (Figure 4.5).

No standard for number of camera-trapping stations, spacing between cameras, or duration of surveys exists for documenting carnivore presence or conducting species' inventories (Kelly 2008). Camera placement and spacing is flexible and often includes targeting likely areas with more cameras, while not surveying unlikely areas. Many studies use a minimum of ~1000 trap nights per study site, but variable objectives and detectability of the target species affects trap nights needed. The lower the detectability of target carnivores, the more trap nights are needed. Increasing camera saturation can decrease the total number of trap nights needed to detect target carnivores (Wegge *et al.* 2004).

Camera trapping was first used in conjunction with capture–recapture models to estimate abundance and density for tigers (Karanth 1995; Karanth and Nichols 1998) and then modified for other boldly marked felids (e.g. Silver *et al* 2004; Maffei *et al.* 2005; Di Bitetti *et al.* 2006; Dillon and Kelly 2007) and carnivores with ear tags (Thompson 2007) or uniquely identifiable ear streamers (Bridges *et al.* 2004a). The technique can even be used for subtly marked species, such as pumas (Kelly *et al.* 2008) and red foxes (Sarmento *et al.* 2009), albeit with more constraints and lower confidence. Alternatively, Rowcliffe *et al.* (2008) treated

Fig. 4.5 Photographs from remote cameras: (a) R. Felix Jean and A. Vonjy Arindrano (WCS/Madagascar) conducting a camera check and demonstrating double documentation of date, camera number, and station number on placard and time embedded on digital image. Information is also recorded on a data sheet and input into a computer database. (b) Non-target species, such as tamanduas (Mountain Pine Ridge, Belize), are often caught on remote cameras and can reveal interesting behaviors. Non-target species (even humans) can prove valuable for biodiversity surveys and potentially can be linked to presence or trapping rates of target species. (c) Bears are notorious for damaging remote cameras, as in this photograph taken near Mountain Lake Biological Station, Virginia. Every remote camera study should plan for sufficient cameras to replace those that are stolen, vandalized, damaged, or malfunction. (d) Remote film camera captures two jaguars in Hill Bank, Belize demonstrating differences in coat patterns that make individual identification possible for mark–recapture studies.

contact rates between cameras and animals using an "ideal gas" model to scale trapping rate linearly with density.

Survey design for estimating abundance using remote cameras is an active area of research (Chapter 5). Most studies use a fixed grid with a minimum of 20 stations with 2 cameras per station, at a spacing that ensures that each individual has a reasonable probability of capture. Capture histories are constructed for individual animals photographed at each site and data analyzed with closed capture models. Because camera grids are often different sizes and can change shape in longitudinal studies, abundance must be converted to density to make comparisons.

Unfortunately, estimating the effective trap area is a sticky problem. One can calculate half the mean maximum distance moved (MMDM) between camera locations among all individuals re-photographed at least once (Karanth and Nichols 1998), and apply this as a buffer around the trapping grid. Wilson and Anderson (1985) provide an entry to the extensive literature on calculating densities by MMDM approaches. New analytical approaches that estimate density directly through spatially explicit capture–recapture models avoid the potential pitfalls of the *ad hoc* mean maximum distance-moved approaches (e.g. Efford 2004; Gardner *et al.* 2009). Comparative analyses suggest that MMDM models can substantially overestimate density compared to spatially explicit capture–recapture models (Obbard *et al.* 2010; Gerber *et al.* 2011).

For long-term, longitudinal camera-trap studies on naturally marked individuals, the Holy Grail is to estimate survival and recruitment. To date, few studies have reached this goal. Karanth *et al.* (2006) used 9 years of data from remote cameras on 74 individual tigers to estimate abundance, population growth rate, survival, recruitment, temporary immigration, and transience using Pollock's "robust-design" (Pollock 1982; Pollock *et al.* 1990; Chapter 5).

The potential to use remote cameras for large-scale carnivore distribution studies is tremendous. Indeed, many countries are required to monitor biodiversity under directives such as the Convention on Biological Diversity (Mace and Baillie 2007). Pettorelli *et al.* (2009) combined camera-trap surveys across 11 sites in Tanzania, East Africa, with ecological niche factor analysis (ENFA; Chapter 10) to reveal distributional and habitat use patterns for 23 carnivore species. ENFA techniques (Hirzel *et al.* 2002) use presence-only data to determine habitat features that promote species' presence. An advancement from the ENFA approach is the occupancy-based approach, which also reveals habitat-use patterns and predicts carnivore occurrence across a landscape, by explicitly modeling detectability as a function of species and environmental variables (MacKenzie 2005; Chapter 10). Thorn *et al.* (2009) used baited camera-traps to estimate brown hyaena (*Hyaena brunnea*) occupancy in South Africa and Linkie *et al.* (2007) combined remote cameras and occupancy to gain information on sun bears (*Helarctos malayanus*).

Many studies have used remote cameras to assess finer scale habitat-use patterns, including use of existing trails by carnivores (Dillon and Kelly 2007; Harmsen *et al.* 2010; Davis *et al.* 2011). Trail systems funnel carnivores past cameras, facilitating photo-captures but potentially biased estimates of density. Alternatively, a researcher can establish a trail system for camera trapping in trailess areas, as carnivores are likely to begin using these paths (Maffei *et al.* 2004). Some carnivores, such as coyotes, however, may be wary of baited cameras on trails (Sequin *et al.* 2003). Other studies have used geographical information systems to extract land-cover data from circular buffers surrounding camera traps (Kelly and Holub 2008; Davis *et al.* 2011).

While limited in scope, camera traps can give insight into carnivore behavior, particularly for describing activity patterns for members of a single species (Bridges *et al.* 2004b; Vanak and Gompper 2007) or of sympatric carnivores studied simultaneously to gain insight into coexistence (Grassman *et al.* 2006a; Chen *et al.* 2009; Di Bitetti *et al.* 2009; Harmsen *et al.* 2009; Lucherini *et al.* 2009). Stevens and Serfass (2008) used remote cameras to examine group composition, seasonality, and activity patterns of river otters at latrines. Hunter (2009) used remote *video* cameras to record the responses of predators to taxidermy models of striped skunks and gray foxes, and learned that carnivores use both coloration and body shape to recognize and to avoid noxious species. Bolton *et al.* (2007) used a digital infrared camera system to monitor predation events at the nests of ground-nesting lapwings (*Vanellus vanellus*) and tree-nesting spotted flycatchers (*Muscicapa striata*). Finally, remote cameras are used extensively to identify species and use rates of highway crossing structures, such as under- and overpasses (Clevenger and Waltho 2000).

4.2.4 Data collection, handling, and analyses for endocrine studies

Techniques that measure steroid metabolites excreted in urine or feces provide an avenue for noninvasive research on the physiology of free-ranging carnivores. Urine is often impractical to procure from wild, free-ranging carnivores but could be feasible for endangered species in *ex situ* conservation programs. Steroids in urine are a reliable indicator of ovulation in carnivores (Dehnhard *et al.* 2006; Durrant *et al.* 2006).

Metabolic studies in captive carnivores show that adrenal and gonadal steroid metabolites are excreted at measureable concentrations predominantly in feces (Brown and Wildt 1997; Young *et al.* 2004). Scats should be handled with gloves and stored individually in resealable plastic bags or polypropylene tubes. The scats should be homogenized prior to lab analysis to ensure representative hormonal concentrations in the sample, since hormones and metabolites may be unevenly distributed. Moreover, after defecation, microflora present in a scat can produce enzymes that further metabolize steroids, altering concentrations. Freezing is recommended to arrest microbial and enzymatic activity, even if the samples have been dried or hormones extracted, unless metabolite stability is confirmed beforehand (Möstl and Palme 2002; Lynch *et al.* 2003; Millspaugh and Washburn 2004). Freeze/thaw events should be avoided. Degradation of steroid hormones in feces can be caused by ultraviolet light, humidity, and temperature. Additionally, degradation rates are influenced by time, diet, and species-specific intrinsic intestinal flora (Touma and Palme 2005; Schwartz and Monfort 2008). These factors may cause fecal glucocorticoid metabolites (FGMs) to increase through time (Washburn and Millspaugh 2002) or decrease (Pelican *et al.* 2007). To address the possible effects of environmental exposure on scats over time, a pilot degradation study under expected environmental conditions is essential *prior* to field collection, especially because usually there is no reliable way to age scats in the field. For example, a recent study on captive jaguars found that FGMs remained relatively constant for 4 days after defecation giving researchers a 4-day cyclical rotation for collecting scat samples in the field (Mesa, Kelly, Brown, unpublished data).

Because metabolism of steroid hormones differs for each species, laboratory procedures should be validated for the target species *prior* to a field study. Both type and concentration of metabolites will be unique for each species and type of biological sample. Validation involves three components. First, endocrine physiology can be evaluated either by stimulating endocrine organs (e.g. adeno-corticotropic hormone, ACTH challenge) or by monitoring physiological events (e.g. acclimatization, recovering from surgery, estral cycles, etc). Subsequently, a metabolite analysis using high-performance liquid chromatography (HPLC) or mass spectrometry facilitates the selection of candidate immunoassays for validation and provides information about metabolism and gut transit time. The use of captive individuals under controlled settings is strongly recommended. Second, hormonal extraction procedures are required to solubilize steroid hormones present in feces, usually through agitation or heating extraction. Hand agitation is practical for field conditions. Heating extraction ("boiling") is the best method of extraction and is often used to corroborate agitation methods. Third, immunoassay selection is based on both affinity of the antibody to the desirable metabolite (obtained from HPLC analysis) and cross-reactivity of the antibody with other hormonal metabolites (Millspaugh and Washburn 2004; Young *et al.* 2004; Palme 2005; Touma and Palme 2005; Keay 2006).

A stress response can be classified as either acute or chronic, depending on the length of exposure, intensity of the stressor (i.e. threat), and the ability of the individual to find a physiological balance (i.e. acclimatization or acclimation). The acute stress response is highly conserved phylogenetically across vertebrate taxa (Romero 2004); thus the mechanism is considered adaptive (Boonstra 2005; Nelson 2005). In fact, acute stress allows an organism to modulate its metabolism, redirecting resources from innate processes, like digestion, growth, immune function, and reproduction, to counter an immediate threat (Nelson 2005). When a stress response is sustained over extended periods of time, the organism is thought to be experiencing chronic stress, also known as distress.

Moberg (1985) noted that the neuroendocrine response to stress has the greatest potential to indicate the impact of stress on an animal's overall well-being. Wildlife endocrinology has demonstrated correlations between physiological responses of individuals to anthropogenic disturbances in ecosystems although connections to fitness or population dynamics are limited. For example, Barja *et al.* (2007) showed a direct correlation between unregulated tourism and the level of FGMs in European pine martens. FGMs in elk (*Cervus canadensis*) and wolves correlate with the intensity of winter snowmobile activity in Yellowstone National Park (Creel *et al.* 2002), though this response had little to no negative effect on the population dynamics of wolves. In a more complex example, dominant female meerkats (*Suricata suricata*) during pregnancy employ stressful evictions to suppress reproduction among subordinates (Young *et al.* 2006, 2008). Subordinates have higher concentrations of FGMs and reproductive down-regulation via decreased conception rates and increased abortions. Dominant females benefit by diminishing competition for limited care among their own and subordinate litters, and through lowering the chances of infanticide by the subordinate females (Young *et al.* 2006).

Reproductive status in wild captive animals has been monitored for decades through noninvasive endocrine sampling (Schwarzenberger 2007). Sex determination also can be achieved using fecal gonadal steroids (Barja *et al.* 2008).

Noninvasive endocrine monitoring has management implications for carnivore translocations, reintroductions, and rehabilitation. While no studies currently exist for carnivores, noninvasive fecal endocrine monitoring during translocation and after release can document time required to return to pre-translocation FGMs (Franceschini *et al.* 2008).

4.3 Combining noninvasive and traditional approaches

4.3.1 Comparative approaches among noninvasive techniques

Comparative studies have helped to increase efficiency and to identify which noninvasive techniques work best for a particular species. Long *et al.* (2008a: tables 12.1 and 12.2) provided a list of survey methods and their attributes for North American carnivores.

Several comparative studies designed specifically to assess noninvasive techniques for carnivores are instructive. Harrison (2006) found that a scat dog produced 10 times the number of bobcat detections as did remote cameras, hair-snares, and scent-stations, but the dog was the most expensive and time-intensive technique. This study did not, however, compare methods for determining the number of individual bobcats identified, which could be achieved through remote camera identification and DNA analysis of feces and hair. Long *et al.* (2007) used a scat-detector dog, hair snares, and remote cameras to survey carnivores in the northeast USA. All three techniques detected black bears but hair snares did not detect fishers or bobcats. They also found the scat-detector dog technique to be the most expensive but it yielded the highest detection rate, rendering it the most cost-effective in the long run. Gompper *et al.* (2006) compared track plates, remote cameras, snowtracking, and scat surveys and found that no one particular technique was best for all species within their carnivore guild. Track plates detected more small carnivores, such as martens and weasels (*Mustela* spp.), and were equivalent to camera traps for midsized carnivores, such as raccoons, fishers, opossums, and domestic cats. Cameras were efficient for bears, while scat surveys and snowtracking were the best methods for coyotes.

Using remote cameras, sign (scat, scrapes, tracks, scent marks) and molecular scatology McCarthy *et al.* (2008) found that low capture and recapture rates of snow leopards (*Uncia uncia*) with remote cameras caused capture–recapture estimates of abundance to be unreliable. Molecular scatology held promise, however, and sign surveys could be most efficient once corrected for observer bias and environmental variance. Tiger (*Panthera tigris*) abundance estimated from genetic capture–recapture models closely matched that from camera traps (Mondol *et al.* 2009).

In short, the choice of noninvasive techniques must be tailored to the target species and study objectives.

4.3.2 Combining traditional with noninvasive approaches

Traditional and noninvasive methods have been compared occasionally using carnivores. Using scat surveys, scent-stations, baited camera-trapping, and live trapping in baited box traps to estimate carnivore species' richness in the Mediterranean, Barea-Azcon *et al.* (2007) found that scent-stations and scat surveys were most efficient logistically and economically over a large spatial scale. They detected genets, however, only with scent-stations and baited cameras and box traps were best for wildcats.

Several studies have simultaneously used radio-telemetry and camera trapping to estimate carnivore densities. Two showed that camera trapping can grossly overestimate carnivore densities (Soisalo and Cavalcanti 2006; Dillon and Kelly 2008), while another found high congruence of the two methods (Maffei and Noss 2008). Balme *et al.* (2009a) used track counts and remote cameras to estimate the size of a known population of radio-collared leopards (*Panthera pardus*) in South Africa. The most accurate estimate of the known population came from camera trapping data, when it was supplemented by movement data from radio-telemetry.

Traditional camera-trapping methodology, however, did not result in gross overestimates, and track counts provided some reliable results.

4.3.3 Data quality and integrity in noninvasive surveys

Noninvasive techniques open doors to sampling carnivores in ways never imagined 30 years ago. The ease of sampling provides an opportunity for involving masses of untrained volunteers, with the potential for creating an unprecedented volume of carnivore data from the field. This strength of noninvasive sampling is also a weakness, as it creates novel problems in maintaining data quality and integrity. Fundamentally, this means that extra steps must be taken in the field to ensure high data quality. For example, training of volunteers and availability of detailed protocols, are essential in any large-scale noninvasive survey.

For camera surveys, an additional necessary step includes double, or triple documentation of camera station locations, dates, and researchers present (Figure 4.5a). Trigger each remote camera in a survey during camera setup, and also during each camera checking, with a placard that reads the station location and date (at minimum), even if this information is already embedded in memory card (Figure 4.5).

Similarly, noninvasive genetic studies require extra checks on data quality, but can still suffer if field collectors fail to follow protocols. The National Lynx Survey, a 3-year noninvasive study to determine lynx distribution across 16 states, provides an instructive example (Mills 2002, 2007). Several hundred personnel initiated the placement and checking of more than 21 000 hair rub pads, following detailed protocols provided by the principal investigators. The study was a success in the ambitious scope of sampling across the species' range in the USA, in that 80% of hair samples collected could be identified to species. Nevertheless, a few personnel threatened the integrity of the entire study by mislabeling samples (Thomas and Pletscher 2002). Fortunately, the study had in place essential checks at both the field and lab level, including the critical design feature that hair collection was only the first step in evaluating lynx presence; follow-up snowtracking and trapping efforts were built into the study to separate actual lynx populations from fur-farm escapees, transient individuals, or mislabeled samples.

Noninvasive approaches draw from cutting-edge advances in molecular genetics, biostatistics, population biology, endocrinology, and epidemiology. Carnivore ecologists must learn more scientific disciplines, in greater depth, than ever before for appropriate application of noninvasive sampling. Though daunting in complexity, the promise of greater understanding of carnivores via noninvasive sampling should be seen as a rallying call across disciplines. Never has there been a more challenging, but also exciting and productive time to study carnivores.

5

Humane and efficient capture and handling methods for carnivores

Gilbert Proulx, Marc R. L. Cattet, and Roger A. Powell

To be effective, conservation and management programs for carnivores require a good understanding of the animals' biology, ecology, behavior, and habitat requirements. To gather scientific information essential to the development of such programs, it is often necessary to capture, handle, and mark animals, a controversial and complex activity (Proulx and Barrett 1989) requiring special skills to minimize negative effects on individuals and populations, and to maximize scientific gains. For this, researchers should use methods that are consistent with codes of ethics and guidelines published by professional societies and countries (Table 5.1). They should continuously improve capture procedures and equipment to work more effectively and more safely for both animals and people (Powell and Proulx 2003). Research design should minimize both potential short-term and long-term effects of capture (Seddon *et al.* 1999, Cattet *et al.* 2008a), and deal with non-random sampling that may affect population structures (Banci and Proulx 1999).

Here, we discuss trap types, sets, and efficiency, and describe humaneness criteria that we use in the selection of specific carnivore traps. We review use of drugs as a primary method of capture through chemical immobilization, but also as a means to support mechanical capture methods by reducing stress and pain. Our approach results in some redundancy but minimizes confusion because different techniques can be used for the same group of carnivores, similar traps and anesthetics may be used for different mammals, and methods that meet performance criteria for one species may not for others. Lastly, we summarize complications that can occur with capture and handling, methods of humane killing, and techniques for restraining and marking carnivores.

Table 5.1 *Non-exhaustive list of codes of ethics and guidelines on animal welfare (all websites were accessed in were accessed in August 2011).*

Code of ethics/Guidelines	Organization/ Country	Reference
Guidelines for the Capture, Handling, and Care of Mammals as Approved by the American Society of Mammalogists	ASM Animal Care and Use Committee 1998	http://www.mammalsociety.org/committees/animal-care-and-use (last accessed 1 July 2011)
Guidelines for the Treatment of Animals in Behavioral Research and Teaching	ASAB/ABS 2000	http://www.animalbehavior.org/ABSHandbook/animal-behavior-society-handbook/27ABSASABGuidelinesForTheTreatmentOfAnimalsInBehavioralResearchAndTeaching
CCAC Guidelines: the care and use of wildlife	CCAC 2011	http://www.ccac.ca/Documents/Standards/Guidelines/Wildlife.pdf
Directive 86/609/EEC on the Approximation of Laws, Regulations and Administrative Provisions of the Member States Regarding the Protection of Animals Used for Experimental and Other Scientific Purposes.	European Union 1986	http://eur-lex.europa.eu/LexUriServ/LexUriServ.do?uri=CELEX:31986L0609:EN:HTML
Animal (Scientific procedures) Act in the United Kingdom	HMSO 1986	http://www.archive.official-documents.co.uk/document/hoc/321/321-xa.htm
Ethical Principles and Guidelines for Experiments on Animals	Swiss Academy of Medical Sciences 2005	http://www.scnat.ch/downloads/Ethik_Tiervers_Nov05_e.pdf
Animal Welfare Act	United States Department of Agriculture	http://www.nal.usda.gov/awic/legislat/awa.htm
Ethical Principles and Guidelines for the Use of Animals	National Research Council of Thailand 1999	http://ird.sut.ac.th/newsite/Form/pet_ethic_eng.pdf
Endangered Species Act	United States Fish & Wildlife Service 1973	http://www.fws.gov/laws/lawsdigest/esact.html

5.1 Mechanical capture methods

5.1.1 Traps and sets

Restraining traps allow captured animals to be released and include cage traps, foothold traps, foot, neck and body snares, and nets (Appendix 5.1). Killing traps include neck snares, and snap, planar, rotating-jaw, and killing box traps, and submarine traps (Appendix 5.1).

Diverse sets exist to capture carnivores (Appendix 5.2). Trap design, preparation, and sets affect trapping efficiency (target captures/trap-night; Boggess *et al.* 1990), and selectivity (number of non-target species). The tripping force of the trigger must match the size of target animals. For example, by setting pan tension on foothold traps at 1.4–1.8 kg, kit foxes (*Vulpes macrotis*) may be excluded from traps set for coyotes (*Canis latrans*, Phillips and Gruver 1996). A trap with a light tripping force may capture carnivores of all sizes and not be efficient due to low selectivity. To increase trapping efficiency, parts of a trap may be modified to control access to the triggering system. For example, a bionic trap with a 6-cm high bait cone aperture will capture small carnivores, such as minks (*Neovison vison*, Proulx and Barrett 1991a), but will restrict access by larger carnivores, such as fishers (*Martes pennanti*, Proulx and Barrett 1993a). The shape and size of triggers can discourage some carnivores from entering a trap. For example, the efficiency of C120 Magnum rotating-jaw traps to capture American martens (*Martes americana*) is higher with one-way four prong triggers, where the central prongs are shorter than the outside ones (Barrett *et al.* 1989), than with pitchfork triggers with four long prongs of equal length (e.g. Naylor and Novak 1994) that interfere with martens' movements (Pawlina and Proulx 1999).

The position of traps in sets also may affect capture efficiency. For example, lynxes (*Lynx canadensis*) can be properly killed by a blow to the neck by placing rotating-jaw traps at least 23 cm above ground and centered in line with bait at the back of a cubby (Proulx *et al.* 1995). With traps set too low, lynxes try unsuccessfully to go over the trap or lose interest in the bait. With a trap set higher but not centered in line with the bait, a lynx may reach for the bait with a front paw, inadvertently firing the trap on its limb.

A trap must be sited carefully to capture carnivores efficiently without causing undue injury. An animal caught in an EGG trap set in a hole dug into a stream bank can injure itself by wrapping the trap anchor cable around something solid and pulling on the captured foot (Hubert *et al.* 1996). Injuries can also occur when foot-snared canids, felids, and ursids become entangled in surrounding vegetation (Mowat *et al.* 1994; Powell 2005).

Baited sets use food or scent to draw target animals to a trap, while trail (i.e. blind) sets are placed where target animals are expected to travel on their own (Powell and Proulx 2003). While baited traps may have higher capture rates for carnivores, they also attract non-target animals.

5.1.2 Trapping efficiency

Trap models and sets, baits and lures, trappers' experience, weather, and biological variables affect trap efficiency (Pawlina and Proulx 1999). Weakened springs (Gruver *et al.* 1996), distorted components (Warburton 1982), and poorly made traps (Linhart *et al.* 1986) affect trap performance. Traps of different generations or manufacturers may have different components. For example, even though the Novak and the Fremont foot snares are similar in design, the latter is markedly more efficient in capturing coyotes (Skinner and Todd 1990). Red foxes (*Vulpes vulpes*) may smell rusty or oily traps, discover traps that move when a fox steps on jaws or springs, and shy from a set that does not provide a clear view (Krause 1989).

Whether baits and lures increase capture efficiency is either unknown or variable for many conditions. Baits compete with odors of natural foods to attract carnivores (Linhart and Knowlton 1975; Humphrey and Zinn 1982). Scent lures may mimic pheromones (Carde and Elkinton 1984) or stimulate curiosity. Their effectiveness is affected by weather, as well as the physiological condition of target carnivores and the animal that is the source of the scent (Pawlina and Proulx 1999).

Trapping efficiency changes with a trapper's experience. Trappers may require a 1-year acclimatization period before becoming proficient with new trapping devices (Skinner and Todd 1990; Pawlina and Proulx 1999).

Weather may interfere with, or enhance, trap operation and affect the behavior of the target species. For example, frozen soil may affect rubber-padded foothold traps set for coyotes more than unpadded ones (Linhart *et al.* 1986) and wind direction affects food detection by dingos (*Canis familiaris dingo,* Joly and Joly 1992).

Finally, biological variables affect capture efficiency. If traps are located diffusely over large areas, they may be absent from small home-ranges (Gehrt and Fritzell 1996). If males and females have home ranges of different size, trap density will affect the sex ratio of captured animals (King and Powell 2007), and when changing resources lead to changes in the sizes of home ranges, capture efficiency changes (Smith *et al.* 1994). Also, animals of different sex often behave differently towards traps and sets (Gehrt and Fritzell 1996). Some animals become trap-shy after initial capture, while others become trap-happy (Pawlina and Proulx 1999). Resident or dominant individuals may intimidate intruders or subordinates with their scent marks, affecting capture rate (Pawlina and Proulx 1999). Finally, life-history condition may affect capture. For example, adult coyotes may be captured more often when rearing pups (Sacks *et al.* 1999).

5.1.3 Humaneness

Killing and restraining traps used to capture carnivores should be humane and either cause unconsciousness as quickly as possible or hold animals with minimal injury and stress.

For state-of-the-art killing traps, we adopt the following criterion, established by Proulx and Barrett (1994):

> Criterion I: at a 95% confidence level, humane killing traps should render $\geq$ 70% of target animals irreversibly unconscious in $\leq$ 3 minutes.

Powell and Proulx (2003) showed that, despite solid technical advances in trap research and development that meet Proulx and Barrett's (1994) criterion, recently developed standards (CGSB 1996, European Community *et al.* 1997) had not completely incorporated those technical advances. Also, instead of adhering to humane trapping standards, the United States developed its own best-management practices on the basis of technical, economical, and social criteria (International Association of Fish and Wildlife Agencies 1997). Nevertheless, Proulx and Barrett's (1994) criterion for killing traps still is the best-defined, objective, and published criterion consistent with state-of-the-art technological development.

For restraining traps, Tullar (1984), Olsen *et al.* (1986), Hubert *et al.* (1996), and others (summarized by Proulx 1999a) developed injury-scoring systems, most of which correspond with pathological changes in captured animals. Over the years, the number of injury classes has increased and, while early scores were based solely on the injuries of captured limbs, more recent injury-scoring systems also include whole-body trap-related trauma (Proulx *et al.* 1993a; Hubert *et al.* 1996). In all systems, injuries that have the potential to decrease the survival of released animals were identified and a 50-point threshold was used to separate humane restraining devices from unacceptable ones (Proulx 1999a). Although captured animals experience behavioral and physiological changes (Kreeger *et al.* 1990b; Proulx *et al.* 1993a; Seddon *et al.* 1999; Cattet *et al.* 2003, 2008a), to date no objective scoring system for restraining traps integrate these changes with physical injuries (Proulx 1999a), at least in part because interpreting such responses is not straightforward (Dawkins 1998). On the basis of Proulx *et al.*'s (1993a) live-trapping tests with raccoons (*Procyon lotor*) in enclosures, and Powell and Proulx's (2003) humane criterion, which does not specify a maximum time of restraint, we adopt the following standard for restraining traps:

Criterion II: at a 95% confidence level, humane restraining traps should hold $\geq$ 70% of animals for $\leq$ 24 hours with $\leq$ 50 points scored for physical injury.

We recommend that this standard be used for the live capture of carnivores, because it will exceed recent national and international standards, which are not as rigorous and fail to integrate state-of-the-art technological advancements.

All killing and restraining traps should be monitored within a 24-hour period to minimize pain and discomfort. Reducing the time that animals spend in foothold traps greatly reduces injuries (Proulx *et al.* 1994). Unless traps can be visited easily, in person, and multiple times daily, they should be equipped with a monitor (Nolan *et al.* 1984; Marks 1996; Larkin *et al.* 2003; Ó Néill *et al.* 2007) that allows false positives but not false negatives, and that notifies a researcher when battery power is low or when a trap has misfired (Powell and Proulx 2003). Remotely monitored traps must, nonetheless, be visited regularly for maintenance; animals avoiding capture may disturb a trap site and render the set ineffective. The mere fact that animals are dead when kill-traps are checked is not evidence that traps are humane, especially if traps are checked only once every 24 hours (Proulx and Barrett 1989). Without knowing a priori whether traps generate enough energy to kill target animals, whether traps consistently strike animals in appropriate locations for a quick kill, and how long trapped animals remain alive, assuming that traps are humane, can lead to undue suffering.

5.1.4 Traps and sets for specific carnivores

Both restraining and killing traps can contribute significantly to research on evolution, ecology, animal behavior, physiology, parasitology, genetics, and other disciplines. The choice of restraining vs. killing traps depends, at the least, on research hypotheses and goals, research design, and study site (Powell and Proulx 2003). Because restraining traps allow the release of trapped animals, they should be used when species-at-risk and pets may be captured. When non-target captures are unlikely, using a restraining trap to capture a target carnivore, only to kill it later (to collect a sample, for example), may be less humane than using a quick-killing trap (Powell and Proulx 2003). Keeping animals alive may be required, however, to avoid freezing or decomposition of tissues to be sampled (Kreeger *et al.* 1990b).

Common sense dictates choosing traps that maximize both selectivity and efficiency (Pawlina and Proulx 1999). Selective, efficient traps minimize the capture of non-target species or individuals, thereby increasing the rate of data collection and reducing the overall impact of the research on the ecological community in the study area. Thus, within the constraints of research design, choose traps based on selectivity, efficiency, and state-of-the-art trapping technology based on

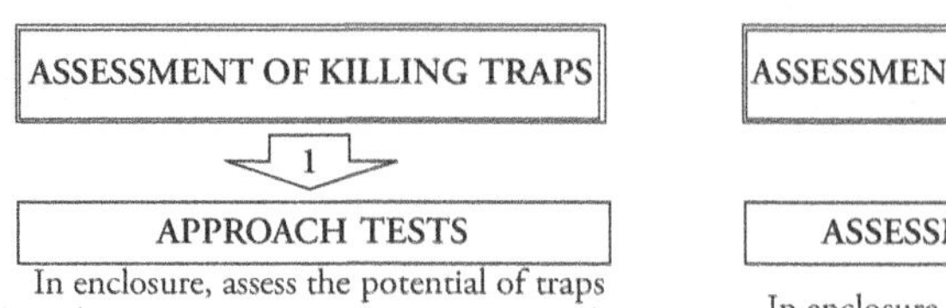

In enclosure, assess the potential of traps (wired in a set position not to cause injury) to strike properly ≥ 5/6 animals in a vital region (Proulx et al. 1989a, Proulx and Barret 1991b).

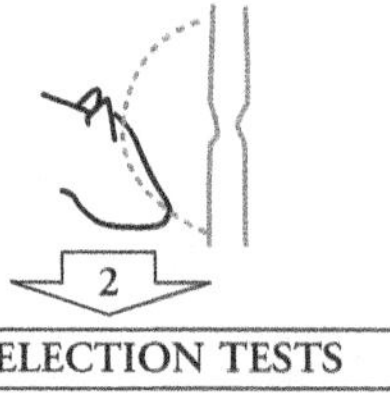

2

PRE-SELECTION TESTS

Assess the potential of traps to render ≥ 5/6 animals immobilized with ketamine HCl irreversibly unconscious in ≤ 3 min. Traps that fail the pre-selection tests are not allowed to go further. The muscles of anaesthetized animals are more relaxed than those of non-anaesthetizeed animals and offer less resistence to the striking bars than conscious animals that are fighting the trap (Proulx et al. 1989a, Proulx and Drescher 1994).

3

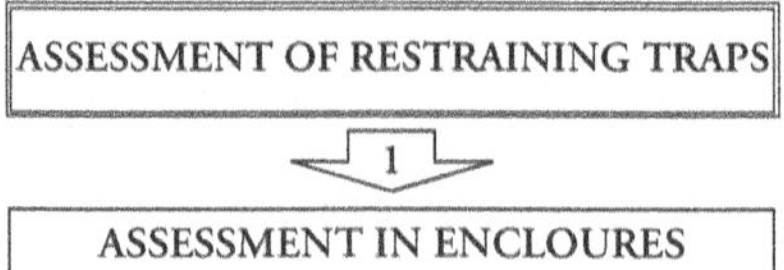

ASSESSMENT IN ENCLOURES

In enclosure, trapped animals are kept captive for ≤ 24hrs, euthanized, and necropsied. Assessment based on a cumulative scoring system to assign points to captured limb and body (Proulx et al. 1993a).

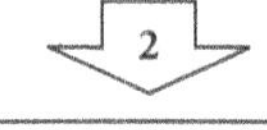

FIELD TESTS

Test traps either alone (comparing capture efficiency to data reported for other traps) or in a comparison with commonly used traps (Pawlina and Proulx 1999).

KILL TESTS

Assess the potential of traps to render 9/9 non-euthanized animals irreversibly unconcious in ≤ 3 min. On the basis of the normal approximation to binomial distribution, at a 95% confidence level, traps would render ≥ 70% of target animals irreversibly unconscious in ≤ 3 min.

Fig. 5.1 Sequential series of biological tests used to assess the humaneness of killing and restraining traps (after Proulx and Barrett 1991b; Proulx *et al.* 1993a).

humaneness. Both efficiency and humaneness must be properly evaluated through sound, scientific protocols (Proulx 1999a), preferably peer-reviewed and published. We evaluate traps here on the basis of published data about capture efficiency and humaneness.

Proulx and Barrett (1991b) described a sequence of biological tests to develop and to evaluate killing traps (Figure 5.1). These tests were carried out in simulated environments (along with mechanical evaluations of trap properties), and they led to the development of most of the state-of-the-art killing traps identified here. Proulx *et al.* (1993a) developed a protocol to assess restraining traps (Figure 5.1), where animals are left in the trap for $\leq$ 24 hours, unless there is evidence of serious trauma. Enclosure tests must be followed by field tests to assess humaneness and capture efficiency fully (Pawlina and Proulx 1999).

The performance of a trap in the field depends on how the trap is set and monitored.

When using killing or restraining traps that are efficient and humane, follow these rules:

1. Do not modify trap size, shape, components, materials, or power, which are essential to achieve a humane kill.
2. Do not modify trigger shape or operation, which affect both humaneness and capture efficiency.
3. Replicate sets that have been used in the assessment of humane traps.
4. Visit traps <24 hours (but preferably <12 hours) after setting them (a) to kill animals that may be seriously injured but are still alive in a killing trap or (b) to release animals captured in restraining traps. No matter how humane a restraining trap is, if it is not visited at short intervals, animals will be injured.

Responsible professionals must strive continuously to improve traps to work more efficiently, more selectively, more humanely, and more safely for both animals and people. Changing the properties of traps, however, may affect the humaneness and capture efficiency of models that meet our criteria. Therefore, modified traps should be re-evaluated.

Finally, safety to the researcher should be kept in mind when developing and assessing traps. In most cases, a locking device can be installed to stop springs from firing or trap jaws from closing. If a trap cannot be safely handled without some safety device, it should not be used.

For this chapter, we reviewed traps on a species-specific basis to describe how and when they have been used, and to determine their advantages and limitations (Appendix 5.3). Wherever possible, we provide examples for all carnivore families, but information about humaneness and capture efficiency for trapping devices is often lacking. On the other hand, by matching size and behaviors of carnivores, one can often predict which trapping device is most likely to be effective for the capture of a species for which little information exists. In general, humane and capture-efficient killing traps and cage traps are available for small- and medium-sized (<5 kg) carnivores. Large carnivores must be captured in rubber-padded foothold traps or cage/box/log traps. Raccoons should be captured in EGG or cage traps. For the majority of canids, foothold traps, foot snares or neck snares are the only devices that are efficient and humane. Small felids (cat, bobcat, lynx, and others) may be captured in cage traps. Foothold traps and foot snares can be effective and humane for all felids and ursids.

Injury caused by cage traps has not been adequately evaluated (Proulx 1999a). In general, cage traps meet Criterion II for the capture of all carnivore species, as they appear to cause less trauma than other restraining techniques (White *et al.* 1991). Their capture efficiency varies among species, and is often lower than other restraining traps with canids (e.g. Muñoz-Igualada *et al.* 2002; Shivik *et al.* 2005). Regardless of the species being trapped, special precautions must be taken to ensure the well-being of captured animals. Carnivores may break teeth and cut their mouths while biting wire-mesh walls (Rust 1968; Belant 1992). Cages with small mesh holes or with solid walls usually are superior to mesh with large holes that allow animals to catch their muzzles. Cage traps without insulated nest boxes and bedding should not be used when temperatures drop below –20°C, or if researchers cannot check traps daily. Warm, dry bedding (e.g. raw wool with natural lanolin) in live-traps can reduce mortalities (Powell and Proulx 2003). Traps should be concealed and covered with vegetation to protect carnivores from direct sunlight, rain, and large predators.

5.2 Use of drugs for capture and restraint of carnivores

Drugs are powerful tools used for capture and restraint of carnivores, and to relieve pain and stress.

5.2.1 Drug access, storage, and handling

Regulations for drugs vary considerably from one country (and states within countries) to the next. Information can usually be obtained by consulting with a local veterinarian working with wildlife or zoo animals.

Although some studies have shown that potency and safety persists well past expiration dates for some drugs (Kreeger *et al.* 1990a; Kreeger and Arnemo 2007), we strongly recommend using non-expired drugs for capture of free-ranging carnivores, to minimize unpredictable variations in drug response and to be fully compliant with regulations. Drug manufacturers provide instructions for appropriate storage of their products with attention to factors such as temperature, humidity, and light exposure.

Storing drugs in a secure and safe place is important to prevent theft for illicit use (Woodward 2005), and to enable accurate inventory in order to know when to order fresh stocks and to dispose of old ones. A running inventory record should have standard information for purchases and use (Cattet *et al.* 2005). Capture records that document drug use and animal response on a case-by-case basis should be maintained to evaluate the effectiveness and safety of drug protocols.

The handling of drugs for use with carnivores requires training and experience. Pay particular attention to human safety to prevent personnel from being exposed inadvertently to veterinary drugs through contact with skin or mucous membranes (eyes and mouth; Kreeger and Arnemo 2007). Persons involved in the capture of carnivores should complete a creditable course in wildlife chemical immobilization, and have current training in basic first aid and cardiopulmonary resuscitation, prior to working with drugs. Personnel who use drugs and associated equipment for the capture of carnivores must have a clearly written emergency action plan in case of human exposure to drugs or capture-related injuries to personnel (Nielsen 1999; Cattet *et al.* 2005). Most physicians are unfamiliar with drugs used in wildlife capture, so communication beforehand will save valuable time in an emergency.

5.2.2 Selection of drugs for use in carnivores

More than one drug may be effective in a given species and availability of drugs changes over time as new products are released and old products are discontinued. In addition, use of drugs in wild animals is often "extra-label" or "off-label" (i.e. use of a specific drug does not follow conditions specified on the label, including specified species, dose, and method of administration). Conditions for extra-label use may be obtained from a wildlife or zoo veterinarian or found in peer-reviewed scientific literature. Regardless of source, the efficacy and safety of a drug based on empirical evidence in a target species should be the primary consideration in selecting a drug protocol. Many published reports describe the effectiveness of a specific drug for capture of a given species, but fewer reports evaluate safety based on the physiological responses of a species to a drug (e.g. vital rates, blood gases, adverse effects). Safety should not be ignored.

Drug effectiveness and safety must be considered when selecting an injectable drug for chemical immobilization (Table 5.2). Because no single drug meets all considerations, different drugs are often combined to attain many desired characteristics, while at the same time eliminating undesired effects (Grimm and Lamont 2007; Kreeger and Arnemo 2007). Many of these combinations include anesthetic drugs (e.g. ketamine, tiletamine), which cause loss of consciousness, with sedatives or tranquilizers (e.g. xylazine, medetomidine, zolazepam, acepromazine), which improve anesthesia in various ways including increased muscle relaxation and pain control. Some immobilizing drugs are used in conjunction with an antagonist drug that is administered to counteract the effects of anesthesia at the conclusion of handling or if complications arise during immobilization. Immobilizing drugs with antagonists are generally preferred because removing effects of anesthesia (1) permits mitigation of anesthesia-related physiological complications, (2) reduces likelihood of injury or death during recovery, and (3) decreases time spent by personnel monitoring recovery (Kreeger and Arnemo 2007).

Adjunctive drugs are used to support immobilization and are generally administered following capture and immobilization. These drugs include the dissociative anesthetic ketamine, which is often administered as a "top-up" to maintain immobilization, even when it is not a component of the immobilizing drug. The

Table 5.2 *Characteristics of ideal drugs for anaesthesia and euthanasia.*

Anaesthesia	
Advantages for the animals	Advantages for the users
Mixes safely with other drugs (i.e. no loss of potency or formation of by-products), and does not react with dart material.	Has low toxicity in humans should accidental exposure occur.
Is safe for pregnant and lactating animals, and nonirritating following intramuscular or intravenous injection.	Rapidly degrades *in vivo* to inactive, non-toxic metabolites (i.e. no harmful effects to humans consuming meat from drugged animals).
Is effective in small volumes (i.e. high potency), and has a wide margin of safety between effective and toxic doses (i.e. accidental overdosage is unlikely to have harmful effects).	Has low potential for human abuse as a recreational drug.
Causes rapid immobilization and loss of consciousness with minimal fear or memory of capture.	Is readily available on the market (i.e. commercial supplier exists, and access is not limited by regulation).
Causes minimal depression of cardiovascular and respiratory function, and produces muscle relaxation.	Is reasonably priced.
Causes minimal inhibition of swallowing reflex.	Is highly water soluble and stable in solution.
Causes good control of pain (analgesia) at immobilizing dosages.	Has a long shelf life.
Is reversible by administering an antagonist drug.	
Causes behavioral effects during induction, immobilization, and recovery that are predictable and safe.	
Rapidly degrades *in vivo* to inactive, non-toxic metabolites (i.e. no harmful effects to drugged animals, predators or scavengers).	
Euthanasia	
Causes rapid loss of consciousness and death without causing pain, distress, or anxiety.	Reliability.
Effects cannot be reversed.	Has low toxicity in humans should accidental exposure occur.
Widely compatible with different species, age, health status, and numbers of animals.	Safe to use in different environments, e.g. urban setting vs. remote field location.
Safe for predators or scavengers that consume the drugged carcass.	Compatibility with subsequent evaluation, examination, or use of tissue.
	Is readily available on the market.

advantage of ketamine for this purpose is that it is metabolized quickly and is less likely to prolong recovery, than will administering more of the initial immobilizing drugs (Cattet *et al.* 2005). Although generally not regarded as a drug, medical-grade oxygen is also a valuable adjunctive "drug" that can be used to prevent and treat several common complications (e.g. hypoxia, hyperthermia) associated with capture and anesthesia (Read *et al.* 2001; Arnemo and Caulkett 2007). Oxygen can be administered intranasally in the field without much difficulty and with minimal training using a lightweight aluminum cylinder (D- or E-type), a pressure regulator, and silastic tubing.

Drugs to relieve pain and stress should be considered for use with carnivores captured either with or without chemical immobilization (CCAC 2003). Aside from obvious concern for the welfare of captured animals, pain and stress can affect their behavior in ways that affect research results (Powell and Proulx 2003; Cattet *et al.* 2008a). Many drugs are available to provide pain relief (analgesia) and to reduce stress for wildlife. These include local anesthetic drugs, opioids, and non-steroidal anti-inflammatory drugs for pain relief (Machin 2007), and sedatives or tranquilizers for reducing stress (Arnemo and Caulkett 2007). Long-acting tranquilizers can be valuable for reducing stress in wildlife that must be translocated or maintained in captivity (Read 2002; Flick *et al.* 2007).

Table 5.3 lists some of the commonly used immobilizing drugs for different carnivore families. Detailed information on specific protocols, including dosages, can be found in extensive reference lists compiled by Kreeger and Arnemo (2007), and West *et al.* (2007).

5.2.3 Methods to administer drugs

Drugs can be delivered to wild carnivores via a variety of methods and equipment (Appendix 5.4) and no one method is suitable for all animals at all times. The choice of delivery method should be based on the behavior of the target species, the circumstances for drug administration, and the user's experience. The goal is to administer drugs in a safe (for personnel and animal alike) and effective manner (Cattet *et al.* 2005). Researchers seeking detailed information on use of equipment and on equipment manufacturers should review books by Nielsen (1999), Kreeger and Arnemo (2007), West *et al.* (2007), and Fowler (2008).

5.2.4 The value of knowledge and experience

Beyond knowledge of drugs and methods of administration, capture personnel must have knowledge of, and experience with, the target species to ensure that captures are effective, consistent, and safe (Nielsen 1999; Fowler 2008). For chemical immobilization, one must be able to visualize where thick, superficial

Table 5.3 *Immobilizing drugs for use with carnivores. "H" denotes relative use of drug is high compared to other drugs used for animals in the Family indicated; "M" use is moderate; "L" use is low.*

Family	*Immobilizing drugs with antagonists*[1, 2] *(Drug/antagonist)*	*Immobilizing drugs lacking antagonists*[1, 3]
Ailuridae (2)[4] (giant panda, lesser panda)	H: Ketamine-xylazine/yohimbine M: Ketamine-medetomidine/ atipamezole	H: Tiletamine-zolazepam
Canidae (28) (dogs, foxes, jackals, wolves)	M: Ketamine-xylazine/yohimbine M: Ketamine-medetomidine/ atipamezole L: Butorphanol-medetomidine/ naltrexone and atipamezole L: etorphine-promazine/ diprenorphine L: fentanyl-xylazine/naltrexone and yohimbine	H: Tiletamine-zolazepam H: ketamine-acepromazine L: Ketamine-promazine L: ketamine-midazolam
Eupleridae (2) (Malagasy civet, Malagasy ring-tailed mongoose)		H: Tiletamine-zolazepam
Felidae (27) (cats, lions, leopards)	H: Ketamine-xylazine/yohimbine M: Ketamine-medetomidine/ atipamezole L: Ketamine-medetomidine-butorphanol/atipamezole and naloxone L: tiletamine-zolazepam-medetomidine/atipamezole L: tiletamine-zolazepam-xylazine/ yohimbine or atipamezole	H: Tiletamine-zolazepam L: Ketamine L: ketamine-acepromazine
Herpestidae (3) (mongooses)	L: Ketamine-xylazine/yohimbine	H: Tiletamine-zolazepam M: Ketamine L: Ketamine-acepromazine
Hyaenidae (4) (aardwolf, hyenas)	M: Ketamine-xylazine/yohimbine L: Etorphine-xylazine/diprenorphine and yohimbine	H: Tiletamine-zolazepam L: Ketamine-acepromazine
Mephitidae (4) (skunks)	M: Ketamine-xylazine/yohimbine	H: Tiletamine-zolazepam H: ketamine-acepromazine
Mustelidae (22) (badgers, ferrets, otters, weasels)	H: Ketamine-medetomidine/ atipamezole H: ketamine-xylazine/yohimbine L: Ketamine-medetomidine-butorphanol/atipamezole and naloxone	H: Tiletamine-zolazepam M: Ketamine M: ketamine-acepromazine L: Ketamine-diazepam L: ketamine-midazolam

Family	*Immobilizing drugs with antagonists[1, 2] (Drug/antagonist)*	*Immobilizing drugs lacking antagonists[1, 3]*
	L: tiletamine-zolazepam-xylazine/yohimbine or atipamezole L: etorphine-xylazine/diprenorphine and yohimbine L: fentanyl-xylazine/naltrexone and yohimbine L: fentanyl-diazepam/naltrexone	
Procyonidae (5) (coatimundi, raccoon, kinkajou)	M: Ketamine-xylazine/yohimbine M: ketamine-medetomidine/atipamezole L: Tiletamine-zolazepam-xylazine/yohimbine or atipamezole	H: Tiletamine-zolazepam H: ketamine-acepromazine L: Ketamine-acepromazine
Ursidae (7) (bears)	H: Tiletamine-zolazepam-medetomidine/atipamezole M: Tiletamine-zolazepam-xylazine/yohimbine or atipamezole M: ketamine-xylazine/yohimbine M: ketamine-medetomidine/atipamezole L: Etorphine/diprenorphine	H: Tiletamine-zolazepam
Viverridae (11) (civets, genets, binturongs)	M: Ketamine-xylazine/yohimbine L: Ketamine-medetomidine-butorphanol/atipamezole and naloxone	H: Tiletamine-zolazepam L: Ketamine-acepromazine

[1] *Kreeger and Arnemo (2007) and West* et al. *(2007) provide information on species-specific drug use including dosages, cautionary comments, and appropriate references.*
[2] *Immobilizing drugs are typically combined prior to injection, hence "ketamine-xylazine." Antagonist drugs are generally administered separately, hence "atipamezole and naloxone."*
[3] *Although antagonists are available for diazepam, midazolam, and zolazepam, they are seldom used because of their short duration of effect.*
[4] *In parentheses is number of species for which drug use is reported in scientific literature.*

muscles lie beneath skin and fur, because drugs are typically administered to wild carnivores by injection, often using remote drug-delivery equipment (blowpipes, modified pistols or rifles, and darts; Kreeger and Arnemo 2007). Injection into other tissues, such as fat or bone, will likely prolong or prevent capture and increase the potential for complications. One must be able to distinguish between normal and drug-induced behavior to monitor the effectiveness of a given drug dose and, if necessary, to determine if more drug is required. Once an animal is anesthetized, it is necessary to monitor its vital signs, as well as its level (or depth) of anesthesia, and to recognize when adverse physiological responses are developing (Cattet *et al.* 2005; West *et al.* 2007). Much of this knowledge can only be gleaned through extensive hands-on experience, not through the pages of books and reports. Nonetheless, attention to current literature and participation in creditable courses in wildlife chemical immobilization will help improve the value of field experiences.

5.3 Identification, prevention, and treatment of medical emergencies associated with capture

Wildlife capture is often unpredictable and relatively uncontrolled. As a result, the potential for medical emergencies is ever present, whether it is injury sustained during capture or adverse physiological response to drugs or restraint (Appendix 5.5). Emergencies or complications can develop at any time between capture and release, and sometimes days to weeks following release (Cattet *et al.* 2005). During capture, an animal can be injured by the trap (Powell 2005; Cattet *et al.* 2008a), through impact or injection by darts (Valkenburg *et al.* 1999; Cattet *et al.* 2006), or while being pursued (Cattet *et al.* 2003). While restrained in a trap, animals can injure themselves while attempting to escape (Proulx *et al.* 1993a; Powell 2005), be injured by other animals (Hooven *et al.* 1979; Craft 2008a), or develop adverse physiological conditions as a consequence of stress, extreme ambient temperatures, or lack of water (Cattet *et al.* 2003). With chemical immobilization, emergencies can arise with inappropriate use of drugs or failure to monitor physiological function (vital signs) of anesthetized animals (Cattet *et al.* 2005). Following release, animals may develop complications as a delayed effect of their response to capture (e.g. exertional myopathy, Cattet *et al.* 2008b).

5.3.1 Homeostasis, stress, distress, and treatment of medical emergencies

Preventing medical emergencies is better and easier than treating them. Effective prevention, however, depends on a sound knowledge of factors that can cause complications and how animals respond to them (Appendix 5.5). Normally, animals actively maintain a relatively constant internal environment (i.e. body temperature, acid–base balance, body water content, etc.) in the face of changing external conditions, such as weather, food availability, and activity. This homeostasis is an essential requisite for many biological processes, including reproduction, growth and development, and immunity. Factors that threaten or disturb homeostasis are called stressors, and the behavioral and physiological responses required to maintain homeostasis are collectively termed the stress response (Figure 5.2) (Hofer and East 1998; Moberg and Mench 2000). If the stress response is effective, homeostasis is maintained and biological processes continue unabated.

When biological processes are disrupted, however, as a result of a prolonged or excessive stress response, the resulting state is termed distress. Manifestations of distress include impaired reproduction, suppression of immune function, stunted growth, and reduced ability to mount an effective stress response in future. At its extreme, distress results in death. Reducing the occurrence and intensity of potential stressors in capture and handling will help prevent distress. Treating medical emergencies in free-ranging wild animals is often difficult, and sometimes impossible. Wild animals are not compliant patients, thus drugs are required to ensure that an animal remains immobilized, or at least sedated, during treatment. Further, effective treatment may require follow-up care over a period following

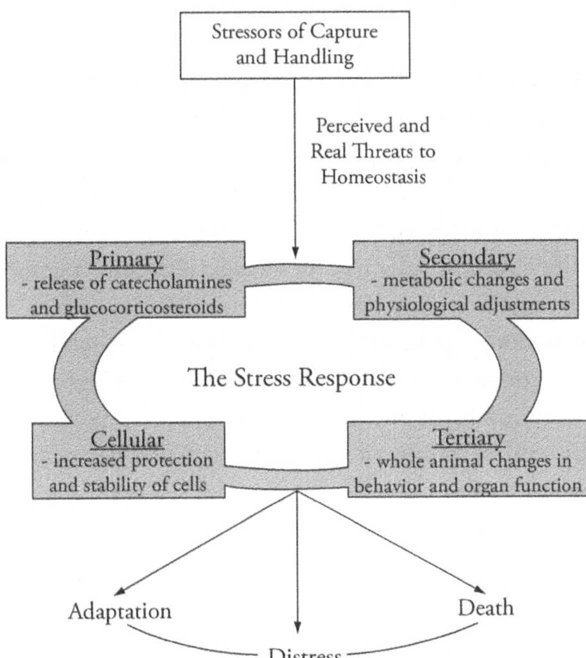

Fig. 5.2 Diagram illustrating the stress response that follows when an animal perceives a threat to homeostasis. The perception may be psychological, physical, physiological, or a combination of types. The overall effectiveness of the stress response—adaptation, distress, or death—is affected by the number, intensity, and duration of stressors.

initial treatment (e.g. to change bandages, to remove sutures or to administer medication). With non-captive, wild animals, follow-up care is not an option. Typically, an animal is released, possibly never to be seen again, and one hopes for the best. These difficulties underscore the importance of placing emphasis on prevention rather than treatment (Appendix 5.5). One must be able to recognize emergencies, to have proper treatment materials on hand, and to have appropriate training and skills required to provide treatment (Appendix 5.5). Researchers seeking more detailed information on treatment of medical emergencies should review books or technical manuals by Nielsen (1999), Cattet *et al.* (2005), Kreeger and Arnemo (2007), and Fowler (2008).

5.3.2 Necropsy

Animals that die during or following capture should be necropsied (Chapter 13; Cattet *et al.* 2005). If an animal dies as a direct result of capture procedures, the capture and handling protocol should be reviewed carefully and minutely, and possibly revised, to ensure similar deaths do not occur in future. If an animal dies as a consequence of concurrent disease combined with the stress of capture, necropsy findings will help to assure continued confidence in the capture protocol and may provide new information regarding the health of the species.

In the field, appropriate tissue samples should be collected and frozen or fixed in 10% buffered formalin for submission to a veterinary pathology facility (Chapters 4, 6, and 13). Appropriate tissue samples should include brain, lung, heart, liver, kidney, spleen, lymph nodes, and muscle. Capture personnel should refer to a wildlife necropsy manual for details regarding required equipment, techniques, and sampling procedures (Chapter 13; Munson 1999; CCWHC 2010). Documentation should include a detailed history and digital images of the field necropsy to assist the veterinary pathologist diagnosing the cause of death. Alternatively, under some circumstances, it may be desirable to arrange shipment of the entire carcass to a veterinary pathology facility for detailed necropsy (Chapter 13).

5.4 Euthanasia

Euthanasia is the humane killing of animals, characterized by minimal pain and distress (AVMA 2007). Minimal pain means the dying animal experiences little sensation of pain because its cerebral cortex, the area of the brain that controls thought, memory, sensation, and voluntary movement, has been rendered non-functional by drugs, concussion, or oxygen deficiency (hypoxia). Minimal distress means the dying animal has not had the opportunity to respond to its situation in a way that is harmful to itself. In the context of wildlife capture, untreatable pain and distress may provide the basis for deciding to kill an animal, so euthanasia by strict definition may not be possible (Drew 2006). Nonetheless, the killing should be as humane as possible.

In addition to severe untreatable injury, other situations can arise during capture and handling when researchers must consider euthanizing an animal; for example, when an animal poses an immediate threat to capture personnel, to the public, to other wildlife (e.g. risk of spreading a serious infectious disease) or to the environment (e.g. an invading species). Consequently, capture personnel must be familiar with acceptable methods of humanely killing wild animals, and must have appropriate equipment (including drugs) on hand to perform a kill quickly. Killing an animal humanely requires appropriate training and experience with the required techniques, restraint of the animal to be killed, and selection of proper drugs (Table 5.2) using criteria that consider humaneness and user safety (AVMA 2007). Specific attention must be given to how best to restrain a wild carnivore prior to killing it (AAZV 2006). Use of "gentle restraint" methods advocated for domestic animals are likely to be ineffective and dangerous with wild animals that are injured or already distressed by being captured. Sedative- or anesthetic-type drugs should be used for these situations, and drugs should be administered by a method that is safe for personnel and minimizes distress in the animal.

Specific consideration should also be given to the human psychological response to killing an animal (AVMA 2007). The decision to kill an animal is sometimes difficult. Uncertainty or differences in opinion often arise regarding the potential impact of a severe injury on an animal's future welfare. Furthermore, a euthanasia decision determined by peripheral factors (e.g. policies or regulations), rather than the condition of the animal, may be controversial. Decision criteria and methods for euthanasia should be discussed and understood by all team members prior to starting trapping.

Appendix 5.6 provides information on acceptable methods of euthanasia for wild carnivores. Some methods, such as exsanguination (bleeding out) or intravenous administration of potassium chloride, are only regarded as acceptable if the animal is killed while deeply anesthetized. Methods regarded as unacceptable, regardless of circumstances, include a blow to the head for animals >1 kg body weight, carbon monoxide, chest compression, drowning, hypothermia (or rapid freezing), and use of neuromuscular blocking agents, such as succinylcholine (AAZV 2006, AVMA 2007).

Table 5.4 *Restraining techniques to handle captured animals.*

Technique	Description	Examples
By hand with a catching pole (pole snare) or a forked stick	Wear gloves—grasp the animal firmly at the base of the neck with one hand, and the hips with the other hand. Hands and arms should be kept above the back of the mammal to avoid claws (Jones *et al.* 1996).	Wolves—Kolenosky and Johnston 1967 Jackals—Rowe-Rowe and Green 1981 Red foxes—Henry 2004; Craft 2008b Pumas—Davis *et al.* 1996 Badgers—Proulx, unpubl. data
Portable cushion	Use cushion to break the fall of anaesthetized animals.	Pumas—McCrown *et al.* 1990
Squeeze cage	This is a cage equipped with a squeeze panel (wire mesh, wood, netting, compact cloth) to hold an animal firmly against the side of the cage for anaesthesia.	Fishers—Buck 1982; Frost and Krohn 1994 River otters—McCullough *et al.* 1986
Wire mesh cone	This is used to handle animals <1500 g (Taber and Cowan 1969; Powell, unpubl. data; Proulx, unpubl. data).	American marten—Bull *et al.* (1996) Long-tailed weasel—Proulx, unpubl.data
Cloth, mesh or heavy plastic bags	May be used during anaesthesia.	Red fox—Zabel and Taggart 1989 Polecats—Forman and Williamson 2005

Carcasses of animals killed while anesthetized, or killed by barbiturate overdose, should be disposed of by deep burial or incineration to prevent secondary toxicity of scavengers (AAZV 2006).

5.5 Restraining and marking techniques

In the absence of, before or immediately after, anesthesia, captured animals must be restrained safely, so as to minimize physical injury and stress (Table 5.4).

Temporary or permanent marks should be as painless as possible and should not affect the animals' behavior or health (ASM Animal Care and Use Committee 1998). Marks must be matched to research objectives and must be appropriate for a carnivore's sizes, body shape, future growth, and behavior (Powell and Proulx 2003). Many short-term, long-term, and permanent markers have been developed for mammals, but few have been tested with carnivores (Appendix 5.7).

5.6 Designing effective trapping programs for carnivores

Carnivores have cognitive maps of where they live, and they do not use space within their home ranges randomly (Chapter 9; Peters 1978; Powell 2000; Proulx 2005). Therefore, setting traps randomly or uniformly across the landscape will likely be less productive than setting traps at special habitat features. Aside from trap and set characteristics, a variety of biotic and abiotic factors affect capture success. To develop an effective trapping program, one must know how a carnivore is associated with the vegetative and physical structures of a study area, and the sizes of home ranges of males and females, adults and juveniles, females with or without young, and dominant and subordinate animals. Traps may be spread so that each individual of a population has one trap within its home range (to trap as many different individuals as possible) or so that each individual has many traps within its home range (to recapture each individual many times; Powell and Proulx 2003). Home range and population sizes may be related to food patches (Macdonald 1983; Fuller *et al.* 1992), but also to intra- and interspecific competition (Rosenzweig 1966; Marker and Dickman 2005; Moorcroft *et al.* 2006). Finally, weather change can affect carnivore activity (Zielinski 2000). Understanding species-specific factors that may affect capture success dictates how and when traps should be set in the field to meet a program objective. The distribution of traps will vary from one species to another but, in all cases, trapping programs should be developed with spatio-temporal schemes that are compatible with the biology of animals.

5.7 Animal welfare

While animals have been captured for centuries by human populations evolving with their environments, today, capturing and handling carnivores is specialized work and must be conducted with scientifically sound protocols and high standards of animal use and welfare.

Researchers should apply Russell and Burch's (1959) "3 Rs," *Replacement, Reduction, Refinement*, to the use of animals in field research. Although the 3 Rs are well-established principles in the field of laboratory animal science, many wildlife researchers are unfamiliar with them and their implementation in wildlife research. This unfamiliarity may be explained, in part, because the goals of wildlife research often value the welfare or needs of populations, communities or ecosystems over the welfare of individual animals (CCAC 2008). Nonetheless, the welfare of individual wild animals is of concern because:

1. Animals (target or non-target species) may be injured during capture or handling.
2. Sampling or marking of captured animals may involve invasive procedures.
3. Wild animals are likely to be intensely stressed during capture because they are not conditioned to human handling.
4. Wild animals may conceal capture-related injuries (from researchers) that could have serious consequences for their long-term survival.
5. Welfare indicators are deficient for many wild species.
6. Peer-reviewed reports on the welfare and research implications of wild animal studies are lacking.

Researchers must consider and implement the 3 Rs to balance the needs of wildlife research and wild animals in accordance with the following definitions (CCAC 2008):

- *Replacement*—Researchers should use animals only if they are unable to find a replacement by which to obtain the required information. Replacement strategies include noninvasive sampling (Chapter 4), collation and use of information already gained, population meta-analyses, population and habitat suitability simulations, and archived tissue samples.
- *Reduction*—Researchers should use the fewest animals needed to provide valid information and statistical inference (Chapter 8). Sample size can be minimized by (1) designing research that yields data appropriate for statistical tests needing small or remotely collected samples (Chapters 4 and 8); (2) using factorial design to explore the effects of several variables in one experiment; (3) using sequential and multivariate statistical methods; or (4) using repeated measured designs (McConway 1992). Reduction also can be applied without compromising animal welfare by maximizing the information obtained per animal (e.g. collection of biological and genetic samples for archiving, Chapter 6), thereby limiting or avoiding, the subsequent use of additional animals. When trapping carnivores, reduction can be applied by designing trapping programs that minimize the likelihood of capturing non-target animals.
- *Refinement*—Researchers should use the most humane, least invasive techniques to minimize pain and distress (Chapter 4). This is the easiest of the 3 Rs to apply in wildlife research. Possible strategies include: (1) assessing and reducing potential sources of harm to captured animals; (2) avoiding methods that raised questions of animal welfare in other studies; (3) using drugs (analgesics) to control pain in invasive procedures (e.g. biopsy, tooth extraction); (4) using noninvasive sampling (Chapter 4) and other sampling not requiring capture to collect biological and genetic samples (e.g. skin samples

by remote biopsy darting; Spong and Creel 2001); (5) minimizing disturbances that can lead animals to abandon home ranges, can pre-empt feeding, can disrupt social structure, and can alter predator–prey relationships; (6) using a minimal (but safe) restraint and the shortest possible handling time; (7) collaborating with manufacturers to produce research equipment least likely to cause pain and distress or to disrupt an animal's normal way of life; and (8) publishing descriptions of refined techniques in the peer-reviewed scientific literature (CCAC 2008).

Researchers and managers can implement and promote the 3 Rs by ensuring that all personnel involved in their capture programs are trained appropriately in field procedures and have undertaken formal training in the concept and implementation of the 3 Rs, and by collaborating in the development and dissemination of training courses, guidelines, and protocols for various species and types of wildlife research (CCAC 2008; Norecopa 2008).

Appendix 5.1 *Trap types used in the capture of carnivores.*

Traps	*Material*	
Restraining devices Cage or box traps	Wire mesh, solid wood, metal, or plastic walls	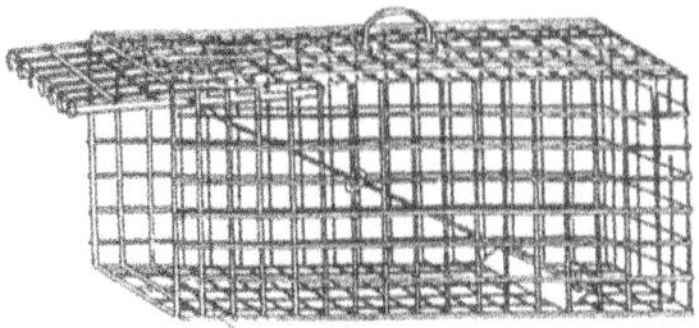
Foothold (leghold) traps	Metal clamping jaws that can be rubber-padded or offset	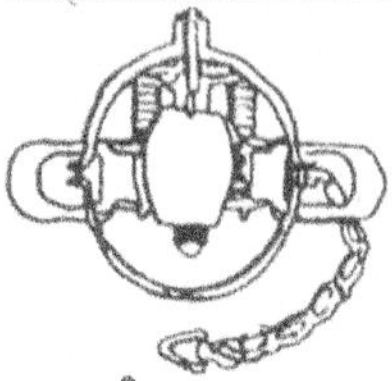
Foot (leg), and neck snares	Metal cable of a single or multiple strands	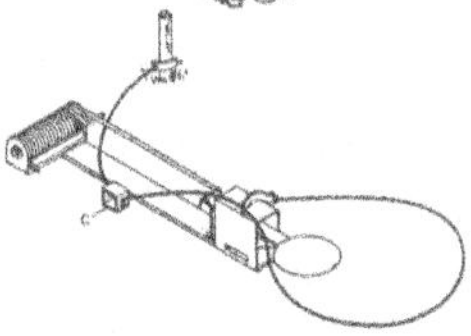
Nets	Nylon mesh	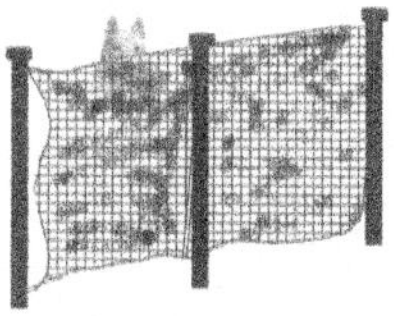
Killing traps Snap trap (mousetrap)	Metal striking bars mounted on flat surfaces	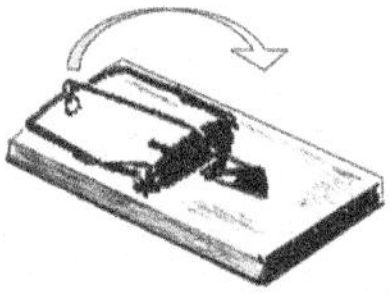
Planar trap	Metal bar	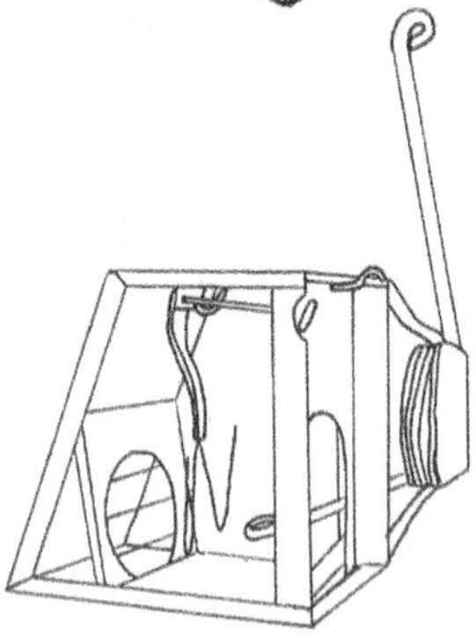

Mode of action	*Species*	*References*
One or two entrances that close when animals step on a treadle or move a triggering device.	Small cages to capture small carnivores to huge structures made of logs or road culverts to trap wolverine- (*Gulo gulo*) to bear-sized (*Ursus* spp.) carnivores.	Powell and Proulx 2003
Jaws open to 180° in their set positions and clamp together to capture animals by a paw or a leg. Some traps have a housing that completely encases a captured limb. All traps are powered by either coil or leaf springs when sprung.	Foothold traps for canids and felids. Traps such as the EGG tap are used for small- and medium-sized carnivores that manipulate and explore with their paws (e.g. raccoon, *Procyon lotor*).	Proulx *et al.* 1993a Proulx 1999a Hubert *et al.* 1999
The energy to tighten the noose around an animal's limb or body is provided by the captured animal or a spring. The cable is equipped with a locking mechanism to prevent the loop from loosening.	Medium-sized carnivores. Neck snares are used to live-trap canids. They hold animals by their necks as if restrained with a leash; a stop prevents the loop from choking animals.	Nellis 1968 Bjorge and Gunson 1989 Proulx 1999a Woodroffe *et al.* 2005b Gese 2006
Drive nets, stretched loosely between two solid objects and supported by poles or branches, to capture carnivores driven by battue and fladry lines, or by helicopters. Hand-held net guns fired from helicopters or all-terrain vehicles.	Medium- and large-sized carnivores.	Beasom *et al.* 1980 Gese *et al.* 1987 Okarma and Jedrzejewski 1997
U-shaped jaw, as in common mouse and rat traps, or a straight bar that closes from 180° onto a flat surface.	Small carnivores.	Powell and Proulx 2003
The spring forms the killing bar and closes in the same plan.	Small carnivores.	Proulx 1999a

(continued)

Appendix 5.1 *Continued*

Traps	*Material*	
Rotating-jaw (body-gripping) traps	Square-shaped metal bars	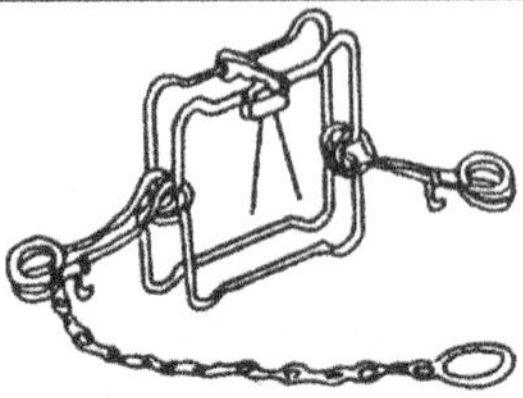
Killing box trap	Metal striking jaw or cable	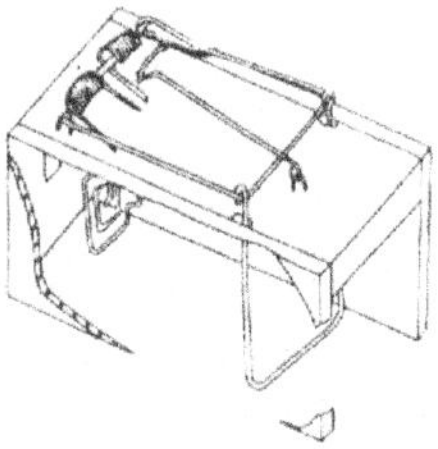
Killing snares	Wire nooses set on land or underwater	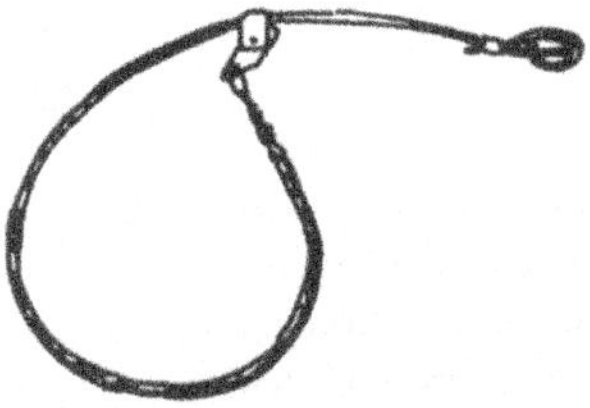
Submarine traps	Cage, rotating-jaw or foothold traps	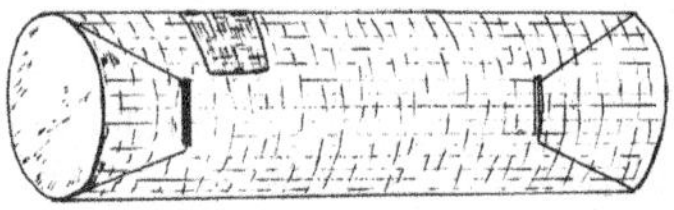

Mode of action	*Species*	*References*
Rotating-jaw traps have two metallic, circular, square, or rectangular frames that are hinged at their center point to operate in a scissor-like action, and are equipped with two torsion springs. Frames rotate and close on the animals upon firing.	Small- and medium- sized carnivores.	Proulx 1999a
Striking jaw or cable set within a box or pipe that is driven by a spring to strike an animal ventrally when the trigger is released.	Although these traps are used mainly for the capture of rodents, they can capture small carnivores such as weasels (*Mustela* spp.).	Proulx 1997, 1999a, unpubl. data.
In manual snares, an animal provides the energy to tighten the noose around its neck. In power snares one or more springs tighten the noose.	Medium-sized carnivores.	Proulx 1999a
Traps are set underwater, or slide underwater. The captured animal may drown or be killed by the trap itself.	Semi-aquatic (e.g. mink, *Neovison vison*) and riparian (e.g. raccoon) carnivores.	Proulx 1999a

Appendix 5.2 *Trap sets that are commonly used for the capture of carnivores.*

Set type	*Description*
Slide wire (drowning)	Foothold trap set in such a way on land, at the edge of water, or in a shallow rill entering a large body of water that it slides into the water upon capture of an animal. A lock stops the trap from coming back up, and the animal is submerged with the trap and drowns.
Channel	A rotating-jaw trap set at the bottom of water channel to capture predators such as minks and otters.
Running pole	A killing trap is set on a pole leaning on a tree trunk. Vegetation placed on top of the trap discourages animals from stepping over the trap to reach the bait, which is between the trap and the trunk. Bait covered with vegetation is less obvious to birds.
Box	A killing trap is inserted and secured in a wire, wooden, or plastic box with one end open and the other covered with wire mesh. Bait is placed behind the trap, at the back of the box near wire mesh. The box may be placed on the ground, on a stump, or on a running pole. Traps set in small boxes with openings at both ends will capture weasels.
Cubby	Teepee-like construction made of logs and branches, a hole dug into a bank, or a rock pile that encloses the trap and bait. The trap is set at the mouth of the funnel-like entrance, which channels the animal toward the bait. For bears (*Ursus* spp.), the back of the cubby should be a large rock or tree that forces the animal to enter the cubby to reach the bait. Large logs should be set on each side of a cubby to direct a bear towards the trap.

	References
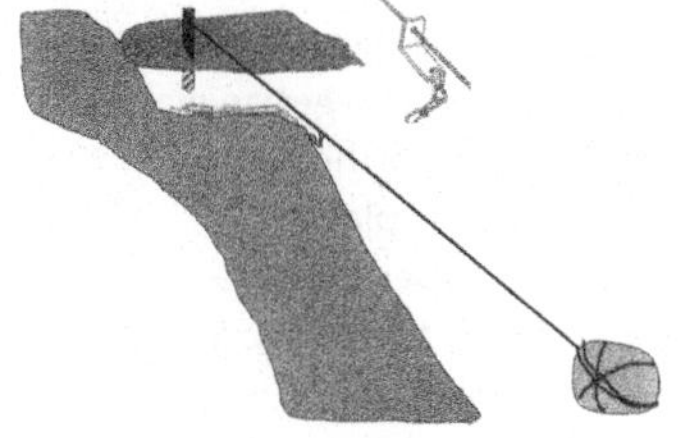	Boggess and Loegering 1985 Hubert *et al.* 1996
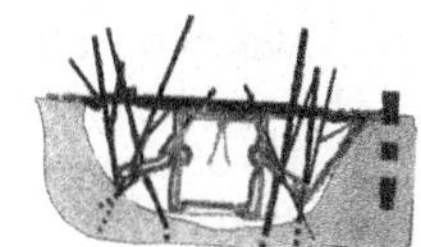	Boggess and Loegering 1985
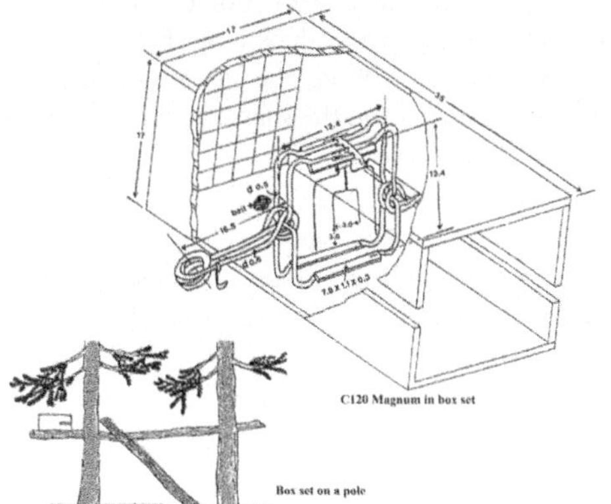	Boggess and Loegering 1985
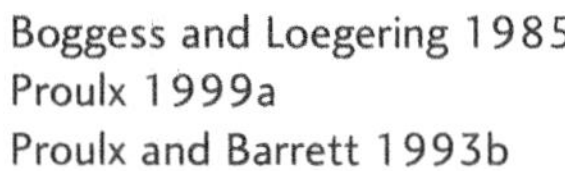	Boggess and Loegering 1985 Proulx 1999a Proulx and Barrett 1993b
	Boggess and Loegering 1985 Proulx *et al.* 1995

(continued)

Appendix 5.2 *Continued*

Set type	*Description*
Trail	A foothold trap or footsnare is set on a game trail. Setting the trap on one side of a log set across the trail forces a target animal to step over the log and land with its full weight on the trap trigger. A two-trap blind set, where a small stick is placed between the traps and at either approach, increases capture rates of some species (e.g. cougar, *Felis concolor*). For bears, a leg snare should be set under a footprint on the bear trail.
Pipe (bucket)	This is a set specifically for bears. The noose of an Aldrich snare is set around a 23-cm long stove pipe or bucket (13-cm diameter) inserted in a 23-cm deep hole in the ground. One side of the pipe has a 6.5-cm long and 2.5-cm wide slot to accommodate the spring throw arm of the snare so the trigger extends through the slot into the center of the pipe. Bait is placed at the bottom of the pipe, below the trigger. The cable loop and the spring throw arm are covered with soil, grass and leaves. When the snare fires, a bear's paw is below the rim of the noose and the snare captures the bear by the leg rather than by the paw.
Tube trap	A rubber-padded snare is placed within a PVC pipe that is 85 cm from the ground between three trees forming a triangle. When a bear pulls on the trigger to reach the bait placed at the back of the pipe, the snare tightens on the leg.
Snare	A manual or power snare is set across a game trail, without bait or scent, or set at the entrance of a baited enclosure (see pen set below). Depending on the size and height of the cable loop, medium- or large-sized carnivores may be captured selectively.
EGG trap	An EGG trap may be anchored to a tree above ground or set in a hole dug into a stream bank within 25 cm of the waterline.

	References
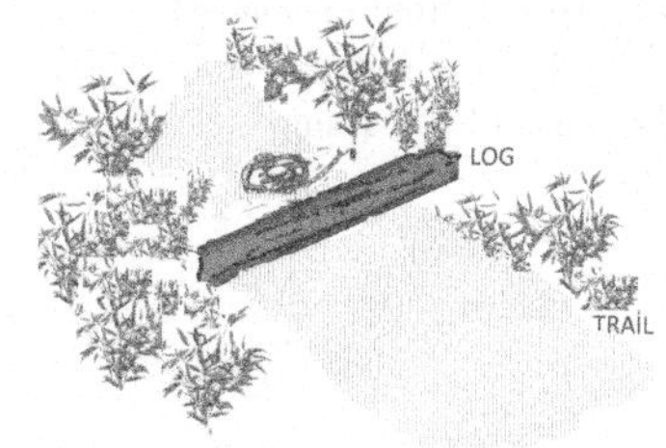	Young and Goldman 1946 Provencher 1969
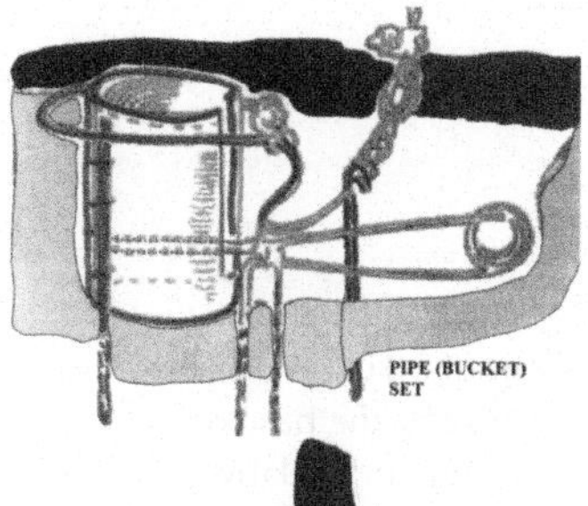	PESCOF 1988 Hygnstrom 1994 Huber *et al.* 1996
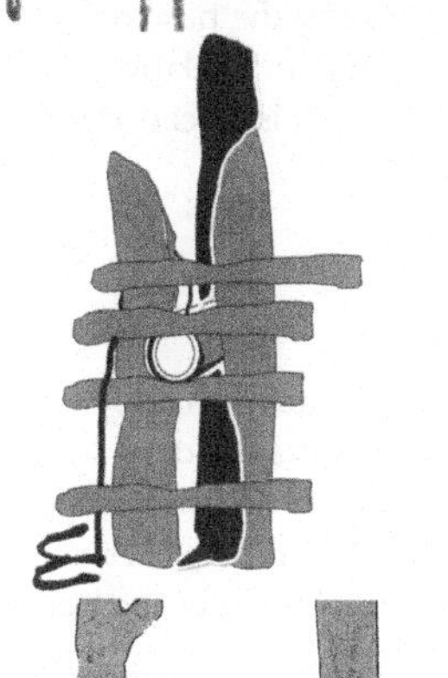	Lemieux and Czetwertynski 2006
	PESCOF 1988
	Proulx *et al.* 1993a Hubert *et al.* 1996

(continued)

Appendix 5.2 *Continued*

Set type	*Description*
Scent post	A scent or lure is placed on a stump, a stick or another prominent object to entice an animal to approach and rub the object. A foothold trap or foot snare set near the base of the stump captures the carnivore.
Dirt hole	A foothold trap or foot snare is set in front of a 10-cm diameter and 20-cm deep hole dug at a 45–60° angle at the base of clump of weeds, small stump or other backstop, in a relatively open area where visibility is good on all sides. Bait is placed at the bottom of the hole and covered with dirt.
Cage trap	A cage trap set uses the trap itself as a self-contained cubby for carnivores that will enter enclosed spaces. Traps should be concealed and covered with vegetation to protect animals from sunlight, precipitations and predators. Bait should be placed behind the treadle or trigger to force the animal to enter the trap and step on the trigger.
Pen	A pen set uses a pen with a single entrance constructed around a burrow system inhabited by a target carnivore. Bait is located outside the pen, in line with the entrance and cage, foothold trap, foot snare, or killing trap is set at the entrance, between the pen and the bait.

	References
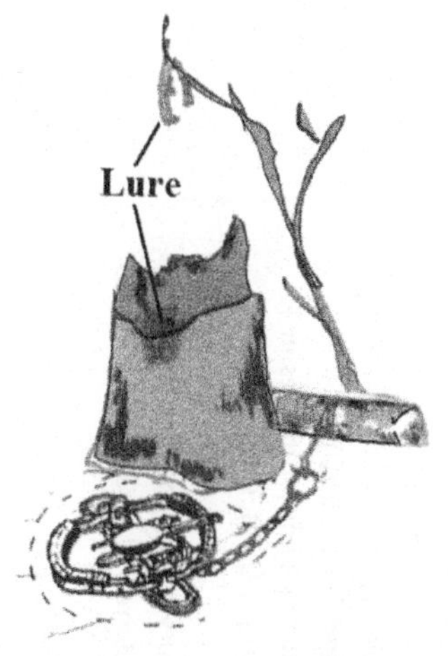	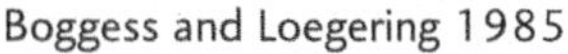 Boggess and Loegering 1985
	Boggess and Loegering 1985 PESCOF 1988 Krause 1989
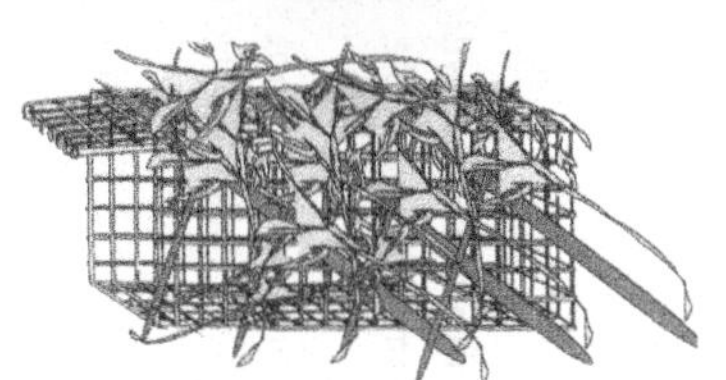	Boggess and Loegering 1985 Powell and Proulx 2003
	Currie and Robertson 1992Proulx, unpubl. data

Appendix 5.3 *Assessment of killing and restraining traps. For details on appropriate trap design and sets, consult the cited references. Criterion I: at a 95% confidence level, humane killing traps should render ≥70% of target animals irreversibly unconscious in ≤3 minutes. Criterion II: at a 95% confidence level, humane restraining traps should hold ≥70% of animals with ≤50 points scored for physical injury.*

Species	*Trap model*	*Performance*	*Meets Criterion I*		*Meets Criterion II*	
			Yes	*No*	*Yes*	*No*
CANIDS						
Coyote (*Canis latrans*)	Manual neck snare and neck snares with Kelly lock, and Gregerson or Denver Wildlife Research Center locks.[1]	Do not meet Criterion I (Guthery and Beasom 1978; Phillips 1996).		√		
	No.3—steel-jawed[2], laminated steel-jawed Sterling MJ600[3], Northwoods[2], and Bridger[4] traps	Do not meet Criterion II (Olsen *et al.* 1986; Phillips *et al.* 1996; Hubert *et al.* 1997).				√
	Novak foot snare[5]	Does not meet Criterion II (Onderka *et al.* 1990).				√
	Nos. 3 and 3½ EZ grip padded foothold traps[6]	Meet Criterion II (Olsen *et al.* 1988; Phillips *et al.* 1996). Can be equipped with tranquilizer tabs (Balser 1965; Berger and Gese 2007).			√	
	Fremont footsnare[7]	Meets Criterion II (Onderka *et al.* 1990). Effective snare cable lock required to minimize escapes (Skinner and Todd 1990).			√	
	0.32-cm diameter cable foot snare with twist-link chain between cam-lock and cable	Based on low frequency of major injuries in 17 animals (Darrow *et al.* 2009), likely meets Criterion II.			√	
	0.32-cm diameter standard cable foot snare with cam-lock, with or without a plastic tube sleeve	Based on mean injury scores and high frequency of major injuries (Darrow *et al.* 2009), does not meet Criterion II,				√
	Collarum neck snare[8]	Meets Criterion II (Shivik *et al.* 2005).			√	

	Cam-Loc[9] neck snare with tranquilizing tab	With 3.2-mm cable, lock stop set at 27 cm, 4-mg diazepam tab meets Criterion II (Pruss *et al.* 2002). Winter capture efficiency comparable or superior to standard neck snares or foothold trapping devices (Pruss *et al.* 2002).		√	
	Net gun[10]	With 8-cm mesh and 13-cm mesh, meets Criterion II (Barrett *et al.* 1982, Gese *et al.* 1987). Aerial net gunning common (e.g. Kitchen *et al.* 1999).		√	
	Cage traps[11] 66 × 51 × 152 cm, 66 × 51 × 183 cm	Used where padded foothold taps and snares illegal, but they are not capture-efficient. Expensive, frequent non-target species, many months of pre-baiting required; because of frequent tooth injuries, may not meet Criterion II (Way *et al.* 2002).			√
	Multicapture, wooden box trap 35 × 35 × 105 cm	Placed at den entrance, meets Criterion II for capture of pups (Foreyt and Rubenser 1980).		√	
Dogs, wild; *dingoes (Canis familiaris dingo)*, feral domestic *(C. familiaris)*, dog hybrids	No. 1½ Victor Soft-Catch™ foothold trap[2]	Meets Criterion II with or without tranquilizer trap device (Fleming *et al.* 1998; Marks *et al.* 2004). As capture efficient as other foothold traps used by trappers.		√	
	No. 3 Victor Soft-Catch™ foothold trap[2]	Meets Criterion II (Meek *et al.* 1995).		√	
	Treadle snare[12]	Meets Criterion II (Meek *et al.* 1995, Fleming *et al.* 1998). More expensive, bulky and awkward than Soft-Catch™ traps. Less capture efficient than commonly used traps. Wire snares must be replaced after each capture.		√	
	Cage trap of metal bars and wire fences, two guillotine doors 1.7 × 1.2 × 0.7 m	Meets Criterion II (Curi and Talamoni 2006). Highly selective for canids.		√	
Fox, arctic (*Vulpes lagopus*)	Sauvageau 2001–08[13]	With baited offset trigger, set on post in frozen ground, portable three-sided wire mesh (or snow) cubby, meets Criterion I (Proulx *et al.* 1993b, 1994).	√		

(continued)

Appendix 5.3 *Continued*

Species	*Trap model*	*Performance*	*Meets Criterion I*		*Meets Criterion II*	
			Yes	*No*	*Yes*	*No*
CANIDS						
	No. 1½ long spring steel-jawed foothold trap[2]	If checked daily, meets Criterion II (Proulx *et al.* 1994).			√	
Manual snares, 0.15 cm cable	If hung directly above dens to capture pups running through the den area, animals processed upon capture, meets Criterion II (Zabel and Taggart 1989).			√		
Fox, crab-eating (*Canis thous*)	Cage trap of metal bars and wire fences, two guillotine doors 1.7 × 1.2 × 0.7 m	Meets Criterion II (Curi and Talamoni 2006). Highly selective for canids.			√	
Fox, grey (*Urocyon cinereoargenteus*)	No. 1½ steel-jawed foothold trap[2]	Does not meet Criterion II (Berchielli and Tullar 1980; Olsen *et al.* 1988).				√
Fox, kit or swift (*Vulpes velox*)	Tomahawk wire cage traps[11] 39 × 39 × 109 cm double-door 31 × 31 × 83 cm single-door Havahart wire cage traps[14] 26 × 26 × 83 cm	If lined with a 3-mm hard board, meet Criterion II (Moehrenschlager *et al.* 2003). Two reverse, double-set traps more capture-efficient than single trap (Kamler *et al.* 2002).			√	
Single-gate, clover-type trap 46 × 46 × 122 cm	Trap is set at the entrance of an enclosure placed over den holes (Zoellick and Smith 1986). Meets Criterion II for the capture of pups.			√		
Fox, red (*Vulpes vulpes*)	Manual snares[15]	Does not meet Criterion I (FPCHT 1981).		√		
	Power snares[15]	Does not meet Criterion I (Proulx and Barrett 1990).		√		
	No. 1½ steel-jawed and No. 3 Victor Soft-Catch™ foothold traps[2]	Do not meet Criterion II (Olsen *et al.* 1988, Kern *et al.* 1994. Seddon *et al.* 1999).				√
	No. 1½ Soft-Catch foothold trap[2]	Meets Criterion II (Olsen *et al.* 1988; Kern *et al.* 1994; Fleming *et al.* 1998; Kreeger *et al.*			√	

		1990b). As capture efficient as equivalent, unpadded traps (Tullar 1984; Linscombe and Wright 1988).		
	Åberg (Swedish) foot snare[16]	Meets Criterion II (Englund 1982).	√	
	Novak foot snare[5]	With blackened cable, appears to meet Criterion II (Novak (1981).	√	
	Belisle foot snare[17]	Meets Criterion II (Muñoz-Igualada *et al.* 2002). Less capture efficient than Collarum foot snare. More selective than cage traps.	√	
	Collarum neck snare[8]	Meets Criterion II (Muñoz-Igualada *et al.* 2002). More capture efficient than Belisle foot snare. More selective than cage traps.	√	
Fox, Ruppell's (*Vulpes rueppellii*)	No. 3 Victor Soft-Catch™ foothold trap[2]	Meets Criterion II, possibly because animals lay relatively quietly in foothold traps (Seddon *et al.* 1999).	√	
Jackals (*Canis mesomelas*, *C. adustus*, *C. aureus*)	No. 3 Oneida Jump trap[1] padded with two layers of cotton mutton cloth	Meets Criterion II when traps checked daily after dawn (Rowe-Rowe and Green 1981).	√	
	Victor Soft-Catch™ No. 1½ foothold trap[2]	Meets Criterion II if animals handled upon capture (Craft 2008b).	√	
Wolf, grey or timber (*Canis lupus*)	No. 4 double long spring, No. 4 jaws offset 2 mm, No. 14 toothed, No. 14 double long spring steel-jawed foothold traps[2]	Commonly used to capture wolves in North America but cause extensive injuries and do not meet Criterion II (Van Ballenberghe 1984; Kuehn *et al.* 1986; Sahr and Knowlton 2000).		√
	Modified Newhouse No.4[18] foothold trap	With one spring only, meets Criterion II (Kolenosky and Johnston 1967).	√	
	Newhouse 14 with 1.8-cm offset toothed jaw[18] foothold trap	May meet Criterion II (Kuehn *et al.* 1986).		

(*continued*)

Appendix 5.3 *Continued*

Species	*Trap model*	*Performance*	*Meets Criterion I*		*Meets Criterion II*	
			Yes	*No*	*Yes*	*No*
CANIDS						
	No. 4 foothold trap with offset steel-jaws[1] with tranquilizer tabs				-	
		With propiopromazine hydrochloride tabs meet Criterion II (Sahr and Knowlton 2000, Chavez and Gese 2006). For use with adults and pups.			√	
	No. 7 EZ Grip Trap[6]	May meet Criterion II (Frame and Meier 2007).			-	
	Aldrich foot snare[19]	Meets Criterion II (Van Ballenberghe 1984; Okarma *et al.* 1998). Difficult to conceal, not capture efficient for trap-shy wolves.			√	
	Drive net	With nylon mesh nets stretched loosely and supported by poles or branches, meets Criterion II (Okarma and Jedrzejewski 1997; Okarma *et al.* 1998; Theuerkauf *et al.* 2003).			√	
	Net gun[10]	Meets Criterion II (Walton *et al.* 2001b; Chavez and Gese 2006). Used to capture adults and pups.			√	
Wolf, maned (*Canis brachyurus*)	Cage trap of metal bars and wire fences, two guillotine doors 1.7 × 1.2 × 0.7 m	Meets Criterion II (Curi and Talamoni 2006). Highly selective for canids.			√	
FELIDS						
Bobcat (*Lynx rufus*)	No. 3 Soft-Catch™ foothold trap[2]	Meets Criterion II (Olsen *et al.* 1988). Can be modified for optimal use (Earle *et al.* 2003).			√	
	Cage trap 38 × 38 × 90 cm	Meets Criterion II (Woolf and Nielsen 2002). More capture-efficient than No. 3 Soft-Catch foothold traps.			√	
Cat, feral domestic (*Felis catus*)	No. 1½ Soft-Catch™ foothold trap[1]	Meets Criterion II (Molsher 2001).			√	
	Cage trap[20] 40 × 40 × 60 cm	Meets Criterion II (Molsher 2001).			√	
Cougar (*Puma concolor*)	Schimetz-Aldrich foot snare[21]				√	

		Meets Criterion II (Logan *et al.* 1999). May cause fewer deaths than treeing animals with trained dogs.				
Lynx, North American (*Lynx canadensis*)	Conibear 330[1]	Used by trappers in North America (CGSB 1996) but does not meet Criterion I (Proulx *et al.* 1995).		√		
	Conibear 330[2] with clamping bars and a one-way four-prong trigger	In simulated natural environments met Criterion I; never field-tested (Proulx *et al.* 1995).	-			
	No. 3 Soft-Catch™ foothold trap[2]	Does not meet Criterion II even with weakened springs (Mowat *et al.* 1994).				√
	Fremont foothold snare[7]	Meets Criterion II (Mowat *et al.* 1994), especially if loosely attached to trees or drag poles and modified to reduce tangling.			√	
	Tomahawk wire-cage traps[11] 66 × 51 × 122 cm, 51 × 38 × 114 cm Home-made wire cage trap 99 × 102 × 122 cm	Meet Criterion II (Mowat *et al.* 1994; Kolbe *et al.* 2003). Large traps may have higher capture efficiency.			√	
Lion (*Panthera leo*)	Aldrich foothold snare[19]	Set in a baited thornbush enclosure, meets Criterion II (Frank *et al.* 2003). After brief struggle, lions lie down.			√	
Tiger, Amur (Siberian) (*Panthera tigris altaica*)	Aldrich foothold snare[19]	Meets Criterion II (Goodrich *et al.* 2001). Effort/capture high.			√	
HYAENIDS						
Hyena, brown (*Hyaena brunnea*)	Aldrich foot snare[19]	Meets Criterion II (Frank *et al.* 2003).			√	
MEPHITIDS						
Skunks, striped, spotted, hog-nosed (*Mephitis mephitis, M. macroura, Spilogale putorius, S.* gracilis, Conepatus mesoleucus)	Foothold traps	Used (Allen 1939; Verts 1960) but limb injuries range from cut skin to fractures to chewed feet; do not meet Criterion II (Novak 1981; Rosatte 1987).				√
	Foothold snares	Do not meet Criterion II (Novak 1981). Poor capture efficiency.				√

(continued)

Appendix 5.3 *Continued*

Species	*Trap model*	*Performance*	*Meets Criterion I*		*Meets Criterion II*	
			Yes	*No*	*Yes*	*No*
MEPHITIDS						
	Wire cage traps	If covered with wood, plastic or cartons meet Criterion II (Larivière and Messier 1999). Opaque-sided traps recommended to reduce stress, facilitate handling (Larivière and Messier 1999).			√	
Wooden box traps	Meet Criterion II (Crabb 1941; Knight 1983).			√		
MUSTELIDS						
Badgers (*Meles meles*, *Taxidea taxus*)	Tomahawk wire cage trap[11] 76 × 25 × 30 cm	Meet Criterion II (Woodroffe *et al.* 2005b). May be less efficient than foothold traps to capture Eurasian badgers (*Meles meles*) (Loureiro *et al.* 2007) but recommended to cull populations in Europe (MAFF 1983).			√	
	No. 1½ Soft-Catch™ foothold trap[2]	Meets Criterion II (Kinley and Newhouse 2008).			√	
	No. 3 Soft-Catch™ foothold trap[2]	Meets Criterion II (Schemnitz 2005).			√	
Fisher (*Martes pennanti*)	Conibear 220[2]	Commonly used by trappers but does not meet Criterion I (Proulx and Barrett 1993b).		√		
	Bionic[22]	With 10 cm aperture cone met Criterion I in simulated natural environments (Proulx and Barrett 1993a). Never field-tested (Proulx 1999a).	-			
	Coon-getter cage trap[23] 37.5 × 37.5 × 90 cm	Meets Criterion II. Relatively efficient, particularly when equipped with radio monitors (Arthur 1988; Frost and Krohn 1994). Serious tooth injuries minimized if no openings larger than 2.5 cm^2 and hard surfaces covered with wood (Arthur 1988).			√	

Marten, American (*Martes americana*)	No. 3 and 4[2] foothold traps in pole box sets	Strike martens in the chest but do not cause major traumatic lesions; do not meet Criterion I (Barrett *et al.* 1989). Energy levels too low (Proulx and Barrett 1994).		√	
	Conibear 120[2]	Commonly used by trappers but does not meet Criterion I (Proulx *et al.* 1989a)		√	
	C120 Magnum[22] with one-way, four-prong pitchfork trigger (central prongs shorter than outside ones)	If in a pole box set, animals single-struck in the head-neck region or double-struck in the head-neck and thorax regions; meets Criterion I (Barrett *et al.* 1989; Proulx *et al.* 1989b). Capture efficiency same as standard Conibear traps.	√		
	Bionic[24]	Judged suitable from work with minks and fishers (Proulx and Barrett 1991a, 1993a). In pole sets, trauma associated with quick loss of consciousness; capture efficiency similar to control traps used by trappers (Proulx 1999b).	√		
	Tomahawk wire cage traps[11] 15 × 15 × 71 cm 20 × 20 × 51 cm	If wrapped in black plastic or connected to wooden nest boxes, meet Criterion II (Bull *et al.* 1996). Relatively capture efficient.			√
Mink (*Neovison vison*)	No. 1½ foothold trap[2] in drowning sets	Does not meet Criterion I (Proulx's 1999a review of Gilbert and Gofton 1982). Note that drowning is not considered to be an acceptable euthanasia method (Ludders *et al.* 1999, 2001).		√	
	Conibear 120[2]	Commonly used by trappers in North America but even upgraded version lacks necessary power; does not meet Criterion I (Gilbert 1981).		√	
	C120 Magnum[22] with pan trigger	The only rotating-jaw trap of its category with the potential to meet Criterion I when it strikes in the head-neck and thorax regions simultaneously (Proulx *et al.* 1990). Captures efficiency similar to traps commonly used by trappers (Proulx and Barrett 1993c).	√		

(continued)

Appendix 5.3 *Continued*

Species	*Trap model*	*Performance*	*Meets Criterion I*		*Meets Criterion II*	
			Yes	*No*	*Yes*	*No*
MUSTELIDS						
Otters (*Hydrictis maculicollis, Lontra canadensis, Lutra lutra*)	Hancock[25] 45 × 59 × 95 cm	Used to capture North American otters (Northcott and Slade 1976; Melquist and Hornocker 1979) but tooth damage, particularly canines, can be excessive and otters may become trap shy (Blundell *et al.* 1999).				√
	No. 1½ Soft -Catch™ foothold trap[2]	Fewer limb injuries than for modified Victor No. 11 double spring trap but no difference in dental injuries (Serfass *et al.* 1996). More research and development needed to minimize serious trauma and to meet Criterion II.				√
	Havahart wire cage traps[14] 40 × 40 × 100 cm unknown 80 × 80 × 140 cm	Recommended by Maxfield *et al.* (2005) for North American otters, and by Perrin and Carranza (1999) for the spotted-necked otters but no data on capture efficiency or tooth injuries.			-	
	No. 11 Victor and Sleepy Creek foothold traps[2, 26]	Meet Criterion II (Blundell *et al.* 1999, Shirley *et al.* 1983).			√	
	No. 3 Victor Soft-Catch™ foothold trap[2] equipped with a trap alarm	Functioning alarms and rapid response (ca. ½ h) necessary to meet Criterion II (Ó Néill *et al.* 2007).			√	
	Nos. 1 and 1½ Victor Soft-Catch™ foothold traps[1] with one spring replaced by a # 2 spring	Set in shallow water, anchored to solid objects, 1-m long chain, monitored daily early morning. Meet Criterion II (Fernández-Morán *et al.* 2002).			√	
Polecat (*Mustela putorius*)	Wire cage traps[6] 15 × 15 × 76 cm	If covered with dry hay for animal welfare so polecats pull at hay instead of trap, meets Criterion II (Birks *et al.* 1994).			√	

Weasel, short-tailed, or stoat (*Mustela erminea*)	Victor No. 1½ steel-jawed foothold trap[2]	Does not meet Criterion I (Trap Effectiveness Research Team 1995).		√		
	Fenn trap[27]	Does not meet Criterion I (Warburton *et al.* 2008). Efficient, used to control stoats and weasels (King 1981).		√		
	DOC traps[28] Models 150, 200, 250	Meet Criterion I (Warburton *et al.* 2008). Conceptually identical to Fenn trap but have six parallel strike bars powered by two coil springs. Set in wooden tunnels, capture as efficiently as Fenn traps (New Zealand Department of Conservation 2008).	√			
Wolverine (*Gulo gulo*)	Foothold No. 4[2] Wooden and metal cage traps	Used to capture wolverines (Hornocker and Hash 1981, Banci 1987, and others) but do not meet Criterion II. Capture efficiency low (Lofroth *et al.* 2008).				√
	Log trap	If bait properly attached, meets Criterion II (Copeland *et al.* 1995; Lofroth *et al.* 2008). Capture efficiency of portable wooden traps and modified round log traps markedly greater than for foothold and metal cage traps (Hornocker and Hash 1981; Banci 1987).			√	
PROCYONIDS						
Raccoon (*Procyon lotor*)	Sauvageau 2001–8[13] Conibear 220[1]	Used by trappers in North America but lack power to meet Criterion I, especially the Conibear 220 (Proulx and Drescher 1994).		√		
	Conibear 160[1]	Used by trappers in North America but lacks power to meet Criterion I (Sabean and Mills 1994).		√		
	Foothold traps with steel or padded jaws[2]	Do not meet Criterion II (Olsen *et al.* 1988; Berchielli and Tullar 1980; Tullar 1984; Proulx *et al.* 1993a; Hubert *et al.* 1996).				√

(continued)

Appendix 5.3 *Continued*

Species	*Trap model*	*Performance*	*Meets Criterion I*		*Meets Criterion II*	
			Yes	*No*	*Yes*	*No*
PROCYONIDS						
	EGG trap[29]	If anchored to a tree above ground, meets Criterion II (Proulx 1991; Proulx *et al.* 1993a). As capture efficient as rotating-jaw traps, avoids non-target species (Proulx 1995). More capture efficient than cage traps (Austin *et al.* 2004). Used in holes in stream banks, more injurious but still more humane than foothold traps in similar sets (Hubert *et al.* 1996).			√	
URSIDS						
Bear, black (*Ursus americanus*), Grizzly bear (*U. arctos*), Malayan sun (*Helarctos malayanus*)	Culvert and barrel trap	If constructed correctly, meet Criterion II (Powell and Proulx 2003). Efficient but cumbersome (Powell and Proulx 2003). May cause less distress and muscle injuries than foot snares (Schroeder 1987; Wong *et al.* 2004; Powell 2005; Cattet *et al.* 2008a).			√	
	Aldrich snare[18]	If set properly and modified with shock-absorbing spring, meets Criterion II (Johnson and Pelton 1980; Kaczensky *et al.* 2002; Powell 2005).			√	
	RL04 tube trap[30] with rubber-padded snare	If set properly, meets Criterion II (Lemieux and Czetwertynski 2006). Use solid anchor for grizzly bears.			√	

VIVERIDS				
Civet. brown palm (*Paradoxurus jerdoni*)	Havahart wire-cage trap[14]	33 × 28 × 107 cm	For animals processed within 30 min after capture, meets Criterion II (Mudappa and Chellam (2001).	√

[1]*Kelley locks: D. Amberg, Morris, Minnesota, USA; Gregerson locks: K. Gregerson, Roundup, Montana, USA; Denver Wildlife Research Center, Denver, Colorado, USA.*
[2]*Woodstream Co., Lititz, Pennsylvania, USA. Current manufacturer: Oneida Victor, Inc., Euclid, Ohio, USA.*
[3]*Sterling MJ600: Glen Sterling, Faith, South Dakota, USA.*
[4]*Montgomery Fur Co., Ogden, Utah, USA.*
[5]*E. R. Steel Products, Barrie, Ontario, Canada.*
[6]*Livestock Protection Co., Alpine, Texas, USA.*
[7]*Fremont Humane Traps, Beaumont, Alberta, Canada.*
[8]*Wildlife Control Supplies, East Branby, Connecticut, USA.*
[9]*Halford Hide & Leather Co., Ltd, Edmonton, Alberta, Canada.*
[10]*Mountain Helicopters Ltd., Taupo, New Zealand.*
[11]*Tomahawk Live Trap Co., Tomahawk, Wisconsin, USA.*
[12]*Glenburn Motors, Yea, Victoria, Australia.*
[13]*Les Pièges du Québec, Enr., St-Hyacinthe, Quebec, Canada.*
[14]*Havahart Woodstream Co., Lititz, Pennsylvania, USA.*
[15]*King snare: Western Creative Services, Ltd., Winnipeg, Manitoba, Canada; Mosher snare: W. C. Mosher, Mayerthorpe, Alberta, Canada.*
[16]*Nordic Sports AB, Kanalgatan 73, S-931 00 Skellefteå, Sweden.*
[17]*Edouard Belisle, Sainte-Veronique, Quebec, Canada.*
[18]*Kirsh Foundry, Inc., Beaver Dam, Wisconsin, USA.*
[19]*Aldrich Animal Trap Co., Clallam Bay, Washington, USA.*
[20]*Manufacturer not identified.*
[21]*Schimetz-Aldrich Spring Activated Animal Care, Sekiu, Washington, USA. Now available from Margo Supplies, Ltd., High River, Alberta, Canada.*
[22]*Originally produced by Woodstream Co. Current manufacturer: Les Pièges du Québec, Enr., St-Hyacinthe, Quebec, Canada.*
[23]*Coon Getter Traps, Miller, South Dakota, USA.*
[24]*Originally produced by W. Gabry, Vavenby, British Columbia, Canada. Current manufacturer: Les Pièges du Québec, Enr., St-Hyacinthe, Quebec, Canada.*
[25]*Hancock Trap Co., Custer, South Dakota, USA.*
[26]*Sleepy Creek Manufacturing, Inc., Berkeley Springs, West Virginia, USA.*
[27]*A. Fenn & Co., Hoopers Lane, Astwood bank, Redditch, Worchestershire, UK.*
[28]*Department of Conservation, Wellington, New Zealand.*
[29]*EGG Trapp Co., Butte, North DAKOTA, USA.*
[30]*Rolland Lemieux, St.-Emile, Quebec, Canada.*

Appendix 5.4 *Methods for administering drugs to wild carnivores.*

Method	Distance from Target Animal	Description	Specific References[1]
Inhalation anesthesia (gas anesthesia)	Close contact	Vaporized drug (volatile anesthetic) delivered directly into the lungs of an animal for absorption by blood and delivery to the brain. Advantages over injectable drugs: finer level of operator control, rapid induction and recovery. Disadvantages: requires field-durable delivery system and dedicated operator to monitor delivery system and animal continuously. Although use in field research is limited, has proven effective and safe with marine and terrestrial carnivores (Heath *et al.* 1996; Mathews *et al.* 2002; Lewis 2004; Potvin *et al.* 2004).	Heath *et al.* 1996 Mathews *et al.* 2002 Lewis 2004 Potvin *et al.* 2004
Hand injection (syringe and needle)	Close contact	Used to deliver drug to restrained or anesthetized animals, to transfer drug to other delivery devices, and to collect blood samples. Syringes available in many sizes (1–60 ml).Needles available in many lengths (16–75 mm) and gauges (14–25 ga).One-time use of disposable syringes and needles strongly recommended to avoid contamination of drugs, blood samples, and animals. Safe practices essential for handling needles and other sharps to prevent "needlestick injury" (the most common cause of accidental human exposure to drugs and animal fluids; Weese and Jack 2008).	See general references for instructions on use of syringe and needle.
Nasal delivery	Close contact	Intranasal drug delivery (drug sprayed into the nostrils) may provide an alternative to intravenous drug delivery in select cases.	Cattet *et al.* 2004 Wolfe and Bernstone 2004

		Advantages: quick and painless, nasal route is immediately available (relative to venipuncture), time to drug effect as rapid as by intravenous delivery (Wolfe and Bernstone 2004). Although use in carnivores not reported, used with xylazine (a sedative-type drug) to reduce stress in wild elk (*Cervus elaphus*) captured by net gun (Cattet *et al.* 2004).	
Oral delivery	Remote	Delivery of drug into the mouth by "tranquilizer tablet" or by drug-laced food (bait). Tranquilizer tablets used with mixed success to calm carnivores captured in foothold traps and snares (Balser 1965; Sahr and Knowlton 2000; Pruss *et al.* 2002) Drug-laced bait used to immobilize captive black bears and brown bears (Ramsay *et al.* 1995; Mortenson and Bechert 2001). Difficulties for drug delivery to free-ranging animals include poor bait acceptance, drug instability, inability to control dosage, risk for non-target animals (Conover 2002). Programs for mass vaccination of wild carnivores with vaccine-laced bait successful (Bachmann *et al.* 1990; Ballesteros *et al.* 2007). Carcasses laced with midazolam used to sedate lions (*Panthera leo*; Stander and Morkel 1991).	Bachmann *et al.* 1990 Stander and Morkel 1991 Ramsay *et al.* 1995 Sahr and Knowlton 2000 Mortenson and Bechert 2001 Conover 2002 Pruss *et al.* 2002 Ballesteros *et al.* 2007
Pole syringe(jab pole)	≤3 m	Extends the distance of operation for injection by syringe and needle. Used to deliver drug to trapped or restrained animals or to give additional drug to animals not completely immobilized but safe to approach. Effective volumes generally limited to <10 ml because animals react quickly to injection. Requires quick delivery through short (≤2.5 cm), large gauge (14–18 ga) needle.	Beltran and Tewes 1995 Grassman *et al.* 2006 Lofroth *et al.* 2008

(continued)

Appendix 5.4 *Continued*

Method	Distance from Target Animal	Description	Specific References[1]
Darts	Determined by type of projector	Used for remote delivery of drug (1–20 ml) to a specific target animal using blowpipe, bow, or dart gun. Many different designs, but basic components are a needle, body (or syringe), plunger, and tailpiece. Discharge of drug by expanding gas from an explosive powder charge, by compressed air, by butane, or by chemical reaction. Some designs and discharge mechanisms more likely to cause injury to target animals (Cattet *et al.* 2006). Needles of different sizes and port configurations have smooth shaft, wire barb, or metal (or gelatin) collar. Barb or collar essential to retain dart securely in free-ranging animal. Can be equipped with radio transmitters to enable location of lost darts or animals (Kilpatrick *et al.* 1996). Can be fitted with skin cutting heads for remote biopsy sampling (Spong and Creel 2001).	Kilpatrick *et al.* 1996 Spong and Creel 2001 Kreeger 2002 Cattet *et al.* 2006
Blowpipe(blow gun)	≤15 m	Effective for restrained animals or free-ranging animals that can be approached closely. Used to deliver small volumes of drug (≤3 ml). Darts propelled quietly and cause minimal trauma. Precise delivery location usually possible. Discrete appearance unlikely to attract public attention when used in urban settings. Long pipes increase accuracy. Risk of drug exposure by mouth. Use in some localities prohibited or requires legal authorization.	Brockelman and Kobayashi 1971 Haigh and Hopf 1976 Ryser *et al.* 2005

		"Powered" blowpipes using compressed air or gas can project darts up to 30 m. Remote-controlled, "powered" blowpipe used to capture Eurasian lynxes (*Lynx lynx*; Ryser *et al.* 2005).	
Long bows and cross bows	$\leq$50 m	Used to deliver $\leq$5 ml of drug to large animals. Commercially-available darts mountable on arrow shafts using adapter. Can project darts long distances, but difficult to prevent trauma caused by high velocity impact (Hawkins *et al.* 1967). Of limited use given wide availability of accurate dart guns that cause less trauma.Modified long bow used to capture lions and leopards (*Panthera pardus*; Stander *et al.* 1996).	Short and King 1964 Hawkins *et al.* 1967 Stander *et al.* 1996
Dart guns	$\leq$90 m	Modified pistols, shotguns, rifles, and custom-designed projectors for propelling drug-filled darts into specific target animals.	Crockford *et al.* 1957 Smuts *et al.* 1977
		Used to deliver drug to trapped carnivores and to capture free-ranging carnivores ($\geq$20 kg) from ground or helicopter. Darts propelled by gas generated from a .22-caliber blank cartridge, by compressed CO_2 or N_2, or by compressed air. Can deliver 1–10 ml of drug for compressed gas/air-powered guns and up to 20 ml for .22-caliber-powered guns. Effective range less for pistols than other types of projectors, and less for compressed gas/air-powered guns than for .22-caliber-powered guns. Darts fired at high velocity can cause injury or death (Valkenburg *et al.* 1999; Cattet *et al.* 2006).	Ballard *et al.* 1982 Kreeger 1999 Valkenburg *et al.* 1999 Cattet *et al.* 2003 Holekamp and Sisk 2003 Cattet *et al.* 2006 Fahlman *et al.* 2008
Remote injection collars	$\leq$25 000 m	Used to recapture animals by remote-controlled injection of drug contained in a syringe mounted on a collar.	Mech *et al.* 1984 Mech and Gese 1992 Jessup 1993 Federoff 2001 Powell 2005

(continued)

Appendix 5.4 *Continued*

Method	Distance from Target Animal	Description	Specific References[1]
		Useful for repeated immobilization of free-ranging animals difficult to dart. Communication with collar greater by air (<25 km) than by ground (<3 km). Used by to recapture wolves and black bears (Mech and Gese 1992; Powell 2005). No longer available commercially.	

[1] *Specific references often chosen as representative examples from a broader selection of literature. Readers seeking additional references or details on use of equipment and manufacturer information should review comprehensive works by Nielsen (1999), Kreeger and Arnemo (2007), West* et al. *(2007), and Fowler (2008).*

Appendix 5.5 *Medical emergencies and complications that can occur with capture and handling of carnivores. Comments provide a definition or description (D) of the emergency or complication and brief summary information concerning clinical signs (S), causes (C), prevention (P), and treatment (T).*

Emergency or Complication[1]	Comments[2]
Physical injury	D: Abrasions (scrapes), contusions (bruises), concussion, dislocation or fracture of bones, lacerations (cuts), and punctures. S: Depends on type of injury—may see blood, reduced mobility, limb irregularities • Animal may vocalize frequently in response to pain. C: Darts • Marking techniques • Prolonged physical exertion (from pursuit or resisting restraint) • Prolonged induction of anesthesia • Rough handling • Sample collection • Self-infliction • Surgery • Telemetry devices (collar, ear tag, implantable) • Traps. P: Animal care approval of capture and handling protocol • Appropriate selection of capture method and equipment for target species • Appropriate training and expertise in capture and handling techniques • Check traps frequently • Perform capture and handling procedures quickly and efficiently. T: Basic first aid for minor injuries • Veterinary care, or possibly humane killing, for major injuries.
Thermal stress (hyperthermia)	D: Increase in body temperature to point where oxygen uptake is insufficient to meet need of oxygen for cellular metabolism. S: Rise in body temperature to >40°C (104°F) • Rapid, shallow breathing (panting) • Rapid heart rate • Warm extremities • Convulsions • Death. C: Concurrent disease • Drug effect (inhibition of thermoregulation) • External heat absorption (solar radiation) • Fear • Improper trap site prevents behavioral thermoregulation • Prolonged physical exertion (from pursuit or resisting restraint) • Species-specific factors (mass, surface-to-volume ratio, insulation) • Weather (high ambient temperature, no wind). P: Avoid capturing animals on hot days • Avoid prolonged pursuit • Check traps frequently • Minimize stress (consider providing sedation) • Monitor rectal temperature frequently, e.g. every 5–10 min • Protect animals/traps from direct exposure to sun • Use immobilizing drugs with effects that can be terminated by administering appropriate antagonist drugs. T: Administer antagonist drug • Administer cold water enema • Administer cold lactated Ringers by intravenous route • Administer oxygen • Apply external cold sources to areas of greatest heat exchange • Immerse animal in cold water • Provide adequate ventilation, e.g. circulate air around animal with a fan • Spray body surface with cold water.

(continued)

Appendix 5.5 *Continued*

Emergency or Complication[1]	Comments[2]
Thermal stress (hypothermia and frostbite)	D: Decrease in body temperature to point where cellular death occurs due to decreased metabolism and/or freezing of tissue. S: Decrease in body temperature to <35°C (95°F) • Shivering • Cold extremities (also firm with frostbite) • Dullness or lack of behavioral responsiveness to stimuli • Decreased heart rate • Shock • Coma • Death. C: Drug effect (inhibition of thermoregulation, decrease in metabolism) • Improper trap site prevents behavioral thermoregulation • Loss of insulation (poor body condition, wet fur) • Prolonged immobility • Prolonged restraint on a cold surface • Species-specific factors (mass, surface-to-volume ratio, insulation) • Weather (low ambient temperature, high wind, precipitation). P: Apply external heat sources, e.g. hot water bottles, warming blankets, chemical heat packs • Avoid capturing animals on cold days • Check traps frequently • Insulate animals from cold, wet surfaces • Monitor rectal temperature frequently, e.g. every 5–10 min • Protect extremities of anesthetized animals from frostbite • Shelter animal from wind and precipitation • Use immobilizing drugs with effects that can be terminated by administering appropriate antagonist drugs. T: Administer antagonist drug *before* animal becomes hypothermic or after it is re-warmed, but not when it is hypothermic • Administer warm water enema • Administer warm physiological saline by intravenous route • Apply external heat sources, e.g. hot water bottles, warming blankets, chemical heat packs • Dry wet fur.
Dehydration	D: Excessive loss of body water. S: Depends on severity of dehydration—may see dryness of mouth (including gums), loss of skin elasticity, sunken eyes, fever, weak pulse, shock, coma, death. C: Decreased water intake • Fever (due to pre-existing illness) • Hyperthermia • Increased water loss (due to panting or persistent vomiting, diarrhea, urination, or bleeding). P: Avoid trapping on hot days • Avoid prolonged pursuit • Check traps frequently • Minimize stress (consider providing sedation) • Protect animals/traps from direct exposure to sun. T: Estimate amount of fluid lost • Administer fluids (lactated Ringer's solution or physiological saline) by intravenous, subcutaneous, or intraperitoneal routes.
Hypoxia (hypoxemia)	D: Decreased availability of oxygen in blood (hypoxemia) or more generally in body tissues (hypoxia) • A common complication of anesthetized animals, less likely to occur in non-anesthetized animals.

	S: Labored or difficult breathing • Blue (cyanotic) mucous membranes • Hemoglobin oxygen saturation (measured by pulse oximeter) $<80\%$ for more than 1 min • Rapid pulse • Unconsciousness • Convulsions • Death. C: Concurrent respiratory disease • Drug-induced depression of respiratory function • Excessive pressure applied to the thoracic cavity (chest) • Obstruction of airways, including nostrils • Regurgitation and aspiration of stomach content. P: Administration of appropriate drug dose • Monitor mucous membrane color and hemoglobin oxygen saturation frequently, e.g. every 5–10 min • Position anesthetized animals correctly • Use proper restraint and handling techniques. T: Administer antagonist drug • Administer oxygen • Correct mechanism causing hypoxia, e.g. reposition body, remove pressure from chest, etc.
Acidosis	D: Disturbance of normal acid-base balance resulting in a blood pH <7.35. S: Rapid breathing • Confusion • Convulsions • Coma • Death. C: Intense or prolonged physical exertion resulting in excessive accumulation of lactic acid (metabolic acidosis) • Obstruction of airways resulting in excessive accumulation of carbon dioxide (respiratory acidosis). P: Avoid prolonged pursuit • Check traps frequently • Position anesthetized animals correctly • Use proper restraint and handling techniques. T: *For metabolic acidosis:* Administer sodium bicarbonate by intravenous route in conjunction with other fluids (physiological saline or dextrose). *For respiratory acidosis:* Assist respiration by artificial ventilation • Establish open airways • Re-position anesthetized animal correctly.
Regurgitation/vomiting and aspiration	D: Passive flow (regurgitation) or forceful ejection (vomiting) of stomach content into the mouth followed by inhalation of regurgitated material into the airways (aspiration). S: Presence of stomach contents in nostrils or mouth • Respiratory distress (gagging, retching, gurgling) • Fever • Death. C: Drug-induced relaxation of stomach sphincter • Improper restraint, handling, or positioning • Recent feeding followed by intense exertion • Stress. P: Avoid pursuing feeding animals • Minimize stress • Position anesthetized animals correctly • Use proper restraint and handling techniques. T: Avoid prolonged immobilization with animal lying on its side • Quickly clear mouth and airway of regurgitated or vomited material • If aspiration has occurred, seek veterinary assistance or consider humane killing.

(continued)

Appendix 5.5 *Continued*

Emergency or Complication[1]	Comments[2]
Shock	D: Failure of blood circulation resulting in ineffective perfusion of tissues. S: Rapid heart rate • Low blood pressure (capillary refill time >2 s) • Shallow, rapid breathing • Bluish pale mucous membranes. C: Concurrent illness • Prolonged physical exertion • Prolonged stress • Severe blood loss. P: Avoid prolonged pursuit • Check traps frequently • Minimize stress (consider providing sedation) • Monitor cardiovascular function (heart rate, pulse, capillary refill time, mucous membrane color, and hemoglobin oxygen saturation) frequently, e.g. every 5–10 min. T: Administer antagonist drug • Administer corticosteroids (dexamethasone) intravenously • Administer fluids (lactated Ringer's solution or physiological saline) intravenously to increase blood volume and blood pressure • Maintain an open airway and provide oxygen • Monitor rectal temperature and keep animal warm.
Seizures/convulsions	D: Disturbance of brain function characterized by involuntary, violent contractions of skeletal muscles. S: Rigid extension of the limbs • Uncontrolled muscle spasms (may be focal or involve whole body) • Increasing body temperature (associated with intense muscular contractions). C: Drug-induced effect, e.g. side-effect of ketamine • Hyperthermia • Metabolic disturbances incited by capture and stress, e.g. hypocalcemia (low blood calcium), hypoglycemia (low blood glucose) • Trauma. P: Administration of appropriate drugs at appropriate doses • Apply same measures as taken to prevent hyperthermia. T: Administer benzodiazepine sedative (diazepam or midazolam) intravenously slowly • Monitor rectal temperature and take appropriate steps to prevent hyperthermia.
Respiratory arrest	D: Cessation of breathing. S: Slow, shallow breathing or cessation of breathing • Blue (cyanotic) mucous membranes • Hemoglobin oxygen saturation (measured by pulse oximeter) <80% for more than 1 min or downward trend in saturation values • Rapid pulse • Unconsciousness • Convulsions • Death. C: Drug-induced depression of respiratory function (possibly as a result of a severe overdose) • Excessive pressure applied to the thoracic cavity (chest) • Obstruction of airways, including nostrils. P: Administration of appropriate drug dose • Monitor respiratory function (respiratory rate and depth, respiratory sounds, mucous membrane color, and hemoglobin oxygen saturation) frequently, e.g. every 5–10 min • Position anesthetized animals correctly • Use proper restraint and handling techniques.

	T: Administer antagonist drug • Establish open airway • Provide artificial ventilation • Administer oxygen • Administer doxapram intravenously.
Cardiac arrest	D: Loss of effective heart function. S: Increased respiratory rate or cessation of breathing • Weak or absent heart sounds or pulse • Low blood pressure (capillary refill time >2 s) • Blue (cyanotic) mucous membranes • Dilated pupils • Cold skin • Unconsciousness • Death. C: Acid-base imbalance • Drug-induced depression of cardiovascular function • Electrolyte imbalance • Hypothermia • Respiratory arrest. P: Administration of appropriate drug dose • Avoid prolonged pursuit • Check traps frequently • Minimize stress (consider providing sedation) • Monitor cardiovascular function (heart rate, pulse, capillary refill time, mucous membrane color, and hemoglobin oxygen saturation). frequently, e.g. every 5–10 min. T: Ensure animal is breathing and, if not, establish open airway and provide artificial ventilation • External cardiac massage (60–100 cycles per min) • Administer epinephrine intravenously.
Exertional myopathy (capture myopathy)	D: A noninfectious disease characterized by degenerative or necrotizing damage to skeletal and cardiac muscles. S: Weakness and loss of muscle coordination • Hyperthermia • Rapid breathing • Rapid heart rate • Dark, brownish urine (myoglobinuria) • Sudden death • Delayed death occurring days or weeks following capture. C: Prolonged physical exertion • Prolonged stress. P: Avoid prolonged pursuit • Check traps frequently • Minimize stress (consider providing sedation) • Use drugs that induce muscle relaxation. T: Administration of sodium bicarbonate and fluids intravenously is sometimes recommended, but treatment is often unsuccessful.

[1] *The likelihood of some emergencies occurring may be determined in part by the method of capture. For example, hypoxia or aspiration of stomach content is much more likely to occur in anesthetized animals than trapped animals.*

[2] *Sources: Nielsen (1999), Cattet* et al. *(2005), Kreeger and Arnemo (2007), and Fowler (2008).*

Appendix 5.6 *Acceptable methods used to humanely kill wild carnivores.*

Method	Comments[1]
Gunshot	Can be used to kill captive, restrained, anesthetized, or free-ranging carnivores. Shooter must be able to make a clean killing shot, using the appropriate firearm and ammunition. Personnel should stand behind the shooter and away from the animal. Large, heavy, slow-moving bullets (e.g. shotgun slugs) are more effective and safer than high-power rifle bullets. Captive, restrained, and anesthetized animals should be shot in the brain, either from the front or side, or in the neck through the vertebral column if the brain needs to be preserved for disease diagnoses, e.g. rabies. For accurate bullet placement, preferable to place the barrel of the gun right on and perpendicular to the skull or neck. For human safety, best location for shooting free-ranging animals is the heart/lung region, rather than the head. Remove lead-contaminated carcasses or body parts from sites where consumption by scavengers can lead to secondary lead toxicity.
Penetrating captive bolt	Can be used to kill anesthetized carnivores. Less risk of injury to bystanders and nearby animals than gunshot. Safe use requires full immobilization of the animal's head, accurate placement of the captive bolt against the skull, equipment that is in good working order, and administration by fully trained personnel. Animals should be exsanguinated (bled out) after the use of a penetrating captive bolt to ensure death. Non-penetrating captive bolts are *not* recommended for humane killing.
Exsanguination (bleeding to death)	Acceptable *only* if animal has been rendered unconscious by drugs or stunning. Can be done quickly and effectively by severing the major arteries leading from the heart by inserting a long-bladed knife into the junction of the base of the neck and shoulder and slicing inwards and downward. Severing of the jugular or femoral veins may also be effective, but will take longer because of slower blood flow. Placing body on incline with head downward may help improve blood flow.
T-61	Can be used to kill anesthetized carnivores. Must be administered intravenously. Mixture of three drugs: embutramide (a general anesthetic), mebezonium iodide (neuromuscular blocker), and tetracaine hydrochloride (a local anesthetic). Not available in all countries, including the United States.
Barbiturates	Can be used to kill anesthetized carnivores. Several euthanasia products contain a barbituric acid derivative (usually sodium pentobarbital) often mixed with local anesthetic agents. Should be administered intravenously, but may also be administered by intraperitoneal or intrathoracic injection in small- to medium-sized carnivores (<50 kg). These drugs are controlled substances in many countries.

Method	Comments[1]
	Effective volume needed to euthanize large carnivores (>150 kg) can be high (>50 ml).
	Animals killed with any barbiturate must be incinerated or buried because of potential secondary toxicity to potential scavengers.
Potassium chloride	Can be used to kill deeply anesthetized carnivores.
	Must be administered intravenously quickly.
	Kills by increasing concentration of circulating potassium in the blood which directly influences the electrical activity of the heart resulting in cardiac arrest.
	Potassium chloride solution is made by adding 300 mg potassium chloride per ml of water (tap or distilled) and shaking vigorously prior to injection.
	Administer at a dosage of at least 50 mg per kg body weight.

[1] *Sources: AAZV (2006), Kreeger and Arnemo (2007), and AVMA (2007).*

Appendix 5.7 *Temporary and permanent marking techniques used to identify carnivores.*

Mark	Technique	General References
Temporary Short-term markers		
Fur clipping and dyeing	Shaving in patterns to reveal underfur of animals. Clipping does not affect body condition, but must be cautiously applied with individuals in poor condition or in cold climates. Hair can be dyed to mark carnivores of all sizes and is particularly useful for mammals with light pelage. Rhodamine B has been used on coyotes (Johns and Pan 1981). Whether clipping or dying fur affects responses to the environment (e.g. physical protection or thermoregulation) or increases visibility of small carnivores to larger predators is unknown.	Ramsay and Stirling 1986 Stewart and Macdonald 1997 Macdonald *et al.* 2004a
Body attachments	Streamers and colored disks of different lengths and color codes attached to a carnivore's body or to ear tags allow identification from distance.	Lentfer 1968 Powell and Proulx 2003
Temporary long-term markers		
Tags	Tags made from metals or plastics of all shapes, sizes, and colors, and stamped with letters, numbers, and short messages, have been affixed to ears, and less frequently to other body parts, such as interdigital webbing (Gorman *et al.* 2006). Ear tags vary in size to accommodate different carnivores of all sizes. Ear tags can be pulled out by animals grooming each other (Stirling 1989) or may catch on vegetation (Hubert *et al.* 1976). Turning metal, crimping ear tags in a mammal's ear so that the clasp is outermost appears to minimize loss (Powell, unpubl. data). Retention of tags varies among species and habitats, but putting ear tags in both ears of an animal and using redundant marking to overcome identification problems are recommended. Ear tags should be lose enough not to interfere with blood circulation and punctures should be treated appropriately to prevent infection and to ensure healing.	Nietfeld *et al.* 1994 Powell and Proulx 2003

Collars and harnesses	Conspicuous metal (copper or brass non-corrosive alloy) or plastic collars with printed instructions, numbered tags, or color codes. Self-collaring collars may be placed like snares along animal trails. Collars may need replacement at regular intervals. Neck collars outfitted with radio transmitters allow researchers to identify animals, to follow their movements, and to investigate population structures and dynamics (Chapter 6).	Sheldon 1949 Zabel and Taggart 1989 De Luca and Ginsberg 2001
Passive integrated transponder (PIT tag)	A PIT tags consists of an electromagnetic coil and custom-designed microchip that emits an analog signal when excited by electromagnetic energy from a scanning wand. The transponder chip is uniquely programmed with an alpha or numeric code, and >34 billion combinations are available. Once inserted under a mammal's skin with a large bore syringe, a PIT tag (2 mm diameter × 10 mm length) can be read by a scanner. PIT tags are relatively expensive and require a specific scanner matched to the tag type to read the identification. They must be read within 10 cm of a wand. A new generation of PIT tags with antennas may be read at greater (up to 0.5 m) distance (e.g. Kurth *et al.* 2007). PIT tags may wander under an animal's skin, especially on large mammals but they are advantageous alternative to ear tags that may be lost during long-term studies. PIT tags have been successfully used with skunks (Neiswenter and Dowler 2007), Eurasian badgers (Rogers *et al.* 2002), ferrets (*Mustela fero, M. nigripes*) (Fagerstone and Jones 1987), Iberian lynxes (*Lynx pardinus*) (Palomares *et al.* 2001), and black bears (Leigh 2007).	Nietfeld *et al.* 1994 Powell and Proulx 2003 Jones *et al.* 2004
Radioactive and chemical markers	A variety of mammals have been marked with radioisotopes as inert implants, as external attachments, and as metabolizable radionucleoides, all of which can be detected in tissues, feces or urine. Rhodamine B is a systemic fluorescent marker that can be detected in hair, whiskers, claws, and other tissues, and it has been used as a qualitative marker of bait consumption in	Nellis *et al.* 1967 Pelton and Marcum 1975 Linn 1978 Savarie *et al.* 1992 Jones *et al.* 2004

(continued)

Appendix 5.7 *Continued*

Mark	Technique	General References
	coyotes (Johns and Pan 1981). Iophenoxic acid is an organic iodine chemical that binds to proteins in the blood, elevating the level of protein-bound iodine. It has been found an effective marker for ferrets, raccoons, coyotes, arctic foxes (*Vulpes lagopus*), and red foxes (Larson *et al.* 1981, Follman *et al.* 1987; Hadidian *et al.* 1989; Ogilvie and Eason 1998).	
Betalights	A betalight is a phosphor-coated glass capsule containing a small quantity of mildly radioactive tritium gas. When the phosphor is struck by low-level beta radiation from tritium, it produces visible light of a characteristic color. It may be used with other markers. They have been successfully used with Eurasian badgers (Cheeseman and Mallinson 1980; Tuyttens *et al.* 2000). They appear to pose no appreciable health hazard to radiation and may function for years.	Rudran 1996
Permanent markers		
Freeze branding	Cryo-branding has been used to mark carnivores. Freeze-branding applies a copper branding iron that is super cooled in liquid nitrogen, or a mixture of dry ice and alcohol, or a commercial refrigerant to a shaved area of the body. It kills the pigment-producing melanocytes of the skin but not the hair follicles, so the hair and skin that regrow in the branded area are permanently white.	Hadow 1972 Day *et al.* 1980 Rood and Nellis 1980 Sasaki and Ono 1994
Tattoos	Tattoos are applied with special pliers or an electric tattooing pencil. Any body part that is relatively free of hair and remains fairly clean can be tattooed, such as upper lips and the groin.	Powell *et al.* 1997 Diefenbach and Alt 1998 Walton *et al.* 2001b Powell and Proulx 2003

Chemical markers	Tetracycline antibiotics chelate with calcium ions in bones and teeth to produce characteristics patterns of fluorescence under UV light. They have been used to mark mustelids, procyonids, canids and ursids.	Linhart and Kennelly 1967 Nelson and Linder 1972 Bachmann *et al.* 1990 Garshelis and Visser 1997
Mutilations	Toe clipping, where the claw and first joint of the toe are removed with dissecting scissors, has been used to mark long-tailed weasels (*Mustela frenata*; DeVan 1982), arctic foxes (Roth 2002), and coyotes (Andelt and Gipson 1980). Toe-clipping could modify the behavior of animals due to pain or reduced function, and could, potentially, reduce success in foraging and competition, resulting in differential mortality. Even though toe-clipping is inexpensive and rapidly applied to very small carnivores, it should be considered only when no other marking method is appropriate. Debrot (1984) and Santos-Reis *et al.* (2004) clipped ears in unique patterns to mark mustelids and viverrids.	Henshaw 1981 ASM Care and Use Committee 1998

6
Carnivores in hand

Kerry R. Foresman

A wealth of information can be collected from an animal in hand, whether it is alive or dead. A particular scientific study may have a specific focus, yet a little time spent obtaining a broader array of information while an animal is "in hand" will be helpful in the future for developing a better understanding of the species. This chapter outlines the most important information to collect and provides the best methods for its collection. A researcher can then develop a database for each carnivore species and the information will be widely useful. Each section of the chapter first discusses information available from a living specimen and then notes additional information that might only be available from a carcass. Table 6.1 summarizes all potential information. The chapter includes numerous, but not exhaustive, examples from published literature. Most important is to collect standard demographic and morphological information.

6.1 Aging

Age of an animal is critical for demographic studies. The variety of morphological changes that occur as a carnivore ages can be used to estimate age. The morphological changes can be grouped into several broad categories: changes in body measurements, dental changes, skull and skeletal changes, and some specialized changes such as eye-lens composition. Knowledge of these age-related changes from known-aged specimens is required, and since nutrition, stress, and disease can influence growth, knowledge about these variables from known populations is helpful.

When carnivores are not aged in calendar ages (months, years), two, and sometimes three, broad age categories are appropriate: juvenile, adult, and sometimes yearling (e.g. North American otters, *Lontra canadensis*, Hamilton and Eadie 1964). For all aging, define terms clearly. For example, the term "subadult" is commonly used, but usually inappropriately. This term correctly refers to an individual who has begun to exhibit many adult characteristics but is still sexually immature.

Table 6.1 *Data to be collected from an animal "in hand."*

Animals Handled Alive	**Additional Information from a Carcass**
Standard body measurements:	
Weight	Skull
Length measurements:	Greatest length
Total length	Condylobasal length
Body length	Zygomatic breadth
Tail length	Palatal length
Hind foot length, width	Canine length
Ear length	Rostral proportions
Snout-to-vent length	Maxillary breadth
Axillary girth	Interorbital breadth
Neck circumference	Postglenoid length
Head length	Length of Maxillary toothrow
Shoulder height	Palatal breadth
Width and length of front and hind feet	Basilar length
Nipple size (width + length)	Diameter of auditory bullae
Muzzle girth	Height of skull
Abdominal girth	Fusion of cranial sutures (degree)
Dental characteristics	Skeleton
Tooth eruption timing	Degree of ossification of long bones
Tooth wear patterns	Baculum length and morphology
Breakage of teeth	
Canine length and diameter	Tissue samples
Radiograph of teeth for ratio of between pulp cavity and overall teeth width	Eye-lens weight
Tooth extraction for sectioning—age determination	Eye-lens protein composition
Reproductive status	
Sex identification	# placental scars (R and L)
Testicular size: length x width, scrotal vs. non-scrotal	# corpora lutea in ovary (R and L)
Vulval morphology—coloration and degree of swelling—height and width	
Nipple size—length and width	
Indication of Past Injuries	
Swelling	Broken long bones, ribs, digits
Broken and/or deformed limbs	Bruising
Pelage characteristics:	
Length of hair and general condition of coat	
Coloration patterns—spotting and distinctive patterns	

(*continued*)

Table 6.1 *Continued*

Animals Handled Alive	Additional Information from a Carcass
Standard body measurements:	
Physiological parameters	
Blood sample (whole blood, serum and plasma)	Additional tissue samples
Tissue samples	Bone marrow index
Hair samples for DNA analyses	
General:	
Detailed photographic portfolio	Detailed photographic portfolio

6.2 Standard body measurements

Body proportions can be used to estimate the age of an animal, show sexual size dimorphism, and may indicate general health and relationships between populations, and taxonomic distinctions. A suite of standard measurements is used for carnivores, with some used only for select species. Since age and sex affect body size, age and sex, if known, should be recorded with the measurements. Both intra- and interobserver errors are common; each investigator should validate the accuracy of his or her techniques with replicate measurements (Blackwell *et al.* 2006). Errors associated with external measurements are more pronounced than for skull measurements, an important factor to consider, especially related to studies on asymmetry (Eason *et al.* 1996; Blackwell *et al.* 2006).

Take standardized measurements. When possible, measure both sides of an animal's body (i.e. both its right and left ear). Develop and maintain databases of this standard information (e.g. de Waal *et al.* 2004; African Large Predator Research Unit, University of the Free State, Bloemfontein, South Africa; polar bears, *Ursus maritimus*, United States Fish and Wildlife Service, Alaska Region 2008).

Models of growth have been developed for several carnivores using a combination of the standard measurements (e.g. polar bears, Stirling *et al.* 1977; Durner and Amstrup 1996; black bears, *Ursus americanus*, Bridges *et al.* 2002; wolves, *Canis lupus*, Mech 2006; brown bears, *Ursus arctos*, Swenson *et al.* 2007) and growth rates are available from captive raised individuals (brown bears, Tumanov 1998; Canada lynx, *Lynx canadensis*, Naidenko 2006). Since diet directly influences growth rates, data from captive animals must be specified as such. Statistical models describe body mass as a non-linear function of age (Swenson *et al.* 2007).

Fig. 6.1 Tripod mounted scale for weighing a brown bear (Pat Owen, NPS, Denali Park and Preserve).

6.2.1 Body mass

Most small carnivores can be weighed with spring scales (e.g. Pesola®), which are available for weights from a few grams (with an accuracy of 0.1 g) to 50 kg (with 500-g subdivisions). Large carnivores (e.g. bears, *Ursus* spp., tigers, *Panthera tigris*) can be weighed using a weighing tripod and available spring or electronic load scales (Figure 6.1). An animal's age, sex, nutritional condition, and season influence body mass (weight), and some carnivores can consume a large amount of food in one meal. Nonetheless, an animal's body weight combined with length is sometimes used as a crude index of the animal's condition. Although a large database exists for most species, recent studies question their validity (Huot *et al.* 1995; Winstanley *et al.* 1999; Pitt *et al.* 2006). Bioelectrical impedance appears to be a more accurate index.

6.2.2 Length measurements

Use a Vernier caliper (for small species), metal tape measure (for small measurements where rigidity is required), or cloth tape measure (for long body

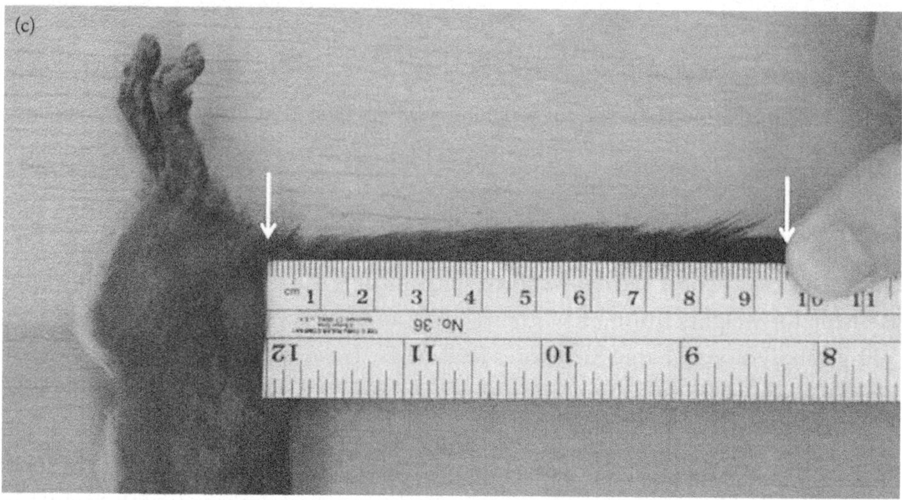

Fig. 6.2 Photos showing measurements on short-tailed weasel (*Mustela erminea*): (a) total length (to distal end of tail vertebrae, *arrow*); (b) body length (to base of tail, *arrow*), (c) tail length (to distal end of tail vertebrae, *arrow*), (d) hind foot length (from heel to tip of longest digit plus claw), (e) ear length (with ruler placed in notch of ear).

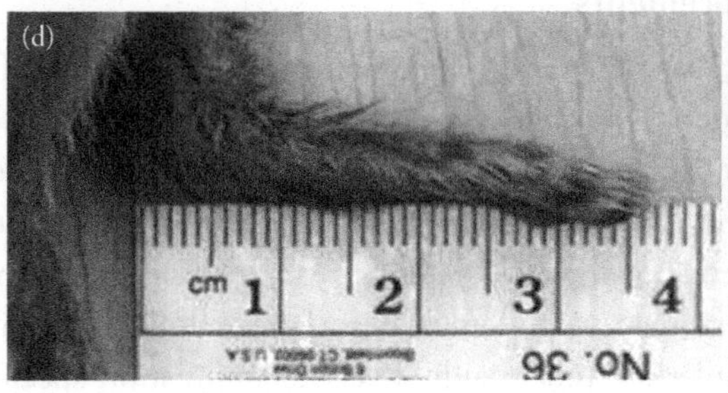

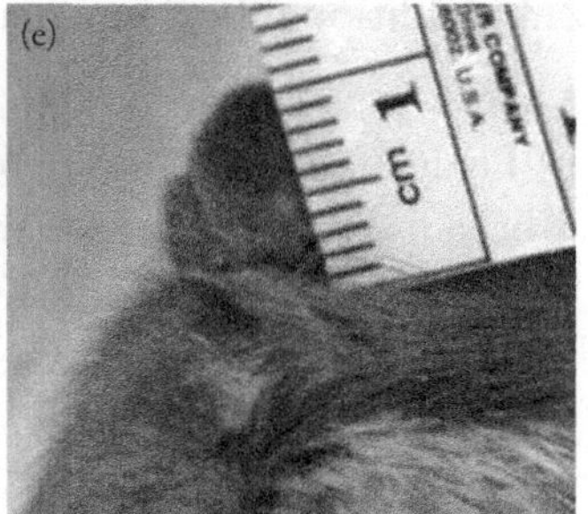

Fig. 6.2 Continued

measurements where body contours must be followed) to take the following measurements to the nearest mm (to the nearest cm in the largest species; Figure 6.2a–e). Take these measurements in the same manner on anesthetized and dead animals.

- *Total length (TL)*—Measure from the tip of the nose to the tip of the last caudal vertebra, following the animal's contours, with the animal recumbent. This measurement is easy with small species (e.g. weasels, *Mustela* spp.). Lay a small- to medium-sized animal on its back to straighten the curvature of the spine. Lay a large carnivore (e.g. African lion, *Panthera leo*; bear) on its stomach and draw a cloth tape measure from the tip of the nose along the body contours from between the eyes, along the midline of the vertebrae to the tip of the last caudal vertebra.
- *Body length (BL)*—Measure from tip of the nose to base of the tail at the notch of sacrum with the animal lying recumbent on its back or stomach as with total length.
- *Tail length (TaL)*—Measure from base of the tail at the rump to tip of the last caudal vertebra.
- *Hind foot length (HF)*—Measure from end of the heal bone to tip of the longest digit, not including the claw (some researchers include the claw in this measurement but doing so imparts added variation since claws can become worn).
- *Ear length (E)*—Measure from notch of the ear opening to tip of the pinna (excluding hair).
- *Snout-to-vent length (SV)*—Measured from tip of nose to anterior edge of anus.

6.2.3 Additional body measurements

- *Axillary girth* – Measure circumference around the chest at the level of the axilla (Figure 6.3). This measurement is strongly correlated with body mass and has been used to estimate weights of some large carnivores when weighing is not possible (e.g. mountain lion, *Puma concolor*, Durner and Amstrup 1996; Charlton *et al.* 1998; cheetah, *Acinonyx jubatus*, Marker and Dickman 2003). Use axillary girth to estimate weight only when a large sample of weights and girths exists for the population under study. Even if a population-specific ratio can be used, compare observed and estimated body masses periodically to check for changes over time in the ratio (Cattet and Obbard 2005).
- *Neck circumference*—Measure around the smallest portion of the neck.
- *Head circumference*—Measure around the largest portion of the head (commonly across the zygomatic arches).
- *Head breadth*—Measure at the maximum width at the level of the zygomatic arches (use calipers). For many carnivores, this measurement correlates better with age of an individual than does body length (e.g. polar bears, Derocher and Stirling 1998; banded mongooses, *Mungos mungo*, Otali and Gilchrist 2004).

Fig. 6.3 Girth measurement of bear (note animal is also on oxygen; photo by Rick Mace).

- *Head length*—Measured from the upper middle incisors to the posterior most projection of the skull on the occipital bones, possibly to the end of a well-developed sagittal crest.
- *Shoulder height*—Measure from the heel of the front foot to the top of the shoulder blade along the contour of the leg. Lay the animal on its side to make this measurement (de Waal *et al.* 2004 noted inaccuracies).
- *Width and length of front and hind feet*—Commonly made on ursids and large felids but even large mustelids. Do not include claws in this length measurement.

6.2.4 Additional measurements, some to estimate age

- *Nipple size*—Combine width + length of the largest nipple (usually inguinal). Measure to the nearest mm. This measurement can often be used to distinguish yearlings from animals in older age classes (e.g. wolverines, *Gulo gulo,* Magoun 1985; wolves, McNay *et al.* 2006).
- *Muzzle girth, girth of abdomen, total length of foreleg, total length of hind leg, foot width*—Commonly collected on cheetahs (Marker and Dickman 2003), but potentially useful for many other carnivores.
- *Ear length and hair length combined*—These measurements, in a mixed regression model, can index age of black bear cubs up to approximately 70 days (Bridges *et al.* 2002).
- *Pelt length*—Measure from tip of nose to base of tail to the nearest cm (Quinn and Gardner 1984; Slough 1996). Take this measurement from pelts that have been stretched and dried. This measurement in used to separate kits from yearling and adult Canada lynxes in population age ratios.

6.2.5 Footpad patterns

Photograph and make prints of footpad patterns. These appear to identify individuals (Herzog *et al.* 2007). Place an animal's foot on an inked pad and print the pad onto white paper: place the animal's foot on a sooted plate and then press it on to contact clear paper to make a clear print, then mount the contact paper on white paper (Figure 6.4).

6.3 Tooth eruption and measurements

The only standard measurement recorded for carnivores during handling is canine length. Canine width and breadth are sometimes measured and, only occasionally, carnassial length.

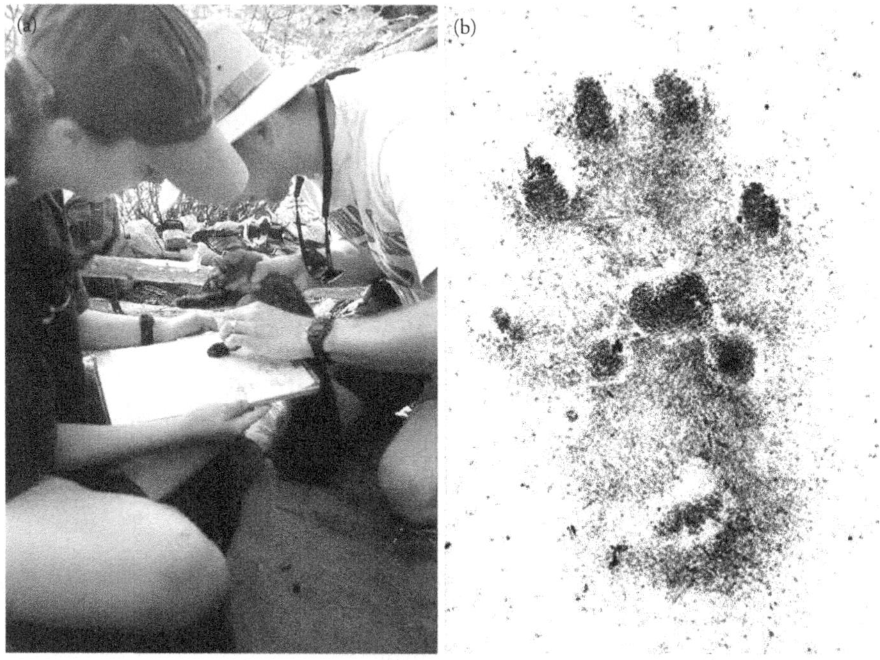

Fig. 6.4 (a) Capturing the footpad structure of an American marten using a sooted plate and contact paper and (b) foot impression.

- *Canine length*—Using calipers, measure from the tip of each canine to the gum line. Record canine identification (upper, lower; left, right).
- *Canine width*—Using calipers, measure each canine from medial to labial at the gum line. Record canine identification.
- *Canine breadth*—Using calipers, measure each canine from front to back at the gum line. Record canine identification.
- *Carnassial length*—Using calipers, measure each carnassial from front to back at the gum line. Record carnassial identification.

6.3.1 Tooth eruption, wear, and age

For many carnivore species, the schedule of tooth eruption and replacement from known-aged animals allows estimates of the ages of young individuals (e.g. Canada lynx, Saunders 1964; red fox, *Vulpes vulpes*, Linhart 1968; brown bear, Rausch 1969; coyote, *Canis latrans*, Silver and Silver 1969; African lion, Smuts *et al.* 1978; bobcat, *Lynx rufus*, Jackson *et al.* 1988; fisher, *Martes pennanti*, Powell 1993b; spotted hyaena, *Crocuta crocuta*, van Horn *et al.* 2003). The timing of tooth eruption and replacement is largely invariant (e.g. polar bear, Hensel and Sorenson 1980), though nutritional condition and disease can affect it.

Patterns of tooth wear patterns, though historically measured, vary tremendously depending on diet (e.g. raccoon, *Procyon lotor*, Grau *et al.* 1970; European badger, *Meles meles*, Harris *et al.* 1992; leopard, *Panthera pardus*, Stander 1997; wolf, Gipson *et al.* 2000; spotted hyaena, van Horn *et al.* 2003). Carnivores that eat a lot of fruit, such as American martens (*Martes americana*, Weckwerth and Hawley 1962), have less tooth wear than individuals of the same species that eat more abrasive foods. One can index age with tooth wear by comparing wear, to wear on teeth of known-age animals from the same population. Dental casts from living animals also provide detail for analyzing tooth eruption and tooth wear (Anderson 1986; Young and Marty 1986; Malcolm 1992; Rasmussen *et al.* 2005).

6.3.2 Pulp cavity measurements and age

Another method, often quick and efficient, used to place individuals into broad age categories (juvenile, yearling, or adult) is to measure the size of the pulp cavity, particularly in canine teeth. As an individual ages, the pulp cavity reduces in size (Figure 6.5; e.g. American badger, *Taxidea taxus*; striped skunk, *Mephitis mephitis*, Fredrickson 1983; coyote, Knowlton and Whittemore 2001).

From an X-ray, from a sectioned tooth or from a ground tooth, measure the maximum width of the pulp cavity and maximum width of a healthy, upper canine and calculate the ratio. Upper and lower canines from the same individual have different ratios, and in some species the upper canine appears most reliable.

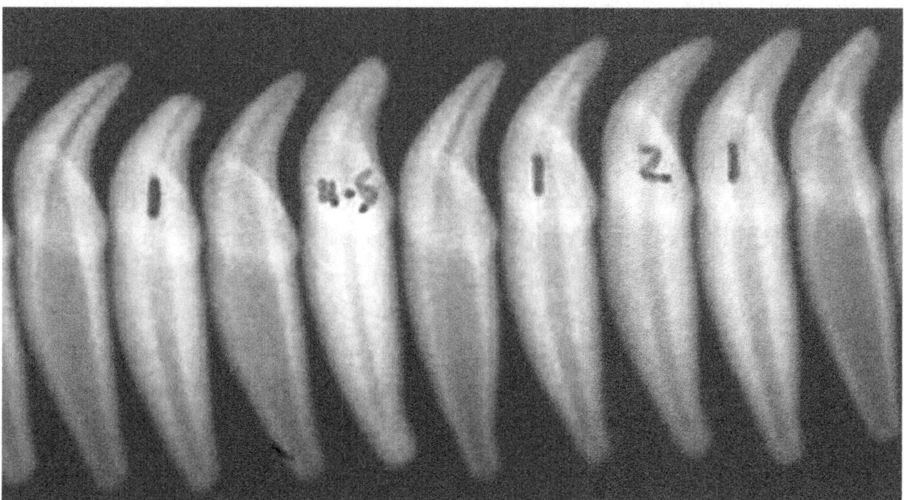

Fig. 6.5 Radiograph of canine teeth of fisher. Notice shrinkage of pulp cavity with age.

The ratios in females are smaller sometimes than those in males of the same age (Knowlton and Whittemore 2001). Pulp cavity ratios are also an effective way to separate the juvenile cohort quickly and inexpensively in a large sample of teeth to be analyzed using cementum analysis.

6.3.3 Cementum annuli and age

The most precise method for aging carnivores is to section and stain teeth histologically to analyze the layering pattern of the cementum (called "cementum annuli"). In carnivores, annual cementum growth exhibits seasonal variation in density and composition, a phenomenon that is most pronounced in more northern latitudes. Cementum developing in the winter stains darkly, while the cementum that grows during the spring and summer stains lightly (Figure 6.6).

Scheffer (1950) first reported this growth pattern and others, such as the Matson Laboratory (Matson 1981), have refined the technique and developed a database of carnivore material providing accurate and precise estimates for many carnivores of many species. For standardization (different tooth types deposit cementum differently), and for the welfare of the animal (so as not to compromise its ability to forage), specific teeth are used for specific species. To alleviate discomfort, teeth should be extracted under one of two regimes. One regime is to use a local analgesia (e.g. Lidocaine), with the additional administration of a longer acting analgesic in conjunction with a broad-spectrum antibiotic (e.g. Ketoprofen and Baytril; Beasley and Rhodes 2007). The other regime is for carnivores that have been live-trapped, which are routinely anesthetized with drugs such as ketamine and xylazine, which themselves possess strong analgesic properties; teeth can be pulled as part of routine handling (Mansfield *et al.* 2006).

For carnivores, the most common tooth taken is the first upper or lower premolar (PM1), a tooth that is often vestigial. In mountain lions (*Puma concolor*), the second upper premolar is preferred (UPM2), while in Canada lynxes and bobcats, a lateral incisor is chosen, since there is no PM1 or PM2. From carcasses, chose the lower canine, since it is large and easily sectioned and aged. Remove the tooth by carefully grasping it with forceps and gently loosening it; insert a dental elevator between the tooth and the gum to loosen the connective tissue around the entire circumference of the tooth, which can then be extracted with a tooth elevator. Take care not to break the root tip. Place the tooth in its own labeled, paper envelop. Simply let the tooth air dry in a cool, dry environment. Teeth stored for an extended time before sectioning should be frozen in a standard, manual defrost freezer. Do nothing more.

For some carnivores, such as American badgers, fishers, red foxes, and Canada lynxes, aging from cementum annuli is highly accurate (95%); for others, such as

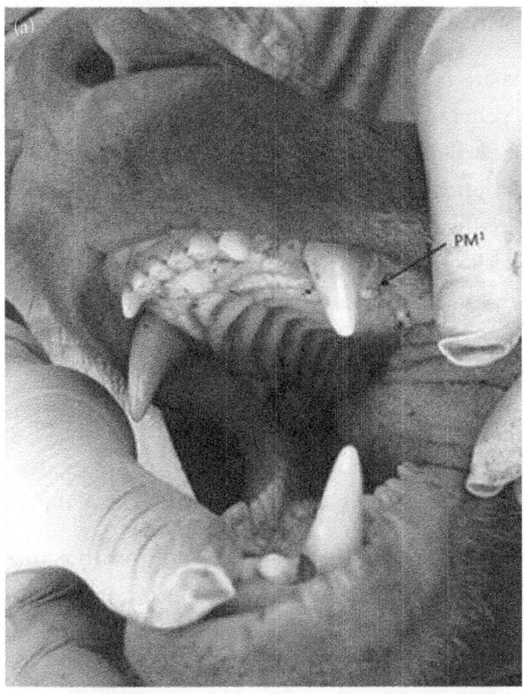

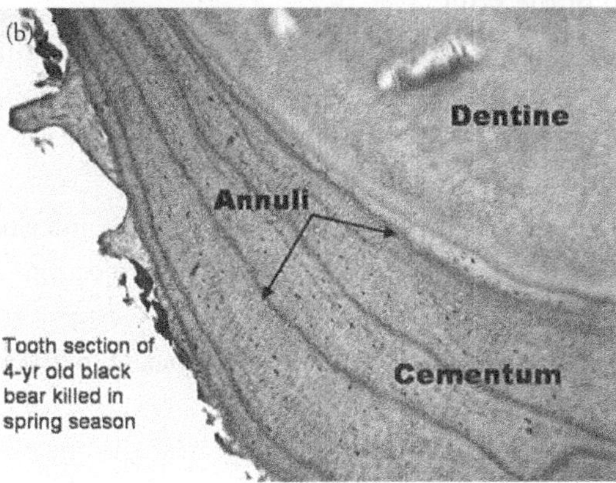

Fig. 6.6 (a) In ursids, pull the upper first premolar for aging (photo—Ben Jimenez); (b) representative tooth section from a black bear illustrating cementum growth rings (from Gary Matson).

mountain lions and sea otters (*Enhydra lutris*), accuracy is low (70–75%); while for most carnivores, accuracy is intermediate (Costello *et al.* 2004). Accuracy is closely associated with the experience of the reader. To age one's own samples, develop a collection of known-age specimens for comparison (Calvert and Ramsay 1998).

6.4 Skull and skeletal measurements

Skull and skeletal materials can provide considerable information related to age of an animal and its general health. Standard measurements are used to determine species and sex for carnivores that exhibit sexual dimorphism, and to assess long-term quality and abundance of food, and long-term stress. Some populations, however, exhibit significant morphological variation and ursids grow throughout their lives (Rausch 1953, 1963), so researchers need to use these measurements carefully. Standardly, these measurements are taken from dead animals but measurements on living animals in the field are now possible using portable X-ray equipment (Schillaci *et al.* 2001).

6.4.1 Skull measurements

Standard skull measurements are usually taken from cleaned material (Figure 6.7).

- *Greatest length (GL) or skull length*—Measure from the anterior-most projection of nasal or premaxillary bones to the posterior-most projection of skull. The posterior-most projection is generally the supraoccipital bone projecting backward at the top of the skull or the occipital condyles projecting backward at the bottom of the skull.
- *Condylobasal length (CBL)*—Measure from the anterior-most projection of premaxillary bones to the posterior-most projection of occipital condyles.
- *Zygomatic breadth (ZB)*—Measure the greatest distance across the outer edges of the zygomatic arches.
- *Palatal length (PL)*—Measure the distance from the anterior-most projection of premaxillary bone to the posterior-most projection of palatine bone measured along the midline axis.
- *Palatal breadth (PB)*—Measure across the palate between the innermost surfaces of the last upper molar teeth.
- *Maxillary breadth (MB)*—Measure the greatest distance between the outer edges of the right and left maxillary bones.
- *Interorbital breadth (IB)*—Measure across the top of the skull between the innermost surfaces of the orbits.
- *Postglenoid length (PGL)*—Measure from the anterior-most projection of the postglenoid process and the posterior-most projection of the occipital condyles.
- *Length of the maxillary toothrow (LMT)*—Measure from the anterior surface of the anterior-most upper premolar to the posterior surface of the last upper molar.

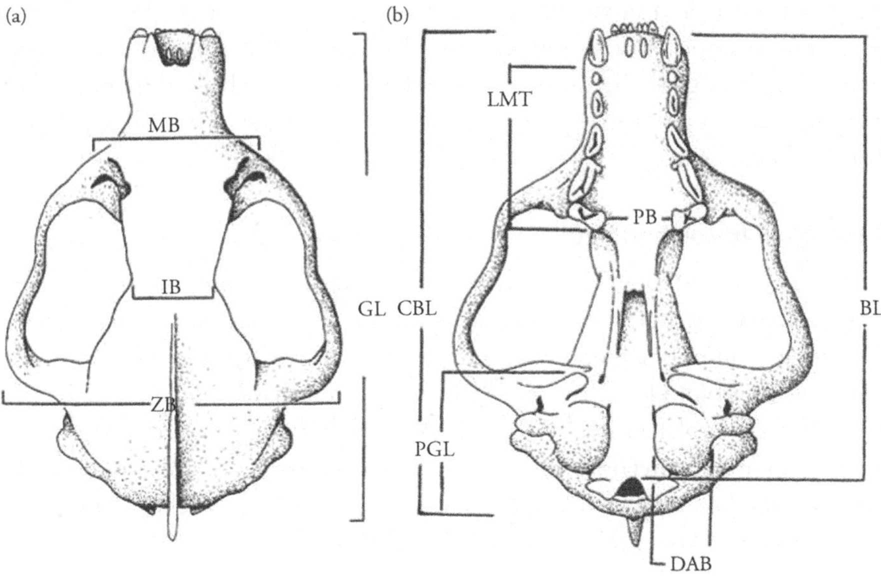

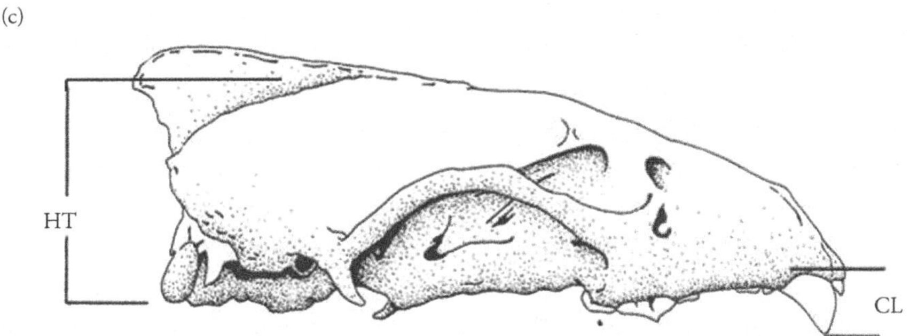

Fig. 6.7 Dorsal (a), ventral (b), and lateral (c) views of a fisher skull illustrating common skull measurements (Foresman 2001; M-molar, P-premolars, C-canine, I-incisors; for all other abbreviations refer to list above).

- *Basilar length (BL)*—Measure from the posterior-most border of the alveoli of the first upper incisors to the anterior-most ventral border of the foramen magnum.
- *Diameter of auditory bullae (DAB)*—Measure across the auditory bullae from the outer to the inner margins at right angles to the long axis of the skull.
- *Height of the skull (HT)*—Measure from the dorsal surface of the parietal bone to the anterior-most point on the ventral surface of the basioccipital bone (omit the sagittal crest if one exists).

Many additional measurements can be taken, depending upon the focus of a study. For example, rostral proportions of skulls in the Family Canidae can be used to distinguish between native and hybrid individuals (Howard 1949, Lawrence and Bossert 1967). Martin *et al.* (2001) provide a broad discussion of these measurements.

6.4.2 Skull fusion and age

The fusion of cranial bones is a function of an animal's age and, therefore, can be used to estimate age, given a collection of known-age skulls. Nutrition can affect this timing, requiring caution in using the estimate of age. Subadult bobcats can be distinguished from adults by viewing closure of the basioccipital–basisphenoid suture (Conley and Jenkins 1969); North American otters can be placed in to five minimum age categories (8–9, 11–12, 20–21, 31–32, 35–36+ months) by closure of the basioccipital–basisphenoid suture, as well as the closure of several additional cranial bones (Hamilton and Eadie 1964).

6.4.3 Skeletal morphology and age

The cessation of growth of long bones and the fusion of the epiphyseal plates can be used to classify individuals into juvenile vs. adult classes of maturity. (e.g. raccoons, Sanderson 1961; North American otters, Hamilton and Eadie 1964; wolves, Rausch 1967; red foxes, Geiger *et al.* 1977).

The baculum, or *os penis*, is a well-developed bone within the urethras of canids, ursids, mustelids, procyonids, and viverids, though vestigial or absent in the felids. The weight, length, and confirmation of the baculum have been used to classify individuals of many species into maturity classes (e.g. gray foxes, Petrides 1950; wolverine, Wright and Rausch 1955; black bears, Marks and Erickson 1966; North American otters, Stephenson 1977).

6.4.4 Eye lens and age

The lens of the eye, made up of crystalline proteins, grows at a decreasing rate throughout an individual's life with no cellular loss (Bloemendal 1977). Consequently, its weight can be used as an index of age. Age curves exist for several carnivore species (e.g. minks, *Mustela vison*, Pascal and Delattre 1981; dingos, *Canis familiaris dingo*, Catling *et al.* 1991) but this technique is really only useful to distinguish between young and adult age categories. This method requires accurate processing of a fresh lens under controlled conditions (drying and length of fixation; Friend 1967) and the starting date for the growth curve may need to be projected back into gestation (Augusteyn 2007).

As an individual ages, biochemical crosslinking increases among tyrosine amino acids in adjacent crystalline protein chains, leading to a progressively higher ratio of insoluble to soluble proteins (Dische *et al.* 1956). Analysis of these changes in small mammals (e.g. meadow vole, *Microtus pennsylvanicus*, Stump and Anthony 1983) and deer (*Odocoileus virginianus*, Ludwig and Dapson 1977) suggests that this ratio may produce a better estimate of age than lens weight. The method has not been commonly applied to carnivores.

6.5 Pelage and age

Carnivores molt seasonally throughout their life to replace damaged hair, to change the density and length of hairs to provide seasonal insulation, and to change color and color pattern with seasons (e.g. weasels). Initial molts reflect aging from infant to juvenile to adult coats, and the timing of these molts is known for many species (e.g. red foxes, Linhart 1968). Young of many felids carry spots that are lost as the adult coat grows. For example, the spotting pattern in mountain lion cubs becomes less distinct and dappled, by nine months of age and is totally lost by 24 months (Logan and Sweanor 2000). This change distinguishes broad but specific age categories.

Pelage quality indicates an animal's health. Both nutrition (National Research Council 1982; Rasmussen and Borsting 2000) and diseases (e.g. sarcoptic mange; Chapter 13) affect pelt quality.

Photograph unique coloration patterns (including scarring) to identify recaptured animals and individuals photographed using remote cameras (Kelly 2001; Heilburn *et al.* 2006; Jackson *et al.* 2006; Larrucea *et al.* 2007; Chapter 4). Keep a complete photographic history of each individual.

6.6 Sex and reproduction

With the exception of spotted hyenas, in which female secondary sexual organs mimic those of a male (Kruuk 1972b), all carnivores exhibit somewhat similar characteristics in their external genitalia that allow identification of sex. In all carnivores, the anal–genital distance is greater for males than females (Figure 6.8).

Wild carnivores in temperate regions generally exhibit seasonal reproductive cycles. Testes of males swell during the breeding season and the testes may be palpated easily in the scrotum. Testicular swelling always anticipates estrus in females. Measuring the length and width of the testes within the scrotum provides an index of testicular size (L × W) and reproductive condition (e.g. wolverines,

Fig. 6.8 (a) External genitalia of male mountain lion showing long distance between anus and penis (the scrotum lies between these structures and black hairs surround the penis in this species), and (b) external genitalia of female mountain lion where the vulva lies close to the anus (photos by Ken Logan).

Mead *et al.* 1991; polar bears, Howell-Skalla *et al.* 2002; wolves, Mech 2006). Most male carnivores possess a baculum in the penis, which can be palpated.

Female reproductive condition can be assessed by viewing the condition of the vulva, which swells as a female enters estrus. Vulval measurements (length or height, and width) may indicate the timing of estrus (e.g. American martens, Enders and Leekley 1941; wolverines, Mead *et al.* 1991; fennec foxes, *Vulpes zerda*, Valdespino *et al.* 2002). More accurate estrus determination can be made by taking vaginal swabs for microscopic examination (Chapter 12). The size (length and width) and coloration of nipples can also be used to determine whether a female has bred: females who have not bred have small, pink nipples, while those who have bred have large nipples that become dark in coloration (e.g. bears, Jonkel 1993; wolves, Mech 2006).

Reproductive status, estrus, and pregnancy in females can be determined endocrinologically in many carnivores, though not necessarily in canids (Kreeger 2003; Chapter 12). Palpation can often document fetuses but the number of fetuses is generally undercounted (Toal *et al.* 1986); take care not to mistake fecal lumps for fetuses. Ultrasound is being developed for use in the field to estimate litter size (Griffin *et al.* 2003; McNay *et al.* 2006) or stage of pregnancy (Stephenson *et al.* 1995; Canon *et al.* 1997; McNay *et al.* 2006).

Tooth cementum analysis indicates the reproductive histories of female black bears, particularly young bears (Rogers 1978; Coy and Garshelis 1992; Carell 1994; Matson's Laboratory: http://www.matsonslab.com/). The more lightly staining cementum, normally laid down in the spring and summer period, is less evident during a cub-rearing year; a thickened layer is subsequently produced the following year, producing an alternation in cementum thickness coinciding with successive cub-rearing years. This method appears ineffective for brown bears (Matson *et al.* 1999) and other carnivores.

Note reproductive information from carcasses (see Chapter 12). The numbers and sizes of corpora lutea, embryos, or fetuses, and placental scars, approximate litter size and are important reproductive parameters in their own right (Sacks 2005; Elmeros and Hammershoj 2006). Counts of corpora lutea, which form in the ovary from the granulosa cells of the follicle following ovulation, are used routinely to assess reproductive performance, with limitations. Not every follicle ovulated is fertilized, some follicles regress in the ovary and produce structures visibly similar to true corpora lutea (termed accessory corpora lutea), some embryos abort or are reabsorbed, and in some species (e.g. bobcat and Canada lynx), corpora lutea do not degenerate and disappear but remain indefinitely (Anderson and Lovallo 2003). In these species, the most recent corpora lutea can often be distinguished by their light, yellowish coloration (Crowe 1975). Color patterns of placental scars may differ between recent and previous litters (Englund 1970; Lindstrom 1981). Scarring also occurs when fetuses are aborted or resorbed (Lindstrom 1981; Strand *et al.* 1995; Mowat *et al.* 1996; McNay *et al.* 2006).

Mountain lions (Logan and Sweanor 2000) and bobcats (McCord and Cardoza 1982; Rolley 1987) are often difficult to sex, since the penis and scrotum are not obvious to those unfamiliar with these species. The hairs surrounding the penile sheath of male mountain lions are commonly dark and the anal–genital distance is much greater than for females (Figure 6.8; Logan and Sweanor 2000).

See Chapter 12 for more information related to sex and reproduction.

6.7 Injuries

Assess the overall physical condition of all animals handled, and record new and old injuries. Note obvious scars, open wounds, and fractures (fresh, healing, and healed) and their severity. In addition to potentially affecting an individual's fitness, these injuries might be valuable for identifying the animal in the future. Porcupine quills in a carnivore's muzzle reduce its ability to capture prey and to eat. Although quills seldom cause infection (Roze 2009), they occasionally pierce internal organs, leading to death (Coulter 1966; Wobeser 1992; Roze 2009). Old injuries provide valuable information about an individual's fitness. For example, broken teeth, particularly canines, affect a carnivore's ability to capture prey. Carnivores, by their very nature, are prone to injury because they attack prey, which inevitably struggle and fight back. Many carnivore taxa in museum collections have >25% incidence of >1 broken tooth. For African lions, >53% of broken teeth were canines (van Valkenburgh 1988). Scoring injuries (Chapter 5) can help researchers decide how to proceed with an injured, live-trapped carnivore (Olsen *et al.* 1986; Hubert *et al.* 1996; Powell and Proulx 2003).

If one has a carcass, do a detailed necropsy to assess overall condition (Chapter 13) and look for subtle injuries (subcutaneous bruising, minor fractures, broken ribs, bone breaks that might have healed, etc.; Mech 1970; Wobeser 1992).

Re-evaluate capture protocols with every capture to minimize future capture-related injuries (Olsen *et al.* 1986; Proulx 1999a; Powell and Proulx 2003; Powell 2005; Chapters 8 and 13).

6.8 Physiological parameters

A variety of physiological measurements provide invaluable information on a captured animal's condition (Chapters 12 and 13). Take blood and tissue samples to be analyzed to assess nutritional status (Chapters 12 and 13), diet (Chapter 11), endocrinological state (e.g. stress and reproductive hormones; Chapter 12), and diseases (Chapter 13).

6.8.1 Blood

Draw blood samples from the jugular, cephalic, or femoral veins using prepackaged Vacutainer® tubes, which can be purchased in a variety of sizes and preparations (Voigt 2000 reviewed this methodology). These tubes have standardized, color-coded rubber stoppers indicating their uses. Whole blood contains red blood cells and plasma, the liquid component including such materials as fibrinogen, a clotting factor, other proteins, water, and electrolytes Purple tubes contain EDTA, an anticoagulant, for whole-blood collection for DNA analyses. For serological tests, use vacutainers containing acid citrate and a dextrose solution (yellow top). Red-topped tubes contain a clot activator allowing serum to be collected for hormone analyses, and green-topped tubes contain a substance such as sodium heparin to prevent clotting, used for plasma samples. Specialty tubes are available.

As their name implies, a vacuum is predrawn in vacutainer tubes; with a plastic holder supporting a double-ended needle, one first punctures the vein and, as blood begins to flow, one then punctures the vacutainer tube with the opposite end of the needle. Multiple blood samples can be drawn by simply removing one vacutainer when filled and inserting a second.

The veterinary community has developed sophisticated hand-held, point-of-care (POC) analyzers for the standard blood chemistry panel and a variety of other blood parameters (Abaxis Veterinary Diagnostics, Abaxis North America, Union City, California, USA; E-Z-EM, Inc., Lake Success, New York, USA). As this technology develops, it will be applicable for field studies.

6.8.2 Tissue samples

Collect tissue samples whenever the opportunity arises, since tissue samples can be used for genetic analyses, for contaminant analyses, and to develop a tissue databank. When handling living animals, collect hair and small tissue samples from ear plugs (Chapters 4 and 12). Use biopsy needles (commonly 11–16 ga.) to collect other tissue samples. Place hair in a small, labeled paper envelope and allow to dry (Chapter 4); for long-term storage, envelopes can be placed in a box with desiccant or frozen in a manual defrost freezer. Carcasses provide a variety of tissues, including liver, kidney, skeletal muscle, and brain (Sheffield *et al.* 2005; Chapter 13). Freeze soft-tissue samples for DNA analyses in small, snap-top vials. Take care to label each sample accurately and adequately with collection data (date, location, sex, unique animal ID). Place a paper label with replicate information written in pencil inside containers, when possible.

Stable isotope analyses provide information that has been assimilated into tissues over a long period from prey that carnivores have eaten (Chapter 11), highlighting the importance of collecting and storing tissue samples. Coupled with molting cycles, the analyses allow researchers to compare recent versus previous diets, showing seasonal differences in diet (McFadden *et al.* 2006). Furthermore, the turnover rate of these isotopes varies among tissues, because metabolic rates of tissues differ (Tieszen *et al.* 1983). Consequently, by analyzing diverse tissues, one can dissect spatial and temporal dietary differences (McFadden et al. 2006). Hair and bone tissues, for example, have low turnover rates and their isotopic ratios reflect season to annual or even to lifetime dietary information (Jones *et al.* 2006).

Fatty acids constitute a significant percentage of the bodies of prey eaten by carnivores and provide unique signatures of prey eaten. The quantitative, fatty acid signature (QFASA) in fat tissues of carnivores reflects the fatty acids available from prey (Iverson *et al.* 2004, 2006; Thiemann *et al.* 2006; Nordstrom *et al.* 2008), again highlighting the importance of collecting tissue samples.

Bone marrow condition is an index of nutritional state (Chapter 13).

6.8.3 Other samples

Collect fecal samples that live-trapped animals have left in traps or at the trap site, and use a fecal loop to extract fecal material from the colon of anesthetized animals. These samples provide appropriate material for DNA analyses and for nutritional (Chapter 13), dietary (Chapter 11), disease (Chapter 13), and endocrine studies (Chapter 12). Storage requirements depend on the studies of interest (Chapters 11, 12, and 13).

Comb the fur of living and dead animals for ectoparasites, such as fleas, ticks, and lice. Store the organisms in labeled vials. For microscopic work, store in 70% alcohol; for DNA analyses, additional preparation may be necessary (Chapter 13).

Collect endoparasites from fecal samples and carcasses. Depending on the study, collect target tissues (e.g. liver flukes, cysts in liver or skeletal muscle).

Carcasses provide contents of stomachs and gastrointestinal tracts for diet studies. Flush the contents of a stomach or GI tract into a large vial containing 70% alcohol (Chapter 11).

6.9 Bioelectrical impedance

A relatively new method to estimate an animal's overall body composition, and thus its relative condition, is bioelectrical impedance analysis (BIA; Robbins *et al.*

2004; Hwang *et al.* 2005; Pitt *et al.* 2006). One passes a small, alternating electrical current through the animal's body between electrodes attached to the upper lip and to the base of the tail. One measures the resistance to current flow (impedance) with a plethysmograph (Hwang *et al.* 2005). Since resistance is directly proportional to body fat content, one can couple resistance with measurements of total body mass, body length, and chest circumference to quantify body fat and total body water (Robbins *et al.* 2004; Hwang *et al.* 2005). The method requires standardized protocol and calibration for each species. Seasonal changes in fat reserves can be measured in this manner.

6.10 Asymmetry

Carnivores, as vertebrates, are bilaterally symmetrical with their symmetry being tightly constrained through gene expression during development (Mather 1953; Møller and Swaddle 1997). Perfect symmetry, however, is rarely achieved; body measurements, such as the lengths of an animal's two upper canines, are inevitably asymmetrical (Manning and Chamberlain 1993, 1994). The degree of asymmetry in an individual for such a paired trait appears to reflect low developmental control over the trait's expression. The reduced control appears to be due to energetic demands of a stressful environment (Badyaev and Foresman 2000; Badyaev *et al.* 2000) and the greater the environmental stress during development the greater the asymmetry (Sommer 1996).

Recognizing asymmetry is important for two reasons. First, researchers commonly measure structures on only one side of the body for standard body measurements. Since some degree of asymmetry is always present and has no consistency for which side will be larger, measuring only one side of an individual will misestimate the measurement means for that individual, making comparisons between individuals inaccurate. Because both asymmetry and researchers are inconsistent regarding which side is larger and measured, population means are unaffected. Small population samples may have significant sampling error.

Second, since asymmetry reflects environmental stresses experienced by an individual during their development, asymmetry reflects fitness of an individual (Møller and Pomianlpwski 1993; Watson and Thornhill 1994). Therefore, one can evaluate fitness of an individual by measuring and comparing structures on both sides of its body (Badyaev 1998). Wayne *et al.* (1986) and Badyaev (1998) used asymmetry to evaluate environmental stresses experienced by cheetahs and grizzly bears. Measuring multiple traits increases the statistical power to detect developmental responses to stress (Leung *et al.* 2000).

7 Radio-telemetry equipment and applications for carnivores

Mark R. Fuller and Todd K. Fuller

Radio-telemetry was not included in the first comprehensive manual of wildlife research techniques (Mosby 1960) because the first published papers were about physiological wildlife telemetry (LeMunyan *et al.* 1959) and because research using telemetry in field ecology was just being initiated (Marshall *et al.* 1962; Cochran and Lord 1963). Among the first uses of telemetry to study wildlife, however, was a study of carnivores (Craighead *et al.* 1963), and telemetry has become a common method for studying numerous topics of carnivore biology. Our goals for this chapter are to provide basic information about radio-telemetry equipment and procedures. Although we provide many references to studies using telemetry equipment and methods, we recommend Kenward's (2001) comprehensive book, *A manual of wildlife radio tagging* for persons who are unfamiliar with radio-telemetry, Fuller *et al.* (2005), and Tomkiewicz *et al.* (2010). Compendia of uses of radio-telemetry in animal research appear regularly as chapters in manuals (Cochran 1980; Samuel and Fuller 1994), in books about equipment, field procedures, study design, and applications (Amlaner and Macdonald 1980; Anderka 1987; Amlaner 1989; White and Garrott 1990; Priede and Swift 1992; Kenward 2001; Millspaugh and Marzluff 2001; Mech and Barber 2002), and in reviews highlighting new developments (Cooke *et al.* 2004; Rutz and Hays 2009; Cagnacci *et al.* 2010). Some animal telemetry products and techniques have remained almost unchanged for years, but new technologies and approaches emerge and replace previously available equipment at an increasing pace. Here, we emphasize recent studies for which telemetry was used with carnivores.

7.1 General background

We use "radio-telemetry" to refer to the process of obtaining information from a remote animal by use of radio waves. Most commonly, biologists use radio waves that are categorized in the spectral bands of Very High Frequency (VHF = 30–300 MHz) or Ultra High Frequency (UHF = 300–3000 MHz [3 GHz]). In their respective countries, governments authorize the specific frequencies that can be used for wildlife studies. The term band or bandwidth is used to describe a specific range of frequencies (e.g. a 2-MHz band of 164–166 MHz). The signal transmitted to, or from, the animal can be used to estimate its location or to carry data about the animal's motion, condition (e.g. heart rate, body temperature), or its environment (e.g. temperature, atmospheric pressure). Radio-telemetry can be used to gather information that is neither practical nor possible to obtain with other methods from rapid-moving, wide-ranging, and secretive carnivores. Biologists have used radio-tracking to find animals for subsequent observation and to document local movements and estimate space use (e.g. home range, Chapter 9), to map dispersal and migration, and to study resource use and selection by carnivores (Chapters 10 and 11). Estimates of population abundance, survival, and fecundity, and information about causes of mortality, can be obtained using data from radio-marked animals (Chapter 5). Radio-telemetry can be applied to research of intraspecific (e.g. social behavior) and interspecific (predator prey) relationships (Chapter 11).

If radio-telemetry seems like a potential technique for addressing objectives, careful study planning is necessary and one must consider several factors relative to alternative techniques (e.g. Long *et al.* 2008a). The first consideration for using radio-telemetry is its potential effects on marked animals (Murray and Fuller 2000; Withey *et al.* 2001; Table 7.1). Researchers must use an appropriate transmitter design and attachment. A study can fail or produce biased results if radio-marking causes aberrant behavior or physiological stress, increases mortality, or reduces reproduction. Animals require capture and restraint, and perhaps sedation or anesthesia (Cattet *et al.* 2008a), while being radio-marked (Mulcahy and Garner 1999; Agren *et al.* 2000; Arnemo *et al.* 2006), potentially compounding the effects of the telemetry package (Tuyttens *et al.* 2002). Biologists must investigate options for packaging and attaching the package when they discuss equipment with manufacturers.

The expense of using radio-telemetry includes costs of equipment (transmitters, antennas, and receivers), salaries, transportation for field personnel to capture and radio-tag animals, and other labor and transportation (e.g. vehicles, aircraft) costs to locate animals in the field. If data are to be obtained via cell phone or satellite, costs include charges for reception, management, and distribution of data. In addition, effort and new procedures might be required to manage the magnitude of the data accumulated from telemetry (Cagnacci and Urbano 2008; Urbano *et al.* 2010). Pilot studies might be necessary for testing the performance and usefulness of equipment and methods, and to learn about variability in telemetry data

Table 7.1 *Uses of radio-telemetry to study carnivores—considerations.*

	Topic	Species	References
Attachment	capture collar	wolves	Mech and Gese 1992
	collars	weasels	Gehring and Swihart 2000
		ferrets	Biggins *et al.* 2006
		brown bears	Schwartz and Arthur 1999
	collar release mechanisms	wolves	Merrill *et al.* 1998
		black bears	Garshelis and McLaughlin 1998
	expandable implants	black bear cubs	Vashon *et al.* 2003
		otters	Soto-Azat *et al.* 2008
		badgers	Ågren *et al.* 2000
		fishers	Weir and Corbould 2007
		foxes	Fuglei *et al.* 2002
		wolves	Crawshaw *et al.* 2007
	multiple methods	otters	Neill *et al.* 2008
	subcutaneous	polar bears	Mulcahy and Garner 1999
Effects	of collars	badgers	Tuyttens *et al.* 2002
	of handling	cougars	Fiorello *et al.* 2007
		bears	Cattet *et al.* 2008a
Aerial tracking		cougars	Stoner *et al.* 2008
		lynx	Vashon *et al.* 2008
		wolves	Hebblewhite and Merrill 2008
Triangulation		bobcats, bison	Riley 2006
		pumas	Laundre and Hernandez 2008
		red foxes	Van Etten *et al.* 2007
Auto-tracking system		ocelots	Mares *et al.* 2008
Video and Telemetry System			MacNulty *et al.* 2008
Satellite system performance		brown bears	Graves and Waller 2006
			Heard *et al.*2008
			Walton *et al.* 2001a
GPS performance			Frair *et al.* 2004
			Jozwiak *et al.* 2006
			Johnson *et al.* 2002
			Lewis *et al.* 2007
		lynx	DeCesare *et al.* 2005
			Mattisson *et al.* 2010
		brown bears	Sundell *et al.* 2006
			Gau *et al.* 2004
		cougars	Lindzey *et al.* 2001
			Ruth *et al.* 2010
		maned wolves	Coelho *et al.* 2007

quantity and quality, which affect sample size, study design, and interpretation of results (Girard *et al.* 2002; Lindberg and Walker 2007; Mills *et al.* 2008). Finally, the number of animals that must be radio-marked, the spatial scale over which they must be sampled, the duration of the study, and training of personnel, affect the expense of using radio-telemetry.

After identifying the study objective(s) and reviewing relevant literature, correspondence with telemetry vendors (via http://www.biotelem.org; accessed March 17 2011) and biologists who have done similar research with the equipment and species is a critical step in deciding how technology can best be applied to the research question. Biologists must be able to describe the environment in which a study is to take place and how they expect the equipment to function. Biologists should test the performance of their radio-telemetry "system" (transmitters, receivers, antennas, coaxial cables, data loggers; Mills and Knowlton 1989; Merrill *et al.* 1998; D'Eon and Delparte 2005), including field-testing in the topographic (D'Eon *et al.* 2002; Zweifel-Schielly and Suter 2007) and vegetative conditions of the study area (Dussault et al 1999; Di Orio *et al.* 2003; DeCesare *et al.* 2005). When equipment is used in naturally varying environments, and when animal behavior causes variability, it often does not perform to the specifications obtained by manufacturers from controlled laboratory testing (Gau *et al.* 2004; Coelho *et al.* 2007; Hwang and Garshelis 2007; Mattisson *et al.* 2010). When a minimum sample of radio-marked animals is critical, researchers must estimate average operational life and failure rates from a sample of transmitters acquired for the study. Some transmitters will surely fail before their predicted operation life is achieved. The scheduled reception of data can be interrupted (Graves and Waller 2006) by factors such as atmospheric interference, vegetation, terrain, and buildings, thus reducing sample size (Hebblewhite *et al.* 2007). Observation bias results when the probabilities of obtaining scheduled location estimates are not equal along animals' movement paths or within a period of interest, and when measurement error occurs when the true location of the animal is different from the estimated location. Equipment performance can affect any use of radio-telemetry data (Hupp and Ratti 1983; White and Garrott 1990; Land *et al.* 2008) and the application of relatively accurate, regular, Global Positioning System (GPS) and other satellite technology has renewed focus on these issues (Rempel *et al.* 1995; Hurlbert and French 2001; D'Eon 2003; Cain *et al.* 2005; Hansen and Riggs 2008).

Researchers must understand the performance of the telemetry equipment to correctly use and interpret telemetry data (Belant and Follmann 2002; Theuerkauf and Jedrzejewski 2002; Frair *et al.* 2004; Mills *et al.* 2006; Hebblewhite *et al.* 2007; Horne *et al.* 2007; Andersen *et al.* 2008; Frair *et al.* 2010). If the telemetry unit on an animal fails to function, and each animal is a sample unit, then sample size is reduced. If data are received less often than expected, or if the telemetry does not perform consistently among all study conditions (e.g. weather, terrain, vegetation), then the quantity or regularity of data might be insufficient for an analysis that had been planned. Analyses for telemetry data are included in some books (White and Garrott 1990) and manuals (Fuller *et al.* 2005), and new procedures are developed constantly (Coyne and Godley 2005; Sand *et al.* 2005; Tinker *et al.* 2006; Young and Shivik 2006; Beyer *et al.* 2010; Boyce *et al.* 2010; Kie *et al.* 2010; Smouse *et al.* 2010).

7.2 Basic telemetry system

The majority of wildlife radio-telemetry has been conducted with VHF frequencies (e.g. Rhodes *et al.* 1998), but satellite telemetry, at UHF frequencies, is being used more and more often. Most component categories (e.g. transmitter, power source, microprocessor) are used with VHF and satellite telemetry and have no special considerations for carnivores. Table 7.1 lists, among other topics, papers that emphasize components; other tables (Tables 7.2, 7.3, 7.4, 7.5, 7.6) provide examples of how various combinations of components and field procedures have been applied to the study of carnivores. Many papers do not provide detailed information about components and simply name manufacturers. Correspondence with telemetry vendors is critically important for deciding which equipment is best for the species, the objectives, and the environmental setting of a proposed study.

Selection of a transmitter includes designating the radio frequency, signal repeat (aka pulse) rate, signal strength (radiated power), duration of operation, and configuration and mass of the unit. The VHF transmitter unit comprises electrical circuitry, a power source, transmitting antenna, encapsulation, material for attachment to an animal, and if needed, sensors. A 10-KHz spacing among transmitter frequencies to be used on a study area is necessary because frequency drift, transmitter crystal variation, and tuning deviations (e.g. nominal 164.000 MHz, received at 164.005) of a receiver can otherwise result in more than one transmitter being received simultaneously at the same frequency setting on a receiver.

The power source is the main determining factor for the duration of operation for an animal's telemetry unit. Signal power and operational life are tradeoffs with battery powered telemetry because batteries with greater energy capacity add bulk and mass to units. Thus, the mass of a telemetry unit that a species can carry without adverse effects limits the size of the power source, and in turn, limits signal power and operation life. Many electronics designs use microprocessors and low power clocks to conserve power by turning transmissions on and off at prescribed times (called the duty cycle). Photovoltaic solar cells, an alternative to batteries, can

Table 7.2 *Uses of radio-telemetry to study carnivores—movements.*

Topic	Species	References
Dispersal	cougars	Stoner *et al.* 2008
	wolves	Ciucci *et al.* 2009
		Kojola *et al.* 2009
Dispersal/survival	raccoon dogs	Sutor 2008
	ferrets	Byrom 2002
Home range/habitat	honey badgers	Begg *et al.* 2005
	brown bears	Edwards *et al.* 2009
	jackals, 3 spp.	Fuller *et al.* 1989
	leopards	Simcharoen *et al.* 2008
	lynx	Vashon *et al.* 2008
	ocelots	Mares *et al.* 2008
	polar bears	Parks *et al.* 2006
	southern river otters	Sepulveda *et al.* 2007
Modeling		Christ *et al.* 2008
Movements/activities	polar bears	Amstrup *et al.* 2001
		Parks *et al.* 2006
	snow leopards	McCarthy *et al.* 2005
Range use/ecology	cheetahs	Marnewick and Cilliers 2006
	Malay civets	Jennings *et al.* 2006
	lions	Metsers *et al.* 2010
Response to wildfire		Ballard *et al.* 2000
Response to big game hunting		Ruth *et al.* 2003
Spatial ecology	spatial theory	Young and Shivak 2006
	spotted hyenas	Boydston *et al.* 2005
	Darwin's foxes	Jimenez 2007
	stoats	Hellstedt and Henttonen 2006
	golden jackals	Admasu *et al.* 2004
	bobcats, gray foxes	Riley 2006

be used with a capacitor or rechargeable battery, but they have limited applicability when animal behavior precludes regular exposure to sunlight (e.g. nocturnal, dense vegetation cover).

Material for transmitting antennas should be strong and, for many applications, flexible. Often the antenna is covered with tough, tight plastic coating. Whip antennas are most efficient when of optimal length, a quarter wavelength of the transmission frequency. Whip antennas, however, are often positioned close to the animal's body to reduce the likelihood of the animal damaging it or having it snag on objects. Proximity to the body and a less than optimal length reduce the efficiency of an antenna and affect reception distance and rates. Flexible wire, or brass or copper loop antennas can be incorporated in a collar but the tuned

Table 7.3 *Uses of radio-telemetry to study carnivores—interactions and predation.*

Topic	Species	References
Comparative ecology	mongooses	Ray 1997
	jackals	Loveridge and MacDonald 2002
Depredations	large carnivores	Kolowski and Holekamp 2006
Food habits	jaguars	Cavalcanti and Gese 2010
Interference interactions		Linnell and Strand 2000
Kill-site locations	wolves	Webb *et al.* 2008
	lions	Tambling *et al.* 2010
Predation rates	cougars	Anderson and Lindzey 2003
		Ruth *et al.* 2010
	jaguars	Cavalcanti and Gese 2010
	wolves	Demma *et al.* 2007
		Hebblewhite and Merrill 2008
		Merrill *et al.* 2010
	large carnivores	Laundré 2008
	wolves	Sand *et al.* 2005
Species interactions	mesocarnivores	Kowalczyk *et al.* 2008
Survival/mortality	coyotes, swift foxes	Karki *et al.* 2007

Table 7.4 *Uses of radio-telemetry to study carnivores—resource use and selection.*

Topic	Species	References
Denning ecology	brown bears	McLoughlin *et al.* 2002a
Finding dens	Iberian lynx	Fernández *et al.* 2002
	Canada lynx	Moen *et al.* 2008
Habitat differentiation	3 spp.	May *et al.* 2008b
Habitat selection/use	brown bears	Berland *et al.* 2008
		Christ *et al.* 2008
		Martin *et al.* 2008
	Florida panthers	Land *et al.* 2008
	polar bears	Mauritzen *et al.* 2003b
	sloth bears	Ratnayeke *et al.* 2007
	Arctic foxes	Pamperin *et al.* 2008
	red foxes	Van Etten *et al.* 2007
Habitat suitability modeling	cheetahs	Muntifering *et al.* 2006

loop antenna must retain the size and shape delivered by the manufacturer when placed on an animal (Gehring and Swihart 2000). For mammals that break antennas (Biggins *et al.* 2006), the antenna is often placed between two layers of heavy collar material (Schwartz and Arthur 1999; Loveridge and Macdonald 2002).

Table 7.5 *Uses of radio-telemetry to study carnivores—behavior and physiology.*

Topic	Species	References
Activity	Asiatic black bears	Hwang and Garshelis 2007
	brown bears	Gervasi *et al.* 2006
		Kaczensky *et al.* 2004
		Kowalczyk *et al.* 2003
	mountain lions	Janis *et al.* 1999
	Andean bears	Paisley and Garshelis 2006
	badgers	Tanaka 2005
	fishers	Weir and Corbould 2007
Circadian activity	wolves	Ciucci *et al.* 1997
		Merrill and Mech 2003
Communication	river otters	Ben-David *et al.* 2005
Disease	jackals	Rhodes *et al.* 1998
Group living	coatis	Hass and Valenzuela 2002
Heart rate/temperature	Arctic foxes	Follman *et al.* 1982
	red foxes	Kreeger *et al.* 1989
	dogs	Li *et al.* 2008
	mink	West and Van Vliet. 1986
	wolves	Kreegar *et al.* 1990
Parent/offspring	pumas	Laundré and Hernández 2008
Paternity and mating system	wolverines	Hedmark *et al.* 2007
	black bears	Kovach and Powell 2003
Response to human activity	wolves	Hebblewhite and Merrill 2008
		Merrill and Erickson 2003
	3 species	Ruth *et al.* 2003
		Linnell *et al.* 2000
	brown bears	Waller and Servheen 2005
Sociality	river otters	Blundell *et al.* 2002
	raccoons	Pitt *et al.* 2008
Sociality and disease	badgers	Böhm *et al.* 2008
Steroids	sun bears	Schwarzenberger *et al.* 2004
Territoriality	wolves	Demma and Mech 2009
		Jedrzejewski *et al.* 2001

Encapsulating, or "potting," the electronics and power source protects the unit from shock and moisture, and keeps animals from damaging components. Most radio transmitters are potted in acrylics, epoxy resins, or hermetically sealed canisters, and large, long-lived transmitters for large mammals usually are cast in a form filled with potting. Transmitters to be implanted require special encapsulation to prevent rapid penetration by body fluids and to preclude adverse reaction from body tissues.

Table 7.6 *Uses of radio-telemetry to study carnivores—population biology.*

Topic	Species	References
Demography	cheetahs	Marker *et al.* 2003
	sea otters	Tinker *et al.* 2006
Genetic and spatial structure	swift foxes	Kitchen *et al.* 2005
Litter size	black bears	McDonald and Fuller 2001
	eastern wolves	Mills *et al.* 2008
Mortality and habitat	brown bears	Nielsen at al. 2006
Population delineation	polar bears	Amstrup *et al.* 2004
	brown bears	McLoughlin *et al.* 2002b
Population genetic structure	polar bears	Crompton *et al.* 2008
Survival and mortality	tigers	Goodrich *et al.* 2008
	furbearers	Kamler and Gipson 2004
	gray fox	Farias *et al.* 2005
	wolverines	Krebs *et al.* 2004
	wolves	Mills *et al.* 2008

A variety of sensors and options is available from many manufacturers. Sensors that detect body movement (Garshelis *et al.* 1982; Gervasi *et al.* 2006), that detect temperature or ambient temperature, or that incorporate a diode for visual tracking (Tuyttens *et al.* 2002) are available. Activity, temperature, and atmospheric pressure are often conveyed by modulating pulse interval, by which a change in pulse is calibrated to a change in the sensor. The pulse interval can be interpreted manually or with a data logger. A tilt switch can report whether an animal's body is in a particular posture by triggering a slow or fast pulse (Theuerkauf and Jedrzejewski 2002). The integration of accelerometers in transmitter units allows three-dimensional motion sensing (Kappeler and Erkert 2003; Moll *et al.* 2007). Conduct a pilot study to test special designs, using a surrogate or captive animal, if possible. Remember, however, that field conditions can affect the function of telemetry systems.

"Activity" can be interpreted from changes in radio signal strength and consistency (Weir and Corbould 2007) that occurs when an animal moves. Motion can also be conveyed from sensors that change pulse rate when an animal has moved within a short period. Such motion sensors can indicate possible mortality of a marked animal that has not moved for a long period (Hass and Valenzuela 2002; Kamler and Gipson 2004; Mills *et al.* 2008). The sensor can reset the pulse rate if the transmitter moves or can be programmed to remain in "mortality" mode for a long period (e.g. to prevent scavengers or other sources of movement from resetting the mortality signal). Mortality also can be indicated with a body temperature sensor that changes the pulse rate when the temperature drops below a prescribed level. Using temperature and motion sensor data together can provide further evidence of mortality or that the transmitter has become detached from the animal (Bates *et al.* 2003). More than one type of data (e.g. motion and temperature) can be sent from a single transmitter of certain manufacturers.

Data storage options allow information to be logged in the telemetry unit for later downloading to a receiver system via radio transmissions to ground-, air-, or satellite-based receivers, or from a transmitter that has dropped from the animal or recovered by recapture or from a dead animal. This option is very useful when the radio signal is beyond reception range or reception is limited by environmental factors (e.g. water, topography), or when numerous locations are obtained from GPS receivers (see below). Temperature, motion, or pressure data can be time stamped and logged for retrieval. These and many other animal and environmental data are gathered by data logging or biologging (Ropert-Coudert and Wilson 2005). Rutz and Hays (2009) summarized engineering and research activity in equipment development and application, data management, and analyses, and the many questions and topics to which biologging is now being applied.

The materials and methods used to attach the telemetry unit to the animal are very important for ensuring the well-being of the animal and for ensuring the unit remains in place for the study period (e.g. Biggins *et al.* 2006). A collar around the neck is a common attachment method for radio-tagging carnivores. The telemetry unit and collar should be able to withstand attention from the animal (D. Garshelis and P. Ciucci, pers.com.) or from conspecifics during social interaction, should fit the animal's neck contours, distribute the package mass evenly, accommodate swallowing, accommodate seasonal changes in neck size, and minimize interference with the animal's natural movements.

A variety of materials are available for collars, depending upon the size of the animal, desired mass, configuration, durability, and attachment method. For small mammals, a collar can be made of steel cable, elastic, or braided fishing line covered by flexible, hollow plastic tubing. Each telemetry unit should be adjusted to fit the individual. For medium or large mammals (Loveridge and Macdonald 2002), the collar is typically constructed of leather, machine belting, braided nylon, or synthetic dog collar material and secured with adjustable bolts, rivets, or buckles to custom fit each individual. Some collars accommodate growth of young animals or temporary neck expansion by incorporating foam rubber inserts or sewn pleats (Garcelon 1977; Strathearn *et al.* 1984; Jackson *et al.* 1985). Telemetry units can be retrieved for reuse, to obtain data loggers, and to relieve animals of the burden. Mechanisms can be integrated in some collars to anesthetize animals using a remote transmitter (Mech and Gese 1992), or to detach the collar at a preprogrammed time (Sawyer *et al.* 2006). Further development continues to overcome performance deficiencies of automatic release collars (Garshelis and McLaughlin 1998; Kochanny *et al.* 2009) and to adapt mechanisms to the variety of species and collar types (Müller *et al.* 2005).

Radio transmitters can be surgically implanted in body cavities or implanted subcutaneously. Abdominal implants often are used in mammals whose body configuration precludes collar attachment (river otters, Melquist and Hornocker 1979; mink, Eagle *et al.* 1984; sea otters, Ralls *et al.* 1989), or to obtain physiological data (Fuglei *et al.* 2002). Implantation of the entire radio transmitter and antenna in the abdominal cavity can reduce reception range of VHF radios by $\geq$ 50% (Melquist and Hornocker 1979). Subcutaneous implants have been used to radio-track polar bears (*Ursus maritimus*) (Mulcahy and Garner 1999). When a smaller transmitter is adequate (e.g. short duration or low transmission power) or other attachment methods are precluded, radio transmitters can be affixed to ear tags, which typically are used for marking livestock (Servheen *et al.* 1981).

Receiving systems for VHF wildlife telemetry comprise radio receivers, antennas, cables to connect the antenna to the receiver, accessories (e.g. head phones, chargers), counters and decoders, and recording devices. The receiver components can process and convert the transmitter's signal to an audio tone, and can produce signals for processing by demodulators, decoders, or pulse counters.

Simple manual receivers operate within bandwidths of about 50 to 200 KHz, and can be used with about 5 to 20 transmitter frequencies. Receivers are powered by batteries and most have a meter to indicate the supply voltage. Most have a built-in speaker and a jack for connecting earphones. Manual receivers are usually the simplest and smallest receivers used in the field. More complex (and more expensive) receivers can cover from 1 to 45 MHz bandwidths, and include a programmable, automatic frequency scanning capability that can be interrupted when a radio signal is detected. Programmable receivers are useful when many transmitters are in the area of reception, such as when one surveys from an aircraft or when a receiver can be left unattended to record signals automatically.

Recorders, counters, decoders, and data loggers mechanize processing of radiotelemetry signals and data by measuring intervals between pulses, recording changes in signal amplitude, marking the presence or absence of a signal, decoding a signal, or recording data.

Receiving antennas have several basic designs: omni-directional, loop, Adcock or H antenna, Yagi, null-peak system. Biologists should discuss their needs with vendors and clearly describe the environment where the research will take place. Omni-directional ("whip") antennae have a uniform, 360° reception pattern, and have relatively low gain. They are easily adapted for magnetic or bolt-on attachment to vehicles and aircraft, and are used commonly to detect presence of signals over comparatively small areas. Directional receiving antennas have a three-dimensional pattern of power oriented by the element(s) of the antenna. A directional receiving antenna will detect the strongest radio signal when directed toward the signal. Making the antenna more directional generally increases gain, and thus the distance over which a signal can be received.

Elevation of the transmitting and receiving antennas generally increases the reception range. However, "obstacles" (e.g. terrain, moist vegetation, buildings) can block signal transmission and having the receiving antenna close to a person, a vehicle, or even earth's surface can affect reception. Therefore, optimal performance is usually achieved by holding or mounting the receiving antenna high. Hold an antenna above one's head, stand on an elevated place, raise the antenna on a mast above the ground, or receive from an aircraft. Custom VHF telemetry systems can be devised for vehicles (Gilsdorf *et al.* 2008) and for remote data collection (MacNulty *et al.* 2008; Mares *et al.* 2008).

7.3 Radio-tracking field procedures

The performance of equipment can be altered dramatically by animal behavior, topography, vegetation, and climate. Personnel need to be trained, and location accuracy and precision are maximized through careful, consistent procedures, including estimates of location error (White and Garrott 1990; Nams and Boutin 1991; Withey *et al.* 2001). The location of a radio-marked animal can be estimated by triangulation along bearings from two or more receiving sites (O'Donoghue *et al.* 1997). Mech (1983), Samuel and Fuller (1994), and Kenward (2001) detail basic tracking methods.

Homing is a method by which the operator uses antenna directionality and signal strength information to move toward and find a transmitter (or animal). Radio-tracking from aircraft is usually a special case of homing to find animals in a large area. Flying can increase reception range by 10 times or more, and increasing altitude usually increases detection distance. When two directional antennas are mounted facing downward and to opposite sides of the aircraft under the wings or struts, signal strength indicates the side of the aircraft where the animal occurs.

7.4 Satellite telemetry systems

Satellite telemetry allows remote tracking of animals from most places on earth. The Argos system became available in the 1980s (Harris *et al.* 1990a, http://www.argos-system.org/manual/, accessed March 17 2011). This system requires using specialized transmitters, called Platform Transmitter Terminals (PTTs), which weigh ≥5 g. Polar-orbiting satellites receive sensor data and calculate animal locations from Doppler measurements of ultra-high frequency (UHF) transmissions from PTTs. Processed data are distributed to researchers in several formats, including Internet access to data received about 4 hours previously. The cost of data acquisition from Argos depends on the quantity of data received and choice of data distribution options.

PTTs can transmit data from up to eight sensors (e.g. temperature, motion, pressure voltage) and can be programmed to transmit at particular times, including only during predicted satellite passes, and during particular periods (e.g. seasons). Transmission strength can be adjusted to conserve battery power, but this can lead to fewer location estimates per duty cycle (Walton *et al.* 2001a). VHF transmitters can be mounted on collars with PTTs to locate animals on the ground (McLoughlin *et al.* 2002b). Finding PTTs (Bates *et al.* 2003) provides valuable information about the status of marked animals and facilitates retrieval of PTTs for reuse.

Location estimates from Argos are assigned to Location Classes (LC) that provide nominal location accuracy. Biologists must consider if regular location accuracy of ≤1 km is appropriate for their objectives. A pilot study can be useful to assess how satellite telemetry performs under the conditions of a particular project. In September 2010, Argos implemented a new location and error estimation procedure, which can decrease error.

The Global Positioning Satellite System (GPS) is a US Department of Defense array of satellites that transmit signals to GPS receivers. GPS systems allow three-dimensional and frequent locations with accuracy of <30 m. Currently, GPS receivers estimate locations within seconds of receiving signals from satellites. Animal-borne GPS systems integrate the GPS receiver into a telemetry unit (on a collar, for example) including a micro-power data acquisition/controller (MDAC). The MDAC turns the GPS receiver, sensors, and data-transfer components on and off to manage the energy budget of the system, to acquire positions at programmed sampling times, and to store location estimates. Some packages provide an interface for the user to program parameters and to download stored data (Tomkiewicz *et al.* 2010). The GPS receiver requires considerable power, which can limit the operational life of the unit. Even with recent low-power GPS receiver technology, receiver power management usually is necessary to achieve regular receptions and long operational life. Collar units for some animals can be powered by solar arrays.

Data stored on board (SOB) a GPS telemetry unit can be downloaded when the unit is recovered. Storing data on board is dependable but if the unit is not recovered, the data are lost. Store-on-board systems can store about 35 000 GPS positions per megabyte of memory and some GPS subsystems have >6 megabytes of memory. Incorporating programmable release mechanisms or VHF transmitters into a GPS telemetry package increase the likelihood of recovering data (Sager-Fradkin *et al.* 2008). Some GPS telemetry packages incorporate transponders that transmit data via VHF or UHF from the unit when queried (Cavalcanti and Gese 2010; Chadwick *et al.* 2010). Power source restrictions limit the VHF transmission to a narrow bandwidth and data rate, requiring perhaps 7 s to transfer a single GPS position and 45 min to transfer 180 positions. Most GPS systems with data-transfer technology also retain data on board, so that data can be downloaded if the unit is recovered.

Another approach to data transfer is to configure each GPS transmitter package as a separate link in a network. When tagged animals aggregate (e.g. a pride of lions), the links communicate, storing data from each package on all other packages. Retrieval of data from all marked animals requires querying only one telemetry package (Juang *et al.* 2002; Martonosi 2006).

GPS data can be transferred from radio-marked animals via the Global System for Mobile Communications (GSM) mobile telephone-data services (Gervasi *et al.* 2006; Sundell *et al.* 2006; Tambling *et al.* 2010). Regular contact and data recovery from an area where service is available (widely in Europe) allows transfer of all data stored in the memory of each GPS telemetry package. The system also can be used in a two-way manner, allowing users to change on-board collar parameters (Sundell *et al.* 2006). Unfortunately, many vast areas (e.g. much of North America, sparsely populated Africa) lack GSM services.

Data transfer via low earth orbiting (LEO) satellites offers much more comprehensive coverage. The Argos LEO satellites orbit about 850 km above the earth, thereby accommodating low power transmitters (100 mW to 1W) with omni-directional antennas. While it is a global system, interference from the environment in Europe and Mongolia–Pakistan disrupts data transfer presently on the Argos uplink frequency, limiting its use in those places. Argos can relay 24–48 GPS positions per day from units on medium- to large-sized animals. GPS data can be recovered as frequently as daily, but the limited energy budget of most animal tags limits the number of GPS positions that can be transferred. Recent developments allow the GPS data transfer to be used with on-board Argos orbit prediction programs enabling transmitting only during satellite overpasses. This increases data transfer efficiency of the system.

The Globalstar satellite system includes more than 40 low earth orbit satellites designed for telephone communication and data transfer. GPS-Globalstar animal telemetry collars can transmit every GPS location acquired in real time or log and store location estimates for later retrieval. Retrieving GPS data via Globalstar is expensive but a GPS-Globalstar unit can be programmed to acquire a GPS location eight times per day, but to transmit only one GPS location per day or per week, thus conserving battery and costing less for data delivery via the satellite system. The other stored data must be obtained later from the retrieved unit. Wildlife collars using a GPS receiving antenna and a Globalstar patch transmitting antenna are recently available and being used on many projects, but we are unaware of published results. Names of current users can be obtained from North Star Science and Technology and from Vectronic-Aerospace. The greatest limitation is that areas lack land-based receiving stations, which limits Globalstar geographic coverage.

The Iridium 66 low earth orbit satellite constellation provides worldwide, two-way, near continuous coverage for voice and data communications. Iridium is used in oceanographic applications in dial-up and in short burst modes. Opportunities for use with wildlife are just becoming available. Current technology is suitable for larger terrestrial species (Aastrup 2009). Vectronic-Aerospace has been the manufacturer offering data transfer via Iridium.

Other data-transfer possibilities (e.g. data recovery using radio modem technology) are available as are innovations with GPS and other technology (Tomkiewicz *et al.* 2010) that will increase the options available to biologists. For example, inertial navigation devices in GPS collars allow estimation of animal locations on a continuous basis between GPS fixes (Hunter *et al.* 2005; Elkaim *et al.* 2006). Inertial navigation systems suffer from deteriorating accuracy as errors compound over time, but GPS fixes at short intervals can be used to reset the accuracy of the estimated track (Hunter 2007). Advances in battery technology, decreases in GPS power requirements, and increases of on-board memory capacity lead to capabilities to track a diversity of species and to know where animals are almost continuously. To ensure having the most appropriate equipment, discuss options with animal telemetry manufacturers and vendors.

7.5 Radio-telemetry applications for carnivores

Many insights into carnivore ecology and behavior have been garnered using radio-telemetry. Animal movements (Table 7.2) are usually studied over time frames of days to years, and often are quantified in terms of distance traveled or total area covered. Ground-based, VHF telemetry has been used extensively to delineate and describe home ranges of animals whose movements are somewhat restricted (e.g. honey badgers, Begg *et al.* 2005), and satellite telemetry has revealed the limitations of ground tracking for some species (e.g. snow leopards, *Uncia uncia*, McCarthy *et al.* 2005). For wide-ranging species, such as wolves, aerial VHF telemetry has long been used to follow resident packs, but packs that depend on migratory prey have been monitored more effectively using satellite telemetry (e.g. in the Arctic, Walton *et al.* 2001a). Similarly, dispersal of smaller carnivore species may be monitored from the ground with VHF transmitters (e.g. ferrets, *Mustela furo*, Byrom 2002) but for large species, GPS telemetry has proved invaluable for studying connectivity among distant populations, specific travel routes, and movements after settling into new territories (e.g. wolves in Finland, Kojola *et al.* 2009).

Investigations of predator interactions with prey and other animals have been enhanced by use of telemetry (Table 7.3), particularly in locating kills and quantifying predation rates (e.g. VHS and GPS telemetry of cougars, *Puma concolor*, Laundré 2008; Knopff *et al.* 2009) and in identifying how interactions affect distribution (GPS and VHF telemetry of wolves and elk, *Cervus Canadensis*, Proffitt *et al.* 2009). Telemetry has provided instantaneous locations and observations and, with new data-retrieval capabilities from satellites, data are very accurate and nearly in real time. Simultaneous VHF telemetry studies of several species also have facilitated in-depth understanding of competition, demographics, space use, and prey use (e.g. jackals, *Canis* spp., Loveridge and Macdonald 2002).

Resource use and selection have been much studied using telemetry because of the volume of locations that can be accumulated (Table 7.4). Many studies have related carnivore distributions to a variety of vegetative or topographic "habitat" variables, sometimes to gain general insights into various habitat requirements and sometimes for specifically targeted insights (e.g. denning of brown bears, McLoughlin *et al.* 2002a). Data have been incorporated into models used to enhance restoration efforts or to minimize conflicts (e.g. VHF telemetry of cheetahs, Muntifering *et al.* 2006).

Behavioral and physiological parameters, otherwise unobtainable, can be monitored via telemetry from wide-ranging animals (Table 7.5). Activity patterns and time budgets have been identified for cryptic, tropical carnivores (e.g. VHF telemetry of Andean bears, *Tremarctoc ornatus*, Paisley and Garshelis 2006), and paternity assignment has been shown to correspond well with overlapping ranges of male and female wolverines (*Gulo gulo*, Hedmark *et al.* 2007) derived from telemetry. Clear identification of social patterns of species not easily observed is possible with telemetry (e.g. river otters, *Lontra canadensis*, Blundell *et al.* 2002), and they can give insights into sociality and disease transmission (e.g. badgers, *Meles meles*, Böhm *et al.* 2008).

Radio-telemetry has contributed to advances in carnivore population biology (Table 7.6). Denning sites and thus litters of young can be found rather quickly using ground-based or GPS telemetry. Unbiased estimates of survival and mortality of individuals leads to better understanding of population status and limiting factors (e.g. ferrets, Byrom 2002; tigers, *Panthera tigris*, Goodrich *et al.* 2008).

Population estimates are more refined because assumptions of population closure can be monitored, and capture–recapture estimates can be assessed in light of having known, marked individuals within the sampling unit (e.g. brown bears, Miller *et al.* 1997).

Research of carnivore biology is fraught with challenges. Radio-telemetry has facilitated gathering much information about many carnivore species that is otherwise not available. Telemetry has been used with carnivores since the earliest application of radio-telemetry to wildlife studies, and many of the techniques from 50 years ago continue to be used with great success today. In the last 20 years, however, advances in electronics and space-age technology have given innovative engineers and manufacturers new material, which they use to provide biologists with new equipment with which new methods are developed. Radio-telemetry equipment is now available with more options from more vendors than ever before, and new capabilities become available at a rapid pace. To know what is available and to apply it best to meet objectives, biologists must have detailed dialogues with manufacturers. Research and conservation of carnivores will surely benefit from continued use of radio-telemetry.

8

Estimating demographic parameters

Ken H. Pollock, James D. Nichols, and K. Ullas Karanth

Recent assessments present an alarming picture of ongoing carnivore population decreases worldwide and highlight the urgent need for targeted population assessments as a basis for mounting appropriate conservation responses (Ceballos *et al.* 2005; Schipper *et al.* 2008). On the other hand, carnivore research and management must address "problem carnivores" that threaten human interests (Treves and Karanth 2003), and the regulated harvest of carnivores as furbearers. Furthermore, investigators are interested in demographic processes simply to satisfy scientific curiosity. A sound understanding of demographic processes is critical to the success of all such endeavors. In the real world, demographic processes are "spatially explicit" in the sense that they are influenced by location-specific ecological features and human impacts. These site-specific ecological processes help us understand demographic processes better. Reliable assessments of different "state variables" such as carnivore numbers or distribution at any location or time, and of "vital rates" that drive changes in these state variables (Williams *et al.* 2002b), form the subject of this chapter. In our view this "parameter estimation" is at the core of gaining reliable knowledge of how carnivore populations function across time and space.

Understanding demographic processes typically requires the measurement and estimation of the following specific parameters (Thompson *et al.* 1998 and Williams *et al.* 2002b provided detailed explanations and definitions of these terms):

1. Population abundance (or its common expression as density) and rates of change in abundance over time.
2. Survival, mortality, recruitment, immigration, emigration, and related parameters that are vital rates that drive changes in abundance.
3. Spatial distribution patterns, habitat occupancy and rates of change in these variables.
4. Additional parameters involving rates of movement and exchanges of individuals among these populations if >1 population is involved.

8.1 Combined challenges of carnivore ecology and survey logistics

Estimating demographic parameters for carnivores poses unique problems not usually encountered when studying many other animal groups. Carnivores typically occur at low population densities. This is particularly true of large and medium-sized obligate flesh-eaters in the *Felidae, Canidae, Hyaenidae*, and even some *Mustelidae*. Furthermore, large and medium-sized carnivores move long distances daily, have large home-ranges, extensive seasonal movements, and long dispersal distances. Consequently, substantial logistical effort is usually inevitable for obtaining reasonable sample counts, whatever the type of survey process being employed (Karanth *et al.* 2010).

Additional challenges are posed because many carnivores are nocturnal, elusive or wary because of human persecution. Thus, visually counting them is usually impossible, compelling investigators to use indirect detections (photos, tracks, calls, scats, etc.) in field surveys (Long *et al.* 2008a; Thompson 2004). For example, a well-developed, rigorous approach for estimating animal abundance based on visual detections, distance sampling (Buckland *et al.* 2001), can be used to estimate carnivore abundance only where carnivores can be detected visually (Ruette *et al.* 2003; Hounsome *et al.* 2005; Ogutu *et al.* 2005; Aars *et al.* 2009), or sometimes with capture data (Corn and Conroy 1998). As animals need not be identified individually, distance sampling does not require "marking" of individuals. Generally, however, this method cannot estimate vital rates, although overall rate of population change can be estimated from temporally repeated data. We do not address distance sampling here, and direct researchers to the studies cited above.

Many carnivore species live in remote or inhospitable habitats (e.g. snow leopards, *Uncia uncia,* in the Himalayas, polar bears, *Ursus maritimus,* in the Arctic, jaguars, *Panthera onca,* in Amazonia). Consequently, lack of qualified field survey personnel, roads, vehicles, electric power, communication, and even physical danger posed by the terrain, climate, or wild animals, may pose formidable challenges to carnivore surveys. In some cases, "loyalty" to local traditions of surveying carnivores, such as some spoor-based *ad hoc* "census" techniques employed in India, Russia, and Africa, impede introduction of modern survey techniques based on concepts of population sampling and statistical inference. For example, for over three decades (1973–2004), official adherence to the "pugmark census technique" in India (Karanth *et al.* 2003 provide details) prevented the introduction of modern tiger (*Panthera tigris*) monitoring approaches, until local tiger population extinctions forced a public outcry.

Because estimating carnivore population parameters is challenging, researchers usually rely on methods that require individual identification of "marked" animals, which are then "detected" and counted in some manner during field surveys using capture–recapture sampling (Williams *et al.* 2002b; Amstrup *et al.* 2005). Although the capture–recapture approach is well-developed, both theoretically and empirically (Williams *et al.* 2002b; Thompson 2004; Amstrup *et al.* 2005; Royle and Dorazio 2008), for carnivores the key challenge lies in individually "marking" the study animals in the field. Unlike fish, birds, and rodents that are often easily captured and tagged, carnivores are difficult to capture, handle and mark (Long *et al.* 2008a; Karanth *et al.* 2010). Furthermore, capture and handling is potentially stressful to carnivores and hazardous to investigators. Overall, logistical difficulties of capturing, sedating, tagging, and recapturing larger carnivores in sufficient numbers generally precludes physical captures for population sampling (Karanth *et al.* 2010).

Therefore, in this chapter we stress the use of noninvasive field methods for sampling carnivore populations to estimate demographic parameters reliably (Chapter 4); invasive methods may be necessary to answer other biological questions (Chapters 5, 6, 7, 9, and 10). At present, reliable, noninvasive methods exist for "capturing and marking" carnivores. The first uses "photographic captures" that facilitate individual identifications from natural marks. The second uses genetics-based individual identifications using DNA that is noninvasively extracted from fecal or hair samples collected in the field (Boulanger *et al.* 2008; Mondol *et al.* 2009). A third approach for individual identifications being explored (Kerley and Salkina 2007) uses trained domestic dogs (*Canis familiaris*) that discriminate among scent signatures of different individual carnivores. Although this third method shows promise, rigorous application is not yet fully developed.

Wherever possible, demographic parameters should be estimated by modeling both the underlying biological process and the observation or survey process that generate the count data (Royle and Dorazio 2008). Thereafter, fit of plausible models can be tested against data generated from carefully designed field surveys. Therefore, the process of a priori modeling and survey design, careful data collection, and testing of plausible models against data *before* generating estimates of demographic parameters are integral.

8.2 Detection probabilities and demographic inference

Carnivore population ecology deals with changes in abundance of animals over time, and with vital rates (survival, reproductive recruitment, immigration, and emigration) that bring about these changes. Inferences about abundance and vital rates must be based on counts of animals from field sampling programs and these counts must then be translated into inferences about the quantities of real interest. This translation is based on the idea that the counts themselves (denote as C) are random variables with expectation given by the product of the quantity of interest (e.g. number of individual animals actually in the area being sampled; denote as N) and a detection or capture probability (probability that a member of N is counted in C; denoted as p):

$$E(C) = Np \tag{8.1}$$

Inference about N thus requires inference about p. For example, if we have some means of estimating detection probability (denote estimate as $\hat{p}$), then we can estimate abundance in the area exposed to our sampling (counting) efforts as (Lancia *et al.* 1994; Pollock *et al.* 2002):

$$\hat{N} = \frac{C}{\hat{p}} \tag{8.2}$$

As emphasized by Lancia *et al.* (1994), the various methods for abundance estimation presented in books, such as those by Seber (1982) and Williams *et al.* (2002b), represent different, sometimes ingenious, approaches to estimating detection probability. The final step of abundance estimation, however, almost always invokes division of the count by this detection estimate, as in Equation 8.2.

The use of so-called index statistics is common in studies of carnivores. In such cases, the counts themselves are used as indicators of abundance for comparisons. For example, ratios of count statistics at different times or locations or species are claimed to estimate time trends in abundance, or relative abundance at different places, or relative abundance of different species. Such uses of counts as indices may be reasonable when detection probabilities are very similar over the dimensions of comparison. When detection probabilities differ for different times, places, or species, then these differences in detection probabilities are confounded with true differences in abundance, potentially producing misleading inferences. The problem with indices is that their valid use in comparisons depends on the similarity of detection probabilities, and this similarity cannot usually be assessed without collecting ancillary data needed to draw inference about p. Our recommendation is, thus, to design a study in such a way as to collect the data needed to draw inferences about p, and hence N. If detection probabilities actually do turn out to be similar over the dimension of comparison, then reduced-parameter models or even counts themselves can potentially be used for the comparison (e.g. Skalski and Robson 1992; Williams *et al.* 2002b). If the detection probabilities do differ, however, then these differences are readily accommodated in inference when appropriate estimation methods are used.

Detection probability is relevant not only to estimating abundance, but also to inferring survival, reproductive rate, and movement (Williams *et al.* 2002b). For example, if we mark a sample of animals in year t, we might try to estimate survival by counting the number of these marked animals still alive in year $t+1$. The count of survivors, however, will likely be smaller than the true number of survivors, and an estimate of detection probability allows us to translate the count properly into an estimate of survivors.

In some applications, it is useful to think of detection probability as the product of two component probabilities, Pr (detection | availability) = p_d, and Pr (availability | membership in population of interest) = p_a (Pollock *et al.* 2004). In this case, the notion of availability means that an animal has a non-negligible chance of being detected given that it is a member of the population of interest. Thus,

$$p = p_a p_d. \tag{8.3}$$

In some cases, decomposition of detection probability is possible. During capture–recapture studies of terrestrial carnivores, we may be interested in inferences about the animals exposed to our trapping efforts. For example, we may profess interest in all animals whose range-centers fall inside the boundaries of our trapping grid. During a particular trapping session, however, an animal may temporarily be outside this boundary (with probability $1 - p_a$), and thus not exposed to trapping efforts. Temporary emigration is a label frequently attached to animals that are members of a group of interest but that are not in the sampled area during a particular sampling period. Certain capture–recapture designs and models permit separate estimation of p_a and p_d in the case of temporary emigration (e.g. Kendall *et al.* 1997; Fujiwara and Caswell 2002; Kendall and Nichols 1995). The resulting overall detection probability for an animal alive in the sampling grid is $p = p_a p_d$. Many inference methods permit estimation of overall detection probability, p, but do not permit its decomposition into detection and availability components.

Sometimes range or occurrence of animals over space is of interest for carnivores (e.g. O'Connell *et al.* 2006; Karanth *et al.* 2009). Indeed range size and extent are criteria frequently used in assessments of the conservation status of species (Ceballos *et al.* 2005; Schipper *et al.* 2008). Occurrence over space is also a natural state variable in (1) geographically extensive studies for which detailed demographic analysis is simply not possible and (2) studies of rare and elusive species (including many carnivores; MacKenzie *et al.* 2004a, 2006). In such cases, the proportion of habitat patches or geographic sample units occupied by a species is a state variable of interest.

Estimating this state variable is typically based on visits to habitat patches or geographic sample units, accompanied by efforts to detect the focal species. Therefore, the resulting count statistic is the number of units at which the species is detected. Non-detections can occur for two reasons: either the species is truly absent or the species is present but was not detected. Once again the issue of detection probability is relevant, although this time at the level of the population or species, rather than at the level of the individual organism. Equation 8.2 is still applicable with the count statistic, C, now being the number of locations at which the species is detected and p being the probability that a species is detected at a location at which it is present. As was the case with the state variable of population size, estimation of the vital rates associated with changes in occupancy, rates of local extinction, and colonization, requires inference about detection probability. Occupancy modeling is essentially an extension of capture–recapture modeling.

8.3 Capture–recapture models

In capture–recapture studies, a researcher samples the animals k times, where k is usually greater than 2. At the time of each sample, all unmarked animals are marked in some unique manner, and all previously marked animals are recorded. This means that at the end of the study we have the capture history of every animal captured at least once. Sometimes animals have to be sacrificed on capture or may be negatively affected so they cannot be released, and these "losses on capture" should be considered in the models used.

Traditionally, marks were applied, such as ear tags, neck collars, and leg bands (Seber 1982; Silvy *et al.* 2005) involving physical capture and recapture (mark–recapture). Animals can also be captured once and marked to be visible and potentially resighted (mark–resight). Now natural tags are also being used from DNA finger prints (Waits 2004) or photo-ID based on natural features, and animals are never physically captured. In the DNA case, an animal's hair or scat is "captured," while in the photo-ID case its image is "captured." Sometimes exploited animal populations have been studied by marking animals and having tags returned or reported by harvesters of the animals (tag-return models; Brownie *et al.* 1985; Williams *et al.* 2002b).

Capture–recapture models are of two main types: closed models for short-term studies and open models for long-term studies. "Closed" means that the population under study does not change in any way during the study. Because the population size (N) must be constant and contain the same individuals, a study should be of short duration to satisfy the assumption of closure. Long-term studies require "open" population models to estimate population sizes and demographic parameters (survival, recruitment, emigration, and immigration). Using the "robust" design (Pollock 1982) combines sampling at two temporal scales and is advantageous for a long-term study, where several short-term pulses of sampling are nested within long periods.

8.3.1 Closed models

The two-sample case of a closed model is the famous Lincoln–Petersen population model (Seber 1982; Williams *et al.* 2002b). Assumptions typically made, besides closure, are that tags are not lost, tags do not influence the survival of the tagged animal, and that all animals in a sample have equal capture probability (that is, animals exhibit no trap response and no heterogeneity of capture probability due to inherent differences among the animals). With only two samples, applications may use different capture methods in the two samples. Mark–resight is one possibility, but many others exist. For example, some mammals have been injected with radio isotope "tags" that can be detected later in their scats (Conner and Labisky 1985).

The eight general models allowing for various sources of variation in capture probability (due to time (t), heterogeneity (h), and trap or behavioral response (b) and their combinations, as well as constant) first detailed by Otis *et al.* (1978), as well as their associated software (MARK, CAPTURE), have been the subject of much research (e.g. White and Burnham 1999; Williams *et al.* 2002b; the MARK online book by Cooch and White 2009). An important issue is selection among models. MARK relies on the Akaike Information Criterion (AIC) method (Burnham and Anderson 2002).

Heterogeneity models originally were based on the jackknife (Burnham and Overton 1978) and the coverage method (Chao 1989). Both finite mixtures (Norris and Pollock 1995, 1996; Pledger 2000) and continuous mixtures (Burnham 1972; Dorazio and Royle, 2003) have also been used. Link (2003) emphasized important non-identifiability issues in these heterogeneity models. An alternative, very important approach is to develop heterogeneity models based on covariates (Huggins 1989, 1991; Alho 1990).

Trap response is a very important issue in capture–recapture studies that involve physical capture and recapture of animals. In many cases, animals become trap-shy, though animals may become trap-happy. The common forms of trap-response models (M_b and M_{bh}) basically estimate N based on removals of unmarked animals, because the recaptured animals with their different capture probabilities turn out not to provide any information about population size (Otis *et al.* 1978; Pollock *et al.* 1990; Williams *et al.* 2002b). Hallett *et al.* (1991) evaluated capture–recapture models for closed populations in their studies of raccoons (*Procyon lotor*) and found support for models that incorporated trap response. Wegge *et al.*

(2004), provided some evidence of trap response to camera traps for tigers in Nepal.

Physical removal or catch per unit effort models can be used for pest or harvested species. One can use M_b and M_{bh} of program CAPTURE for these analyses, if the effort is equal on each sampling occasion (Otis *et al.* 1978; Williams *et al.* 2002b). Catch per unit effort models are a generalization of the M_b model for cases where the sampling effort is unequal across sampling periods (Seber 1982; Gould and Pollock 1997a, 1997b).

An important problem is the conversion of the population size estimate to an estimate of density, which involves defining an appropriate study area size. For trapping grids, the area used is sometimes just the area covered by the traps. Since this approach likely overestimates density, various methods of estimating edge effects have been developed. One approach is to add a buffer strip around the trap array, with the width of the strip similar to half the average home-range width of the animals. A second approach is to use nested subgrids, and a third approach is to use assessment lines (Otis *et al.* 1978 and Williams *et al.* 2002b provide more details). These approaches for estimating density work reasonably well in some practical situations. These are, however, *ad hoc* and involve at least two steps in the estimation process. A fourth approach, trapping webs (e.g. Anderson *et al.* 1983; Buckland *et al.* 2001), uses a specific configuration of traps, combined with ideas from distance sampling, to estimate density directly from capture data. This approach requires substantial effort (a large number of traps) and has seen only one use for carnivores: a density estimate of mongooses (*Herpestes javanicus*) in Antigua (Corn and Conroy 1998).

Recently, spatially explicit closed capture–recapture models have been developed, based first on simulation (inverse prediction) approaches (Efford 2004) and later using likelihood (Borchers and Efford 2008) and hierarchical Bayes (Royle and Dorazio 2008; Royle and Young 2008) approaches. The latter methods have been applied to photographic capture–recapture data on tigers (Royle *et al.* 2009a, 2009b; Royle and Gardner 2011). The spatially explicit models require that each detection history of an individual animal includes not only whether or not the individual was captured at each occasion, but also spatial data on locations of captures. Capture probabilities of animals at specific traps are assumed to be a function of the proximity of an animal's home-range center to the trap. The two substantive advantages of spatially explicit capture–recapture models are that the approach estimates density formally and directly (in a single step), and that the approach deals explicitly with an important potential source of heterogeneity in capture probabilities induced by animal location. This latter point concerns the likelihood that animals with ranges in the center of a trapping array are more likely to be captured than individuals with ranges near the periphery of an array. These new, spatially explicit models incorporate such variation in capture (i.e. detection) probability explicitly, thus dealing with one of the more important sources of heterogeneity of capture probability among individuals.

8.3.2 Open models

In long-term studies, where the population is "open," in addition to population sizes the goal is to estimate population losses (mortality, emigration) and gains (recruitment, immigration). The traditional Jolly–Seber model (Jolly 1965; Seber 1965) revolutionized the analysis of open population studies. This model can be considered to have two parts: the Cormack–Jolly–Seber model (Cormack 1964) for estimating apparent survival and capture probability based on recaptures only; and the full Jolly–Seber model for analyzing first captures and recaptures and estimating recruitment processes, population sizes, apparent survival, and capture probabilities (Seber 1982). In the Cormack–Jolly–Seber model (Cormack 1964; Burnham *et al.* 1987; Lebreton *et al.* 1992; Pledger *et al.* 2003, 2010), covariates can influence both the capture probability and apparent survival parameters.

Jolly–Seber models do not estimate true survival in most studies, but apparent survival (ϕ):

$$\phi = SF \tag{8.4}$$

Apparent survival is the product of true survival probability (S; the complement includes only mortality) and the probability that an animal shows fidelity (F) to the sampled area. One cannot estimate these two parameters separately from capture-–recapture data alone, but one can by combining a capture–recapture study with a telemetry study, which we consider in Section 8.4.

The two modern approaches to estimating recruitment and apparent survival are the super population model (Crosbie and Manly 1985; Schwarz and Arnason 1996; Williams *et al.* 2002b) and the temporal symmetry model (Pradel 1996; Nichols *et al.* 2000; Williams *et al.* 2002b). Both allow richer stochastic representations of important recruitment processes than the original Jolly–Seber formulation.

Using a standard, open model for a population with unequal capture probabilities, due to heterogeneity or trap response, can cause serious biases in estimates of population size (Pollock *et al.* 1990). Violation of the assumption that temporary emigration does not occur (animals move out of the study area and may then return) is also important. Two possible types of temporary emigration are random (an animal's probability of being a temporary emigrant *does not* depend on whether it was a temporary emigrant in the previous period) and Markovian (i.e. an animal's probability of being a temporary emigrant *does* depend on whether it was a temporary emigrant in the previous period). General, open models handle random, temporary emigration, by redefining capture probability as the probability of not temporarily emigrating multiplied by the probability of being captured, given the animal is present. (This is the same as Equation 8.3 because the probability of an animal being available for sampling is the probability that it is not a temporary emigrant). Population size estimates refer to all the animals alive at that time, whether they are temporary emigrants or not. The general open population models cannot handle Markovian temporary emigration.

The standard methods apply to a single population but sometimes one wishes to consider a meta-population system. Multi-state models (Arnason 1972, 1973; Hestbeck *et al.* 1991; Nichols *et al.* 1992; Brownie *et al.* 1993; Schwarz *et al.* 1993) allow transition probabilities between state r and state s:

$$\Phi^{rs} = S^{r}\psi^{rs}, \tag{8.5}$$

where the probability of survival in state r is S^r, and the movement probability is ψ^{rs} between states r and s.

8.3.3 Robust design models

For a long-term study, many advantages come from using the "robust" design, where several short-term pulses of sampling ("secondary periods" that usually assume closure) are nested within long periods ("primary periods" during which the population is open). The traditional closed robust design was developed to use models that permit unequal catchability, due to heterogeneity and trap response for abundance estimation (Pollock 1982; Pollock *et al.* 1990).

Important advances in modeling robust design data include inference about temporary emigration, including Markovian temporary emigration (Kendall and Nichols 1995; Kendall *et al.* 1997), and separation of *in situ* reproduction and immigration using a two-age robust design model (Nichols and Pollock 1990; Pollock *et al.* 1990). Genetic assignment can assist with this separation (Wen 2009; Wen *et al.* 2010). Finally, the open robust design allows entries and exits between the secondary periods within each primary period, instead of complete closure (Schwarz and Stobo 1997; Kendall and Bjorkland 2001).

8.3.4 Natural individual tags

Natural features of animals and DNA "finger prints" are natural tags (Chapter 4). Many carnivores have stripes, spots, and blotches on their coats that can be used to identify individuals unambiguously. Such markings are clearest in felids, but even striped hyenas (*Hyaena hyaena*), African wild dogs (*Lycaon pictus*), Ethiopian wolves (*Canis simensis*), and several smaller carnivores, can be identified individually. Images from normal photography, videography, and "camera traps" can be used but they must be clear and unambiguous (Karanth and Nichols 2010; Karanth 2010). Change in marks as animals age must be documented (Forcada and Aguilar 2000). To prevent loss of data, construction of capture histories should be based on photos of both flanks (Karanth and Nichols 2010).

The complexity of coat patterns and the resulting time and effort needed to identify individuals manually can be significant. For carnivores with complex patterns, species-specific, pattern-matching software, such as EXTRACT-COMPARE (Hiby *et al.* 2009), can identify photos for subsequent, visual comparisons of top-ranked matches. Such software can sometimes match photos from very different perspectives, relying on three-dimensional models of shapes of specific species (Hiby *et al.* 2009).

For DNA finger prints, DNA from a hair or scat sample is used to identify individuals (Chapter 4; also see review by Waits 2004). If the genetic markers used lack the variability to distinguish individuals, then two distinct animals can have the same apparent genotype ("shadow effect"), and the population size will be underestimated. Errors in genotypes due to laboratory scoring errors, contamination, or PCR amplification errors, usually lead to overestimation (Creel *et al.* 2003), unless special capture–recapture models that allow for error are used (Wright *et al.* 2009). Extreme care is required in field collection and laboratory procedures to avoid contamination. Fecal and hair DNA methods may provide information on sex and reproductive status of individuals. Collecting hair and feces can circumvent problems of animal trap-wariness, and equipment theft and vandalism, which may affect camera-trap surveys.

The possibility of combining both kinds of natural tags could prove very useful. For example, visual marks could be used to identify adults, while DNA marks could be used on young animals that have not yet developed stable, visible, natural marks.

When using individual tags applied by investigators, the assumption that animals are not misidentified has rarely been tested. With natural tags (either based on DNA or unique natural marks), this assumption is acknowledged, leading to modeling efforts to handle such errors, especially for closed populations (Yoshizaki 2007; Yoshizaki *et al.* 2009; Yoshizaki *et al.* 2011; Link *et al.* 2010). More research is needed to refine the models and to extend them to open population models and to the robust design.

Photographic or genetic capture–recapture studies must be designed carefully to include species' biology and logistical constraints. Considerations for field design include the period over which demographic closure is assumed in relation to expected population turnover rates in the species, covering a sufficient number of home ranges to sample ample individuals, requisite field skills for locating good trap sites or finding scat deposits, and geographic closure (Williams *et al.* 2002b; Karanth and Nichols 2010; Royle *et al.* 2009, 2010). As with traditional capture–recapture analyses, photographic and DNA studies require multiple sampling occasions.

8.3.5 Design of capture–recapture studies

Make certain that the precision of parameter estimates will be adequate. Otis *et al.* (1978) provided good guidance for closed models, Pollock *et al.* (1990) for open models, and Williams *et al.* (2002b) more generally. The best strategy is to use computer simulation or expected value approximations (e.g. Burnham *et al.* 1987; Pollock *et al.* 1990) based on anticipated parameter values, based on prior knowledge. Computer simulations can generate random capture histories, using random number generators with anticipated parameter values. Expected value approximations require estimating numbers of animals exhibiting each detection history and using these expectations in software such as MARK (White and Burnham 1999). These approaches are accessible even to biologists who have not worked with simulations.

Capture–recapture analyses are based on underlying models of the sampling and demographic processes. Therefore, studies must minimize violations of assumptions as much as possible. For just one simple example, if Markovian temporary emigration is a concern, then the robust design could be used, as that design permits this kind of temporary emigration, whereas the standard open models do not. Pollock *et al.* (1990) and Williams *et al.* (2002b) discussed design-based approaches to minimize assumption violations.

8.4 Telemetry mortality models

8.4.1 Survival models

The staggered entry Kaplan–Meier method is used widely to estimate survival using telemetry data (Pollock *et al.* 1989a, 1989b). This method allows animals to be added to the study while it is in progress and to be "right censored" if animals leave the study area or lose their radio tags. The standard models assume that right-censoring is independent of animal fate, a reasonable assumption in some cases. In some cases, however, disappearance of an animal may be associated with death, for example, in cases involving a predator, scavenger, or road-kill.

In program MARK, the "known fates" option can perform these standard analyses and can be generalized to allow covariates to affect the survival rates. As the name implies, known-fate modeling assumes that the investigator follows the radioed animal closely, knows whether it lives or dies between each pair of potential sampling periods, and detects it at each sampling period in which it is alive. Although radio-tags often result in high detection probabilities, these probabilities are seldom 1. In such cases, detection data for radio-tagged animals should be analyzed using capture–recapture models. Because radios often permit knowledge of animal fate, combined live-recapture and dead-recovery models (Burnham 1993; Williams *et al.* 2002b) provide a natural framework for dealing with carnivore telemetry data with imperfect detection (Hostetler *et al.* 2010). For telemetry studies with animals that are radio-collared, released, and never seen again, perhaps suggesting immediate emigration from the study area, transient models (Pradel *et al.* 1997) developed for traditional capture–recapture data are useful.

8.4.2 Combining telemetry and regular mark–recapture models in one overall analysis

Remember that for standard, open capture–recapture models, true survival and site fidelity cannot be separated (Equation 8.4: $\phi = S F$). To separate these components of apparent survival, one can include telemetered animals in capture–recapture studies (Powell *et al.* 2000; Nasution *et al.* 2001, 2002). Also, recall that we considered multi-state capture–recapture models that decomposed transition probability into survival and movement components (Equation 8.5):

$$\Phi^{rs} = S^r \psi^{rs} \tag{8.6}$$

Adding telemetered animals increases one's ability to estimate the movement probabilities ψ^{rs} between states r and s. Models designed to incorporate both capture–recapture and telemetry data are an important area for future work.

8.5 Occupancy models

8.5.1 Single-season models

The proportion of habitat patches or geographic sample units occupied by members of a species is important in conservation biology. Occupancy is also important for questions about species' range and may be the only state variable for which inference is possible for rare and elusive species. Some of the very first occupancy models were developed for carnivores (Nichols and Karanth 2002) and recent models (MacKenzie *et al.* 2002; 2006; Royle and Dorazio 2008) are applicable for many carnivores. The key issue in occupancy estimation is that typical "presence–absence" surveys do not provide valid inference because locations where animals are not detected are ambiguous regarding presence. Non-detections result from actual absence and from presence without detection. An approach to resolving this ambiguity is to use replicate visits and the resultant detection history data. If sample units are sampled at multiple points or small units of space, and the spatial replicates are selected randomly and with replacement, then the patterns of species' detections and non-detections across the replicates provide the data needed to estimate occupancy.

The occupancy models used in this situation are similar to capture–recapture models. One models detection histories as functions of parameters that describe the process that generated the data and that are relevant to ecological questions. Specifically, single-season occupancy data are modeled using: (1) occupancy parameters that represent the probability that a patch or sample unit is occupied by members of the focal species, and (2) detection parameters that represent the probability that at least one individual is detected during a sampling occasion, given members of the focal species are present in the sample unit.

These single-season occupancy models (e.g. MacKenzie *et al.* 2002, 2006; Royle and Dorazio 2008) all assume that sample units are closed to changes in occupancy over the period in which the replicate visits are conducted. So a unit must be either occupied or not for all visits in a season. This assumption initially appears problematic for carnivores with large range sizes, but may be relaxed. If a sample unit is included in the home range of one or more individuals, and if at least one individual has some non-negligible probability of being located in the sample unit at all of the visits, then this violation of the closure assumption is not a problem. Instead, detection probability is reinterpreted as the product of the probability of presence in the sample unit and the probability of detection, given presence (again the situation reflected in Equation 8.4). Detection probability estimates can be decomposed into these components under some designs (e.g. multiple detection devices at a sample unit, Nichols *et al.* 2008). In such cases of "random" use of the sample unit by a species, the occupancy parameter now reflects use rather than strict presence during the entire season.

The basic occupancy models assume homogeneity of occupancy and detection parameters for all sample units. This assumption can be relaxed, too, by grouping sample units into strata, by collecting covariate information on each unit, by using mixture models for detection probability, or, finally, by modeling patch-level detection probability directly as a function of individual-animal detection probability and number of animals in the sampling unit (Royle and Nichols 2003; MacKenzie *et al.* 2006; Royle and Dorazio 2008). Royle (2006) discussed the implications of heterogeneous detection probabilities in occupancy models.

Standard occupancy models assume no false-positive errors (e.g. an animal track misidentified to wrong species). This assumption can be met by never recording a detection without absolute certainty. Initial work on models to try to deal with false positives (Royle and Link 2006) suggests that such errors make the inference problem much more difficult. New models use ancillary data as an effective approach to dealing with false positives (Miller *et al.* 2011).

Spatial replication requires that spatial replicates within the sample unit be selected randomly and with replacement. A special case of spatial replication that is frequently used for carnivores that travel along trails involves surveying a length of trail for sign (tracks, scats) and breaking the length into segments that are treated as replicates. Because a single animal can travel along the trail for distances corresponding to several segments, the replicates may not be independent. Modeling animal presence at a segment as a spatial Markov process has been useful for large-scale tiger surveys in India (Hines *et al.* 2010; Karanth *et al.* 2011).

Finally, single-season occupancy studies have been extended to include multiple states, where "state" refers to some other attribute of the sample unit relevant to members of the target species (e.g. relative abundance, reproductive success, disease presence). An example is characterizing the state of a site as unoccupied (state 0), occupied with no successful reproduction (1), and occupied with successful reproduction (2). An investigator records the state based on extra information (e.g. if there is evidence of successful reproduction at the site). Evidence of reproduction is taken to mean that the sample unit is in the reproductive state with certainty. Non-detections and detections with no evidence of reproduction are characterized by state uncertainty. Models have been developed to permit estimation of the probabilities of patches being in these different states (Royle 2004; Royle and Link 2005; Nichols *et al.* 2007). A special case of the multi-state model deals with two species and characterizes patches as containing members of neither species, species A only, species B only, or both species. MacKenzie *et al.* (2004a) developed models for this situation and defined interaction parameters associated with lack of independence of occupancy itself and of detection probabilities. This model has seen few applications (Bailey *et al.* 2009) and is potentially useful in carnivore studies focusing on interspecific competition.

8.5.2 Multi-season models

Inferences about occupancy dynamics are obtained by surveying patches over several time periods (called "seasons") and using single-season sampling methods within each year. This approach yields data in the form of a robust design (Pollock 1982). The vital rates that bring about changes in occupancy are probabilities of local extinction and local colonization. The explicit dynamics models of MacKenzie *et al.* (2003, 2006) and Royle and Kery (2007) model occupancy dynamics as a Markov process in which occupancy state in season $t + 1$ depends on occupancy state in season t. Define ψ_t as probability of occupancy in season t, ϵ_t as the probability that an occupied site in season t is unoccupied in season $t + 1$, and γ_t as the probability that an unoccupied site in season t is occupied in season $t + 1$. The fundamental model for occupancy dynamics is then:

$$\psi_{t+1} = \psi_t(1 - \epsilon_t) + (1 - \psi_t)\gamma_t \tag{8.7}$$

If we shift to the view that the occupancy parameters reflect expected proportions of occupied sites, the above expression simply indicates that the proportion of occupied sites in season $t + 1$ is given by the sum of the proportion of sites occupied in season t that do not go extinct locally and the proportion of sites unoccupied in season t that become colonized by $t + 1$.

Multi-season occupancy models permit direct study of the processes underlying occupancy dynamics, whereas inferences based on the single-season models are about occupancy pattern. Investigating the influence of covariates on local rates of extinction and colonization is especially useful. For example, meta-population theory typically posits negative relationships between patch size and local extinction probability, and between distance to nearest source patch and local colonization probability (e.g. Hanski 1999; Hanski and Gaggiotti 2004). Such assumed relationships can be tested formally using occupancy modeling.

Trends in occupancy over time, important for conservation, can be modeled in multiple ways (MacKenzie *et al.* 2006). Multi-state models have also been extended to multiple seasons (MacKenzie *et al.* 2009), permitting, for example, inferences about effects of the presence of one species on the probabilities of colonization and extinction of a potential competitor.

Multi-season models require the same assumptions as single-season models. In addition, they assume similarity of local extinction and colonization probabilities across all sample units. This assumption can be relaxed through the use of site-specific covariates (e.g. associated with habitat) to model these extinction and colonization probabilities. Autologistic models can also be developed (Royle and Dorazio 2008), permitting colonization and extinction of a focal sample unit to be a function of the occupancy of neighboring patches. Of course, occupancy of neighboring patches is not a standard covariate, as it is only partially known because of the detection issue. Such models can be implemented using standard likelihood approaches or Bayesian computational approaches based on Markov Chain Monte Carlo (MCMC, e.g. see Royle and Dorazio 2008).

8.5.3 Software and study design

Computer program PRESENCE (Hines 2006) was developed specifically to fit single and multi-season models to detection/non-detection data. Many of these models can also be implemented in MARK (White and Burnham 1999). PRESENCE is the best source of newly developed models, although new models are usually implemented in MARK within a year of their appearance in PRESENCE. Royle and Dorazio (2008) showed how to implement a variety of occupancy models in WINBUGS (Spiegelhalter *et al.* 2009).

Several key investigations focus on the design of occupancy studies (MacKenzie and Royle 2005; MacKenzie *et al.* 2006; Bailey *et al.* 2007). Occupancy study designs require decisions about allocation of effort between more sample units and more replicate visits to each unit. For single-season studies, MacKenzie and Royle (2005) and MacKenzie *et al.* (2006) provided an expression (and associated table) for computing the number of replicate visits needed to attain a specified level of precision and a number of good suggestions about study design. Multi-season studies require attention to the design issues relevant to single-season studies (e.g. replicate visits per season, selection of sample units), as well as additional considerations. Program GENPRES (Bailey *et al.* 2007) aids in the design of multi-season occupancy studies, using simulation or expected value approaches to address sample size needed for adequate precision, model identification, and designs needed to detect changes in occupancy probabilities or covariate relationships for extinction and colonization.

8.6 Probability sampling of carnivore tracks to estimate population density

Animal tracks have been used as an index of abundance (Section 8.1), and abundance can be estimated using track counts. The method has been used for lynx (*Lynx canadensis*), wolves (*Canis lupus*), and wolverines (*Gulo gulo*) and is based on probability sampling of animal tracks in snow (Becker *et al.* 2004 detail the method).

Becker *et al.* (2004) referred to the method as "transect intercept probability sampling." Either aerial or ground counts of animal tracks may be used. The probability of encountering a track from a group of animals along a transect is estimated from the length and orientation of the track and is then used in a Horvitz Thompson estimator for a population total (Thompson 2002; Chapter 5). Data are augmented by counting tracks for a sample of radio telemetered animals.

8.7 Final thoughts

Because carnivores are increasingly threatened, biologists and managers invest great effort and resources to study their ecology and behavior, and are quick to adopt innovative equipment, field practices, and other physical tools. In contrast, researchers often do not apply adequate thought to hypotheses, to survey design, and to analyses of data. Because carnivores live at low densities and are difficult to study, this inadequate attention to research questions, design, and analyses often results in sparse or poorly collected datasets, which are then routinely shoe-horned into standard statistical packages, which may not be appropriate at all for the research goals. Often, the novelty of a new tool—an expensive camera-trap—dazzles an investigator to the point of forgetting that valid inference from camera-trap data requires a large number of traps and animal capture events.

Whether data collected in the field are "low-tech" counts of carnivore spoor or "high-tech" counts that come from camera-trap photos, DNA, or radio-telemetry, appropriate methods must be used to estimate demographic parameters. Carnivore science and conservation will advance much more rapidly if investigators apply the same healthy curiosity towards the newer approaches to model and estimate demographic parameters that they normally exhibit towards material tools and field craft.

Detection probability must be understood before one can interpret count statistics that arise from field sampling. Study design must include a plan for dealing with that fact that field methods detect only an unknown fraction of the animals present in any sampled area. One approach to dealing with detection is to assume constancy over the dimension of comparison (e.g. time, space, species), but this approach relies on untested assumptions that, when false, can completely invalidate conclusions. The preferred approach is to collect data that permit inference about detection probabilities and their potential variation over time, space, and species. This approach typically involves simultaneous modeling of the detection process and the ecological processes of interest and provides a basis for inference about variation in detection probability, permitting an informed decision about how best to treat these parameters in the analysis. If model selection and model-based tests lead to the inference that detection probability can indeed be viewed as constant over a dimension of interest, then a reduced-parameter model (obtained by equating detection parameters) can be used for inference.

Detection probability is important for making inferences about both state variables (e.g. abundance, occupancy) and vital rates (e.g. survival rate, rate of local extinction). Abundance, or population size, has been the traditional focus of studies in animal population ecology. Estimating abundance and associated vital rates (survival, reproduction, immigration, emigration) typically requires substantial field effort, often resulting in focus on a specific population and geographic location. A host of important questions in animal population ecology must typically be addressed with this state variable and its vital rates. Occupancy studies focus on the proportion of an area occupied by members of a species. This state variable is not equivalent to abundance, and the two quantities carry different types of information about a species. Occupancy studies are most appropriate for addressing macroecological questions (e.g. about range and range dynamics) at large geographic scales. They typically require less intensive field sampling than studies focused on abundance, and hence can be carried out over larger areas. Thus, the two state variables, abundance and occupancy, are not equivalent, and the selection of focal state variable for a particular study depends on the specific question(s) being addressed, which may depend on the resources available. Clear thinking about study objectives and hypotheses is critical.

The inference methods outlined in this chapter represent the current state-of-the-art, and rapid changes are expected. We anticipate substantial advances regarding spatially explicit capture–recapture modeling and methods that integrate data from different sources. Great potential exists to extend spatially explicit capture–recapture modeling to open capture–recapture models covering multiple seasons (Gardner et al. 2010; Royle and Gardner 2011). Such extensions will permit inference about dynamics of individual animal ranges over time, a topic for which only ad hoc approaches are now available. To date, most studies have used a single approach to data collection (true capture and recapture, telemetry, capture–recapture using photos, or using DNA identification) and subsequent inference. In the few studies where multiple methods have been used (e.g. radio-telemetry and camera traps), the data typically have been treated separately for analysis or combined in two-step approaches (Soisalo and Cavalcanti 2006; Dillon and Kelly 2008). Great potential exists for developing joint likelihoods that combine data and corresponding models from multiple sampling approaches. Combining data from camera traps and radio-telemetry in a single analytic framework (e.g. Powell *et al.* 2000 and Nasution *et al.* 2001, 2002 for capture–recapture and telemetry) is a natural step. Combining detection history data from camera traps and genetic analysis of scats also has great potential. Even in the case where individuals detected by the two approaches cannot be linked (i.e. individuals may or may not appear in both data bases), key parameters (abundance, survival, movement) could be estimated with greater precision using data from both sources simultaneously (e.g. Wen 2009; Wen *et al.* 2010).

9

Movements, home ranges, activity, and dispersal

Roger A. Powell

Mammalian carnivores move to find and to capture food, to avoid competitors, to avoid predators, to find mates, and to scent mark and otherwise communicate with conspecifics. Thus, moving is important. Yet, moving is energetically expensive, especially to capture live prey, making each move important. Carnivores are not nomads, moreover, and they do not move at random. They confine their movements in space and, through time, they develop home ranges, which are emergent properties. Home ranges for carnivores can be small, <1 ha for some weasels (*Mustela nivalis*, Nyholm 1959; Lockie 1966), or very large, 100s of km^2 for wolverines (*Gulo gulo*, Vangen *et al.* 2001) to 1000s of km^2 for polar bears (*Ursus maritimus*, Amstrup *et al.* 2000). Sometimes, a carnivore leaves its home range to find mates, to find food, or to find a new place to live, which can become its new home-range. Thus, movement is fundamental for carnivores and is basic to their way of life; good research investigating the movements and home ranges of carnivores can provide important insight into how carnivores live and why they do what they do.

The conceptual problems of understanding an animal's home-range must be faced before one can estimate or quantify that home range. Without knowing what one wishes to estimate, no estimate can be made. Equally important, the hypotheses being tested must be clear, data must be appropriate for the hypotheses, and home-range estimates must be appropriate for the hypotheses and data. We may never find completely objective, statistical methods that use field data to yield biologically significant information about animal's home-ranges (Powell 1987). Nonetheless, we must develop methods that are as objective and repeatable as possible while being biologically appropriate.

9.1 Research design

To be most productive, research requires good design based on good a priori hypotheses (Hayne 1978; Hurlbert 1984; Reynolds-Hogland and Mitchell 2007b; see Chapter 10). Researchers need to develop relevant, non-trivial hypotheses based on previous research and on relevant theory. Good, biological hypotheses should be developed from the synthesis of past research and theory. Good research seeks to learn *why* animals do what they do.

Biological hypotheses are relevant to why animals behave as they do and are based on biological knowledge. Use biological null hypotheses as the nulls for statistical tests (Brosi and Biber 2009). Thus, if the best biological information available predicts a particular outcome, then that outcome should be the null hypothesis for a statistical test because it is the expected outcome, even if it is not the outcome associated with "no effect." This approach is consistent with "Morgan's Canon," which argued against highly anthropomorphized interpretation of animal behavior but not against appropriate, biological interpretations and biological null hypotheses.

For example, large sexual dimorphism in body size is the norm across the Mustelidae. Were a new mustelid discovered in the future and, at first, only females collected, one would hypothesize that the yet-to-be-discovered males would be larger than the females. That should be the null hypothesis for statistical tests because large sexual dimorphism is expected. Now, if a few males are collected and are no larger than females, and the appropriate statistical analysis rejects the biological null hypothesis of large sexual dimorphism, then myriad, exciting hypotheses come to mind regarding why: sexual selection on males is not large, or the habitat is so unproductive that monogamy is required to raise young. In contrast, what would the result be for this scenario using the encrusted, traditional statistics and the null hypothesis of "no effect"? The result would be failing to reject the null. Besides not helping to get one's paper published in *Nature*, this result has two potentially dire problems. First, accepting the null hypothesis is generally not exciting and could cause one to miss the excitement of having found a mustelid without sexual dimorphism, and all the associated hypotheses to test. Second, had result been caused by a small sample size and insufficient power, then one might believe incorrectly that the new species is not sexually dimorphic when, actually, we do not know yet. Note that biological hypotheses are appropriate for frequentist (traditional parametric and non-parametric statistics), Bayesian, likelihood and information-theoretic approaches. Sober (2008) provided a solid background for choosing the approach best for given research hypotheses. Sober also discussed practical problems that arise when biological and statistical nulls differ.

Because a home range is an emergent property of movements, movements and home ranges should be used in research designed to answer different questions and to test different hypothesis. Home-range analyses should be used to understand how individuals space themselves on a landscape, how individuals place their home ranges on the landscape with respect to the distributions of resources, how that spacing changes with changes in neighbors and resources, how individuals might evaluate landscapes, and other questions that extend beyond simply the composite of all movements. Analyses of movements and individual locations should be used to address questions related to travel, to use of specific resources by individuals, to short-term responses of individuals to competitors, to short-term changes in specific resources, and for other questions that concern locations and specific movements. Being hierarchically related, home ranges and movements lend themselves to hierarchy theory to investigate how biological properties emerge across different spatial and temporal scales (Reynolds-Hogland and Mitchell 2007b).

Being enamored with new technology need not lead to the best research. Most research on carnivores is done remotely, using some system of telemetry (see Chapter 7). Direct observations, however, should be used whenever possible. Direct observation allows continuous data collection; shows the study animals on their landscape; allows multiple types of data to be collected simultaneously; allows observations of behaviors not under study; allows study of cooperative, dominance, and other within-group behaviors; and lends itself to serendipity. Tracking animals in snow (not completely direct) shows complete movement paths, including behaviors related to very small-scale aspects of habitat, and lacks only precise timing of behaviors. Telemetry should not supplant methods of direct observation, when the latter are appropriate. Often, however, direct observations limit data collection to one or a few individuals at once, while telemetry can allow data collection on many animals at once. Choice of a telemetry system must be matched to research questions and to the hypotheses being tested (see Chapters 2 and 7). Using relatively inexpensive VHF (very high frequency) transmitters on many animals is usually appropriate for answering population-level questions, while using expensive GPS (geographical positioning system) collars on a few animals is better able to provide data needed to answer questions about individual behavior.

The study of movements and home ranges of carnivores requires information about the landscapes on which carnivores move and establish home ranges. To understand movement, one must know not just how far and when an animal moved, but how the local habitat quality affected movements, whether prey densities caused the animal to linger or to travel through, whether local habitat features facilitated hunting, or whether a local competitor depressed prey availability.

9.2 Movements

All measures of location, activity, and movement lend themselves to testing pertinent hypotheses related to time of day, season, foraging, and use of resources, satiation or nutritional balance, proximity of competitors, predators or members of a social group, and more. Overlap of movement trajectories (paths) provides insight into social behavior, especially when paired with data on proximities and overlap of home ranges.

Animal locations can be estimated by observing animals directly, by capturing animals repeatedly in traps or photos, by following tracks, by collecting samples remotely, (for example, scats, hair) by using telemetry, or by using dogs to track scent (see Chapters 4, 5, 6, and 7). From knowing locations of animals, one can plot their movements from location to location. All animal locations have the potential for error (even GPS locations and maps). Location error affects one's ability to estimate movements.

Turchin (1998) reviewed animal movements and clarified many terms. He defined *movement* as "the process by which individuals are displaced over time," allowing passive transport. Although ecologists and conservation biologists study largely the self-propulsion of animals, passive transport can be important, such as a mother moving her offspring, a carnivore transporting parasites (see Chapter 13), or a polar bear being moved by drifting sea-ice (Mauritzen *et al.* 2003a). A *path* is the "complete spatio-temporal record" of an organism during a period of observation. A *move* is the path segment between two consecutive stopping places (rest sites, for example, or small food patches). And a *step* is the path segment, or the change of location, across a fixed period of time (as between two telemetry locations). These definitions (Turchin 1998), though not universally adopted, are used more consistently than the following terms related to movement. While *duration, direction, distance*, and *speed* of a movement are obvious, a *bout*, the period spent in a single behavior, has proved nearly impossible to define cleanly (e.g. Nams 2006). Path shape has been termed *tortuosity*, *sinuosity*, and *circuity*, and has been quantified in numerous ways and, therefore, lacks both a universal name and a universal way to quantify (reviewed by Favreau 2006). *Displacement* is the distance between an animal's locations at two times. Paths become *routes* when they are reused regularly or repeatedly. Paths and routes can connect sites with resources. Routes can become permanent for an individual or group, and can be used sequentially over time by different individuals and generations (e.g. Andersen 1991). *Dispersal* is usually a one-time behavior, usually during adolescence but sometimes during adulthood, when an individual leaves its mother's home-range or its established home range, to establish its own, new home-range. *Migration* is a repeated behavior, usually done annually in response to seasonal changes in resources. Many animals migrate to reach a site where resources will be relatively abundant during the coming season; some carnivores migrate by following the migrations of prey.

Depending on the hypotheses to be tested, location and movement data can be quantified in many ways. When location data are randomly or evenly collected, numbers of locations in different areas estimate time spent in those areas. Duration measures those time periods directly. Numbers of locations or duration may not correlate with importance of areas to an animal, however. Some resources, a water hole, for example, may be critically important but need be visited only for short periods (used seldom but important). Other resources, a resting site, for example, may be used for a long period but be no more valuable than dozens of other rest sites (used long but not important).

Distance, speed, and time, though obviously related through the defining equation for speed (speed = distance/time), provide different insights into behavior. Distances between telemetry locations are always biased low, because a researcher measures straight lines where an animal may have wandered. Location estimates close in time yield distance and speed estimates that are less biased than those spaced far in time because animals have less chance to wander. Analyses using distance and speed must be robust to such biases. Tracks in snow provide precise measures of distance but not of time or speed. Speed can sometimes be indexed from stride length or gait, but one must be careful to understand the assumptions of such indices. Speed can be measured with added data from telemetry (Powell 1979).

Path shape can be indexed from the ratio of actual distance travelled and straight line distance (Powell 1978; Bell and Kramer 1979; Spencer *et al.* 1990), from relative turn angles across a step (Bovet and Benhamou 1988; Socha and Zemek 2003; Benhamou 2004) or from other measures of angles and distance (Goss-Custard 1970; Caro 1976; reviewed by Calenge *et al.* 2009). Fractal analyses provide insight into important characteristics of movements and how movements build home ranges (Christ *et al.* 1992; Gautestad and Mysterud 1993, 1995; Loehle 1990; With 1994; Bascompte and Vilà 1997; Gautestad *et al.* 1998; Nams and Bourgeois 2004). Although Turchin (1996) argued that use of fractals is inappropriate across scales unless movements are clearly self-similar on different scales (e.g. Atkinson et al 2002), he noted that the fractal dimension, D, does provide a measure of path sinuosity. Fuller and Harrison (2010) showed that D can be used to reveal scale-dependent decisions for use of habitat.

In contrast, Benhamou (2004) argued that D should not be used to index sinuosity because it does not address distances of moves. His caution is pertinent only when one reports sinuosity in isolation.

Activity is the state of being active and is usually measured as the proportion of time spent active. Many telemetry packages now carry various add-ons that record whether an animal is active or not, from variation in signal strength, to activity switches, to accelerometers that provide three-dimensional information about an animal's posture and orientation (see Chapter 7).

Measures of movement rate, activity, and sinuosity of movements show different things and, generally, none should be used in isolation. A female black bear (*Ursus americanus*) with cubs moves at a moderate speed, is active for long periods at a time, but remains within a local area (small displacement) because her movements are highly sinuous (moderate path length, short moves, $D \geq 2$; Powell *et al.* 1997). Breeding male black bears, in contrast, switch between moving at high speeds over long distances nearly in straight lines (long path, large displacement, $D < 1$) while seeking estrous females and moving at slow speeds for long periods along sinuous paths while accompanying estrous females (moderate path length, moderate moves, $D > 1$). No single measure alone describes these movements appropriately; use several in conjunction.

9.3 Home range

Although Darwin (1861) noted that animals have home ranges, Burt's (1943: 351) definition of a mammal's home-range is the foundation of the general concept used today:

> ...that area traversed by the individual in its normal activities of food gathering, mating, and caring for young. Occasional sallies outside the area, perhaps exploratory in nature, should not be considered part of the home range.

This definition is clear conceptually but it is vague on points that are important for quantifying animals' home-ranges. Burt gave no guidance concerning how to quantify "occasional sallies" nor how to discern "the area" out from which the sallies are made. The vague wording implicitly and correctly allows a home range to include areas used in diverse ways for diverse behaviors. Members of two different species may use their home ranges very differently, with very different behaviors but, for both, the home ranges are recognizable as home ranges, not something different for each species.

Nonetheless, I am no longer convinced that Burt's definition is the best way to envision an animal's home-range. Burt's definition is limited to where an animal travels and visits regularly; it does not include areas that the animal knows but visits only rarely at critical times. Animals can be familiar with areas that they do not use regularly. During an autumn with a failed acorn crop, one female black bear and her cubs left what my research team and I considered to be her home range (from multiple years of study) to visit a ridge 10s of km away (Powell *et al.* 1997). From the direct line of this female's travel, I strongly suspect that she knew that distant ridge and knew of its ability to produce food in years when other places did not. If my interpretation of this female's behavior is correct, then I consider that distant ridge to have been part of her home range in other years, because she knew it and knew that she could use it to obtain food during a period of food shortage. An arctic fox (*Vulpes lagopus*) may be familiar with a 100-km^2 area yet use regularly only a small portion with concentrated food (Frafjord and Prestrud 1992). The fox can sample remotely the areas with potential to have food and determine which areas actually do have food at any particular time. Some carnivores can smell and hear over a kilometer and see a few hundred meters or more. They do not need to be physically in a specific small area to know whether berries are ripe, or whether deer have been bedding there regularly, or whether the vole density is high. Thus, an animal's home-range includes critical areas that it may not visit regularly and includes areas sampled remotely to learn if they have important resources.

In addition, Burt's definition is not clearly from the animal's perspective. It is *post hoc* and descriptive from a researcher's perspective. To understand best why an animal uses its home range as it does, and to understand what it gains from its home range, one needs to understand, as well as possible, how the animal itself perceives its home range.

Why do mammals maintain home ranges? Logically, home ranges provide the benefit of knowing where resources can be found. Models of animal movement predict that animals in heterogeneous environments benefit from "knowledge" of their environment (Saarenmaa *et al.* 1988; Folse *et al.* 1989; Turner *et al.* 1994; With and Crist 1996; South 1999; Stillman *et al.* 2000; Moorcroft and Lewis 2006; Dalziel *et al.* 2008; Van Moorter *et al.* 2009; Spencer in press). Such knowledge can affect an animal's fitness. Adult or juvenile carnivores in familiar territory have lower mortality and higher reproduction than juveniles dispersing through unfamiliar landscapes (Blanco and Cortes 2007; Gosselink *et al.* 2007; Soulsbury *et al.* 2008).

Gaining knowledge of a home range requires time, leading to site fidelity, and site fidelity has been used to define whether an animal has established a home range (e.g. Spencer *et al.* 1990; Swihart *et al.* 1988). Unfortunately, such definitions sometimes fail to define home ranges for animals that exhibit true site fidelity, when model assumptions do not match animal movements (Powell 2000).

Mammals create spatial maps using their hippocampus (O'Keefe and Nadal 1978; Peters 1978; Fyhn *et al.* 2004; Sargolini *et al.* 2006; Leutgeb *et al.* 2007; Kjelstrup *et al.* 2008; Pastalkova*et al.* 2008; Solstad *et al.* 2008) and hippocampus size varies with relative selection pressures on cognitive mapping abilities and spatial memory (Krebs *et al.* 1989; Jacobs and Spencer 1994; Galea *et al.* 1996; Clayton *et al.* 1997). Animals plan movements and their hippocampus are sensitive to where they find themselves within their environments (Kjelstrup *et al.* 2008; Solstad *et al.* 2008). In addition, an animal's movements should depend on its nutritional state; resources with low travel costs or that balance the diet should have added value. I showed that fishers appeared to rank prey by energy return (Powell 1979), consistent with optimal foraging models (Charnov 1976a, 1976b; Pyke 1984); those ranks should not change through time but should lead to predictable changes in diets as prey populations change. Thus, a carnivore should update its map for areas with low-ranked prey only when high-ranked prey are rare.

A carnivore has an instantaneous concept of its cognitive map, changing as the animal learns new characteristics of its environment. A researcher, in contrast, learns of changes in an animal's cognitive map only by identifying changes in how the animal uses space over time. A researcher deduces an animal's home-range from locations of the animal and from other data collected over the time required for the animal to visit representative parts of its home range (but see Doncaster and Macdonald 1991). Thus, for most research, a home range must be defined for a specific time interval, e.g. a season, a year, or possibly a lifetime (Fieberg and Börger in press). The longer the interval of time, the more data can be used to quantify the home range, but also the more likely that the animal has changed its cognitive map since the first data were collected. Consequently, certain time frames make no sense for a home range (daily home-range, for example, and perhaps lifetime home-range, in the sense that no animal still uses all sites at the end of its life that it used early in its life).

A related problem is whether animals define boundaries for their home ranges. Researchers must include some radius of familiarity, or perception, around an animal's locations or travel routes in data analyses. If they do not, home range is reduced, *reductio absurdum*, to the places where an animal actually placed its feet, which, clearly, is not satisfactory. The edges of home ranges can be diffuse (Gautestad and Mysterud 1993, 1995), making the area of a home range undefined. This ambiguity does not reduce, however, the importance of a home range to its animal. And for analyses, even crudely estimated areas for home ranges provide insights into animal behavior and ecology. Nonetheless, many carnivores seldom use the edges of their home ranges and, except for territorial carnivores, and carnivores that use physical features of the landscape dictate boundaries (Powell and Mitchell 1998), many carnivores may actually have no real boundaries for their home ranges. After all, they spend the vast majority of their time elsewhere. Consequently, researchers must remember that the boundaries they calculate may mean absolutely nothing to the animals themselves. In addition, area may be the least interesting statistic of home ranges, because it probably means nothing to many animals themselves. Cognitive maps may not have clear edges.

In the end, a carnivore's cognitive map of its home range must allow it to make decisions that affect its fitness, such as where to hunt next for food, how to reach that hunting site while minimizing chances of becoming someone else's prey, and what parts of its home-range overlap with that of a potential mate.

Therefore, I propose that a home range is that part of an animal's cognitive map that the animal chooses to keep up-to-date with the status of resources (including food, potential mates, safe sites, and so forth) and where it is willing to go to meet its requirements. Mammals can sample and update many resources remotely using at least smell, hearing, and sight.

How can we gain insight into an animal's cognitive map, that is, into its own perception of its home range? Aldo Leopold (1949: 78) wrote: "The wild things that live on my farm... frequently disclose by their actions what they decline to divulge in words." We must use the actions of the animals to gain insight into their home ranges, which requires good research design. The rewards of such research will be new knowledge of how carnivores use space related to resources, how such use relates to sex, maturity, experience, social status, nutritional and physiological condition, etc., and, therefore, why carnivores use space as they do.

9.4 Territories

A territory is an area within an animal's home-range to which the animal has exclusive, or perhaps priority, use. The territory may be all, or just a part, of the animal's entire home-range. Thus, a territory is a special type of home range or a particular part of a home range.

An animal maintains a territory only when it has a limiting resource, a resource that is in short supply and that limits reproduction or survival and, thus, population growth (Brown 1969). In consequence, territorial behavior does not limit populations but, rather, limits on individuals within populations stimulate territorial behavior. Theoretical work (Brown 1969; Fretwell and Lucas 1970; Maynard Smith 1976; Watson *et al.* 1979) has shown generally that territoriality, at most, regulates populations proximally.

Most carnivores that maintain territories do so year-round, and food is the most common limiting resource (Rogers 1977, 1987; Sandell 1989; Palomares 1994; Powell *et al.* 1997). For territorial individuals, territory size tends to vary inversely with food availability (Powell *et al.* 1997). For solitary carnivores, the limiting resource for males may be females during the breeding season (Erlinge and Sandell

1986; Sandell 1989; Powell *et al.* 1997). For cooperatively breeding animals in general, theoretical (Powell 1989) and empirical (Jenkins and Busher 1979; Woolfenden and Fitzpatrick 1984; Walters *et al.* 1988, 1992; Powell and Fried 1992) work suggest that the limiting resource is usually some aspect of breeding and living space, such as European badgers' sets (*Meles meles*, Kruuk 1989; Doncaster and Woodroffe 1993) or, for wolf packs (*Canis lupus*), the land within the territory with its prey (Mech and Boitani 2003). Whether individuals, mated pairs, or families defend territories appears to depend not just on the productivity of limiting resources, but also on predictability and fine-grained vs. coarse-grained patchiness (Bekoff and Wells 1981; Macdonald 1981, 1983; Kruuk and Parish 1982; Macdonald and Carr 1989; Powell 1989; Doncaster and Macdonald 1992). Wolff (1989, 1993) warned that food may appear superficially to be the limiting resource when the limiting resource is actually offspring.

Individuals may maintain territories in only part of a species' range, female black bears, for example (Rogers 1977, 1987; Garshelis and Pelton 1980, 1981; Powell *et al.* 1997), and may abandon territorial behavior if the limiting resource increases in abundance. Territorial behavior by female black bears can be predicted from variation in the productivity of food (Powell *et al.* 1997). Such flexibility exists because a territory must be economically defensible (Brown 1969) or economical to maintain (Spencer in press). For territories that are defended, when productivity of the limiting resource is very low, then the costs of territorial behavior are not returned through exclusive access to that limiting resource; when productivity is high enough that the resource can be shared, then the costs are an unneeded expense (Carpenter and MacMillen 1976; Powell *et al.* 1997). If productivity increases, an animal can decrease territory size. If the environment is patchy, however, reducing territory size might require the loss of a large resource patch that will, in turn, drop the territory's resources below the minimum required. Under these conditions, home-range overlap should be allowed (Powell *et al.* 1997). An individual or social group might maintain exclusive access only to those parts of its home range with the most important resources, as exhibited by some coyotes (*Canis latrans*, no overlap at den sites and central, forested areas, Person and Hirth 1991). Alternatively, an individual might allow territory overlap with a member of the opposite sex, intrasexual territoriality (Powell 1979, 1993a, 1994; Sandell 1989). Members of species with intrasexual territoriality generally exhibit large sexual dimorphism in body size, males are polygynous, and females are selectively polyandrous; females raise young without help from males; and the large body sizes and large territories of males may be considered a cost of reproduction (Seaman 1993; Yagamuchi and Macdonald 2003). For species that affect food supplies mostly through resource depression (i.e. have rapidly renewing food resources, such as prey that become wary when they perceive a predator and later relax; Charnov *et al.* 1976a, 1976b), the cost of intrasexual territoriality appears to be minor because the limiting resource renews. This cost may be imposed on females by males (Powell 1993a, 1994).

Thus, territorial behavior is not necessarily a species' characteristic. If environmental conditions leading to territorial behavior remain stable for long periods, however, flexible behavior can be lost as behaviors dependent on territoriality are selected and become fixed. Whether territorial behavior is flexible remains to be tested in most carnivores.

If defended, territories are usually defended with scent marks, calls, and displays (Kruuk 1972b, 1989; Peters and Mech 1975), which are safer and more economical than tooth and claw, and are evolutionarily stable (Maynard Smith 1976; Lewis and Murray 1993). Calls and overt displays are immediately transient and disappear immediately. Scent marks change over time because different components have different volatility. Thus, scent marks provide information on multiple timescales. Individual identification, dominance, or social status, and time since last visit, may all fade from a scent mark at different speeds. Scent marks may also be visual marks (or environmental displays); for example, raised leg urinations on snow in winter and hind-leg scrapes by wolves (Peters and Mech 1975).

Members of many mammal species, but few if any carnivores, defend individual territories against all conspecifics. Many solitary carnivores defend only intrasexual territories (Powell 1979; Sandell 1989; Johnson *et al.* 2000). Members of some other species defend territories as mated pairs (coyotes, Person and Hirth 1991), and still others as extended family groups (red foxes, *Vulpes vulpes*, Macdonald 1979b; wolves, Mech and Boitani 2003; meerkats, *Suricata suricata,* Madden *et al.* 2009), sometimes containing non-family members (Mech 1970; Rood 1986; Kruuk 1989; Madden *et al.* 2009). Where area-specific knowledge improves foraging success or where home-range overlap with conspecifics leads to resource depletion or depression (Charnov *et al.* 1976a, 1976b), avoiding areas of home-range overlap may lead to territories without need of defense (Spencer in press). If such conditions do not change over evolutionary time, selection should favor behaviors that promote minimal home-range overlap, behaviors such as scent marking or calls. Consequently, such advertisement of territories need not be territory defense.

One might hope that territorial behavior would eliminate the problem of home-range boundaries being vague or meaningless, yet we are not so lucky. The wolves that Peters and Mech (1975) followed responded to the scent marks of neighboring wolf packs by marking at high rates. The alpha male of a pack (the dominant male and father of subordinate pack members) even intruded a couple hundred meters into the territory of a neighboring pack to scent mark. Consequently, a territory boundary became a fuzzy space a few hundred meters wide with an amalgam of scent marks, and not a distinct boundary. Distinct boundaries of territories may at times be no easier to identify than are boundaries of undefended home-ranges. Apparent overlap of territories seen in telemetry data may be due to telemetry error but, more likely, are real.

9.5 Estimating animals' home-ranges and territories

Quantifying an animal's home-range uses data that describe the animal's use of space to deduce, or to gain insight into, the animal's cognitive map of its home (Peters 1978). A home-range estimator should delimit where an animal can be found with some level of predictability, it should quantify the animal's probability of being in different places, and it should quantify the importance of different places to the animal. The data on the animal are usually observations, indirect or remote locations, or tracks; remote locations include those gained from telemetry, cameras, track plates, hair snares, and other sources of genetic identification. Because we have, at present, no way of learning directly what a free-living animal perceives as its cognitive map of its home, we have no perfect method for quantifying home ranges.

Many methods for quantifying home ranges provide little more than crude outlines of where an animal has been located. Drawing the smallest convex polygon possible that encompasses all known or estimated locations for the animal (minimum convex polygon or MCP, Hayne 1949), is conceptually simple. Problems with the method, however, are myriad (see my review, Powell 2000). Most importantly, to construct a minimum convex polygon, a researcher discards 90% of the information included in the data he worked so hard to collect and keeps only the extreme data points. More than any other, this method emphasizes only the unstable, possibly imaginary, boundary properties of a home range and ignores the internal structure of the home range and the central tendencies, which are more stable and are important for critical questions about animals. Minimum convex polygons can, however, be useful to define availability for third-order resource selection (Horne *et al.* 2009).

For questions that relate to understanding *why* or *how* an animal has chosen to live where it has, estimators are needed that provide more complex "pictures." An animal's cognitive map will have incorporated into it the importance of different areas. The most commonly used index of that importance is the amount of time the animal has spent in the different areas in its home range. For some animals, however, this index does work (remember water holes and rest sites). No standard approach exists to weight use of space by a researcher's understanding of importance but a number of options exist, such as weighting locations by times between locations and weighting locations by time periods (Katajisto and Moilanen 2006; Fieberg 2007a).

From location data, most home-range estimators produce a "utilization distribution" describing the intensity of use of different areas by an animal. A utilization distribution is often confused with a "utility distribution," a concept borrowed from economics that assigns a value (the "utility") of including a place within an animal's home-range (Ellner and Real 1989). A utilization distribution is derived statistically from animal locations and describes how an animal has used space (estimated from data), whereas a utility distribution is unknown and is probably related to an animal's cognitive map. Utility distributions need not be probability density functions, though they usually are and can easily be transformed to probability density functions.

A utilization distribution, calculated as a probability density function, describes the probability that an animal has been in any part of its home range (Hayne 1949; Calhoun and Casby 1958; Jennrich and Turner 1969; van Winkle 1975; White and Garrott 1990), which provides one objective way to define an animal's "normal" activities. One can arbitrarily but operationally define "home range" as the smallest area that accounts for a specified proportion of its total use. Most biologists use 0.95 (i.e. 95%) as their arbitrary but repeatable probability level: the smallest area with a total probability of use equal to 0.95 is defined as an animal's home-range (the area for "normal" activities). A strong statistical argument exists for excluding some small percentage of the location data, of the utilization distribution, or both: extremes are not reliable and tend not to be repeatable. This argument does not specify, however, that precisely 5% should be excluded. Using 95%, home ranges may be widely accepted because it appears consistent with the use of 0.05 as the (also) arbitrary choice for the limiting *p* value for judging statistical significance. Indeed, the 95% home-range for an animal depends on sample size. Using two different datasets, especially with different sample sizes, both collected from a single animal during a specified time period, will produce two different, though similar, 95% utilization distributions (Figure 9.1; Fieberg 2007a; Kochanny *et al.* 2009; Fieberg and Börger in press).

Some biologists assume that exploratory behavior ("occasional sallies") is excluded by calculating 95% home-ranges, but inspection of real data suggests that this assumption is wrong. Exploration often produces small clumps of locations with probabilities of use >5%. An alternative approach for eliminating "occasional sallies" is to exclude from analyses any isolated locations or clumps of locations. This approach is more likely to eliminate explorations. The clumps

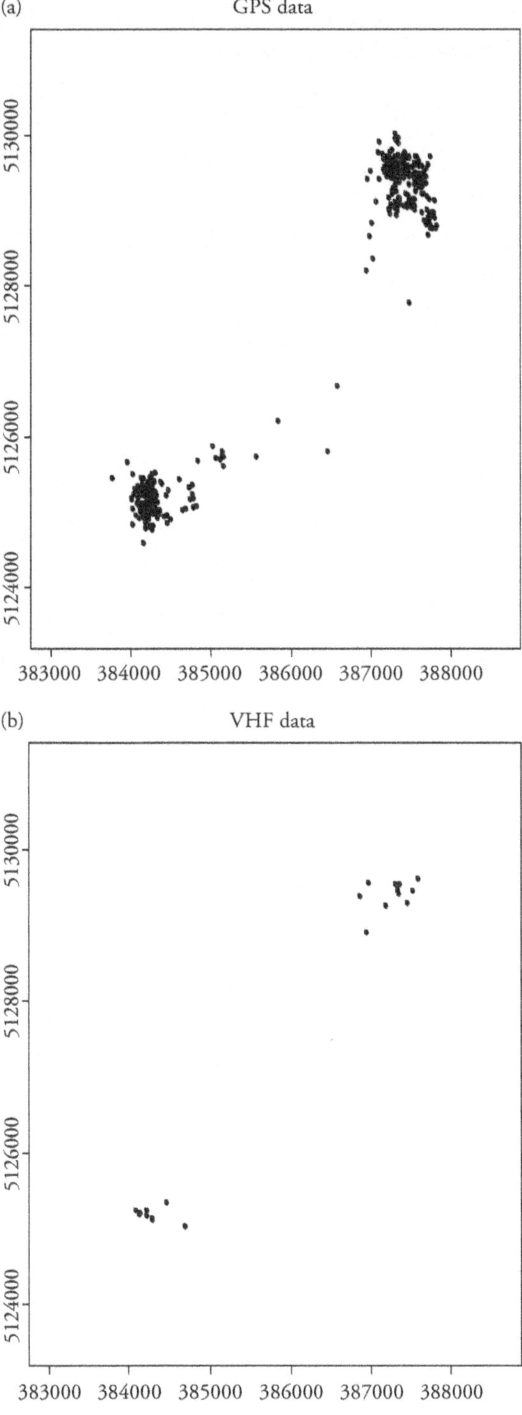

Fig. 9.1 (a) Diurnal locations of an adult female white-tailed deer (*Odocoileus virginianus*) collected by Global Positioning System (GPS) telemetry 15 February to 12 May 1999, Camp Ripley, Little Falls, Minnesota. (b) Locations of the same deer and the same time, collected using very high frequency (VHF) telemetry. (c). Home range of the deer depicted as 95% contours from a fixed kernel density estimator applied to the GPS data (wide, grey lines) and VHF data (think, black lines). From Fieberg and Börger (in press).

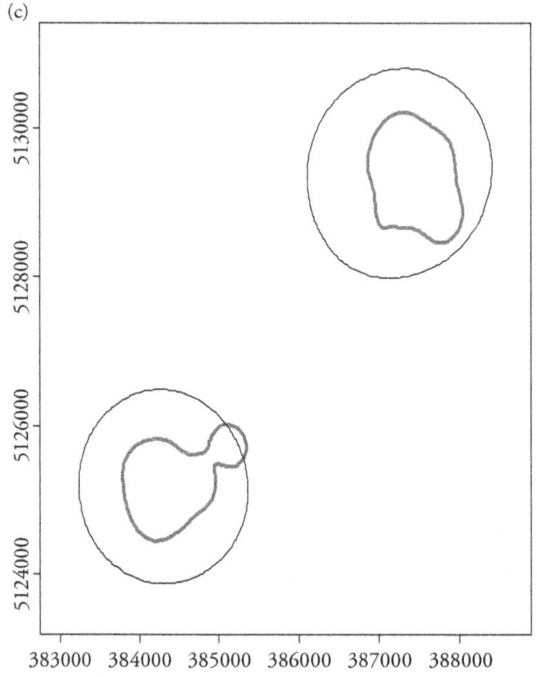

Fig. 9.1 Continued

might not constitute 5% of the data, but is not necessarily a problem. This approach requires researchers to inspect their data closely, which is a good thing.

Data used to estimate a utilization distribution are sequential locations of an individual animal and are seldom independent samples of the true distribution. Lack of independence is not great problem for most analyses (Lair 1987; Powell 1987; Andersen and Rongstad 1989; Gese *et al.* 1990; Reynolds and Laundrè 1990; Fieberg 2007b). In addition, data that are not statistically autocorrelated are, nevertheless, always biologically autocorrelated anyway, because animals use knowledge of their home ranges to determine movements.

The simplest way to build a utilization distribution is to superimpose a grid on one's study area and represent a home range as those cells in the grid having an animal's locations, each cell having a density as high as the number, or proportion, of times the animal was known or estimated to have been within that cell (e.g. Horner 1986; Zoellick and Smith 1992). This approach may seem outdated, given the many sophisticated home-range estimators available; nonetheless, its simplicity and its direct use of data make it a good choice for many analyses. It should be considered for GPS telemetry data because it avoids many problems associated with kernel home-range estimators. Doncaster and Macdonald (1991) estimated the home ranges of red foxes as a retrospective count of the grid cells known to be being visited at any one time. This approach is equivalent to treating the cells as marked individuals for a mark–recapture study and estimating home-range size (population size of the cells) from a minimum number known alive approach (Krebs 1966). Calculating back to any time, a fox's home-range included those cells that had been visited before that time and that would be visited again. This approach allowed Doncaster and Macdonald to follow foxes' home-ranges as they drifted across the landscape. More sophisticated survival estimators could be applied to estimate the rates at which cells were lost from home ranges and new cells added (Doncaster and Macdonald 1996). With this approach, "occasional sallies" are easily identified as those cells visited only once.

Choosing cell size is a major problem for most analyses using grids (Vandermeer 1981). For data on animal locations, cell size should incorporate, in some objective way, information about error associated with location estimates for telemetry data, information about the radius of attraction for trapping data, information about the radius of an animal's perception, the scale of habitat data, and knowledge of the appropriate scale for the hypotheses being tested. For some comparisons, cell size must be equal for all animals, for some others, cell size relative to home-range size must be equal. Because cell size is related to the scale of the behaviors being studied, changing cell size can change results of analyses (Lloyd 1967; Vandermeer 1981). The smallest cell size may be dictated by telemetry error (especially with VHF telemetry) but the ecological processes of the hypotheses being tested dictate the appropriate cell size. Choosing cell size is an application of hierarchy theory.

Kernel density estimators are used widely now to estimate home ranges (Laver and Kelly 2005). These estimators produce unbiased density estimates directly from data and are not influenced by grid size or placement (Silverman 1990). A kernel estimator produces a utilization distribution in a manner that can be visualized as follows. On an x–y plane representing a study area, cover each location estimate for an animal with a three-dimensional "hill," the kernel, whose volume is 1 and whose shape and width are chosen by the researcher. The utility distribution is a surface resulting from the mean at each point of the values at that point for all kernels. In practice, a grid is superimposed on the data and the density is estimated at each grid intersection as the mean at that point of all kernels. Kernel-generated utilization distributions are, basically, smoothed grid cell distributions.

The width of the kernel, called the "band width" (or "window width" or "h"), and the kernel's shape might hypothetically be chosen using location error, the radius of an animal's perception, and other pertinent information. Band width can be held constant for a dataset (fixed kernel); or band width can be varied (adaptive kernel), such that data points are covered with kernels of different widths ranging from low, broad kernels for widely spaced points to sharply peaked, narrow kernels for tightly packed points. Although adaptive kernel density estimators were expected, intuitively, to perform better than fixed kernel estimators (Silverman 1986), they tend to overestimate use of peripheral areas (Seaman and Powell 1996). Kernel shape has little effect on the output of the kernel estimators, as long as the kernel is hill-shaped and rounded on top (Silverman 1986), not sharply peaked (deduced from Gautestad and Mysterud, personal communication). Kernels with infinite tails overestimate use of peripheral areas. No objective method exists at present to tie band width to biology or to location error, except that band width should be greater than location error (Silverman 1986).

Choosing band width is a critical yet difficult aspect of developing a kernel estimator for animal home-ranges (Silverman 1986). The optimal band width to minimize statistical error is known for data that are approximately bivariate normal, but animal location data seldom approximate normal distributions. For distributions that are not normal, band widths can be chosen using least squares cross-validation (Seaman and Powell 1996) and maximum likelihood cross-validation (Horne and Garton 2006a). Alternately, *ad hoc* methods for choosing band width can be based on the biology of the animals being studied, such as the speed that animals can move (Katajisto and Moilanen 2006) or on a researcher's goals, such as making home ranges continuous (Kie *et al.* 2002; Berger and Gese 2007; Jacques *et al.* 2009; reviewed by Kie *et al.* 2010). Choosing kernel width involves a tradeoff (Fieberg 2007b). Narrow kernels reveal small-scale details in the data, and, consequently, tend also to highlight measurement error (telemetry error or trap placement, for example). Wide kernels smooth sampling error but also hide local detail. The result is that, to reduce error, either type of cross-validation tends to choose large band-widths for small datasets, which leads to extensive smoothing (Figure 9.1). Large datasets for the same animals produce similar utilization distributions but with less smoothing, while very large datasets can produce highly disjunct utilization distributions that are under-smoothed. Finally, different software packages use different algorithms for least squares cross-validation and, therefore, often choose vastly different band widths for a single dataset.

For some questions, the optimal band-width should be chosen for each home range, while for other questions, all home ranges may need to be estimated using the same band width. Using a single band-width for all home ranges reduces over-smoothing of small datasets and under-smoothing of large datasets. In the end, band width must be chosen to fit the hypotheses being tested, the datasets, and other research goals.

Seldom are distributions of the use of space by animals, especially carnivores, so "standard" as to allow researchers to apply software packages for kernel home-range estimators blindly. When animals reuse den sites repeatedly, locations at den sites should be removed from datasets before utilization distributions are estimated, then reinserted with appropriate probabilities assigned. Kernel estimates of home ranges sometimes incorporate areas obviously not used by the animals being studied. Lakes, for example, may be avoided or not used at all by some terrestrial carnivores, yet will be included in home ranges, especially if the research animals forage close to the water. One can reset probabilities of use to 0 for non-habitat and recalculate probabilities for areas that can be used.

Local convex-hull estimators are an important alternative to the widely used kernel estimators, especially when use of space has sharp boundaries (Getz and Wilmers 2004; Getz *et al.* 2007). As originally presented, convex-hull kernels were constructed from the $k - 1$ nearest neighbors of a focal location (Getz and Wilmers 2004). In more recent developments, kernels are constructed from all points within a fixed radius of each location (Getz *et al.* 2007). Local convex-hull estimators outperform traditional kernel estimators when habitat boundaries that affect movements are distinct but not when true use of space varies in a smooth manner (Lichti and Swihart 2011).

Brownian bridge estimators are the best-known estimators that do not ignore the time-sequence information that is available with most data on animal locations. All other estimators assume that all location data points are independent and that time-sequence information is irrelevant (Johnson *et al.* 2008, however, have a model of "use" that incorporates serial correlation of location data.). Brownian bridge estimators add ridge-shaped kernels between consecutive location estimates, with the heights, widths, and shapes of the tunnels and hills, dependent on the time and distance between locations (Bullard 1999; Horne *et al.* 2007a). The ridge-shaped kernels are calculated from an animal's distribution of travel speeds and the distribution of potential random routes between two locations. When locations are independent, a Brownian bridge estimator adds nothing to the traditional kernel estimator it modifies.

When data are so abundant that many locations are autocorrelated, allowing many Brownian bridges to be built, then traditional kernel estimators perform nearly as well as Brownian bridge estimators to estimate home ranges. On a simulated landscape with several peaks in resource abundance, I programmed detailed, constrained random movements of animals that were attracted to the resource peaks, yielding utilization distributions that resembled real home-ranges of carnivores. Simulations represented a year with 3-min steps; the animals slept and foraged and avoided non-habitat. I then simulated telemetry datasets by sampling the movements at intervals averaging from once every 6 days to once every 3 min for a year. Estimated home-range sizes were extremely similar for each sampling interval, as were measures of error from the true home-ranges (Figure 9.2). For small sample sizes (long sampling intervals), the Brownian bridge estimator actually averaged slightly larger mean squared error than the standard kernel estimator, while the Brownian bridge estimator showed only minuscule improvement over standard kernel estimator at large sample sizes. Horne *et al.* (2007) had similar results using data for black bears. These estimators are most useful when research goals include identifying specific travel routes, such as road crossings and migration (Horne *et al.* 2007; Sawyer *et al.* 2009).

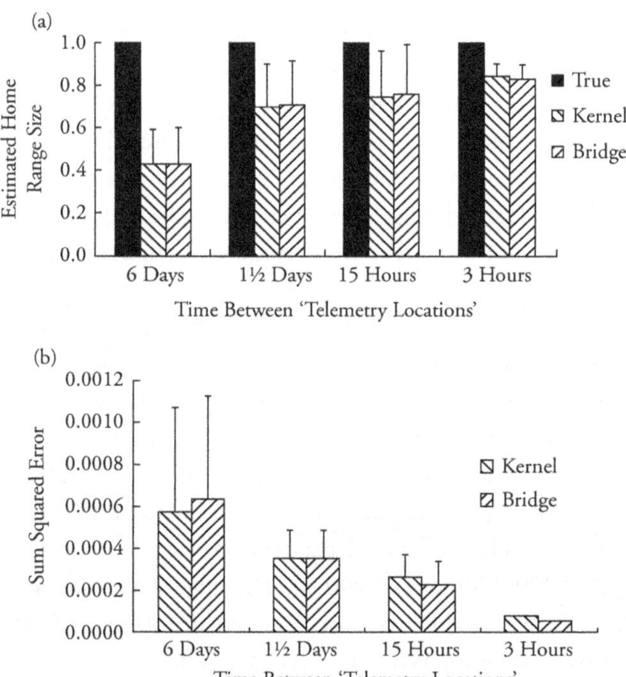

Fig. 9.2 Brownian bridge estimators add little to most home range analyses; home-range area and utilization distributions differ little. Both become more accurate with larger samples. The figure relates to simulations that represent a year with 3-min steps; animals slept and foraged and avoided non-habitat. (a) Mean areas of true home-ranges, of home ranges estimated using a fixed kernel estimator, and using a Brownian bridge estimator. Home ranges calculated as 95% contours from simulated telemetry samples taken once every 6 day, every 1 day, every 15 h and every 3 h. (b) Mean squared error for the kernel and Brownian bridge home-ranges (compared to the true home-ranges).

All the estimators discussed so far estimate the probability that an animal will be in any part of its home range but do not estimate how important that part of the home range is to the animal. For researchers asking questions related to time or for researchers studying animals for whom time and importance coincide, this limitation causes no trouble. This limitation is a problem for researchers interested in the underlying importance of habitats or landscape characteristics when time and importance may not coincide. If time and importance do not coincide, kernel estimators (and all other estimators) will not estimate importance accurately.

Model-supervised kernel smoothing (Matthiopoulos 2003) allows researchers to incorporate other information into kernel estimates of home ranges. This approach calculates a weighted combination of two estimators: one, a standard kernel estimator for an animal's home-range and the other a model of space-use using other information (such as boundaries of lakes, habitats known to have high densities of food, distinctive travel routes). Maximum likelihood cross-validation calculates the relative weights of the two estimators. Horne *et al.* (2008) developed a model with a similar goal of incorporating other information but that modifies the kernel estimator. By incorporating information about important resources, the final utilization distribution will be better than that from the tradition kernel density estimator (Matthiopoulos 2003). Using model-supervised smoothing requires programming skills because no estimator is publicly available.

Another approach to addressing the importance of different places, is to map a landscape depicting different resource qualities, such as food, habitat components that facilitate prey capture, den sites, or the landscape of fear (Brown *et al.* 1999). One can then test for correlation between these resources and time spent where the resources are of different quality, or test which landscapes, or combination of landscapes, best predicts time spent in different places. This approach does not, however, identify limiting resources that need be obtained infrequently.

Mechanistic approaches to home ranges incorporate resources and constraints into models of animal movements, thereby predicting home ranges (Moorcroft and Lewis 2006). The models incorporating the most critical resources and constraints predict home ranges closest to those documented for the target animals. Mechanistic models of coyote territories in Yellowstone National Park, 1991–93, show that prey abundance, topography, and the presence of other coyotes were critically important to how coyotes use space (Moorcroft *et al.* 2006; Moorcroft and Lewis 2006). An alternative mechanistic approach is to model the value of resources for a carnivore, then predict home ranges incorporating those resources using an optimal foraging approach to use of space (Mitchell *et al.* 2002; Mitchell and Powell 2003b, 2007). Black bears in the Southern Appalachian Mountains, USA, establish home ranges that minimize the area needed to meet food requirements, allowing for modest resource depression by other bears. This approach does not model the distribution of movements but, rather, the optimal space to occupy, given that movements are restricted by the cost of travel.

Another approach is to build home ranges (utilization distributions) using other resources as currencies (Powell 2000, 2004; Powell and Mitchell in press). For example, build a utilization distribution that represents not time but, for example, represents how an animal spreads energy expenditure across its landscape, or how it spreads energy gain, or the potential to capture prey. To build alternate home-ranges, one can place a kernel on each animal location but weight the kernels by, for example, rate of energy expenditure for the animal when the location was collected. Ultimately, one would like to build a distribution of fitness gained (Powell 2004). This approach does have the potential to identify critically limiting resources that are seldom used, but only if the biology of the study animal is well known.

With this myriad of potential home-range estimators, how can one choose which to use? Often, the hypotheses being tested will limit the choices. In addition, information-theoretic approaches can be used to choose the best estimator (Horne and Garton 2006b based on the hypotheses.

9.6 Home-range cores, overlap, and territoriality

9.6.1 Home-range cores

Particular parts of an animal's home-range must be more important than other parts. In general, foods and other resources are patchily distributed (e.g. Curio 1976; Goss-Custard 1977; Frafjord and Prestrud 1992; Powell *et al.* 1997) and, therefore, those parts of a home range with high densities of critical resources ought, logically, to be more important than areas with few resources. Trying to identify as the core that part of an animal's home-range that is most important has a long history (e.g. Burt 1943; Kaufmann 1962; Ewer 1968; Samuel *et al.* 1985). Identifying cores, if they exist, could be important for understanding home ranges.

Understanding home-range cores has two parts (Powell 2000). First, a core is used more heavily than the apparent "clumps" of heavy use that appear from random use of space. This first part leads to the second: a core, therefore, can not be defined by an *ad hoc* level of use (i.e. a 25% home range, or all areas used more than average) but must be determined by the animals themselves. The literature is filled with definitions of cores that are *ad hoc* or subjective. Were an animal to use its home range randomly or in an even fashion, the home would have a core by these definitions. The consequence is that using an *ad hoc* definition can lead to spurious results and lead a researcher to identify behavior with a random pattern as being significant. To have a biological meaning, a core must have a definition that has a mechanistic biological interpretation, not an *ad hoc* definition.

Seaman and I (1990; Powell *et al.* 1997; Powell 2000) introduced a technique for identifying cores that is objective, not arbitrary, and that allows the animals themselves to define their cores (Box 9.1). Our technique is based on the logic that has been used to identify behavior bouts (Fagen and Young 1978; Slater and Lester 1982). Bingham and Noon (1997) used the same method and Harris *et al.* (1990b) may have, but their explanation is not clear.

This criterion for a home-range core clearly identifies the most intensively used areas within an animal's home-range, and it allows the data (i.e. the animal) to determine the boundary between core and periphery. The criterion is objective and intuitive, showing that some animals have large cores and some small (Seaman and Powell 1990; Powell 2000).

Box 9.1 Home-range cores.

To identify an animal's home-range core, plot each value for each cell of the home range onto an x–y graph. Each cell's y-value is its value for cumulative percent home-range (100%, 95%, 90% down to 0%). That is, take a home range plotted as percent home-ranges and each cell's y-value is the value of the contour line that runs through the cell. Each cell's x-value is its probability of use, expressed as a percent of the maximum probability of use. Thus, the cell with the highest probability of use has x-value 100%, and the cell with the least non-zero probability of use has x-value of nearly 0. Both x- and y-axes on the graph range from 0 to 100 (Figure 9.3). Were an animal to use space within its home range randomly, all points would fall on a straight line going from $x = 0$ for the 100% home range and $x = 100$ probability at 0% home range with slope = −1 (Figure 9.3a). If use of space by an animal is clumped, however, the points will sag below the line of random use (Figure 9.3b) and if use of space is even (all areas used with equal intensity), the points will remain as a high plateau from x equals 0 probability of use up to x equal some large probability of use and then plummet (Figure 9.3c).

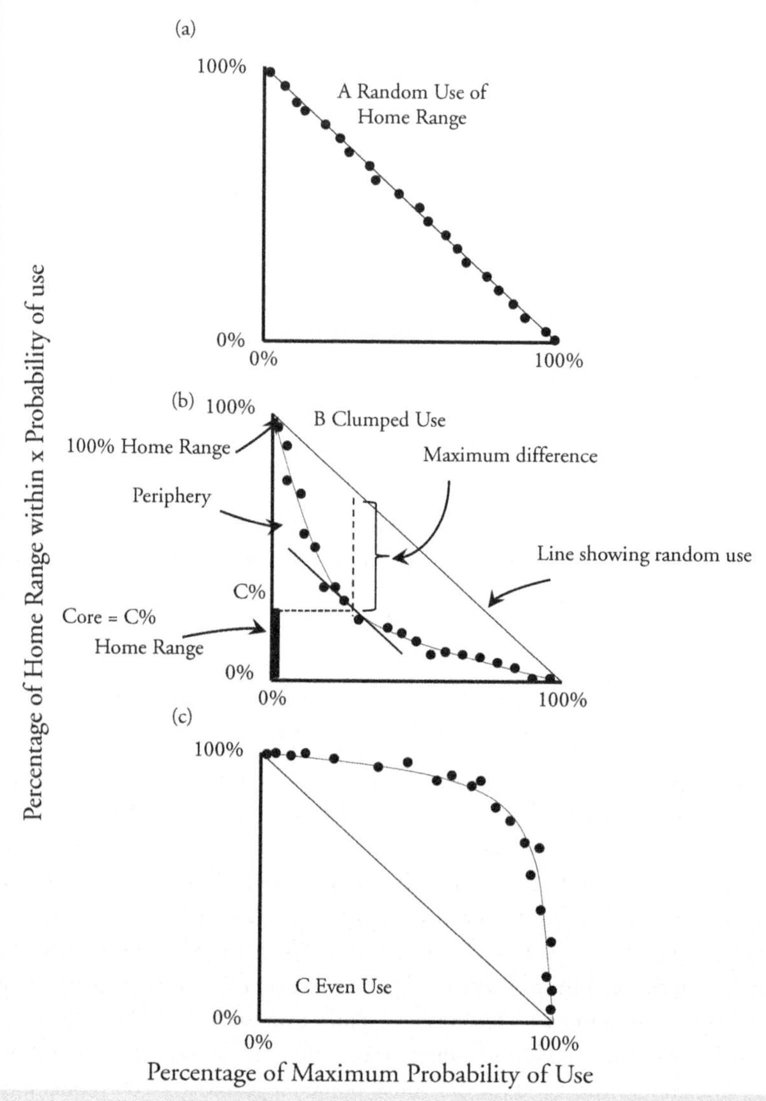

Fig. 9.3 The concept for calculating home-range cores. For an animal's home-range, whether an animal's use of space is random, clumped, or even affects the relationship between probability of use and the percentage of the home range with that percentage of use or greater. The *x*-axis is the probability of use for areas within an animal's home-range calculated as the percentage of the maximum probability of use (maximum value of the utilization distribution). The *y*-axis is the percentage home range (100% home range, 95% home range, etc.). (a) Relationship for random use by an animal of the area within its home ranged; (b) for clumped or patch use of the home range; and (c) for even or over-dispersed use of the home range.

Box 9.1 *Continued*

When use of space by animals is clumped (Figure 9.3b), Seaman and I defined an animal's core as containing the points close to the *x*-axis (after Fagen and Young 1978; Slater and Lester 1982). Those points of the home range mapped onto the steeply descending slope along the *y*-axis are used least and constitute the periphery of the home range. The curve can be divided into its two parts at the point whose tangent has slope = −1 (i.e. whose tangent is parallel to the line for random use). This is also the point on the home range curve that is furthest from the line with slope = −1 (Figure 9.3b). Plots of actual data that do not yield smooth curves can be fit with smooth curves (Figure 9.4b).

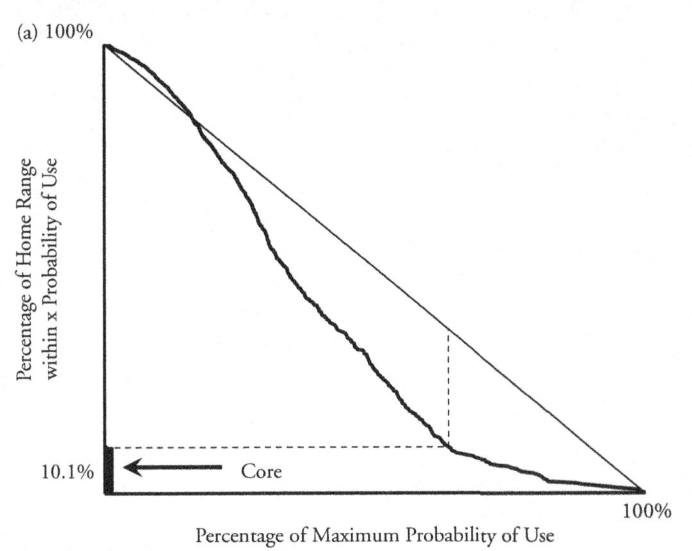

Fig. 9.4 Home-range curves for adult male black bear 12 (a) and adult female bear 61 (b) both from 1985, studied in the Southern Appalachian Mountains of North Carolina, USA (Powell *et al.* 1997). The dotted vertical line shows the maximum difference between the line representing random use to the curves for bear home-ranges. The dotted horizontal lines show the percent home-ranges that correspond to the core for each bear. The core for male 12 corresponded about to his 10% home range, while that for female 61 corresponded about with her 20% home range.

Box 9.1 *Continued*

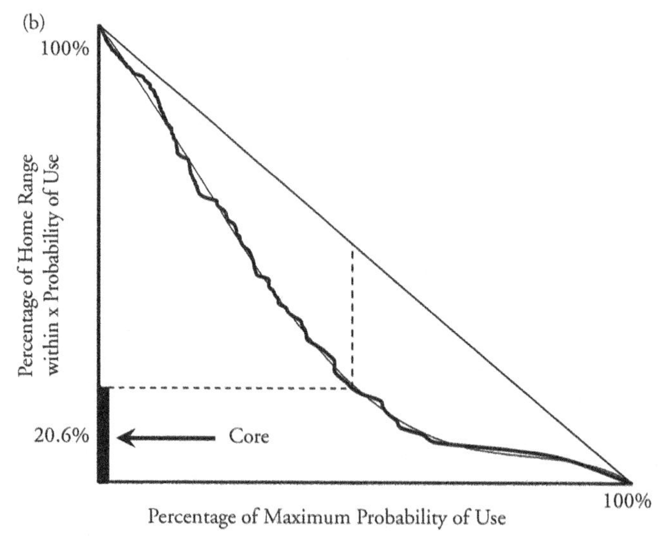

Fig. 9.4 Continued

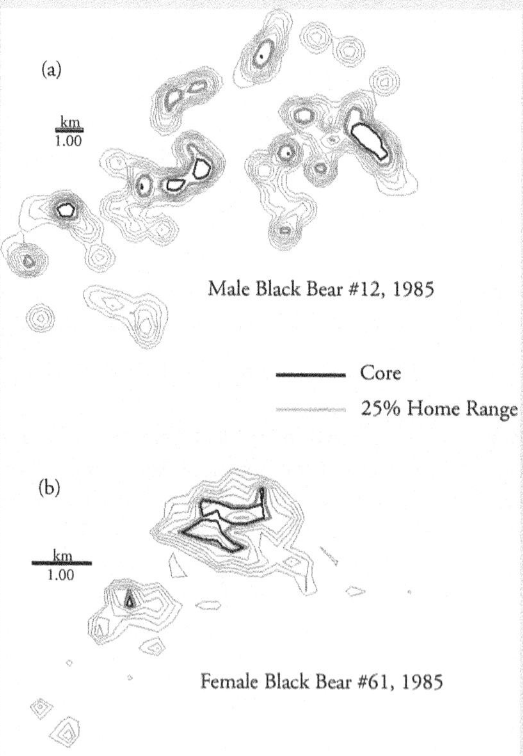

Fig. 9.5 Contour plots of the home ranges of adult male black bear 12 (a) and adult female bear 61 (b) both from 1985, studied in the Southern Appalachian Mountains of North Carolina, USA (Powell *et al.* 1997). The conventional 25% home ranges, often chosen to represent a home-range core, is shown in heavy grey while the cores calculated as described in this box are shown in heavy black.

Box 9.1 *Continued*

The home-range curves for adult male black bear 12 (Figures 9.4a, 9.5a) and female bear 61 (Figures 9.4b, 9.5b) in 1985, studied by my research team in the Southern Appalachian Mountains of North Carolina, USA (Powell *et al.* 1997), show that different animals will have different sized home-range cores. The home-range core for the male was approximately his 10% home range, while the core for female 61 was approximately her 20% home range. Just as home ranges vary seasonally and yearly for a given animal, cores vary over time as well.

Note in Figure 9.4a that the home-range curve for male bear 12 is above the random use curve for areas with low probability of use, the areas that makeup the far periphery of this bear's home-range. That the home-range curve is above the random use curve means that the bear's use of space in that area was evenly spaced, not clumped. Whether this is an artifact of smoothing using a home-range estimator or whether animals really do use the peripheries of their home ranges evenly and at low probabilities of use deserves investigation.

9.6.2 Home-range overlap

Home ranges of conspecifics often overlap, sometimes extensively. Relatives may often use areas of overlap simultaneously, while non-relatives may avoid simultaneous use. Females with young may avoid areas of home-range overlap with adult males. How animals in a population respond to home-range overlap is important, making measures of overlap important.

In some species, territorial behavior has been documented objectively, especially for many birds. For many carnivores, however, territory defense or responses to scent marks are difficult to document and apparent lack of home-range overlap is the only evidence of territoriality. For such carnivores, home-range overlap must be quantified in an objective manner that weights probability of use of different parts of a home range or territory. Home-range overlap can then be compared statistically among, for example, populations that appear to differ in territorial behavior but for whom territorial behavior has not been manipulated experimentally.

Doncaster (1990) defined two types of overlap, termed static and dynamic interactions. Static interaction is the spatial overlap of two home-ranges, while dynamic interaction involves interdependent movements of the two individuals whose home ranges overlap. These types of overlap can be quantified in several different ways.

9.6.3 Static interactions

Area of overlap is a poor estimate of the effect or importance of home-range overlap. Areas of overlap vary in probability of use and the two individuals may have a large overlap of areas used little by each, or a small but critical overlap of areas used intensively by each. Although Genovesi and Boitani (1997) found that area overlap of minimum convex polygons correlated strongly with weighted overlap, this need not always be the case. Indeed, the biology is most interesting precisely when this pattern is not the case, because it means that the animals are responding both to resources and to each other.

Several indices work well to quantify pairwise overlap of home ranges using probability density functions for each animals' home-range. The first, I_p, is:

$$Ip = \Sigma p_{ki} . \Sigma p_{kj},$$

$$k \in O$$

where p_{ki} and p_{kj} are the independent probabilities that at any arbitrary time animals i and j are in cell k of a study area that is within the area of overlap, O, of the animals' home ranges. I_p ranges from 0 to 1.

This index is the simple probability that the two animals will be in their area of home-range overlap at the same time were they to move independently of each other. Of course, most animals do not move without respect to the movements of other animals. Consequently, static interactions should be studied in conjunction with dynamic interactions.

Spearman's coefficient of rank correlation is a robust index that can be used broadly to index overlap of home ranges (Doncaster 1990). Calculate Spearman's r for the utility distributions of two animals with overlapping home-ranges, or for the frequencies of use of cells in a grid. The index behaves well and nonlinear responses of the index outside of the area of overlap (where one individual has probability of use equal 0) do not affect the overall usefulness of the index.

Because an animal's home-range can overlap with the home ranges of several other animals, all pairwise index values within a study site are not strictly independent. Similarly, for studies that follow some individuals for more than one year, index values for different years may not be independent. Statistical tests must be controlled for both individuals in each pairwise overlap when testing for differences among sites, among years, or among populations of different species.

These indices of overlap, and other similar indices (Hurlbert 1978), can be used to compare overlap between sites, to compare changes in overlap with changes in food or other limiting resources, or to deduce territorial behavior when active defense, scent marks, or calls have not been documented.

9.6.4 Dynamic interactions

Several approaches can be used to quantify and to test whether two individuals affect the behavior of each other. The indices test predominantly for attraction or avoidance and are important to testing hypotheses related to pair-bonds, to the existence of extended family groups, or other kin-related social behavior, such as group territories.

χ^2 and G tests can be used to explore whether two animals located at the same time (approximately) tend to be found together. In a 2×2 contingency table, label paired and unpaired distances as near (animals together or associated) vs. far (not together or not associated). Many telemetry packages now carry proximity loggers that record whether another animal carrying a telemetry package is nearby. Barring such technology, Doncaster (1990) labeled the red foxes he studied as "near" when they were close enough to detect each other; Horner and Powell (1990) labeled black bears as "near" when they were closer than the median distance of telemetry error. Minta (1992) defined "near" as two animals being within an area of overlap at the same time. The N-paired locations are a sample estimating how often the two animals are close together. Take each location for each animal, calculate the distance to the $N - 1$ locations of the other animal not taken at the same time to obtain a sample of $N^2 - N$ distances that can also be divided into near and far, and estimates how often the animals would be near each other if the movements of each were unaffected by the other. A significant χ^2 or G value indicates that the animals attract or avoid each other. Minta (1992) noted that χ^2 statistics behave better than G with small sample sizes.

My coworkers and I used fixed kernel estimates of probability distributions to estimate the probabilities that two animals would be in their area of overlap simultaneously if each used the area without regard to use by the other (index I_p above, Powell *et al.* 1987). I_p can be tested against the actual proportion of time (proportion of location estimates) that either animal spends in the area of overlap with the other. Because the two animals are unlikely to be located the same number of times, but must be in the overlap the same number of times, the proportion of locations in the area of overlap will differ slightly between the two. This approach can not be used to test whether two specific individuals attract or avoid each other but can be used to test whether classes of animals exhibit attraction or avoidance.

Minta (1992) developed further tests for attraction or avoidance to an area of overlap that allow researchers to test for more specific use of areas of overlap and that accommodate more diverse data. He showed how to test whether one animal of a pair is attracted but the other not, how to test for attraction when animals are not always located simultaneously, and how to test for attraction when home ranges are not known but the area of overlap is.

Finally, social network analyses will be a productive approach for understanding how individual carnivores, members of groups, and individuals interact, and how those interactions depend on resources and landscapes (Dalziel *et al.* 2008; Sih *et al.* 2009). This approach will lead to an understanding of, for example, how the interactions between "asocial" male and female carnivores are affected during the breeding season by their interactions before the breeding season and, in turn, how they affect interactions afterward. Network theory has been used to understand relationships within, and among, meerkat (*Suricata suricata*) groups and how meerkat interactions affect the spread of tuberculosis (Drewe *et al.* 2009; Madden *et al.* 2009; Drewe 2010).

9.6.5 Testing for territoriality

For many animals, territory defense is difficult or impossible to document but patterns of home-range overlap can be indexed using either I_p or Spearman's r to deduce territorial behavior. I_p is little affected by cell size, as long as cell size is relatively small compared to home-range size, because probabilities of being in cells are summed over the area of overlap. I_p, however, can be used only to test for differences between, or among, populations or classes of individuals. Spearman's r is sensitive to cell size and, therefore, has the potential to give different results with different cell sizes. Cell size must match the grain or scale of the hypotheses being tested and must be the same for all data being compared. Spearman's r can, though, identify spatial attraction or avoidance on a scale much finer than that of individual home ranges.

9.7 Parting thoughts

Know your animals. Incorporate as much background biology into research as possible.

Research design is paramount for successful research. Match your methods to the hypotheses you are testing. Different approaches and different technologies are needed for different hypotheses. Develop biological hypotheses, not statistical hypotheses. Biological hypotheses can be "tested" (evaluated) using Bayesian, likelihood, frequentist (conventional parametric and non-parametric statistics), and information-theoretic statistical approaches. Sober (2008) provided an excellent overview of the strengths and short-comings of these approaches, as used to test biological hypotheses.

Know your data. Inspect your data. Plot your data. Knowing your data will help you spot mistakes in your analyses.

Know the landscape and habitats (which are different from land cover types, see Chapter 10) in your study area.

Collecting data on movements, activity, and home ranges of carnivores is of little to no value without collecting, or having, data on the landscape, local habitats, prey populations, conspecifics, and competitors, etc.

To compare different datasets, critical methods must be the same for comparisons to be of value. For example, the differences in algorithms used to choose band width by different software packages for kernel home-range estimators prohibits comparisons.

Nearly all the approaches I have recommended in this chapter require raster GIS (geographical information system) datasets (cell-specific data), if GIS software is even used. To learn the probability of an individual carnivore using a specific habitat or land cover, to quantify home-range overlap, and to locate movements within habitats, all require cell-specific data for convenient analysis.

If you let available software define your analyses, then you are letting someone else design your research. Perhaps the designer of the software deserves credit for your results and not you.

10

Carnivore habitat ecology: integrating theory and application

Michael S. Mitchell and Mark Hebblewhite

Few terms in wildlife ecology and conservation biology enjoy jargon status more than the word "habitat." The ubiquity of the word in popular, scientific, and administrative literature suggests a universal definition, yet the diversity of contexts in which it is used clearly indicates little consensus. This conceptual imprecision has strong, but generally unacknowledged, implications for understanding and managing populations of wild animals, particularly for those where human-caused changes to ecosystems threaten viability. Few vertebrate groups better epitomize such populations than carnivores. Yet efforts to quantify what makes places habitable for carnivores are strongly compromised when poorly considered or biologically meaningless definitions of habitat are used.

We agree with Morrison *et al.* (1992), Hall *et al.* (1997), and Sinclair *et al.* (2005) that a definition of habitat must explicitly consider the resources that contribute to an animal's fitness. Describing habitat simply as the places or prevailing conditions where an animal is found is tautological, precluding robust knowledge and effective conservation. Nonetheless, descriptive definitions are overwhelmingly prevalent in the habitat literature. Why? We hypothesize three possible explanations. First, so little is known about an animal's habitat that only the initial steps of the scientific method are available to investigators: observe and hypothesize, the essence of description. Such cases are surely much rarer than the prevalence of descriptive habitat definitions suggests. The second explanation is that scant critical thought has been given to defining habitat because of the challenges of employing the entire scientific method (i.e. testing of hypotheses). In the absence of careful thought, over time such traditions become paradigms by weight of representation, irrespective of their limited scientific or biological merits. A final explanation is that data sufficient for developing rigorous, resource-based definitions of habitat are unavailable. This real-world constraint does sometimes limit the application of even the best of habitat definitions, requiring the careful use of surrogates (e.g. using proportion of hardwoods in the over story as a surrogate for the specific hardwoods that produce hard mast); indeed, every habitat definition we know relies on surrogates. Nonetheless, the uncritical use of surrogates, particularly given the rapid growth of remotely sensed land-cover data, computing power, and the use of sophisticated analytical techniques, has produced a large number of studies whose definition of habitat would seem to be "throw a bunch of conveniently available environmental variables into the statistical hopper and see what pops out."

The prevalence of descriptive habitat definitions not linked to fitness suggests both biological and scientific shortcomings in how we understand and study habitat. Describing *where* animals live is not informative science; for robust understanding that can lead to effective management and conservation, we need to know *why* animals live where they do (Gavin 1991). For many species, including a large and growing number of threatened carnivores, the consequences of poor understanding or misguided conservation are real and strongly negative. Knowing why an animal lives where it does is not just an academic exercise; we must bring the best science possible to bear on problems that may ultimately prove insoluble if we do not.

This chapter outlines our understanding of how to bring the best possible science to bear on discerning why carnivores live where they do. We discuss the concept of habitat, particularly as it applies to carnivores, whose resources contributing to fitness are often mobile. And we will discuss how habitat for carnivores can be quantified and its use interpreted. Finally, we discuss a study design that uses sound logic and robust analysis to maximize strength of inference. We then review some of the recent advances linking carnivore habitat to populations. We suggest a way of thinking about, and studying, carnivore habitat that will improve the efficiency of learning and the efficacy of conservation.

10.1 What is habitat?

The habitat definitions of Hall *et al.* (1997), Morrison (2001), and Sinclair *et al.* (2005) are based on the classic notion of the ecological niche, whereby animals select the resources and conditions that increase fitness (hence resource selection is distinct from habitat selection). Individuals, populations, and species have habitat and, consequently, habitat cannot occur without the animal. As with habitat, many definitions of the niche exist but Grinnell's original concept includes all subsequent definitions. The niche is a property of a species, includes abiotic and biotic components, is related to fitness, and includes long temporal and large spatial scales. Niche-based modeling has spurred recent investigation into explicit linkages between the niche concept and use of habitat by animals (Pulliam 2000; Soberon 2007; Hirzel and Le Lay 2008). Another important contribution of niche theory to habitat ecology is the distinction between fundamental and realized niches (Hutchinson 1957). An important consequence is that, unless we use experiments (MacArthur 1967), as empiricists we almost always describe the realized niche, or habitat, of a species. Under niche theory, populations have habitat but in Figure 10.1 we can see clearly that habitat is hierarchical from populations to individual foraging decisions by an animal. The concept of the niche is a good starting point for understanding habitat in a way that can be applied across scales.

10.1.1 Potential, sink, quality, source, suitable, or critical? What kind of habitat is it?

Following from the niche-based definition of habitat, habitat cannot just be a geographical description of an area or piece of land. Certain conditions must be

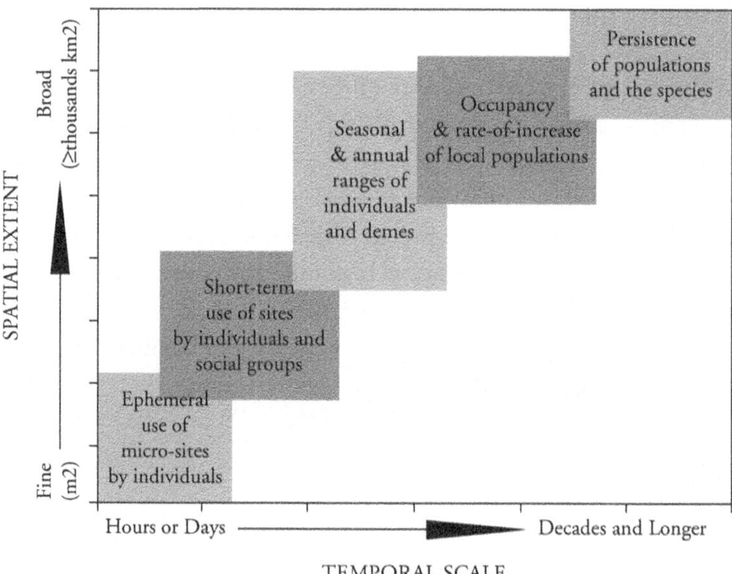

Fig. 10.1 Habitat occurs at multiple temporal and spatial scales; at the 1st order, habitat selection scale of the persistence of the species, equivalent to the species' niche; the 2nd-order, growth of local populations and seasonal and annual ranges of individuals; 3rd order (short-term use of sites by individuals and social groups) and finally; at the 4th order scale, where individuals make microscale foraging or selection decisions. Source: Mayor *et al.* (2009)

present for a species to survive and to reproduce. Hirzel and Le Lay (2008) illustrated the relationship between habitat and its distribution in geographic space (Figure 10.2). This approach to habitat helps us define several confusing terms, such as source habitat, sink habitat, potential habitat, habitat quality, suitable habitat, and critical habitat (Garshelis 2000; Pulliam 2000; Hirzel *et al.* 2002, Soberon 2007). First, presence of animals in an environment does not define habitat because presence alone does not consider survival and reproduction. Thus, environments where animals can occur, but where potential for survival is low and reproduction absent, are sink habitats, and environments with sufficient resources to support high survival and reproduction are source habitats (Figure 10.2). Note that a sink habitat can be critical to a species if residents of a sink habitat emigrate to a source habitat when a source population is low for reasons other than habitat. Environments where members of a species could occur, but presently do not, are

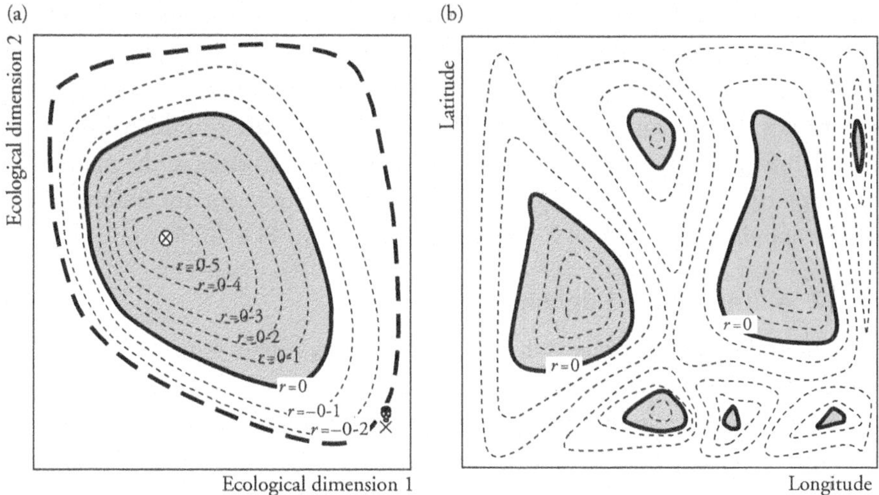

Fig. 10.2 Conceptual diagram of the relationship between habitat as defined by Sinclair *et al.* (2005) and the geographic distribution of that habitat in space adapted from Hirzel and LeLay (2008). (a) represents the relationship between intrinsic population growth rate (*r*) and two ecological dimensions (such as lichen abundance, or snow pack). Shaded areas indicate source habitat where population growth (*r*) is >0 (i.e. the population is growing), and the area inside the solid dashed line is considered sink habitat where species can persist, but only through immigration from an adjacent source. The skull and crossbones represent areas where the species cannot persist. (b) Represents this environmental space translated to geographic space given spatial measurements of the same resources for caribou in space. Shaded areas again represent source habitat where the conditions present are favorable for species persistence. Note that *r* here assumes density independence.

potential habitats. Similar to the fundamental niche, measuring a potential habitat well in field studies is almost impossible.

A habitat is of high quality (i.e. suitability) if individuals can experience high survival and reproduction and, thus, the population has the potential for a high growth rate. Note, however, that neither high nor low rates of survival and reproduction are necessarily reliable indicators of habitat quality where vital rates are density dependent; survival and reproduction could be high in poor habitat that is sparsely populated and low in excellent habitat occupied by a population near carrying capacity. Also, under the niche-based definition, the term "unsuitable habitat" has no logical meaning: all habitat, by definition, is of various degrees of suitability. "Non-habitat" is outside the solid dashed lines in Figure 10.2a, where a species cannot persist. Under this definition, only habitat (where populations can exist, with immigration for sink habitats) and non-habitat (where populations cannot persist) can exist.

Finally, no stand-alone, *biological* definition of critical habitat exists because "critical" implies importance for a specific goal or objective function. For endangered species, the goal is most often making the species non-endangered by reaching some recovery goal, but the target is a socially or politically defined goal. Heuristically, however, we can imagine some smaller subset of the shaded area in Figure 10.2 as being high-quality habitat that is sufficient for maintaining a specific population size, given a geographic area and species' life-history.

10.1.2 A fitness-based definition of habitat

The best understanding of habitat will explicitly relate resources to the survival and reproduction of an animal. This is a conceptually satisfying understanding of habitat because it proceeds from first principles, providing the mechanism that explains why an animal does what it does. If we can understand the potential contribution of each point in space to an animal's fitness based on the resources found there, we can evaluate the decisions an animal makes in its day-to-day activities (i.e. the behaviors that we perceive as habitat use). Mitchell *et al.* (2002) presented such an approach, originally developed by Zimmerman (1992) for black bears (*Ursus americanus*) in the southern Appalachian mountains, presenting habitat as a "fitness landscape" (Box 10.1). The fitness landscape proved highly robust for predicting habitat selection of black bears (Mitchell *et al.* 2002), effects of forest management on habitat use by bears (Mitchell and Powell 2003a), and optimal selection of home ranges based on the spatial distribution of resources (Mitchell and Powell 2007). Mosser *et al.* (2009) linked a fitness surface to habitat variables for long-term studies of lions (*Panthera leo*) in the Serengeti and showed a disconnect between habitats with high lion density and habitats selected because they contributed the most to adult female survival and

BOX 10.1 A habitat suitability model for black bears in the Southern Appalachians

Black bears obtain most of their nutrition from seasonally available vegetation, augmented by colonial insects, carrion, and rare acts of predation. In the southern Appalachian Mountains of North Carolina during spring, black bears eat predominantly grasses, forbs, and a saprophytic parasite of red oaks (*Quercus* sp.), squaw root (*Conophilus americanus*); berry-producing plants during summer; and hard-mast producing trees during fall. Bears in the region eat anthropogenic foods but suffer high rates of mortality near roads.

Zimmerman (1992; Mitchell *et al.* 2002) approached modeling habitat for black bears in the region based on first principles, attempting to quantify those resources and environmental attributes that contributed strongly to survival and reproduction. Zimmerman's approach was based on the habitat suitability index paradigm (Brooks 1997) but departed from it in some important respects. Drawing on published literature, Zimmerman modeled a priori the values of 15 food, denning, and escape resources important to bears (Table 10.1) and combined them into a

Table 10.1 *Habitat components used to calculate a Habitat Suitability Index for black bears living in the Southern Appalachians.*

Habitat Component	Relationship to Fitness of Bears	Method of Sampling
Number of fallen logs/ha	Abundance of colonial insects	Field sampling
Anthropogenic food source	Availability of food from human point sources	Aerial/ground survey
Distance to anthropogenic food source	Costs of traveling to human food source	GIS
Distance between anthropogenic food source and escape cover	Risk of acquiring food from human sources	Topographic maps
Distance to perennial water	Abundance of grasses and forbs in spring	GIS
Percent cover of *Smilax* spp.	Availability of fruit in fall	Field sampling
Percent cover in berry species	Availability of fruit in summer	Field sampling
Presence of red oak species	Availability of squaw root in summer	Forest inventory data/ GIS
Forest cover type	Availability of hard mast in fall	Forest inventory data/ GIS
Age of stand	Productivity of hard mast	Forest inventory data/ GIS

(continued)

Table 10.1 *Continued*

Habitat Component	Relationship to Fitness of Bears	Method of Sampling
Number of grape vines/ha	Availability of fruit in fall	Field sampling
Distance to nearest road	Risk of encountering humans	GIS
Area of conterminous forest not bisected by roads	Risk of encountering humans	GIS
Percent closure of understory	Escape cover	Field sampling
Slope of terrain	Escape cover, availability of caves for denning	GIS
Area in *Rhododendron* spp. or *Kalmia* sp.	Availability of thickets for denning	Aerial photo
Number of trees ≥90 cm DBH/ha	Availability of large trees for denning	Field sampling

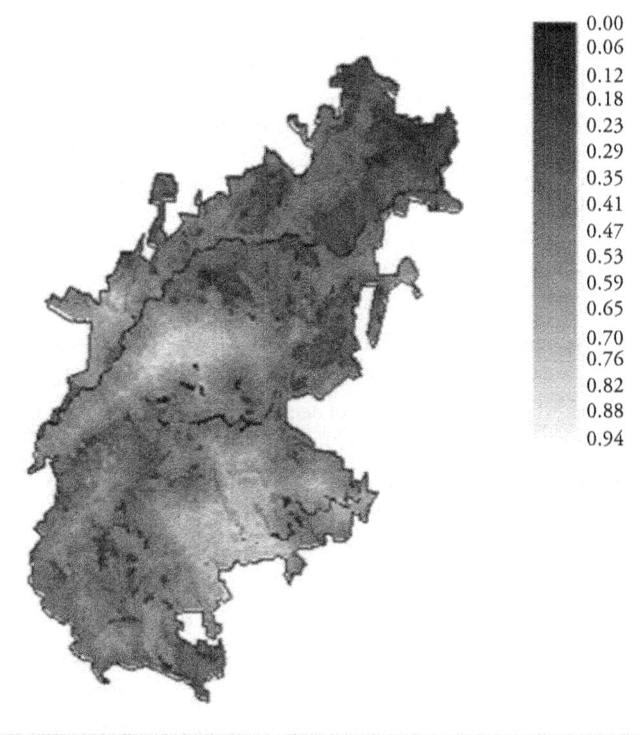

Fig. 10.3 Zimmerman's habitat suitability index (HSI) for black bears in the Southern Appalachians depicted as a fitness landscape for the Pisgah Bear Sanctuary in western North Carolina. HSI values range from 0, poor quality, to 1, high quality.

BOX 10.1 *Continued*

model that indexed habitat suitability on a scale from 0 (poor) to 1 (good; Figure 10.3). In a test of the model using independent data, the index predicted strongly habitat selection by 81 telemetered bears, especially when escape resources were removed (Figure 10.4), and helped to elucidate complex responses of bears to habitat changes caused by forest management (Mitchell and Powell 2005). Used as a currency for individual-based, optimal home-range models, the index facilitated accurate prediction of the home ranges of 100 adult female bears (Figure 10.5; Mitchell and Powell 2004, 2007). Winning no awards for parsimony, the index nonetheless has yet to be improved; sensitivity analyses showed that no variable or suite of variables dominate its predictions and no attempts to reduce the model have resulted in improved predictiveness. The explicit linkages between resources and their value to bears likely contributed strongly to the robustness of model predictions across a variety of applications, making the map of model predictions a "fitness

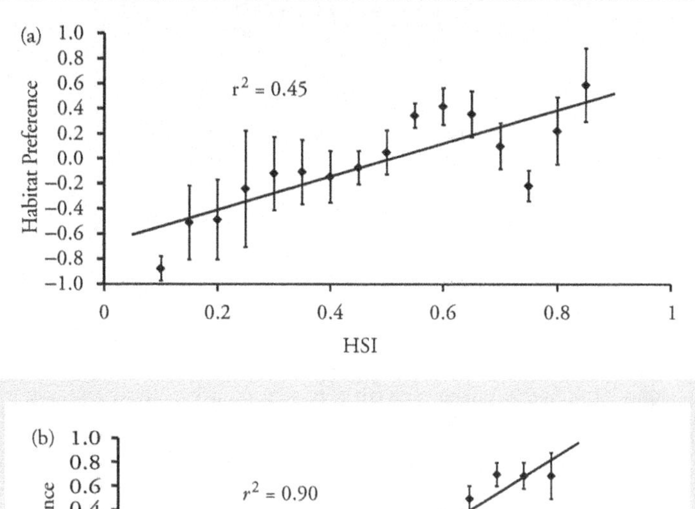

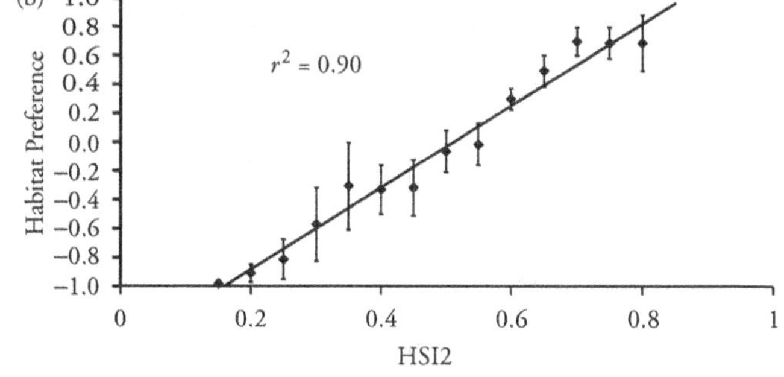

Fig. 10.4 Relationships between habitat use and Zimmerman's HSI (a) and the HSI without escape resources (b) for black bears in the Pisgah Bear Sanctuary, Pisgah National Forest, North Carolina.

BOX 10.1 *Continued*

landscape," i.e. the potential contribution of each point in space to the survival and reproduction of black bears (Powell 2004). The explanatory value of Zimmerman's model, beyond describing habitat selection by bears, highlights the merits of testing hypotheses about fitness-based definitions of habitat.

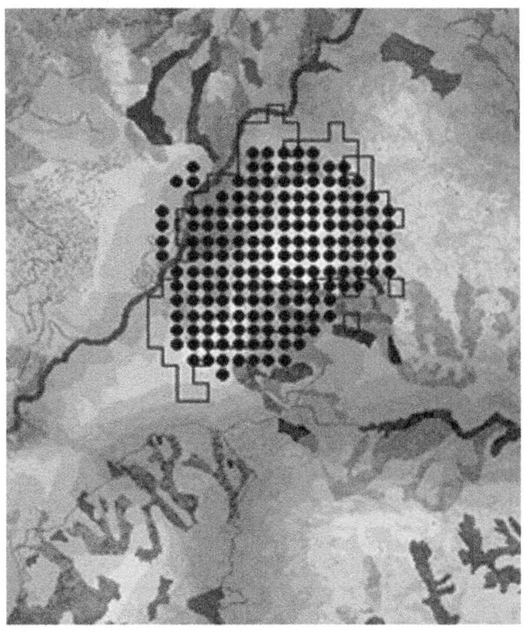

Fig. 10.5 Estimated optimal home range (dots) superimposed over true home-range (outline) of female bear 96 in 1984, Pisgah Bear Sanctuary, North Carolina. Shades of gray depict Zimmerman's HSI values (dark is low quality, light is high quality) that served as the currency for home range optimization.

cub production. This result emphasizes the long-understood, but seldom addressed, flaw in equating density to quality (Van Horne 1983), i.e. where habitat quality is high, a population has the potential to have high density but this potential may not be realized for a number of reasons.

The notion of habitat as a fitness landscape, where the contribution to survival and reproduction of resources at each point in space is made explicit, has conceptual appeal, but in practice can prove a daunting challenge. More often than not, resources, such as specific food types, are difficult to observe or model over the large landscapes that carnivores use, necessitating the use of surrogates. Thus, even in

Zimmerman's (1992) model (Box 10.1), few resources were measured directly. For example, the model used percent cover of berry-producing species as a surrogate for productivity of berries. The use of these surrogates relied on assumptions about their relationship to what they represented, and few ecologists would have difficulty imagining circumstances under which those assumptions could be violated. Nonetheless, as with any model, the relative merits of assumptions can only be evaluated if the assumptions are stated. Contrast how assumptions can be evaluated and tested in a model where fitness-based relationships are explicitly hypothesized with those implicit yet undefined in a model that defines habitat simply as, say, cover types. In the latter case, if a cover type is a perfectly predictive model for the behavior of interest, a researcher or a manager cannot know why it was.

In effect, any mapped habitat model is either implicitly or explicitly a fitness landscape representing a hypothesized or tested relationship between resources available to an animal and how it uses them, whether or not this is recognized. This fact should be dealt with explicitly and from the outset for any habitat model. What fitness relationship is the model intended to represent? Are those relationships operant at the scale of investigation (e.g. it may make no sense to include resources important to reproduction if observations used to build or test the model do not include the breeding season)? Should others have been included and how does their exclusion affect model performance? Under what circumstances could the assumptions of how the model captures fitness relationships be violated? A model that cannot stand up to such scrutiny invites questions about the biology underlying its predictions and, thus, about its usefulness for understanding or managing animals.

10.2 What is carnivore habitat?

Previous research on the habitat ecology of carnivores has focused too much on the environmental variables that predict carnivore presence or density, and not on variables with direct links to carnivore fitness. In the process, studies have often neglected biological first principles defining what it is to be a carnivore. Many carnivores are obligate predators; for these species, habitat definitions must include explicit measures of biotic interactions with prey. Such measures must include abundances and distributions of prey, and environmental characteristics that facilitate capture of prey. Instead, hosts of studies relate occurrence, use, or selection by carnivores to vegetation communities, digital elevation models, remote-sensing variables, and other kinds of spatial variables easily obtained in a geographical information system (GIS) framework. For example, Mace *et al.* (1996) and Boyce and Waller (2003) examined habitat selection by grizzly bears (*U. arctos*) in western Montana as a function of vegetation communities identified

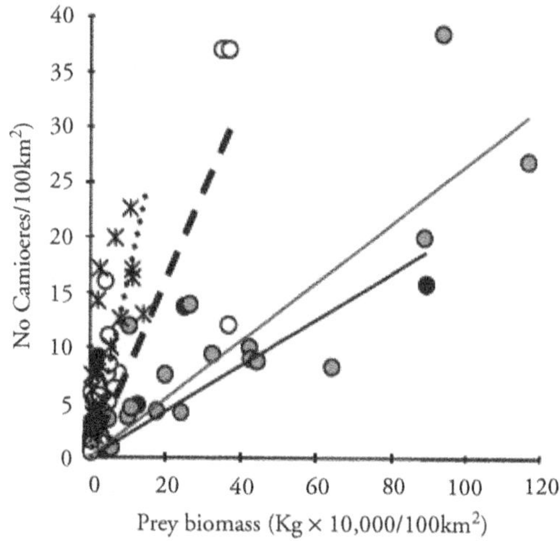

Fig. 10.6 Carnivore density (#/100km^2) as a function of prey biomass for tigers (solid circle and line); lion (gray), leopard (open/dashed), and Canadian lynx (*/—). Source: Carbone and Gittleman (2002).

from Landsat-TM imagery combined with topographic and some human features. Such an approach makes sense for omnivorous carnivores that rely heavily on vegetative resources, and may even be useful at prediction. However, the merits of these approaches are less certain for carnivores that are strongly reliant on mobile and unevenly available prey. For example, vegetation communities and glacial features explained little about use of space by wolves (*Canis lupus*) in the Canadian arctic (McLoughlin *et al.* 2004); or by wolves in the Canadian Rockies (Hebblewhite *et al.* 2005). The assumptions that such variables are surrogates for availability of plant forage for omnivores or of prey for carnivores are often unwarranted and infrequently tested. That these habitat models, convenient to mapping, do not explain carnivore behavior argues strongly for considering prey resources explicitly. Few studies of habitat for carnivores include availability of prey species, fundamental to the persistence of carnivores. The main factor driving densities of obligate carnivores is food, i.e. the density or availability of prey. (Miquelle *et al.* 1999; Fuller and Sievert 2001; Carbone and Gittleman 2002; Fuller *et al.* 2003). The ratio of carnivore to prey biomass scales to the reciprocal of carnivore mass (Figure 10.6, Carbone and Gittleman 2002). We do not argue that non-prey resources are irrelevant. Moorcroft *et al.* (2006), for example, showed that coyotes (*Canis latrans*) avoided scent marks from conspecifics in Yellowstone National Park; even in this case, however, prey density explained much of the variation in coyote movements. Thus, if we wish to define habitat in a manner that helps us understand why obligate carnivores do what they

BOX 10.2 Empty forest syndrome: comparing predictions from Amur tiger habitat models with and without measures of ungulate prey availability

The forests of Asia are "empty" of large ungulate prey for tigers, leading biologists to coin the term "empty forests" syndrome (Karanth *et al.* 2004a; Datta *et al.* 2008). This syndrome occurs when environmental, structural aspects of tiger habitat are present (forests, water, stalking cover) but the most critical factor, large ungulates, are not. The main cause for "empty forests" is overhunting and poaching, which leaves forests depopulated of sufficient ungulate prey for tigers (Miquelle *et al.* 1999; Karanth *et al.* 2004b). In this case-study, we illustrate the effects of empty forest syndrome on predictions from resource selection function (RSF) models developed for the Amur tiger in the Russian Far East. Basic study design is a used–unused design, whereby large units where tigers were detected during intensive winter snowtrack surveys during winter 2005 were compared to unused units using logistic regression. This used–unused design corresponds to a true probability. Miquelle *et al.* (2006) provided details of data collection.

The full analysis was conducted as part of predicting habitat for Amur tigers expanding their range from Russia into the Changbaishan region of NE China (Li *et al.* 2009). Russian biologists tracked Amur tigers in the snow during winter surveys and also collected data on the tigers' main ungulate prey species: sika deer (*Cervus nippon*), roe deer (*Capreolus capreolus*), red deer (*C. elaphus*), wild boar (*Sus scrofa*), musk deer (*Moschus spp.*), and the rare moose (*Alces alces*). Data of similar resolution were not available in the Chinese portion of the study area. Hence, we were interested in quantifying the effect of not including prey availability in RSF models. We developed a two-staged approach to examine the effects of ungulate prey on habitat modeling by (1) developing an "environmental" RSF model, including surrogate environmental variables, such as land cover, elevation, net primary productivity, and snow cover from MODIS satellites (Huete *et al.* 2002), that would be expected to correlate with prey distribution; and (2) developing a "biotic" RSF model that also included track density of the main ungulate prey species. Modeling details were similar to Box 10.1.

The overall biotic RSF model was significant (Likelihood-ratio ratio χ^2 = 125·5, $p < 0.00005$) and demonstrated good model fit (Hosmer and Lemeshow goodness of fit test, test, χ^2 = 8·45, p = 0.35), and had better explanatory power, discriminatory power, predictive capacity than the environmental model (Table 10.2). Moreover, in a model selection sense, the biotic model was over 10,000 times more likely to be a better model compared to the environmental model (ratio of Akaike weights of the two models). Clearly, knowledge of ungulate distribution and relative abundance improved the ability of the model to predict tiger habitat. The biotic model had superior discriminatory ability at predicting tiger habitat as measured by an average ROC, pseudo-R^2, and the k-folds cross-validation procedure

BOX 10.2 *Continued*

Table 10.2 *Amur tiger resource selection function model diagnostics and covariate structure for the best environmental covariate RSF model and the best ungulate RSF model in the Russian Far East during winter 2004/2005. The top habitat and ungulate models are compared using AIC, ROC, pseudo-R², and k-folds spearman rank correlations.*

	AIC	ROC	Pseudo-R^2	*k*-folds
Environmental Covariate RSF Model	594.7	0.71	0.12	0.712
Ungulate RSF Model	531.8	0.89	0.25	0.881

(Table 10.2). ROC scores between 0.8 - 0.98 are indicative of excellent discriminatory ability, echoed with the very high *k*-folds spearman rank correlation of 0.881. The biotic model provided a higher overall classification success for survey units of 72%. Briefly, tiger's selected areas with high densities of sika deer, red deer, and wild boar in the ungulate model. Li *et al.* (2009) provided full details.

We compared the predicted distribution of tiger habitat probabilities between the two models (Figure 10.7). This comparison shows that without taking ungulate densities into account, the environmental model overpredicted the amount of

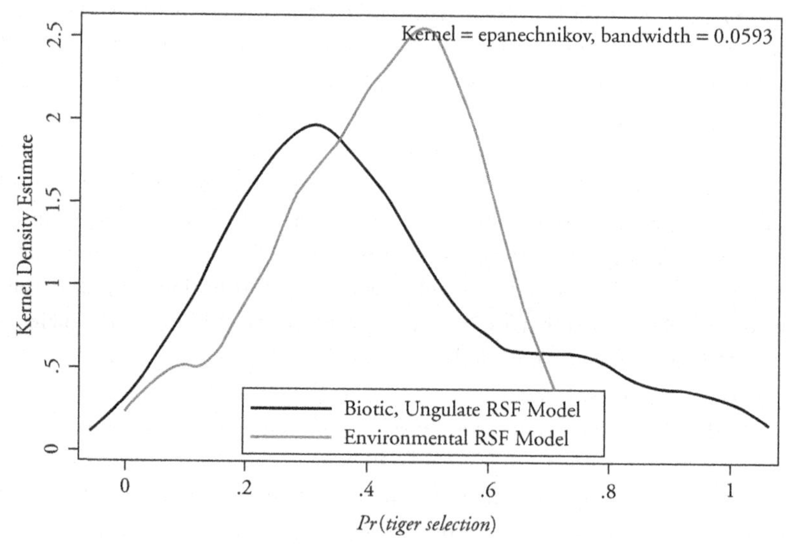

Fig. 10.7 Relationships between the probability of tiger selection and ungulate track counts for red deer, sika deer, and wild boar from resource selection function modeling for tigers in the Russian Far East, winter 2005. Resource selection was assessed at the sample unit scale (135 km²), and the best linear predictions from the logistic regression model from Equation 10.1 are shown against observed sample-unit scale predictions (*Pr*(*tiger selection*)).

BOX 10.2 *Continued*

"high-quality" habitat available for tigers compared to the biotic model. The consequences of this overprediction was a poor Spearman rank correlation between the frequency of tigers and high-ranked categories of tiger habitat (environmental model Spearman rank correlation, $r_s = 0.71$, biotic model $r_s = 0.88$). Therefore, even on the Russian side of the border, environmental covariates were not adequate spatial surrogates for ungulate data, and did not adequately capture the determinants of ungulate distribution and abundance, resulting in an optimistic prediction of the amount of high quality tiger habitat available. Results of our extrapolation of the environmental model to areas without similar prey density data should overestimate the availability of "high" quality tiger habitat in a similar fashion.

Unless we explicitly model the key resources for carnivore—namely, their prey—we risk creating habitat models for carnivores that are overly optimistic and leave out the key fitness-drivers of population dynamics. In the case of tigers, the recent criticisms of Project Tiger in India especially emphasize the critical conservation importance of these mistakes. Many of the tiger reserves created especially for tigers are devoid of the large prey that tigers need, driving tiger densities down to the point where many tiger reserves are devoid of tigers, too.

do, including prey availability should be the critical first step in addressing carnivore habitat ecology.

To be fair, quantifying the availability or abundance of prey across large spatial scales for most carnivore species is difficult. This is the main reason why surrogates, such as vegetation type or land-cover classifications from remote sensing, are often used, despite few tests of these surrogates. Numerous recent studies have attempted to integrate availability of prey resources into habitat-selection models for carnivores, however, and offer great promise for connecting habitat selection to population processes. Hierarchical analyses of habitat selection by Amur tigers (*Panthera altaica*) for the five main prey species available in the Russian Far East showed that it was the distribution of their main prey, not vegetation communities *per se*, that limited tiger "habitat" (Miquelle *et al.* 1996, 1999). Explicitly linking habitat selection by tigers to their ungulate prey (and hence to tiger fitness) made the case for controlling one of the main ecological reasons driving carnivore population decreases—poaching of ungulate prey (Miquelle *et al.* 1999; Chapron *et al.* 2008). Failure to include a biotic definition of habitat is the cause of the "empty forest syndrome" discussed in Box 10.2. This example illustrates the conservation costs of using poor, vegetation-only, definitions of habitat for obligate carnivores, and makes a convincing case for relating fitness directly to prey abundance.

Because many carnivores are threatened or limited by human activity, many studies include the biotic interaction with humans as an important influence on carnivore habitat. Thus, humans reduce habitat, changing the relationship between fundamental and realized niches of carnivores on a landscape. Conceptually, reducing conflict with humans would restore great amounts of "potential" habitat for many carnivore species. Researchers have investigated effects of humans and human developments on many carnivores, often focusing on roads. Gray wolves, cougars (*Puma concolor*), jaguars (*Panthera onca*), Amur tigers, Tasmanian devils (*Sarcophilus harrisii*), grizzly bears, and black bears all show that roads may be important limiting factors in the environments of these carnivores (Thurber *et al.* 1994; Mladenoff *et al.* 1995; Jones 2000; Gibeau *et al.* 2002; Dickson *et al.* 2005; Hebblewhite *et al.* 2005; Carroll and Miquelle 2006; De Azevedo and Murray 2007; Reynolds-Hogland and Mitchell 2007a; Cushman *et al.* 2009). Often, the effects of human persecution depend on context. In National Parks in Alaska and Alberta, for example, gray wolves do not avoid human activity inside protected areas, but show typical avoidance of human activity outside (Thurber *et al.* 1994; Hebblewhite and Merrill 2008). This context dependency explains recent debate about the mechanisms of road avoidance in the Great Lakes region of North America (Merrill 2000; Mech 2007). Thus, simply including human biotic interactions with surrogate variables, such as road density or distances to roads, may not capture the mechanisms of carnivore–human relationships.

10.3 Measuring habitat use and selection by carnivores

At least some of the confusion about habitat-selection studies can be attributed to the bewildering number of ways that carnivore ecologists can design habitat ecology studies: habitat suitability indices, resource-selection functions, resource-selection probability functions, resource-utilization functions, compositional analysis, environmental niche factor analysis, occupancy modeling, classification and regression trees (CART), genetic algorithm for rule-set prediction (GARP), maximum entropy, Mahalanobis distances, and the list goes on. Arguments and confusions within the literature (Boyce *et al.* 1999; Keating and Cherry 2004; Johnson *et al.* 2006) about the nature of statistical tests of habitat selection, while important from a statistical viewpoint, do nothing to remedy the confusion for the practitioner. Rigorous review of the statistical bases for all methods is outside the scope of this chapter. Instead, we review the importance of critical considerations often ignored: question-driven research, theoretical foundations for selectivity, scale-dependency of ecological processes, effects of density dependency, study design, and the relationships between different classes of habitat modeling approaches.

10.3.1 The over-riding importance of questions

Any habitat model is an answer to a specific question about causal relationships between an animal and its environment, whether the question is stated or not. Without well-stated questions about these causal relationships, however, analytical answers have limited or no meaning (much less usefulness). Yet the literature on habitat studies is replete with answers for apparently one unasked question: what is habitat for animal X? This approach presumes that understanding why the animal is where it is, is not important, and the approach does not reveal causation for observed effects. Such descriptive habitat models might pose credible explanations for why animals are found where they are, but until such models are tested, their credibility is unconfirmed and the cause and effect relationships implicit within them are hypothetical. Unfortunately, confirmed causes for why animals exist where they do are critical for conservation and recovery of a species, and underlie habitat-based conservation. The scientific method, fully employed, offers a comprehensive mechanism for understanding cause and effect habitat relationships. Nonetheless, surprisingly few habitat studies make complete use of the hypothetico-deductive logic it embodies. By far the most common approach to modeling habitat is to construct statistical models and to interpret their biological meaning a posteriori (i.e. the first two steps of the scientific method), resulting in the generation of untested hypotheses about causation. To conclude causation from an a posteriori hypothesis is to make the logical error of affirming the consequent (Williams 1997). Until an a posteriori hypothesis is tested using independent data (i.e. the remainder of the scientific method), its credibility and usefulness is no greater than the myriad other, equally credible, a posteriori hypotheses that could have been used to explain the same patterns.

Logically, causation can only be established by testing hypotheses, whereby predictions from hypotheses derived empirically (e.g. from previous observations or studies) or theoretically are compared to observations to determine their capacity to predict empirical patterns; doing so can provide evidence for causation in two ways that differ in their level of logical support. Hypotheses that are supported in classic experiments, where the magnitude of effects are evaluated both in the presence of hypothesized causes (e.g. environmental attributes in the case of habitat studies) and where the causes are known to be absent (i.e. the control), provide evidence of *sufficient causation*, wherein presence of the cause was alone sufficient to produce an observed effect (Williams 1997). A common, but misguided, justification for a posteriori analyses in habitat studies is that causation cannot be established in ecological research because classic experiments are difficult to conduct. Controlled experiments, however, are not the only means to establish causation. *Necessary causation* can be established using observational studies, where the magnitude of effects are evaluated only where hypothesized causes are present and not where they are absent (generally the case for ecological studies); in these cases, a supported hypothesis indicates that the proposed cause produced the observed effect, at least in part, but other possible causes that were not evaluated cannot be excluded (Williams 1997). Whereas establishing necessary causation lacks the inferential strength of finding sufficient causation, it far exceeds the logical rigor of generating an untested hypothesis that establishes no causation at all. Testing meaningful, a priori hypotheses always provides stronger inferences on the cause and effect relationships that underlie habitat selection, than failing to do so.

Few circumstances exist where a researcher should choose to generate hypotheses rather than test them. For the vast majority of habitat studies, the empirical and theoretical fodder for constructing excellent hypotheses is vast, though often neglected in favor of sophisticated statistical approaches to generating a posteriori models. However sophisticated the means for generating an a posteriori habitat model might be, though, what can be learned from such an untested hypothesis is logically limited, compared to what can be learned through the test of any carefully considered a priori habitat model, however simple. We argue that the best approach to developing a robust understanding of carnivore habitat is to do a lot of thinking in advance of collecting a single data point, figuring out what the relevant questions motivating the study really are, and developing hypothesized answers to those questions that can be tested using field observations. Doing this thinking will increase inferential strength for the study; it will also allow effective planning for the data needed, the necessary sample sizes, the hardware required, the analytical framework, etc., needed to maximize study success and effectiveness of conservation applications based on the research.

10.3.2 Why should carnivores be selective?

A fundamental but rarely considered question for those who embark on habitat study is "why do we expect that habitat should predict animal behavior or population dynamics?" The clear answer is that natural selection has shaped behavior of animals to be selective, and that they will generally choose to exploit those places providing the resources that most contribute to their fitness. Without this assumption, no reason exists for quantifying relationships between behavior and habitat. Nonetheless, the theoretical foundations underlying the assumption are often completely ignored or even denied. The discipline of optimal foraging (Pyke *et al.* 1977; Pyke 1984; Stephens and Krebs 1987) is devoted to exploring precisely the fitness-based behaviors assumed in habitat analyses. Habitat research, in fact, is no more than a subdiscipline within optimal foraging, yet habitat studies rarely take advantage of the rich theoretical and empirical background available to them from this field of inquiry.

When applied to habitat selection, the optimal-foraging approach explains both the central tendencies we expect to see (optimizing phenotypes), and the variation around those central tendencies (e.g. error associated with learning through iterated experiences, extrinsic influences of competition, variation in resource productivity, etc.). The challenge common to studies of both foraging decisions and habitat selection is discerning the expression of the optimizing phenotypes amidst the processes that shape and influence them. Generally, habitat studies pursue a straightforward, if simplistic, approach to this question, using proportional use as an indicator of habitat value. This is directly analogous to Charnov's (1976a) model of optimal choice of prey, whereby proportions of different prey types in a predator's diet result from an iterated decision-making process that maximizes profitability of the diet by weighing the benefits of consuming an encountered prey against the costs of capturing it. Note, however, that whereas the prey model is explicit about the economic mechanism determining the proportional representation of prey types in a diet, habitat analyses assume such mechanisms result in disproportionate use of habitat features without specifying what they are; disproportionate habitat-use is thus taken as prima facie evidence of selection. This is a safe assumption when proportional use of habitat characteristics differs from that available and extrinsic factors, such as predation, competition, and population density, have little effect on habitat use. The absence of mechanistic explanations is problematic for forecasting or extrapolating, however, where such factors play important roles.

Using a fitness-based definition of habitat based on resource distribution and productivity promotes quantifying the benefits of selecting habitat characteristics (for a rare example, see Andruskiw *et al.* 2008), but costs and constraints that also influence selection can be more difficult to quantify. Finding the means to identify and to measure these costs and constraints on optimal use of resources is one of the defining challenges for the future of habitat studies. Inevitably, costs and constraints of habitat use will be measured as imperfectly and indirectly as the benefits. Nonetheless, even simplistic measures of costs and constraints offer strong explanatory improvement to habitat models. For example, relatively coarse measures of resource depression, travel costs, and conspecific avoidance have strong explanatory power in predicting how animals balance costs and benefits of habitat use in their selection of home ranges (Mitchell and Powell 2004, 2007; Moorcroft and Lewis 2006; Moorcroft *et al.* 2006; Moorcroft and Barnett 2008; van Beest *et al.* 2010).

10.3.3 The importance of scale

Habitat is inherently scale-dependent (Figure 10.1). When considering scale, most habitat researchers immediately think of Johnson's (1980) nested scales of habitat selection, including 1st order (geographic range of the species), 2nd order (placement of home ranges within the range of the species), 3rd order (use of habitat patches by an individual within its home range), and 4th order selection (foraging within patches). These scales outline a continuum of behaviors producing ecological patterns that depend on the geographical and temporal scales of observation. Johnson's scale, however, is categorical, whereas space and time are continuous. Thus, even within Johnson's scales of habitat selection, observations and therefore inferences can vary strongly depending on the spatial and temporal scales of observation (Figure 10.1). The important context of spatial and temporal windows of observation is often misunderstood or ignored when modeling habitat (Boyce 2006). Because this context drives the robustness and usefulness of habitat models, a researcher needs a strong understanding of the variation in the ecological processes driving habitat selection during observations. Thus, space and time define a window of observation within which ecological processes are often uniquely expressed. Conceptually, this is intuitive: observing an animal during a single day precludes extrapolation of its behavior over a year. Intuitiveness can break down, however, when explanatory patterns at one spatio-temporal scale are absent, completely different, or even reversed at another scale.

Knowing a priori how a spatio-temporal window of observation or application frames what can be learned about ecological processes is challenging. Hierarchy theory (Allen and Starr 1982; O'Neill *et al.* 1986; King 1997) offers a conceptual framework for inferring a priori mechanisms, whereby ecological processes are understood in terms of both lower level mechanisms and higher level constraints. A researcher can begin designing a study by placing his or her question on the spatio-temporal continuum of ecological processes (Figure 10.1) and asking whether the spatial and temporal extents required to answer the question are feasible for study. If not, how could the question be changed so that its answer can be found within an ecological process observable within realistic constraints on time and space? Perhaps this seems obvious, but it can be argued that the spotty predictive record for habitat models (Garshelis 2000) can at least in part be attributed to failure to acknowledge hierarchical structure of ecological systems, whereby decisions by animals influencing their use of habitat at broad temporal or spatial scales were naively modeled using data collected over short time periods and within small spatial extents. Additionally, studying a phenomenon at one scale, and assuming that it scales up or down linearly to another, presumes a perfectly nested hierarchy with no emergent properties across spatial and temporal scales. This presumption is highly questionable in ecological systems (Allen and Starr 1982; O'Neill *et al.* 1986; King 1997). To avoid such outcomes, Reynolds-Hogland and Mitchell (2007b) suggested that habitat studies could be conceptually organized according to hierarchy theory using three intersecting and interdependent axes: time, space, and ecological process of interest. The ecological axis is, by definition, hierarchically organized, such that any point selected along that axis has a correspondingly appropriate intersection point on the time and space axes. A shift in any one of the axes (e.g. a smaller temporal window of observation, or a different resolution for the ecological process) requires concomitant shifts in the other axes.

Results of habitat studies are extremely scale-dependent. Failure to acknowledge and plan for such dependency can result in misleading inferences (Boyce 2006). Understanding the hierarchical organization within an ecological system before attempting to tease out its processes in space and time is essential for successful, applicable habitat research.

10.3.4 Density dependence and habitat selection

The effects of population density on habitat selection are important yet underappreciated (Fretwell 1972; Rosenzweig 1981; Haugen *et al.* 2006; McLoughlin *et al.* 2010). Extending optimal foraging-type models, Fretwell (1972) showed that, for animals foraging to increase fitness, habitat (patch) selection would be affected by the density of conspecifics in a density-dependent fashion. Given the two basic assumptions, that individuals have "ideal" knowledge about the distribution of resources and that they are "free" to move between patches to maximize fitness, as density increases, animals will select patches in a frequency dependent fashion that equalizes realized fitness among individuals. This scenario results in an evolutionary stable strategy, where individuals make the best of a bad situation as density increases and no individual can achieve higher fitness. The density ratio between two patches at ideal free distribution is the habitat "isodar," which reflects differences in demographic quality between habitat patches (Morris 2003a, 2003b). The "ideal free distribution" predicts habitat selection for a wide variety of species (Oksanen *et al.* 1995; Beckmann and Berger 2003; Haugen *et al.* 2006; Griffen 2009). Unfortunately, the ideal free distribution has been tested only once for carnivores (black bears; Beckmann and Berger 2003), yet many studies assume density equates to fitness, clearly not the case under this form of habitat selection. Testing predictions of ideal free distribution theory should help carnivore ecologists understand the mechanisms governing habitat selection, even when animals clearly are neither ideal nor free.

As an extension of the ideal free model, consider territorial animals that are not "free" to move. Here, animals are divided into territory holders and non-territorial animals. Territory holders take the best real estate for themselves and achieve high fitness payoffs, making an "ideal despotic distribution" (Fretwell and Lucas 1970) in which density and fitness are not necessarily equal. The ideal despotic model has predicted spacing of Serengeti lions (Mosser *et al.* 2009), male black bears in California (Beckmann and Berger 2003), wolves in Yellowstone National Park (Kauffman *et al.* 2007), and other carnivores. Unfortunately, few studies, and almost none with carnivores, have examined (or acknowledged) the potential role of density in shaping habitat selection. Habitat selection by carnivores should change in density dependent fashions.

10.3.5 Understanding habitat selection: study design

Selection implies a behavior shaped by natural selection, whereas use is the observed outcome of that behavior. Some research questions lend themselves to understanding patterns of use, such as utilization distributions (Millspaugh *et al.* 2006), analyses of the amount of use (North and Reynolds 1996), and hazard models of resource use rates (Freitas *et al.* 2008). Understanding the process of selection, however, provides the only opportunity to address why or how a particular pattern of habitat use is achieved, particularly given the multiscale nature of habitat. For this reason, we focus on selection, studying the use of resources by an animal and also what resources *could* have been used but were not. Two main different sampling protocols underlie almost all habitat-selection studies: comparing (1) used resources with unused resources, or (2) used resources with available resources. A third design compares unused resources with available resources (Manly *et al.* 2002) but we know of no example of this design.

Used–unused (presence–absence) designs are perhaps the more powerful and straightforward for habitat-selection studies because we can use any number of statistical frameworks to compare attributes of used versus unused units and we can make inferences about utility of habitats from the resultant statistical functions. A common statistical framework for comparison is logistic regression, which uses a binary response variable for used and unused (Hosmer and Lemeshow 2000). When density or counts are modeled, generalized linear modeling (GLM) frameworks, such as Poisson, probit, zero-inflated Poisson, or zero-inflated negative binomial models are used (Guisan *et al.* 2002; Manly *et al.* 2002; Nielsen *et al.* 2005). Common used–unused data include remote-camera trapping (animals are either photographed or not-photographed); vegetation plots where plants are either present or absent, eaten or not eaten; mark–recapture trapping, photographing, and DNA sampling; and aerial surveys where animals are seen or not seen. The key here is that a survey unit was sampled and had a probability of either containing the animal (i.e. p) or not containing the animal ($1-p$), and that sampling had no bias.

When animals occupying a unit may not be observed, resulting in a false absence, then detection probability <1 (MacKenzie *et al.* 2005; see Chapter 4). Occupancy models that explicitly incorporate detection probability into the habitat model are beneficial, especially when detection probability itself is a function of habitat (MacKenzie *et al.* 2005; Hines *et al.* 2010). To estimate probability of detection, repeated sampling of units is required. For example, when a carnivore that is detected in 3 out of 5 surveys of a sample unit, detection probability is 3/5, or 0.6, and the probability is 0.4 that the carnivore occupies units where it was not observed (under a set of assumptions, MacKenzie *et al.* 2005). Marucco and McIntire (2010) used this approach with wolves. If detection probability is constant, or if multiple sampling is not conducted, then the used–unused design reduces to the use–availability design. In this case, relative probability of detection is estimated, which is still extremely useful for conservation and management.

Use–available (presence-only) design only has information about where animals used habitats (Pearce and Boyce 2006). Radio-telemetry studies are perhaps the most common method used to collect such data, and use–available designs are among the most common for analysis of habitat selection. Other studies with the use–available design include studies of animal distributions from museum collections (Pearce and Boyce 2006), aerial surveys where detection probability <1, scat analyses, and track count surveys. Resource selection functions (RSFs) and environmental niche factor analysis (ENFA or niche-factor analysis, Hirzel *et al.* 2002) are used commonly to compare used and available locations. Niche models are identical to RSFs from a study design perspective because used locations are compared to what is available within some defined study area. Thus, distinctions between different use–available designs are often false.

The distinction between a use–available design and used–unused design, however, can sometimes be tricky and often researchers can adopt both designs with the same data. For example, researchers conducted surveys over 10-km^2 grid cells in northern Ontario for wolverines (*Gulo gulo*), recording the presence or absence of wolverine tracks (Krebs *et al.* 2004). Their goal was first to describe the distribution and occurrence (use) of wolverines, yet this rich dataset clearly could be used with habitat-selection models. Both a used–unused design (units with and without wolverine tracks) or a use–available design (units with wolverine tracks versus the entire study area) could be adopted. Moreover, in this case, a used–unused design could be extended to a true occupancy model because sites were surveyed multiple times and detection probability could be estimated. Which study design is the "best" to use in this case? The answer depends on the research question. If knowing a relative probability is sufficient for conservation, then a used–available design is fine. If the true detection probability is needed, then the additional costs of collecting multiple sampling rounds was worth doing.

Within these two broad categories of study designs in habitat-selection studies, data can be collected and inferences applied among populations and individuals levels on at least three levels. Often, researchers collect data on wildlife only at the population level with no information about individual patterns of use, non-use, and availability. Manly *et al.* (2002) called this Design I (see Chapter 11). Common examples include aerial surveys, track or scat transects, distance sampling, diet selection based on scats, 2nd order scale (Johnson 1980) comparisons of resources used within animal home-ranges compared to what they could have used across the whole study area. In this design, animal observations occur at the population level and data include what animals did not use or was available to them. In Design II, inferentially between the population and individual level, resource use by individual animals is recorded but not where individual animals did not occur, or what was available to individual animals. Availability is measured at the population-level. An example includes observing individual bighorn sheep on aerial surveys or distance sampling and comparing their individual use of resources to that which was available to the entire population (Manly *et al.* 2002). For Design III, use and availability or lack of use is known at the individual level. Radio-telemetry is the most common tool for this design. An example is habitat selection by individually snowtracked wolves, compared to availability sampled along individual movement paths (Whittington *et al.* 2005). Costs and benefits of the different study designs depend on the research question and cost. Design I studies are often relatively inexpensive but lack mechanistic insights into why carnivores select habitat.

10.3.6 Using resource-selection functions and other approaches

In the North American literature, resource selection functions (RSFs) have gained prominence in habitat-selection studies (Boyce and McDonald 1999; Manly *et al.* 2002), although they are conceptually identical to niche-factor analyses that compare presence-only data to availability within a fixed study area. Other modeling approaches include maximum entropy models (MAXENT, Peterson and Robins 2003; Phillips and Dudik 2008), habitat suitability index models (Brooks 1997), and occupancy models (MacKenzie *et al.* 2005). Manly *et al.* (2002) defined RSFs as any function that is proportional to the probability of use, so this broad definition encompasses almost all other types of habitat models that could be conceived.

For the most common used–unused study designs, the used and unused sample units are commonly contrasted with logistic regression using the following equation (used–unused):

$$\hat{w}(x) = exp\ (\beta_0 + \boldsymbol{\beta X}) / \Big(1 + exp\ (\beta_0 + \boldsymbol{\beta X})\Big),$$

where $\hat{w}(x)$ is the probability of selection as a function of variables x_n, β_O is the intercept, and $\boldsymbol{\beta X}$ is the vector of the coefficients $\hat{\beta}_1 x_1 + \hat{\beta}_2 x_2 + \ldots + \hat{\beta}_n x_n$ estimated from fixed-effects logistic regression (Manly *et al.* 2002).

In applying the used–unused design, $\hat{w}(x)$is a true probability and is referred to as the resource selection probability function (RSPF).

For the use–available design, the resultant function is a relative probability, and is estimated using:

$$\hat{w} * (x) = exp\ (\boldsymbol{\beta X}),$$

where $\hat{w} * (x)$ is the probability of selection as a function of variables x_n, and $\boldsymbol{\beta X}$ is the vector of the coefficients $\hat{\beta}_1 x_1 + \hat{\beta}_2 x_2 + \ldots + \hat{\beta}_n x_n$ estimated from fixed-effects logistic regression (Manly *et al.* 2002).

In the use–available design, because the true sampling fraction is unknown, the prevalence of use, or the absolute amount of use, cannot be estimated and, hence, the intercept is meaningless.

RSFs have been commonly used to develop a posteriori statistical models to describe habitat (i.e. to generate hypotheses) but they also lend themselves readily to hypothesis testing, as do other modeling approaches. The selection of environmental variables, x_n, for inclusion in RSF analyses implicitly reflects hypothesized contributions of habitat characteristics to selection. Stating these hypotheses explicitly makes clear their biological justification for inclusion in a habitat model; the proximity of coefficients, $\hat{\beta}_n$ to 0 (i.e. whether 0 is included in the confidence intervals for $\hat{\beta}_n$) and their relative magnitude estimated by logistic regression, thus, represent tests of the hypothesized contribution of each variable to habitat selection. Hypotheses about the relative importance of specific habitat features to specific carnivores, and about the importance of combinations of those features, can be tested by evaluating competing multivariate RSF models (using Akaike's Information Criterion, AIC; Burnham and Anderson 2002). Ciarniello *et al.* (2007) demonstrated a novel way of testing hypotheses through cross-validation of RSFs generated for the same species at different study sites. Testing the ability of an RSF generated on one dataset to predict observations for an independent dataset remains the most robust means of using RSFs to test the hypothesized causes and effects of habitat relationships.

Considerable debate about RSFs has centered on the statistical mechanics and the interpretation of functions estimated from use–available designs. The problems are that contamination arises because some available points may actually contain used points and that the overall prevalence in a logistic model with a use–available design is unknown (Keating and Cherry 2004). These problems exist, but so long as the output from a logistic regression based on use–available design is treated as relative, resources or habitat quality can be interpreted validly (Johnson *et al.* 2006; Lele and Keim 2006). The debates, unfortunately, have taken focus away from the ecology of habitat selection (McLoughlin *et al.* 2010). Readers can read the relevant literature (Boyce and McDonald 1999; Manly *et al.* 2002; Johnson *et al.* 2006; Lele and Keim 2006).

Researchers should adopt a particular modeling approach only if it is useful. A researcher must know how a model will be used and have some way of measuring the predictive accuracy, precision, or generality of the model. To measure the latter for used–unused models, typical logistic regression diagnostics apply; for use–available designs, the problems of defining availability renders these approaches suspect (Boyce *et al.* 2002). Regardless, cross-validation, both with internal and external data, is necessary to test the predictive accuracy and utility of a habitat model (Roloff *et al.* 2001; Boyce *et al.* 2002; Johnson and Gillingham 2005; Johnson *et al.* 2006). Cross-validation also provides insight into how robust a habitat models is to aspects of study design, such as autocorrelation, non-independence, multicollinearity, and sample size (Manly *et al.* 2002; Johnson and Gillingham 2005; Gillies *et al.* 2006). Typically, in a k-fold procedure, a researcher divides data into k-partitions and cross-validates the predictive capacity between observed frequency of use and predicted frequency of use across the partitions of the data. This is internal cross-validation because the data used to generate the model is used to test different "versions" of the model. Conceptually, this is similar to evaluating model fit with the coefficient of determination, and gives a measure of how well the data are explained by the model. Boyce *et al.* (2002), however, showed substantial annual variation in predictive ability of RSF models for boreal songbirds, throwing caution on the utility of the "average" year model in predicting distribution over time. In addition, biased datasets may show good internal validation despite being ecologically wrong.

Obviously, a far better way to test generality, accuracy, and precision of a model is to compare model predictions to independent data, i.e. external validation. Independent data can be collected in different years, different study areas, and with different technology (e.g. GPS vs. VHF data). Ultimately, only the test of time reveals how "useful" a particular habitat model is. In perhaps the best example of model validation, Mladenoff's *et al.* (1999) tested a previously developed RSF

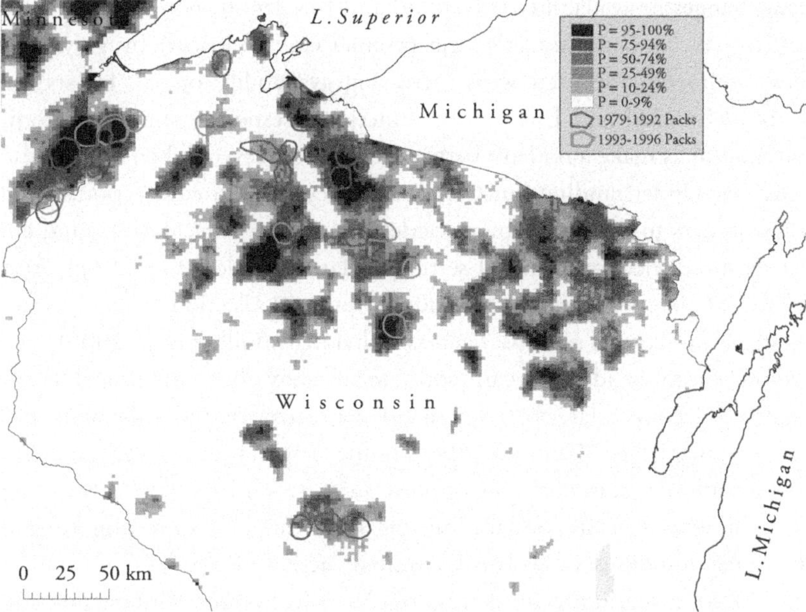

Fig. 10.8 Spatial predictions of gray wolf habitat in the American Midwest by Mladenoff *et al.* (1999) made using data from 1979–92 (wolf pack polygons in white) tested against observed distribution of new packs (black boundaries) observed during 1993–98. Model fit was remarkably high, and the model was able to predict colonization of new smaller patches previously unused by wolves. Source: Mladenoff *et al.* (1999).

for the expanding gray wolf population in the Great Lakes states of the US against new data collected later. The initial RSF model predicted accurately the wolf distribution 5 years later (Figure 10.8). More often, model validation reveals systemic problems with the model, such as poor prediction across individuals, or spatial differences in habitat selection that suggest selection may vary systematically as a function of some biological gradient (called functional responses in resource selection, Mysterud and Ims 1998).

10.3.7 Functional responses in resource selection

One extremely important ecological mechanism is the variation in the strength of selection as a function of availability. Such functional responses in resource selection for *spatial* variables (habitat) may be extremely common in carnivores, and parallel the concept of frequency dependence in *non-spatial* selection of prey, which has been recognized for a long time (Greenwood and Elton 1987). Functional responses address how selection for a spatial resource should change as that resource changes in availability (Mysterud and Ims 1999). For example, selection for oak forests (and, presumably, the productivity of acorns) by gray squirrels (*Sciurus carolinensis*) declines with increasing availability of oak forests on the landscape (Mysterud and Ims 1999). Functional responses should be common whenever animals make a tradeoff between two resources, or when thresholds exist for resources. Understanding functional responses in resource selection, therefore, allow researchers to develop habitat models that are general, flexible, and able to predict resource selection in novel settings (Matthiopoulos *et al.* 2011). Application of mixed-effects models to the study of resource selection enables researchers to investigate functional responses across individuals (Gillies *et al.* 2006).

Two important studies of carnivores relate show how functional responses in resource selection relate to frequency-dependence. As the availability of land increases in polar bears' (*Ursus maritimus*) home-ranges, bears select for ice closer to land, which affords greater hunting opportunities, (Mauritzen *et al.* 2003a), typical of a tradeoff between areas good for hunting and areas good for resting. An analysis of functional responses of wolves to human activity helped Hebblewhite and Merrill (2008) to synthesize conflicting results of wolf–human interaction studies. Previous studies of wolves' responses to roads showed attraction, ambivalence, and avoidance. Such results caused Mech *et al.* (1988) to conclude that wolves showed no consistent responses to human activity. What previous studies had not done, however, was address how selection changed as a function of the availability of human activity. Hebblewhite and Merrill (2008) found that avoidance of human activity by five wolf packs living in different human activity levels, depended on the overall amount of human activity in their pack territories. Packs with little human activity in their territories showed weak or no responses, whereas packs with high human activity showed strong avoidance, especially outside of protected areas. This and other recent examples of wolf–human functional responses (Houle *et al.* 2010) illustrates the power of understanding functional responses to produce syntheses of previous studies and to produce a solid framework for understanding carnivore–human relationships. We expect that carnivores commonly exhibit functional responses in resource selection. The most powerful approach to understanding functional responses is to combine an understanding of frequency dependence in prey selection (Greenwood and Elton 1979) with functional response analysis of spatial selection for these same prey species by a predator.

10.3.8 The importance of defining availability: recent advances from the field of movement modeling

Inferences from habitat-selection modeling with the use–availability design are highly contingent on how availability is defined (Beyer *et al.* 2010). Unfortunately, no biologically objective means of calculating availability exist; researchers can only infer indirectly what resources an animal considers to be available from what it did, compared to what we imagine it could have done. Further, the concept of availability is inherently scale-dependent and depends on the spatial scale at which resource selection is investigated. In other words, no "correct" way exists to sample availability. Many studies have compared telemetry locations for an animal to a set of random locations within its entire home-range (i.e. 3rd order selection; Johnson 1980), making the implicit assumption that animals can move anywhere within their home ranges at any time between successive locations. While this assumption could biologically be true for some highly mobile carnivores (e.g. wolves), it is clearly unrealistic for many others. And, with the growing use of global positioning system telemetry collars (GPS) in carnivore research, assuming that a carnivore can go anywhere within its home range between locations that are mere minutes apart, is unrealistic. Moreover, the debate over the use–available design has confirmed that the way this design had been applied in previous studies has problems. Improved understanding of availability is needed.

Fortunately, GPS technology has helped ecologists define availability somewhat more from an animal's behavioral perspective, and these definitions help circumvent some of the other problems with the use–available design. A study on turtles started it all. Comparing locations of slow-moving wood box tortoises (*Clemmys insculpta*) to random locations across their home ranges made no sense to Compton *et al.* (2002). Consequently, the authors borrowed a statistical method from the biomedical literature, matched-case control logistic regression, and defined availability as the area each tortoise could have reached from each location, based on its history of movements (Figure 10.9). The used and available locations are then compared using a conditional logistic regression model (also known as case-control, paired logistic and conditional logistic regression; Hosmer and Lemeshow 2000). The key here is that each used location is paired against *n* number of cases that represent where the animal could have actually moved (availability). The conditional likelihood of the logistic model takes into account what was available at each step and, consequently, the inferences from the overall model are conditional on the availability at each time step (Aarts *et al.* 2008; Moorcroft and Barnett 2008). This is now the recommended approach for determining availability in the use–available design at the individual level, especially including weights of "available" locations at different distances. Whittington *et al.* (2005) adopted this design to demonstrate that wolves avoid human activity in Jasper National Park, Alberta.

A caveat to this approach, however, is that restricting available points based on movement rates defines the scale of selection under evaluation (Forester *et al.* 2009). On a continuum of infrequent to frequent locations, the decisions being

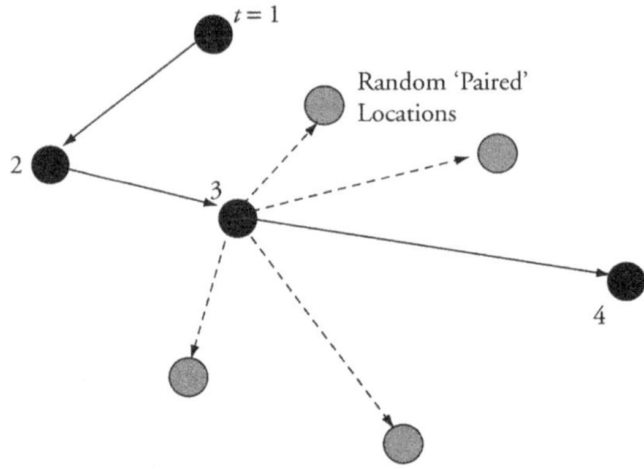

Fig. 10.9 Matched-case control sampling design for use–available study designs with animal-tracking data. Sampled locations (black circles) are paired with biologically realistic samples of "availability" given where the animal could have gone at time $t = 3$ in this example. Random paired available points can be generated from the observed step length from $t = 3$ to $t = 4$, or the empirical step length and turning angle distribution for the vector of animal relocations along the entire path $t = 1$ to 4, in this case.

modeled by this approach transition from Johnson's (1980) 3rd order selection to something on a finer scale, perhaps to Johnson's 4th order. Alternatively, the resources selected by an animal while it moves are environmental characteristics that facilitate safe and efficient travel. This begs an interesting question about how decisions made during movement represent selection in their own right vs. the extent that these decisions are structured and constrained by selection at higher orders. In the first case, decisions made during movement may very well represent an order of selection new to the nested orders traditionally considered, requiring further theoretical development before empirical results are fully understood. In the second case, selection of a travel route may be nothing more than the most convenient way to move between two selected habitats, rendering the notion of habitat selection during movement moot.

10.3.9 Quantifying resources

Two approaches, which can be combined in some cases, exist for examining resource availability at used and unused or available locations. The first is a "macro" approach to measure resource availability at broad landscape scales using spatial GIS models (Franklin *et al.* 2001; McDermid *et al.* 2005, 2009). This approach has been used most successfully for abiotic or loosely biotic variables, such as vegetation cover, topographic variables (digital elevation models), and human-related variables, such as distance to roads or road density.

The second is a more "micro" approach, whereby small-scale habitat covariates are measured at sites used and unused or available by using standard field-based monitoring approaches. For example, Kunkel *et al.* (2004) measured the availability of vegetative cover, tree species, and snow depth along travel routes of wolves, compared to areas where wolves killed ungulate prey, and compared to "random" areas not along travel routes. Values for such micro-variables are expensive and time-consuming to collect but often provide richer mechanistic insights into the factors influencing different stages of carnivore habitat selection, such as hunting, resting, or attacking.

Both approaches usually rely on availability of static or abiotic surrogates that do not reflect what was truly available to an animal, leading to two problems. First, maps of "static" vegetation types do not really reflect availability of resources for most animals, including carnivores. While a static land-cover model using different forest cover types (such as spruce, open conifer, shrubs, grasslands) has some explanatory power as a habitat model, it does not capture what might be important to a carnivore in a dynamic sense. For example, if we accept that prey are a critical biotic resource for many carnivores, grassland land-cover types could have dramatically different "value" to an ungulate over the course of a year, and hence, to a carnivore (Hebblewhite *et al.* 2008). Moreover, many spatial covariates (such as vegetation and snow cover) are temporally dynamic, yet habitat-selection models

BOX 10.3 A prey-based habitat model for gray wolves in Banff National Park

Gray wolves are the most widely distributed terrestrial, mammalian carnivore in the world (Mech and Boitani 2003). They require only the availability of large ungulate prey. As such, wolves are habitat generalists and densities are driven solely by ungulate biomass (Fuller and Sievert 2001), except when limited by human-caused mortality.

In this example, we illustrate including prey availability directly into habitat-selection models. Our goals here are to compare habitat-selection models based on just environmental covariates, to those based on prey availability, to illustrate the insights gained by explicitly considering prey availability, and also the drawbacks of such an approach.

We developed use–available resource selection functions (RSF, Boyce and McDonald 1999) for VHF telemetry locations for 14 wolves during winters 2001–05 in Banff National Park (Hebblewhite *et al.* 2002; Hebblewhite 2005). We estimated 99% kernel home ranges with a 6-km band width. We accounted for correlation within packs using a random effect for each wolf pack. Attributes of

BOX 10.3 *Continued*

used locations were compared to those of available locations using a mixed-effects logistic regression model (Gillies *et al.* 2006) that yielded a relative probability of wolf use of a resource type. We considered two broad types of models: (1) "typical" RSF models as a function of spatial covariates, including topographical variables (elevation, slope) and land cover type derived from LANDSAT imagery (McDermid *et al.* 2009); (2) "biotic" RSF models that explicitly modeled prey availability and distance to high human activity (Hebblewhite and Merrill 2008). We used a previously developed habitat suitability index for prey (Holroyd and Van Tighem 1983). Moose, deer (white-tailed, *Odocoileus virginiana*, and mule deer *O. hemionus*), elk, bighorn sheep (*Ovis canadensis*), and mountain goat (*Oreamnos americanus*) models were considered. Wolf diet was ~50% elk, 30% deer, 10% moose, and 10% other species, such as bighorn sheep and mountain goats (based on biomass, Hebblewhite *et al.* 2004). Thus, we predicted that 3rd order habitat selection within the home ranges of wolf packs would correspond to previous results of diet selection. Despite the importance of this hypothesis, which would allow us to scale up to spatial distributions from simple and easy to collect diet studies, few ecologists have tested the generality of correspondence between scales in carnivore habitat studies.

Covariates were screened for colinearity using a liberal correlation cutoff of $r > 0.7$ (Menard 2002). We used stepwise-AIC model selection to select the top typical and biotic RSF model, and compared the two models using AIC, and predictive capacity using k-folds cross-validation (Boyce *et al.* 2002). Because this model was use–available, using normal logistic regression diagnostics was invalid (Boyce *et al.* 2002).

Comparing the top typical and biotic models illustrates the tradeoffs carnivore ecologists will often face between predictive capacity and ecological understanding with habitat-selection models. The typical covariate model was by far the best model from an AIC perspective, with the biotic model over 66 AIC units "worse" than the typical model. Nonetheless, examination of the models' abilities to predict within-sample wolf telemetry data revealed that the biotic model fared better, explaining 10% better than the typical model. Coefficients for all models were as expected from previous studies on wolves in mountainous terrain (Oakleaf *et al.* 2006; Hebblewhite and Merrill 2008) and the rank order of predictions from the diet of wolves in Banff matched the rank-order of selectivity coefficients from the RSF model (Table 10.3, Figure 10.10). The relative probability of the five ungulate prey species changed as a function of habitat quality, confirming that as diet suggests, wolves avoid goat and sheep habitat, and select for moose, and deer, and elk approximately equally (Figure 10.10).

BOX 10.3 *Continued*

Insights from this biotic-RSF model are limited by the usual restrictions of regression-based studies. These regression models do not demonstrate whether wolves are really selecting elk or deer because these prey species were highly correlated in space; likewise for avoiding goats and bighorn sheep. Wolves could

Table 10.3 *Resource selection function (RSF) model structure and diagnostics for the top competing environmental covariate and biotic covariate models for wolves in winters 2001–05 in Banff National Park, Alberta, in two wolf packs.*

Model	Logistic Model Structure and Coefficients (*K* = number of parameters)	AIC	*k*-folds Spearman rank correlation
Environmental covariate model	*K* = 10, *Pr*(Use) = −0.005*Elevation + 1.67*Burn + 0.67*Water + 1.4*Shrub +0.28*OpenConifer + 0.28*ModerateConifer + 1.44*MixedForest + 1.14*Herbaceous −2.9*Alpine	1587	0.83
Biotic covariate model	*K*= 7 *Pr*(Use) = -0.439*DistHuman −0.23*Sheep −0.48*Goat + 0.72*Elk + 0.30*Moose + 0.66*Deer	1653	0.92

Notes: elevation is in meters; see Hebblewhite & Merrill (2008) for explanations of the landcover covariates; DistHuman is the distance, in kilometers, to high human access, defined by Hebblewhite and Merrill (2008).
* = ×

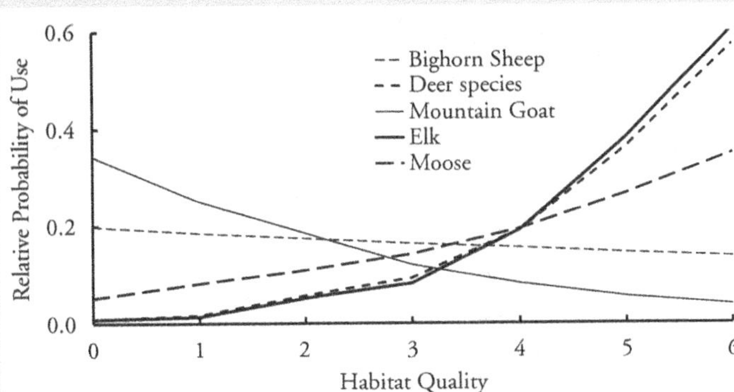

Fig. 10.10 Relative probabilities of use of five ungulate prey species by wolves as a function of relative habitat quality for five ungulate prey species in the Canadian Rockies from Resource Selection Functions.

BOX 10.3 *Continued*

be selecting elk but, because they encounter deer between predictable elk patches, deer could, actually, not be actively selected by wolves (*sensu* Huggard 1993). To tease this apart requires comparisons among wolf packs with different availabilities of prey (e.g. a functional response). Regardless, this example illustrates the exciting biological hypotheses that can be generated if we move from merely trying to predict habitat selection by carnivores to understanding the mechanisms of how prey availability drives carnivores.

have done little to link the spatially dynamic resource selection process to similarly dynamic measures of resource availability (Hebblewhite 2009). The growing access to remote-sensing products that measure the dynamic availability of forage, through indices like the normalized difference vegetation index (NDVI) and snow cover through MODIS satellites, means that future carnivore habitat models should be dynamic measures of resource availability (Hebblewhite 2009).

The second more critical problem is capturing the availability of dynamic, biotic resources, such as prey availability. A growing number of studies do include biotic covariates in habitat models (Miquelle *et al.* 1999; Hebblewhite *et al.* 2005; Heikkinen *et al.* 2007; Webb *et al.* 2008; Basille *et al.* 2009; see Box 10.2). While collecting sufficient data on availability of prey across large areas for many carnivores is difficult, carnivore habitat studies will increasingly include mechanistically measures of prey availability (Box 10.2, Box 10.3). Recent advances in non-invasive monitoring will certainly help. Camera-trapping and snowtracking can collect data on prey and predator simultaneously (Stephens *et al.* 2006).

10.4 Linking habitat selection to population consequences

Numerous authors have addressed the difficult conceptual and empirical challenge of linking habitat selection by individuals to population consequences (reviewed by Fryxell and Lundberg 1997). Here we focus on three empirical approaches with a demonstrable record for carnivore studies and that are perhaps the best scientifically defensible approaches: (1) population extrapolation based on habitat models, (2) combining habitat models with spatial models of mortality risks to develop core and sink habitat maps, and (3) spatially explicit models of population viability. No particular method is necessarily superior but note that data requirements, complexity, and assumptions increase from method 1 to 3.

10.4.1 Habitat-based population estimates

This approach combines habitat modeling with information about population densities to predict the number of animals in a given area, and has strong potential for answering questions about mechanistic links between habitat and population sizes and distributions. The principles behind this approach are first to model habitat selection, relate this habitat model to known abundance in the same area, and then to extrapolate the potential population size and distribution by applying the spatial habitat-selection model and habitat-abundance ratio to a new area (Boyce and McDonald 1999; Johnson and Seip 2008). It was this approach that Mladenoff and coworkers used successfully to predict the distribution and abundance of the recovering wolf populations in the Great Lakes region of the United States (Mladenoff *et al.* 1995, 1999; Mladenoff and Sickley 1998). It was also used to predict the numbers and distribution of grizzly bears in the Selway–Bitterroot ecosystem following potential reintroduction (Boyce and Waller 2003); the grizzly bear distribution and abundance in the Parsnip river area of Northern British Columbia (Ciarniello *et al.* 2007); recolonization habitats and population sizes of recolonizing Amur tigers expanding into NE China from the Russian Far East (Li *et al.* 2009; Box 10.2); and potential habitat and population size for critically endangered Far Eastern leopards (*P. pardus occidentalis*, Hebblewhite *et al.* 2011).

The first step involves developing a habitat-selection model for a particular carnivore species using (ideally) empirical data on the spatial locations of animals. The model should, ideally, have high predictive capacity, good model fit, and be hypothesis driven. One might use an RSF model to obtain the spatial prediction of the relative or absolute probability of use ($\hat{w}(x)_i$ from Equation 10.1) for a particular study area with a known or estimated population size of the focal species ($\hat{N}$). Next, the total predicted "habitat" required for each animal is estimated by dividing the total amount of habitat across the study area by the population size $\sum \hat{w}(x)_i / \hat{N}$. This ratio then provides the habitat/population ratio that can be used to extrapolate population size in adjacent areas, over time, and in different study areas. The assumptions of this approach, which include (1) the right biotic variables driving fitness have been measured, (2) similar selection patterns will exist for spatial variables in both areas, (3) similar landscape configurations exist for available spatial variables in both areas, (4) similar relationships between population parameters and available habitat in both areas, and (5) resource selection results in higher densities in those habitat types (or resource units) that are selected by a species. These are valid assumptions for many theoretical patterns of habitat selection (such as ideal free distribution, Fretwell and Lucas 1970). For an endangered species caught in an ecological trap, where animals select habitats that lead to reduced fitness (Robertson and Hutto 2006), the positive correlation between habitat selection density may break down. This potential problem leads us to the second potential approach to link populations and habitats.

10.4.2 Combining habitat and spatial models of mortality risk

A second approach to link habitats to population sizes is to combine a habitat model, such as the RSF designed above, with a complementary spatial mortality model that allows biologists to relax the assumption that selection = density. This approach entails identifying areas that are selected for high use by a species and identifying areas that cause high mortality and then dividing the area into habitat that can be classified as a sink (selected, high mortality) or source habitat (selected, low mortality), and non-habitats.

In the first example of this approach, Nielsen *et al.* (2004) developed spatial habitat models using resource selection functions for threatened grizzly bears in Alberta, and combined this habitat model with a spatial model of mortality risk for bears developed using spatial locations of mortalities, mostly human-caused (Nielsen *et al.* 2004). They then combined the two spatial models to identify primary sink and source habitats, secondary sink and source habitats, and non-critical habitat for grizzly bears. This model was then spatially mapped for grizzly bears on the landscape, identifying important sink areas for grizzly bears (Figure 10.11). Sink habitats were closely associated with roads and timber harvest. Therefore, Nielsen *et al.* (2006) recommended adopting access management of industrial roads to increase security and habitat quality for grizzly bears (Figure 10.11).

While Nielsen *et al.* (2004) used a large sample size of over 279 spatial mortalities of grizzly bears over 25 years, other recent studies have developed spatial models mortality risk for endangered species using fewer data and complementary approaches. For example, Falcucci *et al.* (2009) developed an integrated occurrence–mortality model for the small brown bear (*U. a. marsicanus*) population in central Italy to identify the "attractive sink" and source habitats. They contrasted bear presence (2544 locations) and mortality data (37 locations) used as proxies for demographic performance. Both Johnson *et al.* (2004) and Schwartz *et al.* (2010) used a landscape-linked Cox-proportional hazards survival model with telemetry locations of grizzly bears over 22 years in the Greater Yellowstone Ecosystem and with 63 grizzly bear mortalities to develop spatial mortality risk models. More carnivore ecologists should use these methods to combine risk and habitat models to define source-sink habitats.

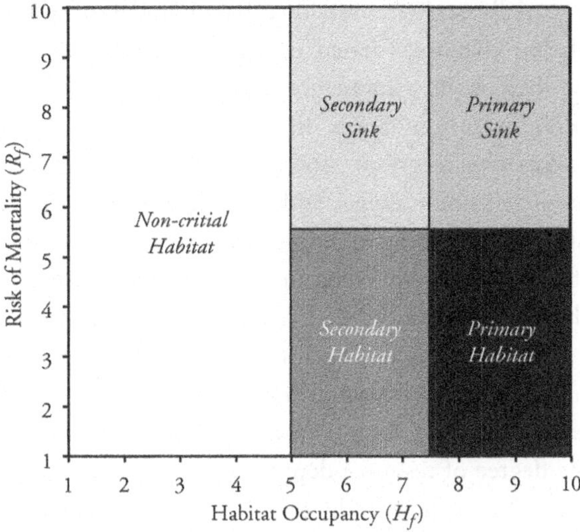

Fig. 10.11 Predicted habitat states for west-central Alberta based on combining habitat quality from an RSF model and spatial mortality risk predictions. Source: Nielsen *et al.* (2006).

10.4.3 Spatially explicit population models

A sophisticated approach to linking critical habitat to population size is to develop spatially explicit, individually-based, population models (Morris and Doak 2002; Carroll and Miquelle 2006; Linkie *et al.* 2006). Population viability analyses (PVA) predict the probability of persistence of a population (Boyce *et al.* 2001; Morris and Doak 2002). Although PVA models have faults (Caughley 1994), they are useful for making relative comparisons between different management or recovery scenarios for engendered species, and often help identify critical knowledge gaps (Brook *et al.* 2000; Holmes *et al.* 2007). Making PVA spatially explicit requires a link between populations and habitats. This link is most often made using simulation models of realistic movements and survival of individual animals on a specific landscape. Spatial PVA accommodate the landscape context, habitat fragmentation, and meta-population structure (Carroll *et al.* 2003b; Linkie *et al.* 2006). The cost of these models, of course, is the requirement of large datasets and the difficulty of parameterizing all required inputs with empirical data. The models present a tradeoff of parsimony versus complexity. Spatially explicit population viability models have been used for tigers (Carroll *et al.* 2003a; Linkie *et al.* 2006), wolves (Carroll *et al.* 2003b), and have even included the effects of climate change for Canada lynxes (*Lynx canadensis*) and American martens (*Martes americana*) in the eastern US (Carroll 2007).

Alternatively, spatially explicit, individual-based models of space use can be used to model population dynamics based directly on landscape characteristics and estimates of basic behavioral parameters. Mitchell and Powell (2004) presented optimality models for home ranges that maximized the benefits of spatially distributed resources over costs of repeatedly visiting resource-bearing patches. These models predicted home ranges and their distribution on a landscape under resource-maximizing and area and minimizing strategies. The resulting spatial distribution of home ranges depended on spatial characteristics of resources and the extent to which animals reduced the value of resources (i.e. resource depression) to conspecifics through consumption or protection. Simulating home ranges on a landscape using these models produced predicted distributions of animals that ranged from ideal free to ideal despotic (Fretwell and Lucas 1970; Fretwell 1972), depending on the degree of resource depression.

Using these spatially explicit, individual-based, home-range models, Mitchell and Powell (2007) showed that black bears living in the southern Appalachian Mountains generally pursued an area-minimizing strategy for selecting their home ranges, with slight levels of resource depression (e.g. Box 10.1, Figure 10.5). This

Fig. 10.12 Change in area of simulated, area-minimizing home-ranges for female black bears in the Pisgah Bear Sanctuary, North Carolina, as a population increases. Simulations were of sequentially established optimal home ranges constructed under an area-minimizing strategy with moderate resource thresholds and low resource depression (Mitchell and Powell 2007), and based on the food component of a habitat suitability index (HSI) for bears in the Southern Appalachians. As more home ranges are added to the sanctuary, area of home ranges increased in size, suggesting that area of home ranges may be useful for understanding population size (*N*). Eventually, no new area-minimizing home ranges could be added to the sanctuary, resulting in a maximum of 52, the estimated carrying capacity (*K*) for the Pisgah Bear Sanctuary. Source: Mitchell and Powell (2011).

finding has very strong ecological implications because resource depression sets a maximum number of home ranges a landscape can support. Thus, home-range models such as these can be used to estimate both the distribution of animals and the carrying capacity (K) of a landscape for those animals, without knowing their abundance. Accordingly, Mitchell and Powell (2011) estimated carrying capacity for adult female bears, K_{AF}, in their study site by sequentially adding simulated, area-minimizing home-ranges to a resource landscape comprising the food component of a habitat suitability index (HSI, Zimmerman 1992; Mitchell *et al.* 2002; Box 10.1) and using behavioral parameters found best to predict home ranges for adult females (Mitchell and Powell 2007). Simulated home-ranges increased in area as the simulated population grew; the point at which no new home-ranges could be added predicted that K_{AF} was approximately 52 (Figure 10.12). Mitchell and Powell (2011) then estimated carrying capacity for all bears (all age and sex classes except cubs), K, by adjusting K_{AF} for the proportion of adult females in the population, yielding $K = 126$ bears. For the 235-km^2 study site, density at carrying capacity was 0.54 bears/km^2, which is only slightly higher than the upper limit of density estimated for black bears living in the nearby and fully protected Great Smoky Mountains National Park (0.35 bears/km^2; McLean and Pelton 1994).

10.5 Conclusions

Research that provides the most rigorous understanding of carnivore habitat scientifically possible is based on asking good questions first and foremost, hypothesizing good answers to these questions based on both theory and empirical evidence, testing the hypotheses by comparing their predictions to empirical data using the best analytical approaches available, and linking selection behavior directly to population consequences. This is a demanding process at all levels: asking good questions is difficult, developing good hypotheses is difficult, mastering rapidly evolving, highly complex analytical techniques is difficult, bridging from behavior to demography is difficult. Pressing management and conservation needs facing carnivores rarely allow the luxury of easier approaches that provide weak to poor inferences, limited scope and generality, and ultimately uncertain applicability (at best).

11

Describing food habits and predation: field methods and statistical considerations

Erlend B. Nilsen, David Christianson, Jean-Michel Gaillard, Duncan Halley, John D.C. Linnell, Morten Odden, Manuela Panzacchi, Carole Toïgo, and Barbara Zimmermann

11.1 Quantifying predators' diets

During the last century, a variety of field techniques have been developed to gain insight into predator–prey relationships. A common starting point for a predator–prey study is to quantify the predators' diet. Although the analysis of diet reveals little of the predation process (an item being ingested through predation or scavenging), diet information helps us understand how predators of a particular species relate to their various potential prey and to their environment. Traditionally, diets have been analyzed using scats, stomach contents, by searching for kills of radio-collared individuals, and by following animals' tracks in the snow or sand. These methods are still important but have been supplemented by recent advances in DNA technology, such as DNA barcoding (Roth *et al.* 2007; Valentini *et al.* 2008) and stable isotope techniques (Crawford *et al.* 2008). These new techniques provide more certain identification of prey species in scat samples and avoid the many problems of visual or microscopic identification (Teerink 2004).

11.1.1 Scat analysis

Scat analysis is used widely to estimate the amounts of different foods ingested by carnivores based on identifying the indigestible parts of animals and plants found in scats (Putman 1984). Standard laboratory procedures (reviewed in Reynolds and Aebisher 1991) allow the identification of prey species from macroscopic, undigested remains, such as teeth, bones, feathers, tissues, and exoskeletons of insects, and from microscopic analysis of hair and of invertebrate exoskeletons found in scats. Items in scats can be identified using classification keys (Day 1966; Teerink

2004) and by comparison with reference materials (e.g. seeds, feathers, skeletons) collected from the study site. When analyzing scats of small- and medium-sized predators, examine also microscopic remains to ascertain the occurrence of chetae, as earthworms can comprise a substantial part of a diet.

Scat analysis is simple, cheap, noninvasive, allows relatively large sample sizes, but presents both technical and interpretation difficulties. Bias and sampling error can arise as early as during scat collection due to the inclusion of scats from non-target species, or due to inadequate study design. Collecting scats at kill sites, along predator tracks, or along trails is particularly prone to bias for predators with large prey because the production of several scats per individual prey may lead to over-representation of a given prey item. If such pseudoreplication occurs, subsampling is useful (Mattson *et al.* 1991), or considering all scats collected at a single kill-site or along one trail as one sample (Marucco *et al.* 2008). Modeling can help in finding the optimal sample size to avoid a lack of power when comparing diet within and among species. Targeted collection of scats is sometimes desired; scats at a den might provide insights into the diets of offspring vs. adults (Lindström 1994; Panzacchi *et al.* 2008a). In addition, recently developed fecal DNA methods allow sex determination from scats (e.g. Hedmark *et al.* 2004), potentially allowing one to compare diets by sex.

A second possible source of bias and error lies in the misclassification of food remains. Some predators ingest a wide variety of animal and plant items, as well as anthropogenic foods, such as garbage, whose remains in scats take the form of macroscopic fractions intermingled with microscopic particles. This bias and error can be reduced through proper training of the lab personnel (Spaulding *et al.* 2000) and by applying the point frame method for identifying prey remains (Ciucci *et al.* 2004). Even when prey remains are identified correctly, their ecological significance should be carefully considered. For instance, hairs of predators of the target species, which occur often in scats, might indicate intraspecific predation, or scavenging, or simply self-grooming. Similarly, it is not possible to conclude whether prey remains in a scat indicate predation or scavenging.

Choice of analytical method is critical. As no unbiased procedures exist, combining techniques is recommended. As an index of how often a predator eats a given prey item, the most simple and time-saving method is the *frequency of occurrence* (FO), which measures the number of scats (n_i) containing remains from food category *i* with respect to the total sample size of scats (N), thus (Leckie *et al.* 1998):

$$FO_i(\%) = (n_i/N) * 100.$$

Even though the FO has the advantage of providing results that can be compared among studies, it has the disadvantage of treating all prey items equally, regardless of size or probability of leaving remains. Hence, it over-represents small items such as insects (Ciucci *et al.* 1996).

A modified version of the frequency of occurrence, the *whole scat equivalents* (WSE), attempts to limit this problem by summarizing the relative volume or weight of each prey category within the sample (Angerbjörn *et al.* 1999). The plot of the FO against the mean volume, which can be calculated by dividing the sum of the volumes of each item in all scats by the total number of scats, helps one visualize the contribution of each item to the total volume of the scats (Kruuk and Parish 1981). Such methods based on volume are usually quick and easy, but they contribute little to understanding the amounts of different foods ingested or the nutritional value of the food.

If one wishes to investigate consumption or gain insight into the nutritional significance of different foods, combine FO with methods that use remains in the scats to estimate the fresh weights of different foods ingested. Methods based on biomass seem to have the greatest potential for estimating the actual bulk consumed, but they tend to be the most time-consuming, as accurate models describing the relationship between prey biomass consumed per scat produced are required (Rühe *et al.* 2008). The proportion of each macroscopic item can either be measured directly, or can be estimated by volume, and then multiplied by the total dry weight of the scat to estimate the dry weight of each food category (Reynolds and Aebisher 1991). Then apply coefficients of digestibility, quantifying the ratio of fresh weight of a given prey to the dry weight of its remains in scats, to estimate the fresh weight consumed (e.g. Jędrzejewski and Jędrzejewska 1992). Coefficients of digestibility can be obtained through rigorously repeated species-specific feeding trials, where predators are fed known amounts of different foods (Lockie 1958; Weaver 1993). The weights of consumed earthworms can be estimated from the number of chetae and gizzard rings (Brøseth *et al.* 1997). Even though coefficients of digestibility can be found in literature for some species, their definitions vary among authors and the combination of different coefficients may lead to significant biases (Reynolds and Aebisher 1991). Estimates of biomass ingested is overestimated using this approach, when the prey are not completely consumed (Ciucci *et al.* 1996).

As all methods for extrapolating from occurrence in feces to biomass or energy consumed are fraught with a range of errors and assumptions, researchers often adjust techniques to their particular studies, thereby producing countless variants of each technique and complicating comparative studies among species or areas. Nonetheless, all above-mentioned methodologies tend to rank prey items similarly when the diet is based on few, large prey (Ciucci *et al.* 1996). Inconsistencies often arise, however, when the diets include a wide spectrum of prey sizes and species (Liberg 1982). Quantifying biases and errors associated with data collection and analysis (see Reynolds and Aebisher 1991) facilitates comparisons among studies.

11.1.2 Analysis of partly digested food items

If a sufficient number of carcasses of the target species are available (e.g. due to hunting or trapping), diet can be investigated by evaluating the contents of stomachs and gastrointestinal tracts (Thompson *et al.* 2009). The method is based on washing and separating the material contained in the gastrointestinal tract with a sieve, and identifying prey remains using classification keys and reference collections. As with scat analysis, FO should be complemented with methods providing estimates of the nutritional significance of foods to the predator. For each food item, multiply indices of metabolizable energy by estimates of the minimum and maximum food consumption per meal, obtained from captive conspecifics, to get estimates of caloric intake. This method accommodates different digestibilities among foods, and therefore accommodates the importance of large items.

An advantage of analysis of stomachs and gastrointestinal tracts is that the sex, age, and body condition of individual predators are usually available. One disadvantage of the investigation of the gut contents is that the results may not be comparable to those obtained with scat analysis; even the contents of stomachs and intestines may differ considerably (Witt 1980). Discrepancies in the outcomes of the three approaches reflect differences in the process of digestion at different stages. Of course, carcasses and scats may suffer different age and sex biases (Cavallini and Volpi 1995). If the carcasses belong to animals killed at bait, gut analyses should be avoided or the results should be interpreted with care.

11.1.3 Snow- and sandtracking

Following tracks in the snow has been an important source of data on predation in northern environments. Snow can be a good substrate to record animal tracks and often allows one to follow continuous behavioral sequences with little bias related to prey detectability. Sometimes one can assign a track to an age or sex class based on size and behavior (raised leg urination, for example, by canids). Following tracks allows a researcher to collect scats and document kills, to investigate hunting behavior (sometimes of known individual predators), and to quantify hunting success (Sand *et al.* 2005). Tracks can be used to monitor predator populations (Chapter 16), based on statistical modeling of track frequencies and distributions (Wabakken *et al.* 2001), or based indirectly on DNA analyses from scats collected along tracks (Kohn and Wayne 1997). Sometimes one can estimate the age of the different stretches of track, facilitating the calculation of kill rates (Pedersen *et al.* 1999) and the predator's functional response (O'Donoghue *et al.* 1998). Snow-tracking combined with radio-telemetry (Chapter 7) improves the temporal precision of track data and identifies individuals leaving tracks. Sand may also provide a suitable substrate for following predator tracks (Bothma *et al.* 1984). The major drawback of ground-tracking methods is their dependence on weather and climatic conditions; for some studies, the difficulty of assigning tracks to individual predators is a problem.

11.1.4 Telemetry-based methods to study predator diet

The development of VHF radio-telemetry methods in the 1960s and 1970s allowed individual-based studies of animal movement and behavior. By using telemetry to locate a predator frequently and to document its movements, activity patterns, and habitat use, a researcher can identify potential kill-sites. Intensive aerial-tracking surveys within given time periods (e.g. Vucetich *et al.* 2002) can be effective but are expensive, constrained by daylight and good weather conditions, and often biased due to low detection rates of kills in closed habitats and during snow-free periods, especially for small prey species. Searches of potential kill-sites often produce prey remains. Because telemetry allows kills to be linked to time, one can calculate kill rates. Such methods remained the standard approach for studying large predators (e.g. O'Donoghue *et al.* 1998; Laundre 2008; Nilsen *et al.* 2009a) until the late 1990s. Data collected by this method may, however, be biased towards large prey items, which are easy to detect and which cause a predator to spend more time at a kill site or revisit the site several times. The use of carcass-searching dogs can reduce this bias. In the late 1990s, the advent of GPS technology provided many new opportunities (Chapter 7). GPS data are more accurate than traditional VHF telemetry data and typically allow the collection of many more location estimates (Sand *et al.* 2005). As a result, the detection of kill sites has improved immensely, producing more reliable measurements of kill rates, even under snow-free conditions when carcasses are hard to find. GPS telemetry has been used predominantly with large-bodied predators (>10 kg) that can carry the weight of a GPS collar and that consume mostly large prey (Figure 11.1).

With respect to studying predation with GPS telemetry, researchers need to develop objective criteria to identify points of interest, especially clusters of positions, to prioritize in an unbiased manner the sites for ground searches for prey remains. Researchers have started to develop rules for identifying kill sites (Anderson and Lindzey 2003; Sand *et al.* 2005; Zimmermann *et al.* 2007). Binomial regression models for presence and absence of large kills at clusters of GPS

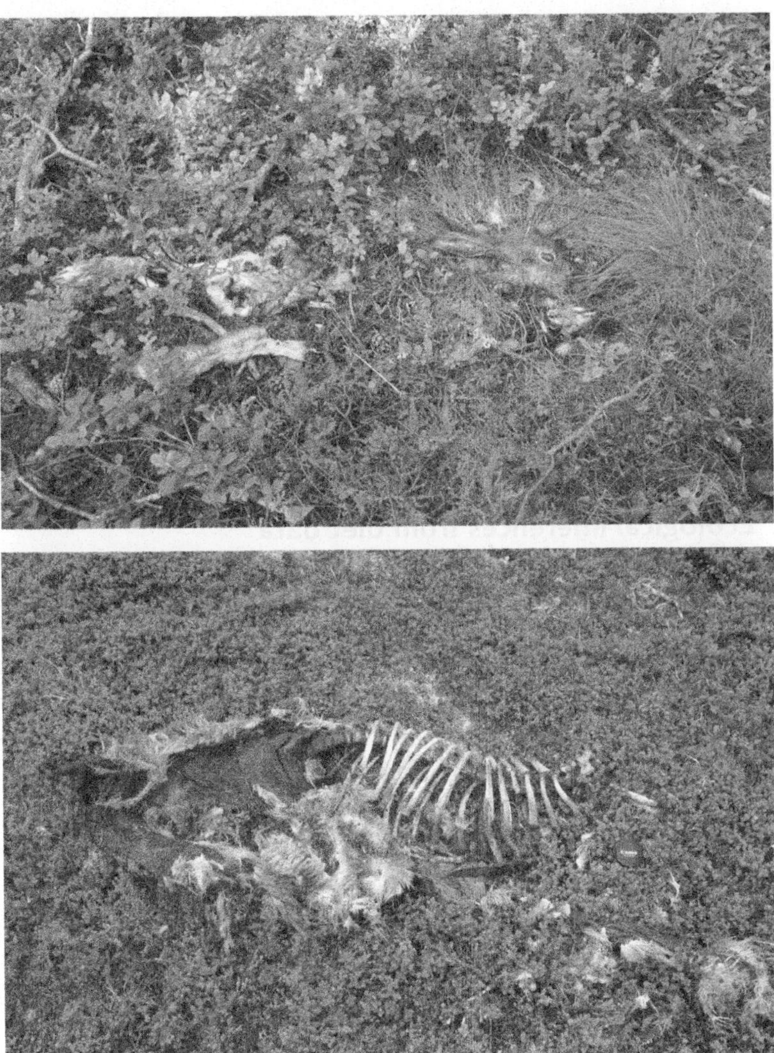

Fig. 11.1 By visiting clusters from radio-collared animals, researchers are able to find the remains of prey animals. Top photo shows the remains of a hare (photo: Robert Needham), whereas the lower photo depicts the carcass from a semi-domestic reindeer (photo: Andrea Mosini). Both prey were located by visiting clusters of tracking positions from radio-collared lynx in Northern Norway.

positions (Anderson and Lindzey 2003; Zimmermann *et al.* 2007), two-step binomial and multinomial regression that estimate the probability of a site holding a large-bodied kill, a small-bodied kill, or no kill (Webb *et al.* 2008), classification trees that divide clusters into kills or non-kills by threshold criteria of predictor variables (Tambling *et al.* 2010), and hidden Markov modeling techniques to distinguish kill sites, bed sites and movements (Franke *et al.* 2006), all hold potential. In multicarnivore systems with predator interference, field visits to clusters are needed to determine displacement and scavenging, which might bias estimates of kill rates based on cluster methods (Ruth *et al.* 2010). Sampling effort required appears to depend heavily on characteristics of both predator and prey; kill sites appear easier to find for large, solitary felids that hunt large prey than predators such as wolves that show less stereotypic behavior when handling kills. Several issues, such as positioning bias in different habitats, still require attention, yet this new technology is opening a range of possibilities that were inconceivable only 10 years ago. Telemetry-based approaches, however, are less useful for finding smaller prey items or multiple or surplus killing events (common for livestock) when the predator does not stop for a significant period to consume the item killed.

11.2 Ecological inferences from diet data

The field methods described above are much utilized techniques in many field studies of predator–prey interactions. However, with these data at hand, one might also ask questions that go beyond simply describing the diet and food habits of the predator—questions that directly relate to the impact of predation on the prey population. We will here focus on methods used to estimate kill rates and functional responses, prey selection, niche breadth, and diet overlap.

11.2.1 Quantifying kill rates and functional responses

Kill rate is defined as the number of prey items killed by a predator (individual or group) within a certain time window. Linking this information to prey density (Holling 1959) or predator and prey density (Abrams and Ginzburg 2000) makes it possible to estimate the functional response of the consumer. Kill rates can be assessed directly by tracking predators on snow or with telemetry during defined sampling intervals with the aim to detect all kills within these periods, can be assessed indirectly by comparing prey densities or mortalities in areas or time periods with and without predators, or can be inferred from scat analysis.

Estimating functional responses based on such field data is not trivial. Estimates of prey (and predator) density must be available (Chapter 5). Obtaining robust density estimates remains a challenge. When data are available to model functional responses, the most commonly used approach is to fit non-linear regression models to the data, assuming Gaussian distributed error terms, and to compare candidate models based on Akaike's information criterion (AIC) (Vucetich *et al.* 2002; Nilsen *et al.* 2009a). Robust model selection is often hampered, however, by the need for large datasets to estimate key parameters in the models (Marshal and Boutin 1999; Nilsen *et al.* 2009a; see also Vucetich *et al.* 2002). If one assumes that prey depletion is negligible within the sampling period, as is usually assumed for large predator–large prey systems, models are usually expressed as kills per time unit (often per month or 100 days for large carnivores). Integrated models may work better when prey are depleted (Vucetich *et al.* 2002).

In addition to the challenge of differentiating among different functional responses, different sampling regimes can yield different estimates for the parameters in a particular functional response model. At Isle Royale, USA, comparing functional-response models for wolves preying on moose across different spatial scales (whole island, per pack, and "mixed scale") resulted in different models being selected for different spatial scales (Jost *et al.* 2005); nonetheless, the selected models included predator dependence and satiation at all scales. Because the number of tracking days needed to obtain robust estimates of kill rates varies for different predator–prey pair systems, the number of tracking days should be included as a part of the general assessment of the robustness of the estimated model. New methods for analysis of GPS telemetry data are continuously being developed, for example, distinguishing between search time and handling time (Merrill *et al.* 2010) allowing researchers to test predictions of optimal foraging models. Search plus handling times sum to the inverse of kill rate. By relating search time to habitat variables and prey densities, attack success and predation risk can be assessed from environmental variables directly. For small predators, obtaining kill sequences by following individuals is complicated because (1) the remains of the kills are hard to find and recover, (2) the predators' small body sizes result in rapid (often within a few minutes) and total consumption, and (3) some species, such as stoats (*Mustela erminea*) and weasels (*M. nivalis*), frequently kill prey in subnivean or subterranean spaces. Thus, inferences based on a predators' diets have been used to estimate predator kill rates and functional responses (Gilg *et al.* 2003, 2006). The basic assumption in these approaches is that a close link exists between a predator's diet and its functional response, so that the relative occurrence of a given prey in the predator's feces, together with knowledge of the predator's metabolism, can be used to calculate prey consumption. In an alternate approach, Sundell *et al.* (2000) manipulated the densities of radio-collared voles (*Microtus*) within large enclosures, measured kill rates of weasels at different prey densities in a relatively controlled manner, and thereby estimated a functional response.

Miller *et al* (2006) inferred about predator functional responses from variation in observed survival rates of prey by integrating formulations of predator functional responses into the estimator of survival. This approach opens the possibility of using data on prey survival to estimate directly one of the key parameters of a predator–prey system. Advances in the modelling of survival probabilites might prove to be a valuable extension to analysis of functional responses, at least in cases where mortality factors are known and mainly caused by predation.

The final aim of diet analysis is often to estimate the total consumption by a predator population. Differences in diet among groups of scats (e.g. collected from animals in different areas or in different seasons) can be detected using χ^2 tests or contingency tables (Wright 2010), or with logistic regression to model the presence or absence of a given item in the scats. Multinomial models developed for capture–mark–recapture data can be used to quantify uncertainty of diet estimates and to assess differences in diet composition when using presence/absence data (Lemons *et al.* 2010). The capture–mark–recapture approach should be more productive when foods are quantified by weight or volume. As the proportions of different food items in a scat are interdependent (i.e. they sum up to one), such data should be analyzed with methods taking into account this interdependency (e.g. compositional data analysis; Reynolds and Aebisher 1991; Aitchinson 1986).

11.2.2 Studying selection—the difference between use and availability

The basis for selection of prey is that individuals of different species, or subgroups within species, yield more energy than do other prey (Krebs and Davies 1993). One generally expects a predator always to try to capture the energetically most valuable prey if encountered, whereas other prey should be captured only if the most valuable prey is rare (Charnov 1976a, 1976b). Apart from experimental work, direct tests of this hypothesis are uncommon, as estimating energy yield and expenditure associated with different prey items is not easy.

Technically, prey selection is a measure of the deviance between the proportion of prey X in the diet, compared to the availability of prey X in the standing population of all prey. Study designs for measuring used and available are grouped into three main categories based on the definitions and assumptions regarding use and availability (Manly *et al.* 2002). In Design I, the resource (prey) use is not recorded for specific individuals, and used and available resources are assumed to apply to the whole predator population in a study area. Individual resource use is recorded in the Designs II and III. For Design II, the availability of resources is estimated at the population level but for Design III the availability of resources is estimated at the individual level. Sampling of scats usually allows only the application of Design I. Since the introduction of fecal genotyping, however, scat analyses may also convey individual-based information, thus allowing use of Design II (Prugh *et al.* 2008). Design III can be applied in telemetry-based studies, if estimates of prey availability are available for each individual predator. Among the three categories of study designs, Designs II and III clearly have the potential to address a wider range of questions. For instance, variation in prey selection with respect to sex and age classes cannot be analyzed using Design I. Taking individual differences into account provides insights into patterns of resource partitioning stemming from individual specialization (Araújo *et al.* 2010), which may have stabilizing effects on population dynamics (Kendall and Fox 2002) and may affect sympatric speciation (Schluter and McPhail 1992).

A common difficulty in all studies of prey selection is defining and estimating prey availability. Prey availability is not directly equivalent to prey abundance (Molinari-Jobin *et al.* 2004). Rather, prey availability is a function of prey abundance, prey antipredator behavior, differential vulnerability of prey of different life-cycle stages, and more. Consequently, prey availability changes across time, not only because prey abundance changes, but also because different age classes are more prone to predation. Prey vulnerably might change regionally (Molinari-Jobin *et al.* 2004; Panzacchi *et al.* 2008b; Nilsen *et al.* 2009b).

Once use and availability of prey have been estimated, several methods exist to measure selection. Manly's selection index compares the relative usage (r_i/n_i) of prey of category i (species, age class or other categorical variables) to the relative usage of all prey $\sum r_j/n_j$ in the environment; Chesson 1978):

$$\alpha_i = \frac{r_i/n_i}{\sum r_j/n_j},$$

where r_j represents the resource of category i that is used, n_i the resource in category i that is available the selection index α_i range from 0 to 1, and a higher value indicates greater selection.

Estimates of both use and availability of prey are susceptible to biases and error that increase the risk of rejecting the null hypothesis of no selection (Type I error). Furthermore, values of Manly's index depend strongly upon which prey species are considered to be available. Including or not including an abundant prey species that is rarely consumed may reverse preference classifications of the other prey species (Johnson 1980). In such cases, using Johnson's (1980) rank index is less sensitive to subjective choices of available resources. This index ranks use and availability of resources and the difference in ranks forms the basis of selection classifications.

In the literature, *use*, *selection*, and *preference* are often applied interchangeably. Use and preference generally differ from selection by being independent of availability. Use refers to a food item being consumed in a specified time period, while preference is the probability that a food item is selected when offered on an equal basis with other items. Accordingly, the existence of a preference requires an outcome of behavior by the predator, whereas the selection index is simply an estimate of non-random association between predators and prey. Selection is not equivalent to importance. A prey type can be highly selected, even if it is relatively rare and, therefore, rarely eaten; accordingly, the importance of this prey for the survival and reproduction of the predator may be minimal compared to common prey that are not selected.

11.2.3 Quantifying food niche breadth and diet overlap

Diet data can be used to investigate and compare the diet breadth in different areas and periods and among different species. Based on the breadth of their diets in different ecological settings, focal species can be placed along the generalist–specialist continuum (Jedrezejewska and Jedrezejewski 1998). This approach also provides insights into the ways that species within ecological communities partition the available resources, and allows inferences on competition and coexistence.

The study of resource partitioning has a long history in community ecology. The term *trophic niche* was coined at the beginning of the twentieth century to describe the position of a given species in the foodweb of a community. In the absence of competition, the observed resource use constitutes a species' *fundamental trophic niche*, while in the presence of competitors the species' niche can be substantially different and it is termed its *realized trophic niche* (Hutchinson 1953). The breadth of the fundamental and realized trophic niches can be measured by a variety of classical diversity indices that are sensitive to both the number of items in the sample and to their relative abundance. One of the most widely used indices for measuring the niche breadth is the *Shannon–Wiener index* (H') (Krebs 1999):

$$H' = -\sum p_i(\ln p_i).$$

where p_i is the proportion of each food item in the sample. H' ranges from 0 (lowest niche breadth) to 1 (greatest breadth).

Alternatively, several studies use the *Levins' index*, B, and the *Levins' standardized index* B_s (Levins 1968):

$$B = 1/\sum p_i^2.$$

where p_i is the proportion of each food item in the sample;

$$B_S = B - 1/n - 1.$$

where n is the total number of food categories identified.

Both the choice of the method for quantifying the diet and the choice of the index for estimating the niche breadth affect one's results. While H' tends to overestimate the importance of rare food items, B tends to emphasize the evenness in the distribution of items (Pielou 1975).

The first method for estimating niche overlap is based on the relative use of different segments of a niche resource axis and on the overlap between species in their use of common segments (MacArthur and Levins 1967). A variety of other indices followed (e.g. Schoener 1971) but one of the most widely used is the index of trophic niche overlap between species j and k developed by Pianka (1974):

$$O_{jk} = \sum p_{ij} p_{ik} / \left(\sum p_{ij}^2 \sum p_{ik}^2 \right)^{\frac{1}{2}}.$$

where p_{ij} is the proportion of the food item i in the diet of species j;. O ranges between 0 (total niche separation) and 1 (total overlap).

Multivariate ordination techniques, such as correspondence analysis, can be used for a graphical visualization of the resource partitioning in the community by displaying the matrix formed species (e.g. predator guild) and food items (e.g. prey guild) in a multidimensional space.

The indices of niche overlap provide useful insights into the resource partitioning within a community at a given time and place. Trophic interactions are spatially and temporally dynamic, however, and food niches converge and diverge as the diversity of resources changes (Schoener 1982). Hence, a repetition of a study in different seasons is often advisable. Also, the existence of trophic niche overlap does not necessarily imply the occurrence of competition, as competition is shaped by a variety of environmental and behavioral parameters (Colwell and Futuyma 1971).

11.3 Using stable isotopes to infer trophic interactions

Stable isotopes, particularly (but not only) those of carbon and nitrogen, can be used to investigate the assimilated diets of organisms and, hence, trophic interactions (Kelly 2000; Post 2002; Newsome *et al.* 2007). Foods vary systematically in their proportions of elemental isotopes: in the case of carbon, ^{12}C and ^{13}C, and for nitrogen, ^{14}N and ^{15}N (Brand 1996). The isotopic "signature" of a consumer's tissues reflects the proportional intake of the isotopes in its diet (DeNiro and Epstein 1978), with correction for preferential uptake and loss of given isotopes in digestion process, known as diet-tissue or trophic fractionation. Trophic fractionation varies for different elements, species, and tissues (Dalerum and Angerbjörn 2005). Marine food-chains tend to be longer than terrestrial chains and baseline $\delta^{13}C$ values differ. Thus, when a predator mixes marine and terrestrial food sources, the typically large differences in both $\delta^{13}C$ and $\delta^{15}N$ make discriminating their proportions in the diet relatively simple (e.g. Ben-David *et al.* 1997).

Using stable isotopes to estimate trophic position relies on trophic fractionation. Generally, $\delta^{15}N$ fractionates much more strongly than $\delta^{13}C$. Generic values for trophic fractionation, typically *ca* +3–3.4$^{0}/_{00}$ for $\delta^{15}N$ and *ca* +1$^{0}/_{00}$ for $\delta^{13}C$ (Kelly 2000; Post 2002), are often used in trophic analyses, but naive application of such values can be a pitfall for the unwary. The isotopic signature of a predator is not usually sufficient to infer trophic position without an appropriate isotopic baseline, and estimates of trophic position are very sensitive to assumptions about the trophic fractionation of $\delta^{15}N$ in particular, and to different methods of generating an isotopic baseline (Post 2002). The average $\delta^{15}N$ value for terrestrial omnivores in the literature is slightly *lower* than that for terrestrial herbivores, probably because of the very high fractionation values reported for ruminant herbivores (Darr and Hewitt 2008) compared to animals with simpler digestive processes. Moreover, different species eating identical diets vary up to 3.6$^{0}/_{00}$ in $\delta^{15}N$ fractionation of the same tissue, i.e. more than a whole trophic level at the usual generic rate (Sponheimer *et al.* 2003). Therefore to conclude, without species- and tissue-specific fractionation being taken into account, that similar $\delta^{15}N$ values in different species of predator (or any animal) indicate trophically similar diets is unwise; or, conversely, that differences in values between species imply a difference in diet or trophic position.

This consideration also renders assessing variation in trophic niches of individual predators within a species more complex than simply reading off a $\delta^{15}N$ value and mapping directly to "trophic level." Although for a given species and tissue, trophic fractionation can be assumed to be uniform in strict carnivores, whose foods are typically similar in protein concentration, uniformity may not be the case for omnivores, such as bears (Phillips and Koch 2002). Different prey species with identical diets may vary in isotope profiles, and prey of differing diets and trophic positions may have similar isotope profiles.

Where (1) baseline isotope values for diet sources (different prey species and isotopically distinct classes within those species) and (2) fractionation values for the predator tissue analyzed are fairly securely known or can be inferred, the comparison of stable isotope values within and among species is a powerful technique for assessing relative trophic niches and trophic niche width (Bearhop *et al.* 2004). A variation of this technique with wide application in predator studies is to use the method to assess the diet at different time periods, either by measuring tissues, such as blood plasma, with very rapid turnover, or tissues that capture the isotopic signature when growing but are thereafter metabolically inert, e.g. hair or horn (Dalerum and Angerbjörn 2005).

11.4 Estimating non-lethal effects of predation

Measuring the effects of carnivores on their prey is often limited to estimating the impact from the killing and consumption of prey (numerical or direct effects), as this is the most obvious effect of carnivores. Carnivores also affect prey survival, growth, and reproduction indirectly by altering prey behavior or physiology (risk effects or non-consumptive effects). In some cases, the risk of predation alone can be the strongest driver of population dynamics, even stronger than the direct effect of predation (Pangle *et al.* 2007). Because of this, risk effects can cascade to lower trophic levels through two pathways (Figure 11.2). For example, the presence of a carnivore can reduce the foraging time of an herbivore releasing local vegetation from both the reduction in herbivore foraging effort (*trait-mediated indirect effect*) and the reduction in herbivore numbers due to the nutritional costs of this antipredator behavior (*density-mediated indirect effect*). Understanding how carnivores shape the risk of predation as perceived by their prey is central to understanding ecosystem function.

Much of our understanding of risk effects comes from experiments with invertebrates and small animals (Preisser *et al.* 2005); yet much of this work has parallels with vertebrate carnivores in natural systems. Most prey animals are capable of detecting carnivores long before an attack and most prey possess several traits for avoiding predation. While physiological stress responses may exist (Boonstra *et al.* 1998), most research has focused on estimating behavioral responses and their costs. Common behavioral responses to the presence of predators are shifts in habitat selection foraging behavior and vigilance (Lima 1998). Although these responses correlate with one another, antipredator responses cannot be assumed to increase with levels of predation risk (McNamara and Houston 1987; Lima and Bednekoff 1999). Unfortunately, we have few examples where the demographic costs of behavioral responses to carnivores have been estimated for any wild prey species. Surprisingly, most research on risk effects with terrestrial carnivores has focused on estimating how changes in prey behavior affect lower trophic levels (e.g. hardwood plant regeneration in Yellowstone National Park following wolf recolonization; Fortin *et al.* 2005). We know little about how carnivores disrupt survival, reproduction, and growth in prey through risk effects (but see Creel *et al.* 2007).

Information on carnivore behavior is crucial to detecting and quantifying relevant antipredator responses in natural systems. Current approaches differ primarily in temporal and spatial scales. A dichotomous index of predation risk can be applied across the sampling space (carnivore-absent and carnivore-present). At the broadest scales, prey behavior, prey nutrition, or the dynamics of lower trophic levels are compared between ecosystems with and without carnivores

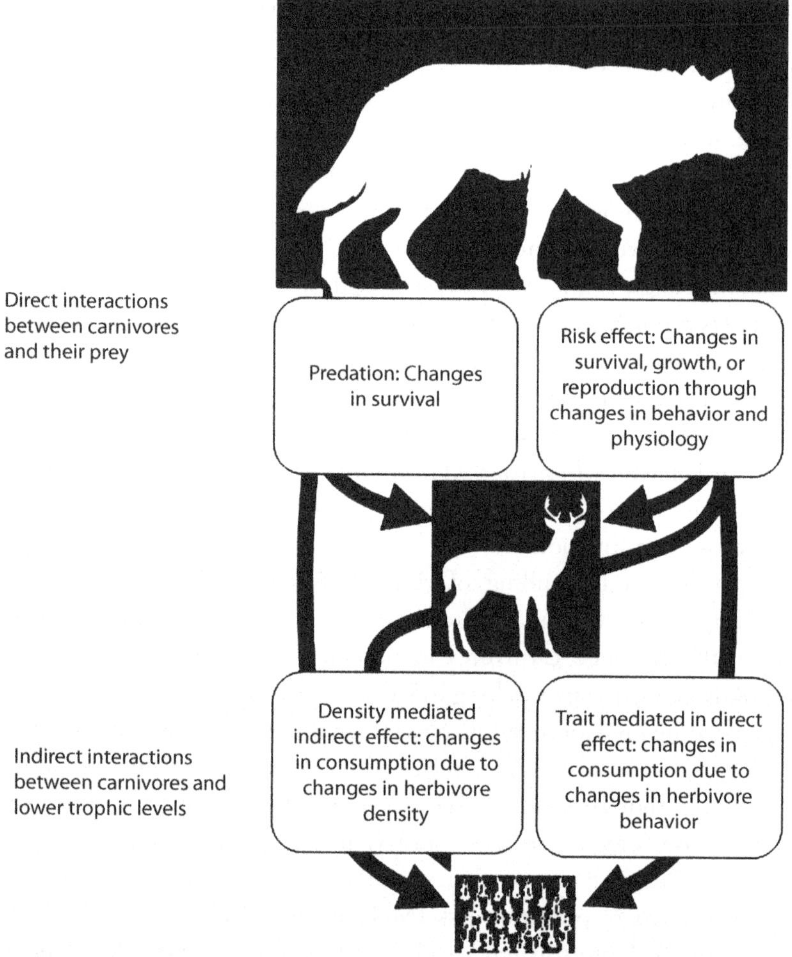

Fig. 11.2 Measuring the total impact of carnivores on ecosystems requires an understanding of the complexity of their interactions. For example, with three species at different trophic levels a carnivore should be expected to affect the demography of its prey through predation and by eliciting the expression of costly antipredator behavior in prey. Lower trophic levels might experience reduced grazing pressure if herbivore numbers are limited by predation, limited by the costs of antipredator behavior, or if herbivore foraging behavior is depressed as part of the antipredator response.

(Wolff and Van Horn 2003), sometimes with the use of predator exclosures (Hodges and Sinclair 2003). Similarly, prey behavior, prey nutrition, or plant dynamics can be compared within an ecosystem but over different states of carnivore status, e.g. prior to, and after, a reintroduction (Mao *et al.* 2005) or before, and after, predator removal programs (Banks *et al.* 1999). At finer scales, particular groups of prey animals within one population or particular sites within a study system may be classified as under the influence of carnivores or not, and compared, e.g. groups near or far from refuges (Frid 1997). These latter methods have detected general relationships between carnivores, prey, and lower trophic levels. Nonetheless, these methods can mask the effects of predators if ecosystems differ widely over space and time. Moreover, many responses to carnivores often occur at fine spatial and temporal scales (Winnie and Creel 2007), indicating that most prey can distinguish variation in predation risk at much finer scales than are usually measured. Hence, the magnitude of the indirect effect of a carnivore could be underestimated or missed entirely at broader scales.

Information on local carnivores can provide finer spatial and temporal resolution of predation risk. Describing gradients of risk rather than dichotomies, such as estimating the carnivore/prey ratio over time and space (Creel *et al.* 2007) or describing spatial variation in local carnivore habitat use (Valeix *et al.* 2009), are becoming more common. These methods rely on efficient carnivore detection and tracking, such as snowtracking, radio-telemetry, or GPS collars.

Most current approaches for estimating the impact of predation risk from carnivores on ecosystems largely assume that a carnivore's proximity to a prey animal confers risk, and that the behavior of a prey animal in the presence of a carnivore reflects its efforts to increase its probability of surviving an attack. Nonetheless, some studies have shown that the behavior of a carnivore (beyond its movement) may also influence how prey determine risk. For example, wolves close to elk affect vigilance of elk, but how recently wolves have made a kill also matters (Liley and Creel 2008). Future research will likely reveal that prey animals are highly attuned to the behavior of local carnivores and adjust their antipredator behavior in response to changes of predation risk at very fine spatial and temporal scales. If so, quantifying the indirect effects of carnivores on ecosystems may require fine scale information on carnivore behavior coupled in time and space to the behavior of prey.

11.5 Some further challenges

While we have learned much about predator–prey dynamics, central aspects of predator–prey interactions are still understudied. Some of these aspects can be addressed using the field methods described in this chapter, other aspects require innovative thinking in the field. In a dynamic predator–prey system, a link between predator consumption rate and predator recruitment, and thus population growth, is generally assumed. Although several studies have documented a positive relationship between prey density and predator density (e.g. Carbone and

Gittleman 2002) and from this inferred a numerical response (Gilg et al., 2003), very few studies have been able to measure the relationship between predator consumption rate and predator population growth rate (Millon and Bretagnolle 2008; but see Vucetich and Peterson 2004). Gaining a better understanding of this relationship will greatly enhance our understanding of predator–prey dynamics. In addition, our knowledge of scramble competition and predator population growth is still very limited. Indeed, models of predator–prey dynamics may make assumptions that contradict behavioral models. For instance, while the hypothesis of economy of resource defense (i.e., territoriality) predicts that at very high and very low prey densities it does not pay to be territorial, many models of predator–prey dynamics assumes that predator interference limits predator number also when per capita food abundance is high. Compared to prey species, the demography and population dynamics of predators are understudied.

In a real and complex world, interactions between prey and predators are indeed complex. The context in which trophic interactions take place contributes to the outcome. We should, therefore, in the future not ask questions such as, "does predation affect prey population dynamics?" but rather seek to identify conditions that determine the relative strength of top-down and bottom-up effects. In some cases, these conditions might relate to the biotic aspects of the system, such as taxonomic differences, differences between hunting styles, and predator–prey body weight ratios, while they might also relate to factors describing the physical environment, such as terrain, landscape productivity, and climate. Further work is needed to integrate the field methods described in this chapter with conceptual models of how predator–prey system ought to behave under these various conditions.

12
Reproductive biology and endocrine studies

Cheryl S. Asa

Historically, data on carnivore reproduction collected during field studies was limited to analyses of anatomy and morphology from animals that were trapped or from carcasses. Examinations of gonads, uteri, and mammary tissue provide some information on age of puberty, reproductive seasonality, pregnancy, and litter size. Nonetheless, because such sampling represents only one point in time for a particular animal, physiological processes are not apparent. For example, one cannot discern from a single anatomical sample whether females of a species are induced or spontaneous ovulators, or whether they are monestrus (that is, have only one ovulation) or polyestrus (multiple cycles until conception). And, despite new technologies, such as ultrasound and endocrine assays, one sample still provides only limited information. These techniques do have the advantage, however, of being noninvasive and of offering opportunities for sequential sampling.

This chapter reviews carnivore reproductive systems and tools available for field studies of carnivore reproduction and endocrinology, and interpretation of results. Despite the diversity of species in Carnivora, available information on reproductive physiology is biased toward canids and mustelids, due to the commercial production of fur from Arctic and red foxes (*Vulpes lagopus* and *V. vulpes*) and mink (*Mustela vison*), which not only made them readily available for study, but also because research affects commercial interests. For a more complete review of mammalian reproduction, see Asa (2010).

12.1 Carnivore reproductive physiology: the basics

12.1.1 Puberty

Puberty is the time when an animal is first able to reproduce, but this is the culmination of a process rather than a discrete event (van Tienhoven 1983), and is accompanied by activation of the hypothalamic–pituitary–gonadal axis, manifest by increased production of testosterone in males, and estrogen and progesterone in females. For practical purposes, puberty can be defined for males as the first appearance of sperm in the ejaculate, although detection of sperm by testicular biopsy or dissection is also diagnostic. For carnivores that have been studied, the first increase of testosterone occurs about 2 months prior to the appearance of sperm in the ejaculate, because of the time required for spermatogenesis and epididymal sperm maturation (coyote, *C. latrans*, Kennelly 1969; domestic dog, *Canis familiaris*, Foote *et al.* 1972; jaguar, *Panthera onca*, Costa *et al.* 2008; ocelot, *Leopardus pardalis*, Silva *et al.* 2010). For females, puberty can be defined as first ovulation but in some species, such as those with induced ovulation (Table 12.1), first estrus might be the more accurate measure, since not every female in estrus may have an opportunity to mate and, hence, to ovulate.

Hormonal markers of puberty include increases in testosterone for males and, for females, increased estradiol associated with estrus and progesterone following ovulation. Gonadal hormones affect many features that can be measured, such as changes in vaginal cytology and vulva size for females, and an increase in testis size and appearance of species-specific secondary sex characteristics in males.

Estimations for the age of puberty from field studies are usually based on the observation of mating or birth of young. Nonetheless, physiological puberty (sperm production or ovulation) can occur without these obvious manifestations since, among males, in particular, not all individuals have an opportunity to mate and most matings are undocumented. Birth of young is an especially conservative measure of puberty, because females may have multiple ovulatory cycles before successfully conceiving and carrying a pregnancy, a phenomenon called adolescent sterility (Spear 2000). Numerous studies of coyotes, for example, have documented the first signs of ovarian and sexual activity during a female's first breeding season, although few females subsequently gave birth (reviewed in Green *et al.* 2002). In coyotes, however, the likelihood of reproducing in the first year appears dramatically influenced by local food availability and density of conspecifics (Nellis and Keith 1976; Windberg 1995), a phenomenon that is likely shared by other species.

Using birth of young as the marker of puberty is also conservative because it ignores social suppression of reproductive behaviors, as has been documented in gray wolves, *C. lupus* (Packard *et al.* 1983). Young wolves may reach physiological puberty (sperm production in males and ovulation in females) at the species-typical age of about 22 months, but be prevented from mating by their parents. In contrast to behavioral suppression, most subordinate dwarf mongooses (*Helogale parvula*, Creel *et al.* 1992), and African wild dogs, (*Lycaon pictus*, van Heerden and

Kuhn 1985) past the typical age of puberty appear to be physiologically suppressed (i.e. suppression of spermatogenesis in males and follicle growth and ovulation in females), although a small percentage of females may be able to ovulate and give birth. Observations of copulation by subordinate male and female banded mongooses (*Mungos mungo*) suggest they are not physiologically suppressed (Rood 1980). Many other reports of reproductive suppression among subordinate members of carnivore social groups have not distinguished between behavioral and physiological suppression (red foxes, Macdonald 1979*a*; silver back, *Canis mesomelas*, and golden jackals, *C. aureus*, Moehlman 1983; Ethiopian wolves, *C. simensis*, Sillero-Zubiri *et al.* 1996).

12.1.2 Seasonal reproduction

In carnivore species of the Arctic and Temperate zones, seasonal control of reproduction is mediated primarily by changes in photoperiod (Turek and Campbell 1979). Some species are considered short-day breeders, because they mate during late summer or fall when day length is becoming shorter. Others are long-day breeders and mate during late winter or spring, after the winter solstice when day length is increasing. Gonadal recrudescence at the beginning of each breeding season is equivalent in most ways to puberty (Ebling and Foster 1989).

In general, seasonal reproduction is characteristic of Arctic and Temperate zone species, although even Tropical species may be affected by seasonal environmental changes, such as rainfall patterns (Lincoln and Short 1980; Bronson 1989). Young are born in late winter or spring in the Temperature and Arctic zones, during the period most auspicious for their growth and survival. Carnivores have relatively short gestations (about 1–4 months; Hayssen *et al.* 1993), which puts mating season in mid to late winter for many species. The factors that select for time of parturition (the ultimate cause, usually climatic factors) differ from those that control the mating period (the proximate cause: activation of the reproductive system; Baker 1938). For species with delayed implantation, however, the time of mating is not constrained by gestation length. This decoupling of birth and mating season from gestation length allows flexibility for otherwise solitary males and females to associate at more opportune times. Furthermore, females can increase their fitness through mate choice, which may have selected for mating to occur when possibilities for female choice or male competition are greatest (Sandell 1990; Spady *et al.* 2007).

Most information on reproductive seasonality comes from the annual distributions of matings or presence of offspring, which are good estimators of female reproductive potential but do not necessarily represent male capacity. That is, males typically produce sperm longer than the period when females in the population may ovulate, effectively bracketing the range of female fertile days. Furthermore, even males of some seasonally breeding species may never cease spermatogenesis completely, although sperm quality or quantity may decline outside the breeding season (e.g. red foxes, Creed 1960; ocelot, margay, *Leopardus wiedii,* and tigrina, *L. tigrinus*, Morais *et al.* 2002). Seasonal reproductive timing for a species appears to vary by latitude (Fletcher 1974), which has been demonstrated for gray wolves (Mech 2002) and red foxes (Lloyd and Englund 1973; Asa and Valdespino 2003).

12.2 Stages of the female reproductive cycle

Of the variety of terms used to describe female reproductive cycles, "ovarian cycle" best encompasses the wide range of patterns and focuses on the recurrent follicular growth and development common to all. For example, induced ovulators can have cycles of follicular growth that are not ovulatory at all if they fail to mate, making the term "ovulatory cycle" inappropriate, whereas "estrous cycle" describes behavioral but not physiological events. In addition, different conventions exist for characterizing the reproductive stages of female cycles. One uses designations that refer primarily to external signs and behavior: proestrus, estrus, and diestrus (sometimes called metestrus in dogs, Jochle and Andersen 1977), with anestrus to describe the non-breeding period of seasonal breeders. The classifications by Conaway (1971) refer to diestrus as pseudopregnancy, because it is characterized by a prolonged period of progesterone elevation similar to pregnancy. Another naming convention refers to follicular and luteal phases to reflect changes on the ovaries, which may be more useful for understanding the underlying endocrine and morphological changes.

The follicular phase of the typical ovarian cycle is characterized by the growth and development of follicles that secrete estradiol, the hormone that stimulates proestrous and estrous behavior and the changes in vaginal cytology and vulva tumescence. Oocyte and follicular growth progress to the mature, tertiary or Graafian follicle with its fluid-filled antrum and culminate in ovulation, involving follicle rupture and release of ova (eggs). Both proestrus and estrus are characterized by elevated estradiol levels; females in proestrus are attractive to males but are only receptive to mating during estrus (Beach 1976). Still, in most species there is not a clear delineation between these two phases, since estradiol progressively increases until the time of ovulation, with what seems to be merely a progressive change that culminates in estrus. In contrast, in dogs and probably other canid species, a pre-ovulatory increase in progesterone clearly separates proestrus from the sexual receptivity of estrus (Concannon *et al.* 1977a; Asa, 1999).

The durations of proestrus and estrus vary widely among species. Perhaps the longest proestrus, determined by sanguinous discharge, has been documented for coyotes (2–3 months; Kennelly and Johns 1976), with wolves next at 6 weeks (Asa *et al.* 1986). Although estrus may be less than 24 hours in some species, in most carnivores it lasts more often from several days to more than a week (Hayssen *et al.* 1993). The increasing progesterone following ovulation terminates the estrous period and initiates the luteal phase, regardless of mating or conception. For induced ovulators that do not mate, however, estrus may recur and appear continuously for a considerably longer period without intervening production of progesterone (Weir and Rowlands 1973).

Although spontaneous ovulation generally is reported for most mammals, induced ovulation occurs in more carnivore species (Table 12.1) than any other taxon, except pinnipeds (Ewer 1973). In spontaneous ovulators, ovulation occurs regardless of the presence of males, whereas in induced ovulators, copulatory stimulation typically is required to trigger ovulation. Nonetheless, species considered to be induced ovulators may, at times, ovulate without mating (e.g. African lion, *Panthera leo,* Schmidt *et al.* 1979; leopard, *P. pardus*, Schmidt *et al.* 1988;

Table 12.1 *Carnivore species with known induced ovulation.*

Species	Reference
Cheetah, *Acinonyx jubatus*	(Czekala *et al.*, 1994)
Lion, *Panthera leo*	(Schmidt *et al.*, 1979)
Leopard, *Panthera pardus*	(Schmidt *et al.*, 1988)
Tiger, *Panthera tigris altaica*	(Seal *et al.*, 1985)
Snow leopard, *Uncia. uncia*	(Schmidt *et al.*, 1993)
Puma, *Puma concolor*	(Bonney *et al.*, 1981)
Clouded leopard, *Neofelis nebulosa*	(Brown *et al.*, 1995)
Ocelot, *Leopardus pardalus*	(Moreira *et al.*, 2001)
Tigrina, *Leopardus tigrinus*	(Moreira *et al.*, 2001)
Margay, *Leopardus wiedii*	(Moreira *et al.*, 2001)
Island fox, *Urocyon littoralis*	(Asa *et al.*, 2007)
Black bear, *Ursus americanus*	(Boone *et al.*, 2004)
Japanese black bear, *U. thibetanus japonicus*	(Sato *et al.*, 2001)
Wolverine, *Gulo Gulo*	(Mead *et al.*, 1993)
Hawaiian mongoose, *Herpestes auropunctatus*	(Hoffmann and Sehgal, 1976)
Ferret, *Mustela furo*	(Marshall, 1904)
Mink, *Mustela vison*	(Hansson, 1947)
Weasel, *Mustela nivalis*	(Deanesly, 1944)
Raccoon, *Procyon lotor*	(Whitney and Underwood, 1952)

mink, Sundqvist *et al.* 1989; clouded leopard, *Neofelis nebulosa*, Brown *et al.* 1995; ocelot, Moreira *et al.* 2001; tigrina, Moreira *et al.* 2001; margay, Moreira *et al.* 2001). In raccoons (*Procyon lotor*), just the presence of a male has been reported to induce ovulation without copulation or even physical contact (Morris 1975). Thus, a continuum of ovulatory mechanisms exists, ranging from exclusive spontaneous ovulation to an absolute need for copulatory stimulation, with most species falling somewhere between those extremes (Jochle 1975).

Following ovulation, the cells comprising the follicle transform to become the corpus luteum (CL), which produces progesterone, the hormone that characterizes the luteal phase or diestrus. If conception has not occurred, the CL regresses and either a new cycle or anestrus ensues. The length of the luteal or diestrus phase varies greatly in carnivores. The typical lifespan of the mammalian CL is about two weeks (Conaway 1971) but in many carnivores it can be considerably longer. In particular, the canid CL continues to secrete progesterone for a length of time equivalent to the species-typical gestation period, hence called pseudopregnancy. Felids and mustelids, which are induced ovulators, also experience pseudopregnancy following a sterile mating (Conaway 1971). The length of pseudopregnancy in ferrets (*Mustela furo*) is equivalent to gestation length (Hammond and Marshall 1930), as in canids. In contrast, in felids the duration is roughly half that of pregnancy (e.g. African lion, Schmidt *et al.* 1979; puma, *Puma concolor*, Bonney *et al.* 1981; leopard, Schmidt *et al.* 1988; snow leopard, *Uncia uncia*, Schmidt *et al.* 1993; cheetah, *Acinonyx jubatus*, Czekala *et al.* 1994; clouded leopard, Brown *et al.* 1995).

12.2.1 Pregnancy

Pregnancy is characterized hormonally primarily by a sustained elevation of progesterone. Other hormones, such as estrogen, prolactin, and relaxin, are important in some species and can be diagnostic. For example, the only hormonal difference between pregnancy and pseudopregnancy in canids is relaxin, which is elevated in the second half of gestation but is absent in non-pregnant females (Bauman *et al.* 2008). Although mean fecal progestagen levels may be higher in pregnant than pseudopregnant canids (e.g. maned wolf, *Chrysocyon brachyurus*, Velloso *et al.* 1998), this difference is not diagnostic because of considerable temporal and individual variability in progestagen concentrations. Relaxin also might be useful in distinguishing pregnancy and pseudopregnancy in species with induced ovulation but, as mentioned above, pseudopregnancy occurs infrequently.

12.2.2 Delayed implantation or embryonic diapause

It is advantageous for all species to have offspring born at a time of year favorable for their survival, which in Temperate zones is usually late winter or spring. For most, the length of gestation determines when mating must occur to time births properly. Some species, however, have evolved a mechanism to adjust the length of gestation so that mating can take place when socially or environmentally advantageous (Mead 1989, 1993; Sandell 1990). That mechanism, called delayed implantation or embryonic diapause, involves the arrest of development at the blastocyst stage before uterine implantation occurs (in European badgers, *Meles meles,* development continues albeit very slowly; Canivenc 1966). Continued, fertile ovulatory cycles are typical for mink early in the period of delay (Hansson 1947; Enders and Enders 1963), which has been described also for European badgers (Yamaguchi *et al.* 2006). Resumption of development and implantation is stimulated by a change in photoperiod and is accompanied by increasing progesterone concentrations (Wimsatt 1975); in some species, such as the mink, a rise in prolactin precedes that of progesterone (Murphy *et al.* 1981). Among carnivores, delayed implantation is common in ursids and mustelids (Table 12.2; (Wimsatt 1975;

Table 12.2 *Carnivore species that experience delayed implantation or embryonic diapause.*

Taxon	Reference
Wolverine, *Gulo gulo*	(Wright, 1963)
River otter, *Lutra canadensis*	(Wright, 1963)
Sea otter, *Enhydra lutris*	(Sinha *et al.*, 1966)
American marten, *Martes americana*	(Wright, 1963)
Beech marten, *Martes foina*	Canivenc in (Renfree and Calaby, 1981)
Pine marten, *Martes martes*	(Canivenc *et al.*, 1975)
Fisher, *Martes pennant*	(Wright, 1963)
Sable, *Martes zibellina*	(Baevsky, 1963)
Stoat, *Mustela ermine*	(Wright, 1963)
Long-tailed weasel, *Mustela frenata*	(Wright, 1963)
Mink, *Mustela vison*	(Pearson and Enders, 1944)
Western spotted skunk, *Spilogale gracilis,*	(Mead, 1968)
Striped skunk, *Mephitis mephitis*	(Wade-Smith *et al.*, 1980)
American badger, *Taxidea taxus*	(Wright, 1966)
European badger, *Meles meles*	(Canivenc, 1966)
American black bear, *Ursus americanus*	(Wimsatt, 1963)
European brown bear, *Ursus arctos arctos*	(Hamlett, 1935)
Grizzly bear *Ursus artcos horribilis,*	(Craighead *et al.*, 1969)
Polar bear, *Ursus maritimus*	(Hamlett, 1935)
Malayan sun bear, *Helarctos malayanus*	(McCusker, 1974)
Giant panda, *Ailuropoda melanoleuca*	(Zhang *et al.*, 2009)

Renfree and Calaby 1981). Delayed implantation is suspected in the red panda (*Ailurus fulgens*), spectacled bear (*Tremarctus ornatus*), and sloth bear (*Melursus ursinus*, Mead 1989).

12.2.3 Seasonal and lactational anovulation

In the anovulatory or anestrus period of seasonal breeders, ovarian steroid levels are low or undetectable. Anovulation also may result from nursing young, termed lactational anovulation or anestrus, which prevents the mother from facing simultaneous energetic demands from gestation and lactation. In African lions, removal or loss of nursing cubs results in rapid resumption of estrous cycles, which is considered an explanation for infanticide by new males that take over prides with females nursing young. Such infanticide allows new males to impregnate females sooner and hasten their own reproductive success (Pusey and Packer 1994). Surprisingly, though, estrous cycles of large felids do not resume after weaning at the species-typical time during the period of prolonged dependence of offspring, which indicates that a mechanism other than lactational suppression is operating. This form of estrous suppression, independent of lactation, appears common at least in the large felids whose young remain with their mothers for a year or more post-weaning (Ewer 1973).

12.2.4 Frequency of ovarian cycles

Females of many carnivore species are monestrous, that is, they come into estrus and ovulate only once annually. In contrast, others are polyestrous and can have repeated estrous cycles until conception occurs. Polyestrus species can be continual or seasonal breeders, but the definition of monestrum is not as clear. Some authors use monestrum to describe the cycles of all canids, regardless of the number of ovulations per year. Because the canid cycle is so long (e.g. the long obligate pseudopregnancy), the classification of cycles in the non-seasonally breeding canid species deserves review.

12.3 The endocrinology of stress

The term "stress" is used commonly to mean psychological stress, but the endocrine system evolved to cope with the physiological, as well as psychological, stress responses of an organism facing a "fight or flight" situation. The major responses are to mobilize glucose to fuel muscle activity, to shunt blood to skeletal muscles and the heart, and to relax the bronchi to facilitate oxygenation of blood. This acute stress response is integrated primarily by epinephrine, and perhaps norepinephrine, released from the adrenal medulla acting through α- and β-

adrenoceptors in appropriate target tissues. Sustained, or chronic, stress progresses through several phases, characterized by Selye (1951) as the General Adaptation Syndrome. Chronic stress responses are mediated by adrenal glucocorticoids, either cortisol or corticosterone, depending on the species.

Cortisol is the major circulating glucocorticoid in carnivores. As their generic name suggests, glucocorticoids, such as cortisol, play a basic physiological role in glucose metabolism, but they also influence blood glucose levels during stress (Möstl and Palme 2002; Hadley and Levine 2007). Although cortisol is clearly involved in the psychological stress response, cortisol concentrations vary in response to other factors as well. For example, glucocorticoid levels increase during hunting, courtship, and copulation (Möstl and Palme 2002). Even baseline levels of glucocorticoids in the non-stressed, homeostatic state show a clear circadian rhythm of release, so sequential samples should be collected at the same time of day. As steroid hormones, glucocorticoids are excreted in urine and feces, primarily conjugated as glucuronides or sulfates, but they can also be measured in saliva and hair, as well as blood.

Monitoring glucocorticoid levels provides insight into the effects of potentially stressful events or situations, such as human disturbance, severe weather conditions (e.g. storms, flooding or draught), or responses to translocation or reintroduction. No clear relationship exists between glucocorticoid concentrations and social status. Both dominant and subordinate individuals can have high cortisol levels, although recent data from free-ranging carnivores suggest that levels are elevated more often in dominant individuals (Creel 2005). Another perspective on this apparent paradox is that social stability is typically less stressful than instability, as represented in the "challenge hypothesis" or the concept of allostasis (Wingfield *et al.* 1990; Wingfield 2005). Thus, when evaluating cortisol data, the broad social context should be considered.

Sustained adrenal activation, regardless of the cause, can interfere with reproductive processes through negative feedback on the endocrine cascade that culminates in stimulation of gonadal steroid production, ovulation, and spermatogenesis (Pottinger 1999). Persistent elevations of glucocorticoids can also have deleterious effects on health, such as compromised immune response, muscle wasting, and even symptoms of diabetes mellitus.

12.4 Endocrine studies and sampling strategies

Two basic classes of hormones exist: steroid and peptide hormones (Tables 12.3 and 12.4). In the past, blood samples were required for hormone measurements. As part of natural metabolism, however, most hormones are excreted in urine or feces

Table 12.3 *Peptide hormones of possible interest in carnivore research.*

Hormone	Action	Present in
Follicle stimulating hormone (FSH)	Stimulates follicle growth Initiates spermatogenesis	Females Males
Luteinizing hormone (LH)	Indicative of ovulation Stimulates testosterone	Females Males
Prolactin	Supports lactation May support parental care	Females Males and females
Relaxin	Marker of pregnancy in canids and felids	Pregnant females

Table 12.4 *Steroid hormones of possible interest in carnivore research.*

Hormone family	Hormones	Action(s)	Present in
Glucocorticoids	Cortisol, corticosterone	Regulate glucose and responds to physical and psychological stress	Males and females
Estrogens	Estradiol, estrone	Stimulate estros behavior, vulval swelling, cytological changes in vaginal epithelium	Females (low levels in males)
Progestagens	Progesterone, pregnenolone	Support gestation (also elevated during luteal phase in anticipation of conception)	Females
Androgens	Testosterone, androstenedione, dihydro-testosterone	Support spermatogenesis, secondary male characteristics and male-typical behavior	Males (low levels in females)
Thyroid hormones[1]	T3, T4	Stimulate metabolic rate in response to cold; necessary for successful reproduction	Males and females

[1] *Thyroid hormones are not truly steroid hormones, but are similar in structure and are excreted by similar pathways.*

(or both) and can be extracted and measured with standard endocrine assay techniques, such as radioimmunoassay (RIA), enzyme immunoassay (EIA), or chemiluminescence. Hormones may be present in their native form, as a metabolite, or conjugated to a glucuronide or sulfate, which makes them soluble in water (urine). Saliva contains hormone levels that generally reflect blood concentrations (Cook 2002) but can be challenging to collect. Because peptides are digested in the gastrointestinal tract, only steroid hormones can be extracted from feces. Steroid hormones can now also be extracted from hair shafts (Koren *et al.* 2002), adding another possible source of endocrine information.

Among the steroid hormones (Table 12.3), testosterone is the most abundant circulating androgen in males, but androstenedione may also be important. For females, estradiol is the most common and bioactive estrogen, but estrone may be at higher concentrations in urine and feces as an estradiol metabolite. Cortisol is the best-known product of the adrenal cortex, but the adrenals of most species also secrete sex steroids, such as dehydroepiandrosterone and estrone. Corticosterone, rather than cortisol, is the major glucocorticoid in some mammals (e.g. rodents) and in birds, and may be present at high concentrations in feces as a cortisol metabolite in carnivores (Young *et al.* 2004).

Peptide hormones (Table 12.4) of interest include luteinizing hormone (LH) and follicle-stimulating hormone (FSH) that, despite their names, are present in males as well as females. LH stimulates the final stages of follicle growth and ovulation in females but in males stimulates testosterone synthesis. FSH is responsible for early stages of follicle growth in females but in males is necessary, in particular, for initiation of spermatogenesis at puberty or onset of seasonal reproductive function. Other peptide hormones relevant to reproductive processes include relaxin, which is elevated in late pregnancy in all carnivore species studied to date, and prolactin, which has many roles in both males and females. As its name implies, prolactin is essential to milk production in females, but it appears to be associated with parental care in males as well as females.

Endocrine analyses provide information on gender, age class, breeding seasonality, reproductive status, pregnancy, and sometimes can be related to social status and stress. Gender cannot be deduced from the mere presence of estrogen or testosterone in a sample because both hormones are present in both males and females. Likewise, LH, FSH, and prolactin are produced by the pituitaries of both males and females. Nevertheless, in general, testosterone is higher in males and estrogen higher in females, so a ratio of the two hormones in a sample is usually diagnostic. In contrast, progesterone is typically found only in females and at especially high concentrations during pregnancy.

The clearest delineation of age class by hormone analysis is between adolescence and adulthood. Gonadal hormones, as well as LH and FSH, should be low or undetectable in prepubertal animals, although seasonally breeding species also have low concentrations outside the breeding season. For old-age classes, complete reproductive senescence and cessation of ovarian hormone secretion in otherwise healthy animals has only been documented in primates (Walker and Herndon 2008). In general, however, gonadal hormone production wanes with age in both males and females of most, if not all, species, but this decline is likely to remain within the range of variation detected in middle-aged individuals and, therefore, is not diagnostic.

The hormone changes associated with pregnancy are perhaps the most dramatic and, thus, the clearest to interpret. Progesterone concentrations may be elevated by as much as one or more orders of magnitude higher during pregnancy. Primarily a product of the placenta, relaxin is only present in pregnant females after embryo implantation. Relaxin is the only hormone that distinguishes pregnancy from pseudopregnancy in canids (Steinetz *et al.* 1987; Carlson and Gese 2007; Bauman *et al.* 2008). It can also be diagnostic of pregnancy in felids (de Haas van Dorsser *et al.* 2006), although felid pseudopregnancy, which is the result of an infertile mating or occasional cases of spontaneous ovulation, is uncommon. In contrast, the spontaneous ovulation of canids is followed, when conception does not occur, by an obligate pseudopregnancy, resulting in a challenge for pregnancy diagnosis. Species that undergo delayed implantation or embryonic diapause present an even greater challenge, however, because no pregnancy-specific hormones are elevated during the period of diapause. For example, although relaxin was detected in serum and urine of giant pandas (*Ailuropoda melanoleuca)*, variable and confounded results make it a questionable indicator of pregnancy in this species (Steinetz *et al.* 2005). Increases in prolactin and progesterone, associated with implantation and resumption of development, mark the beginning of the gestational period in species with delayed implantation (Lopes *et al.* 2004).

Lactation can be detected easily via manual expression of milk but requires capture and handling. The hormone prolactin is elevated during lactation and declines gradually until weaning, but prolactin can also be elevated in individuals that have not been pregnant and that are not lactating (Kreeger *et al.* 1991). In Temperate zones, annual increases in prolactin are mediated by changes in photoperiod that affect males as well as females of many species (Lincoln *et al.* 2003) and, thus, are not diagnostic of reproduction and lactation.

12.5 Sample collection

Hormones can be found in blood, urine, feces, saliva, and even hair. Choosing which substance to sample depends on research questions and hypotheses, access to animals for handling, and options for processing and preserving the samples. Following sample extraction, the assay protocols are similar, yet each sample type has advantages and disadvantages. A general discussion of endocrine sampling for studies of reproduction and stress in mammals can be found in Hodges et al. (2010).

12.5.1 Blood

Blood samples provide the best measure of the amount of a hormone reaching target tissues at the time the sample is drawn. The great disadvantage to using blood samples is that they require animal capture and restraint. Single samples can be collected during capture for procedures such as radio-collaring, but sequential sampling to monitor dynamic physiological processes is usually impractical. Because virtually all hormones are secreted in pulses, and many exhibit circadian rhythm of release, one blood sample may not reflect the average for a day. For some conditions, e.g. pregnancy, a threshold concentration may be established so that any concentration above that level can be considered a positive diagnosis. If samples are collected across successive days to look for patterns, circadian variability can be avoided by sampling at the same time each day. When collecting samples for cortisol analysis, working quickly is critical, since capture and restraint cause release of cortisol within minutes. Both steroid and peptide hormones circulate in blood; levels may be lower, however, than in urine or feces, where they can become concentrated, possibly making detection during assay more difficult.

Blood samples must be centrifuged soon after collection to separate red blood cells from the serum or plasma, which then must be frozen until assay, preferably in an ultralow rather than conventional frost-free freezer, since the warming cycles that prevent frost buildup can compromise sample integrity. Perhaps the greatest disadvantage to blood sample collection is that it requires animal capture and restraint. Single samples are possible during capture for procedures such as radio-collaring, but sequential sampling to monitor dynamic physiological processes may be impractical.

12.5.2 Urine

A major advantage of urine as a source of hormones is that noninvasive sampling is possible. Urine can be collected from scent-marking sites, but the likelihood of contamination with urine from other animals is high. Alternatively, a scent-marking station with a single-use or washable collection vessel can be used, ideally if the animal can be observed marking, not only for individual identity but also to ensure that the sample represents only one animal. Another possibility, if urine is sufficiently concentrated, is to blot a urine mark with filter paper (Shideler *et al.* 1995). Urine also can be collected during restraint or anesthesia by manual expression of the bladder, bladder catheterization, or cystocentesis, a procedure in which a needle is inserted transabdominally into the urinary bladder and urine withdrawn by syringe. For species that scent mark with urine, collection of sequential samples become (or can be) more practical, although identifying individual samples still can be challenging.

A further advantage of urine relative to blood sampling is that urine samples average hormone pulses across hours. Nonetheless, protocols for daily sampling of urine should specify sample collection at approximately the same time each day to control for circadian rhythms of excretion. Samples should be frozen or dried on filter paper soon after collection to avoid hormone breakdown. It is necessary to correct for urine dilution before calculating hormone concentrations (Bonsnes and Taussky 1945). Creatinine is secreted into urine at a relatively regular rate, regardless of the volume of urine production, so it can be used to index hormone concentration.

As with blood samples, both peptide and steroid hormones are excreted in urine. Steroids, however, are likely to be conjugated as glucuronides or sulfates, which requires an additional step to cleave the conjugate if the assay antibody only recognizes the native hormone. This step can be avoided by using assay antibodies that recognize the conjugated forms as well. Because urine is more likely to contain metabolites of the steroid hormones, rather than the native forms, instead of assaying progesterone, for example, a common urinary metabolite such as pregnanediol glucuronide may be measured. Domestic cats (*Felis catus*), and probably wild felids, excrete steroid hormones predominantly in feces, with such small amounts in urine as to be virtually undetectable (Shille *et al.* 1990). Differential excretion in urine versus feces has not been established for most species.

Because steroids excreted in urine (and in feces) may be metabolites rather than the native forms, biological validation is needed to demonstrate that the hormone being measured does relate to the phenomenon being studied. For example, changes in estrogens and progestagens should correspond to periods of estrous behavior or parturition when studying reproduction. Another validation method is to administer a hormone challenge. This approach is most commonly used in adrenal/stress studies to document measurement of the appropriate glucocorticoid. Injection of ACTH, the pituitary hormone that stimulates the adrenal cortex, should be followed by increased glucocorticoid (Eiler and Oliver 1980; Keay *et al.* 2006), which may be in the form of cortisol, corticosterone, or a metabolite. Similarly, GnRH can be used to stimulate LH, which then stimulates gonadal hormones such as testosterone (Asa *et al.* 1990). Release of estradiol or progesterone in response to GnRH, however, depends on whether the ovaries contain follicles or corpora lutea at the time of treatment, so is more useful in males. In addition, capture is a stimulus for cortisol release, with samples collected at timed intervals as after ACTH injection (Wingfield *et al.* 1997).

12.5.3 Feces

Fecal samples are simple to collect, making sequential sampling practical, and contamination with feces from another animal is not as likely as with urine, except for animals that defecate at latrines. Another potential confound can occur for animals that may urine-mark on feces, which can contaminate samples with excreted hormones from other individuals or from the same individual at another time. Feces can be collected from the ground or directly from the rectum from a restrained or anesthetized animal. Trapped animals often defecate in traps, but judging the age of the sample is important if traps are checked only once daily.

Because sampling does not require animal handling, feces are a good choice for studies that measure glucocorticoid levels as an indicator of stress (Keay *et al.* 2006). Even when restraint is required to collect a sample, glucocorticoid levels in feces deposited at the time of capture do not reflect the stress of capture, because of the lag time between the increase in glucocorticoids in blood and their appearance in feces. This lag, which reflects gut passage time, can vary among species and individuals, but 12 to 48 hours is common (Schwarzenberger *et al.* 1996). Time of day for collecting fecal samples is less important than even for urine, since feces are typically produced less frequently and represent hormone excretion across the time since last defecation. Like urine, hormone concentrations in feces average the pulsatile and even to some extent the circadian changes reflected in blood samples, depending on the frequency of defecation.

Because distribution of steroid hormones in feces may not be uniform (Brown *et al.* 1994; Millspaugh and Washburn 2003), a sample should be mixed thoroughly before taking an aliquot. Hormone concentrations in samples change over time due to metabolism by bacteria, temperature, and humidity, so care must be taken to collect fresh samples. Methods for preserving fecal samples include freezing in either a conventional freezer or in liquid nitrogen, drying, or extracting in an organic solvent for storage until assay. Species differ in the direction of change in hormone concentration depending on storage method (Hunt and Wasser 2003), complicating selection of the most appropriate method (Khan *et al.* 2002; Lynch *et al.* 2003; Millspaugh and Washburn 2004). Ideally, the best field-storage method should be determined for each study. In general, however, immediate freezing at –20°C or lower, drying, or extraction in ethanol, are acceptable. For international shipping, some countries require that potential pathogens in fecal samples be killed using, for example, ethanol, formalin, acetic acid, or autoclaving.

In contrast to the advantages, fecal samples have several drawbacks. First, only steroid hormones and their metabolites can be measured, since peptide hormones are broken down in the digestive tract. Even for steroids, calculating an accurate concentration that reflects blood concentrations of the hormone being measured in impossible. No method exists for indexing excretion rate in feces, such as creatinine in urine. Instead, concentrations of hormones are calculated relative to the weight of dry matter, or undigested material, in the sample, which vary from sample to sample. This variability can be especially dramatic in large carnivores that may not eat daily. Seasonal changes in diet that may contain different percentages of fiber also confound interpretation of results (Wasser *et al.* 1993), as do changes in food intake for hibernating animals.

12.5.4 Saliva

Salivary concentrations of both peptide and steroid hormones correlate well with blood levels (Cook 2002) and collection of samples has been relatively successful in a number of domestic and captive wildlife species. However, collecting saliva from free-ranging animals is a challenge under field conditions.

12.5.5 Hair

A recent method for assessing steroid hormone production has come from the need to test athletes for use of performance-enhancing androgens, commonly (and erroneously) referred to simply as "steroids." Hair samples reveal hormone levels during the period of hair growth and document drug use during that time. For non-human animal research, measuring cortisol, another steroid hormone, in studies of stress is more common. This approach can be more useful than point sampling with blood or excretory products, since it reflects patterns over a longer period of time (Koren *et al.* 2002). A difficulty for field sampling is determining when hair growth began (or ended). One way to control that variable is to shave an area and capture the animal again after the desired period, shave again, and analyze the steroid content of that new growth (Davenport *et al.* 2006). For animals that shed seasonally, using either the shed hair or new growth provides a rough approximation for the timing of hair growth. Domestic dogs have variable cortisol content of hairs that contain different color pigments (Bennett and Hayssen 2010). Consequently, hormone content of hair may vary by season for species with varying coat color (weasels, *Mustela* spp); hormone levels taken at the same time may not be comparable for animals with different hair color, and one should use hair of the same color when sampling species with patchy color (e.g. African wild dogs).

Unlike hair collection for genetic studies, the follicle cells are not necessary for hormone extraction; steroid hormones appear to be incorporated in the hair shaft as it grows. The amount of hormone being released from sebaceous glands surrounding the hair follicles is of concern, though, since hormones in the glandular secretions typically coat the exterior of hair shafts. Various methods have been tested for washing the exterior before hormone extraction, so as not to remove the target hormone in the hair shaft interior (Davenport *et al.* 2006).

12.6 Non-endocrine techniques for studying reproduction

Although not practical for all field research projects, standard measures of reproductive function may be appropriate and feasible in some circumstances or to answer particular questions. Some traditional methods and measurements are commonly used when animals are trapped or when examining carcasses; others are more technologically advanced, such as ultrasound.

12.6.1 Males

Testis measurements (Chapter 6) generally correlate with sperm production. Ultrasound allows more accurate measurements of testis size than is possible using calipers, and can also be used to assess testicular activity (Eilts *et al.* 1993). Semen can be collected from anesthetized males using electroejaculation (Green *et al.* 1984; Wildt *et al.* 1984; Wildt *et al.* 1986; Brown *et al.* 1989; Farstad 1996; Kojima *et al.* 2001; Johnston *et al.* 2007) and sperm can be retrieved post-mortem from the testis or epididymis (Pérez-Garnelo *et al.* 2004).

12.6.2 Females

Vulva measurements reflect reproductive condition (Table 12.5), since estradiol stimulates vulval swelling in most females in proestrus and estrus. Single smears of vaginal epithelial cells are diagnostic of estrus in species where cytological patterns have been established, but sequential samples revealing relative changes are more informative (African lion, Liche and Wodzicki 1939; red fox, Bassett and Leakley 1942, Boue *et al.* 2000; raccoon, Sanderson and Nalbandov 1973; coyote, Kennelly and Johns 1976, Carlson and Gese 2008; raccoon dog, *Nyctereutes procyonoides*, Valtonen *et al.* 1977; wolf, *Canis lupus*, Seal *et al.* 1979; brown hyena, *Hyaena brunnea*, *Ensley et al.* 1982; black-footed ferret, *Mustela nigripes*, Hillman and Carpenter 1983; North American otter, *Lontra canadensis*, Stenson 1988; steppe polecat, *Mustela eversmanni*, Mead *et al.* 1990; cheetah, Asa *et al.* 1992; wolverine, *Gulo gulo*, Mead *et al.* 1993; fennec fox, *Vulpes zerda*, Valdespino *et al.* 2002; giant panda, Durrant *et al.* 2003; bush dog, *Speothos venaticus*, DeMatteo *et al.* 2006a; gem-faced civet, *Paguma larvata taivana*, Liu *et al.* 2007; sun bear, *Helarctos malayanus*, Frederick *et al.* 2010). In members of the genus *Canis*, signs of sanguinous uterine discharge are indicative of proestrus and estrus (Table 12.5), and some degree of non-sanguinous uterine and vaginal discharge is common in most female mammals during those periods.

Ultrasound can be even more useful in females than in males, since it can detect follicles and corpora lutea on the ovaries as well as embryos *in utero*, long before manual palpation is effective in diagnosing pregnancy. In addition, an experienced

Table 12.5 *External signs of proestrus or estrus.*

	Reference
Vulval swelling	
Raccoon, *Procyon lotor*	(Whitney and Underwood, 1952)
Marten, *Martes americana*	(Enders and Leekley, 1941)
Stoat, *Mustela erminea*	(Gulamhusein and Thawley, 1972)
European polecat, *Mustela putorius*	(Hammond and Marshall, 1930)
Black-footed ferret, *Mustela nigripes*	(Hillman and Carpenter, 1983)
Wolverine, *Gulo gulo*	(Mead *et al.*, 1991)
Gem-faced civet, *Paguma larvata taivana*	(Liu *et al.*, 2007)
Red fox, *Vulpes vulpes*	(Mondain-Monval *et al.*, 1977)
Fennec fox, *Vulpes zerda*	(Valdespino *et al.*, 2002)
Bush dog, *Speothos venaticus*	(DeMatteo *et al.*, 2006b)
Raccoon dog, *Nyctereutes procyonoides*	(Valtonen *et al.*, 1977)
Black-backed jackal, *Canis mesomelas*	(van der Merwe, 1953)
Sanguinous Discharge	
Coyote, *Canis latrans*	(Kleiman, 1968)
Wolf, *Canis lupus*	(Seal *et al.*, 1979)
Black-backed jackal, *Canis mesomelas*	(van der Merwe, 1953)
Raccoon, *Procyon lotor*	(Whitney and Underwood, 1952)
Gem-faced civet, *Paguma larvata taivana*	(Liu *et al.*, 2007)

ultrasound technician can count the number of embryos or fetuses and, depending on the species, may be able to estimate the expected time of parturition (Boue *et al.* 2000; Clifford *et al.* 2007). Newer battery-powered instruments can be taken into the field easily. Unfortunately, current ultrasound resolution is not effective in detecting placental scars, the traditional method for counting the number of fetuses in past pregnancies.

Although not yet incorporated into a field study as an indicator of proestrus or estrus, ample evidence exists from several domestic species that an increase in activity is associated with increased estrogen (Gerall *et al.* 1973; Roelofs *et al.* 2005). This increased locomotion can be monitored by simple pedometers and with radiotelemetry units that distinguish degrees of movement (Williams *et al.* 1986). Temperature telemetry can also be used to monitor reproductive conditions, since progesterone concentrations correlate with basal body temperature. Basal body temperature increases around the time of ovulation, when progesterone rises and falls, and just before parturition when progesterone concentrations decline (Christie and Bell 1971a). In domestic dogs, basal body temperature, which falls by about 1°C approximately 24 hours before parturition, is often used to indicate of impending birth (Concannon *et al.* 1977b).

12.7 Gamete preservation and assisted reproduction

Although not historically of importance, management of small, endangered, and threatened populations may benefit from gamete retrieval and preservation, particularly from genetically valuable individuals. Methods applied to domestic and captive species (Holt and Pickard 1999; Farstad 2000; Andrabi and Maxwell 2007) may in some cases be appropriate and advantageous for free-ranging animals.

For males, semen can be collected by electroejaculation under anesthesia when animals are trapped for radio-collaring or other procedures. Likewise, sperm can be retrieved directly from the testes, epididymides or vas deferens of males that have recently died (Anel *et al.* 1999; Pérez-Garnelo *et al.* 2004; Ganan *et al.* 2009). Before cryopreservation, semen requires addition of a diluent or extender to protect sperm cells from damage during freezing and thawing (Hammerstedt *et al.* 1990; Parks and Graham 1992; Holt 2000). Extenders have been tested for many carnivore species. Semen cryopreservation can be accomplished in the field using liquid nitrogen vapor, such as from a "dry shipper," an insulated container that absorbs liquid nitrogen into its lining to prevent spillage during transport.

Collecting and preserving female gametes is considerably more challenging. First, oocytes are much more difficult to collect than sperm. Methods include an ultrasound-guided needle to aspirate oocytes directly from ovarian follicles or surgical removal of ovaries for better access to follicles and higher oocyte yields (Brogliatti and Adams 1996). Second, traditional semen-freezing techniques do not succeed with oocytes. A new, practical technique has proven successful with a number of species. Oocytes can be vitrified, which entails gradual dessication via successively more concentrated media baths followed by plunging into liquid nitrogen (Vajta and Kuwayama 2006). This technique is too specialized for field application, but special cases, where genetically valuable females may be lost, as during disease outbreaks, may warrant consideration of heroic measures. Removal of ovaries for shipment to a lab within 10 hours may be adequate for oocyte rescue and preservation (e.g. Mexican wolf, *Canis lupus bailey,* Boutelle *et al.* 2011). Cryopreserved sperm can be incorporated into the population via artificial insemination (AI), which is the simplest assisted reproduction technique (ART). Other ARTs include *in vitro* fertilization (IVF), intracytoplasmic sperm injection (ICSI), embryo transfer, and even cloning. Intracytoplasmic sperm injection might be justified in cases of post-mortem sperm retrieval when sperm quality or vigor is compromised by initial stages of degradation. Given profound species' differences and the poor success rate of other assisted reproduction techniques in most species, AI is the technique most likely to be useful for field applications.

Although a variety of methods exist for introducing sperm into a female's reproductive tract, the greater difficulty is determining the appropriate time for insemination; the lifespans of sperm and egg are finite, and an egg is only present for fertilization around the time of ovulation. Fortunately, estrus and ovulation can be induced for timed insemination with an increasing number of protocols, e.g. gonadotropins and GnRH agonists (Asa *et al.* 2006), that may make AI more practical in the field.

The prospects for using cryopreserved female gametes are limited, since they require *in vitro* maturation followed by *in vitro* fertilization and embryo transfer into a recipient female. *In vitro* maturation and *in vitro* fertilization must take place, of course, in a properly equipped lab, but embryo transfer could potentially be accomplished with an anesthetized female in the field. Alternatively, cryopreserved gametes could be used more easily in captive animals whose offspring could supplement free-ranging populations as an indirect method of improving genetics.

12.8 Control of reproduction

Reproductive management includes both enhancing and controlling which animals breed and how many young are produced. Concerns for animal welfare and failure of lethal means to reduce problematic populations effectively have led to interest in using non-lethal methods to control populations (Asa and Porton 2005; Porton 2005). Non-lethal methods include permanent sterilization or reversible contraception and can be accomplished either surgically or chemically (Asa 2005; Kirkpatrick and Frank 2005; Asa and Porton 2010). Gonadectomy or chemical destruction of gonads do prevent an individual from reproducing, but these methods also eliminate gonadal hormones (testosterone in males, estradiol and progesterone in females). Gonadal hormones stimulate sexual behavior and also secondary sex characteristics (e.g. a male lion's mane) and other aspects of sociosexual behavior, such as territory maintenance. Furthermore, if all males in a population are not sterilized, those remaining fertile are likely to inseminate more females than they otherwise might, compensating for the ones eliminated from the breeding pool. In contrast, males vasectomized, either surgically or chemically, continue to guard territories and females. Studies have shown that vasectomized coyotes and wolves maintain pair bonds despite failure of the pair to produce young (Spence *et al.* 1999; Bromley and Gese 2001).

Vasectomy is contraindicated in species with induced ovulation, since sterile copulations are followed by pseudopregnancies with elevated progesterone that can cause overgrowth of the uterine endometrium, predisposing the female to potentially life-threatening infection (pyometra) or development of tumors. Carnivores appear to be more vulnerable than ungulates and primates to the potentially deleterious effects of progesterone or synthetic progestins used as contraceptives (Munson *et al.* 2005). Canids that have repeated pseudopregnancies, with associated high levels of progesterone, have higher incidence of pyometra than females that regularly give birth (Devery 2010). Currently, no safe, effective non-surgical method of permanent sterilization exists for female carnivores. Tubal ligation, which is somewhat less invasive than ovariectomy, subjects females to repeated estrus and pseudopregnancy associated with elevated concentrations of progesterone. Ovariectomy eliminates production of both estradiol and progesterone but will also likely interfere with formation of pairs and maintenance of pre-existing pair bonds (Asa 1996).

The GnRH agonists provide safe, effective, and reversible contraception for both male and female carnivores, but the length of efficacy may not be more than one or two years, necessitating retreatment for longer periods of suppression (Bertschinger *et al.* 2001; Boutelle and Bertschinger 2010). Their disadvantage is that they produce a condition similar to gonadectomy, wherein testicular and ovarian steroids are suppressed, altering behavior.

13

Investigating cause-specific mortality and diseases in carnivores: tools and techniques

Greta M. Wengert, Mourad W. Gabriel, and Deana L. Clifford

Conservation of carnivores entails an understanding of ecology and life-history requirements of species, which is essential for identifying the factors limiting or threatening populations. Because population growth and persistence relate to mortality and fecundity, identification of threats requires thorough characterization of cause-specific mortality or causes of low fecundity, along with representative sampling to ensure an unbiased perspective on the relative importance of specific causes. Disease, as a potential cause of mortality or influence on reproductive success, can be easy to overlook as a contributing cause of population decline. For this reason, a working knowledge of disease processes at individual and population levels, and basic methods to study these processes, are vital to biologists and managers involved in carnivore conservation. In this chapter, we describe several tools and techniques for collecting biological data and samples to study cause-specific mortality and the presence of pathogens and disease in carnivores, and emphasize the necessity for thoroughness and representativeness. We offer a variety of ways to analyze mortality and disease data through epizootiological studies and modeling programs. Lastly, we present a variety of commonly used disease-intervention programs for the prevention or control of detrimental pathogens in carnivore populations.

13.1 Determining causes of mortality in carnivores

Studies of sources of mortality should accomplish two objectives: thorough understanding of the direct and indirect reasons for the death of each individual, and the relative frequency and importance of each contributing cause of mortality to the population as a whole. The latter requires representative sampling, a sufficiently large sample size, and knowledge of whether sources are additive or compensatory, in addition to accomplishment of the first objective. Here we focus on the techniques for finding, recovering, and analyzing carcasses effectively, in itself a challenging task, but stress that this information translates to effective management only if causes of mortality are evaluated in terms of their effects at the population level.

13.1.1 Locating dead animals to determine cause-specific mortality

Continuous monitoring, or at least daily monitoring, of transmitter-tagged animals is the most effective technique for locating dead animals quickly enough to determine the cause of death and understand predisposing factors. Such monitoring, however, requires extensive live-trapping (Chapter 5) to outfit many animals with transmitters, and is expensive, requiring extensive person-hours of ground telemetry, frequent flights to locate animals, or expensive satellite-transmitter systems that report at least daily.

Most wildlife transmitters with a VHF telemetry component and GPS/Argos telemetry packages can be equipped with mortality sensors (Chapter 7). Sensors cause VHF units to pulse at different rates, indicating whether an animal has moved or not for a pre-set period of time, allowing a researcher to locate a dead animal reasonably quickly. Alternatively, some researchers use activity sensors if studying activity patterns is a research goal. Though not as effective as mortality sensors, repeated "inactive" signals suggest possible mortality, prompting a researcher to locate an animal and verify death vs. extended inactivity.

Alternatively, non-probability sampling, or convenience sampling, is used commonly to find dead animals. Unfortunately, such opportunistic sampling with no a priori sampling design is characterized by large sampling error and commonly misrepresents the relative importance of causes of mortality in the population (Nusser *et al.* 2008). Thus, if determining cause-specific mortality is a main objective of a carnivore research program, a sampling design is mandatory.

Direct interaction with personnel of public agencies, including transportation departments, state and national parks, and game-management agencies, can provide a wealth of information on the locations or final dispositions of carcasses found by their personnel (Knight *et al.* 1988). Animal-removal services or depredation programs also provide carcasses and information on locations of carcasses. Additionally, local wildlife-rehabilitation centers and veterinary clinics may generate information from the public on morbidity and mortality of wild carnivores. Regularly scheduled contact with a agency and other pertinent personnel may yield samples with less bias than completely opportunistic sampling.

For research on road mortality, schedule systematic road-kill surveys. Surveys of equal, or representative, lengths along roads of different classes throughout a study area allow inference about the rates of road-induced mortality on roads of different classes by individuals of different age classes and sex (if age and sex data are collected; Clarke *et al.* 1998).

13.1.2 Handling dead animals and important precautions

Approximately 75% of emerging human infectious diseases originate in animals (Taylor and Latham 2001). Therefore, researchers should be familiar with the common routes of disease transmission and with the clinical signs of pathogens that can be passed between animals and people (zoonotic pathogens) before starting research. Accordingly, when handling carcasses, the following basic precautions should be observed to minimize risk of researcher exposure to diverse zoonotic pathogens, including some that are quite debilitating and sometimes fatal, like rabies. First and foremost, wear personal protective equipment when handling dead and live animals. Gloves (latex, nitrile, or vinyl) provide protection from most biological agents and some chemical agents. Masks prevent the transmission of airborne zoonotic pathogens (Wong *et al.* 2009), especially important when performing field necropsies. Double-bag a carcass with plastic bags that are thoroughly sealed, airtight, and well-labeled. If possible, keep the carcass on ice, or at least cool, until it can be refrigerated. If it cannot be sampled or necropsied within 24–36 hours, freeze it. Store carcasses and tissues in an area that is not used by humans or domestic animals; especially avoid freezers and refrigerators used to store food. Researchers should also carry an easily accessible informational card advising medical personnel that zoonotic diseases should be considered in differential diagnoses, in case a researcher cannot do so him- or herself.

Whenever possible, tag a limb of the animal with relevant information; at a minimum include the date and time, location where the carcass was collected (latitude and longitude, if possible), the collector's name and contact information, and other important reference information (i.e. project name, affiliations, etc.). Munson (2006) detailed the safest, most effective methods for collecting, storing, and shipping carcasses for necropsy and diagnostic testing (also consult www.nwhc.usgs.gov/mortality_events/shipping_instructions.doc).

13.1.3 Field-data collection at mortality sites

In human forensics cases, thoroughly documenting the details of the scene is essential. Likewise, when investigating the cause of mortality in wildlife, follow strict protocols to investigate a mortality site. Upon discovery of a deceased animal, leave the carcass and site undisturbed until fully photographed. Key characteristics to record include: whether the carcass was cached by a predator; whether the carcass was intact, dismembered, or partially consumed; whether the carcass was dragged; and other notable features of the site. For example, the position or location of a carcass may suggest whether the animal secured itself in a sheltered location due to extreme morbidity before death. If interested in linking habitat features to mortality risk, especially in the case of predation (Kunkel and Pletscher 2000; Hebblewhite *et al.* 2005), record information on habitat and terrain in detail. Write complete directions to reach the site, flag the location to facilitate return to the site, if necessary, and record the location with a GPS unit.

13.1.4 The clinical necropsy

More often than not, the reason for death of an animal is not obvious at first examination, and multiple factors may have interacted to cause death. To understand the underlying factors that contribute to mortality, both at the individual and population levels, conduct clinical necropsies. Ideally, field biologists or wildlife managers can collaborate with veterinary pathologists experienced with wildlife to conduct systematic, thorough necropsies. Because we strongly suggest that carnivore researchers studying mortality make every attempt to have carcasses investigated in a pathology laboratory, we highlight fundamental components of a necropsy in order to inform researchers observing or assisting with necropsy.

Hemorrhage is easily identified during necropsy and is used to determine whether injuries occurred ante- or post-mortem. Photograph all areas of hemorrhage such that the pathologist, if not present, can later characterize the nature of the hemorrhage. In addition to gross examination of the tissues, the pathologist collects samples from all major organs for in-depth histological investigation, since lesions at the cellular level may be associated with an animal's death or morbidity prior to death. Occasionally, immunohistochemistry, a method that detects antigens within cells, is employed by the pathologist to confirm or rule out infection by certain pathogens. Staining techniques (e.g. Gram staining, acid-fast staining) also help detect and identify pathogens. Clinical necropsies also allow biopsy of tissues for diagnoses of non-infectious diseases, such as cancer, which can be critical for the health of some species (Tasmanian devil, *Sarcophilus harrisii,* McCallum *et al.* 2007; island fox, *Urocyon littoralis*, Vickers *et al.* 2007). Pathologists typically work with toxicologists, who screen for abnormal levels of heavy metals and presence of toxins within tissues, such as anticoagulant rodenticides or other pesticides. Gross necropsy is also an opportune time to remove a tooth for cementum annuli analysis, to estimate the age of the animal (Chapter 6).

13.1.5 When clinical necropsies just aren't feasible—a quick guide to field necropsy

Sometimes, logistics or budgets simply don't allow for a full clinical necropsy by a wildlife pathologist. Whether financial constraints or remote field sites prohibit a full necropsy, field biologists can perform field necropsies. Obtain training from a wildlife pathologist to ensure safe and thorough necropsy procedures. Photograph and sample all tissues for later lab analysis and secure collaboration with a pathology laboratory beforehand to receive and analyze tissue samples. Split each tissue sample between formalin and plastic whirlpacks on ice, then freeze the whirlpacks as soon as possible. Collect blood, nasal, and ocular exudates, ectoparasites, and fecal samples. Document and record all collections and abnormal observations (write on a necropsy form or voice-record). Consult and use the online manual of necropsy methods (Munson 2006; http://www.vetmed.ucdavis.edu/whc/pdfs/necropsy.pdf). Box 13.1 contains a minimal list of tools and supplies needed for a field necropsy.

13.1.6 Field and laboratory investigation of intraguild predation

Intraguild predation in carnivore communities can be a frequent cause of mortality (Ralls and White 1995; Mills and Gorman 1997; Moehrenschlager *et al.* 2007; Thompson and Gese 2007). When intraguild predation is suspected, photograph the immediate surroundings and all bite wounds or obvious injuries to the dead animal. Measurements can be taken of the bite wounds, though only punctures in bone can be truly diagnostic for identifying predator species (Lyver 2000).

Molecular analyses are proving quite useful for determining predator species. Until recently, these methods were restricted to identifying livestock predators (Williams *et al.* 2003; Sundqvist *et al.* 2008), but in research on fishers (*Martes pennanti*) and American martens (*M. americana*), intraguild predators have been determined through sampling saliva around bite wounds and extracting DNA (Wengert *et al.*, unpublished data). Rub sterile, polyester-tipped swabs within bite wounds and clip fur surrounding wounds. Store these samples in airtight vials, and freeze at or below –20°C. Arrange collaboration with a genetics laboratory ahead of time, so that the laboratory personnel can determine the most appropriate genetic protocols for a particular project. For example, if the researcher is studying intraguild predation on small carnivores, such as weasels or small foxes, a variety of predators ranging from raptors to bears must be considered.

A somewhat less accurate method for identifying predators is use of molecular techniques on feces left at carcasses (Ernest *et al.* 2002; Onorato *et al.* 2006). Though this evidence is often only circumstantial, it can infer predator identity

Box 13.1 Minimum list of supplies and equipment a researcher should have available during field necropsy (Munson 2006)

Personal protective equipment

- Gloves (rubber, latex, or nitrile)
- Rubber boots or plastic foot protectors
- Scrubs or coveralls
- Mask (to cover mouth and nose)
- Eye protection

Necropsy equipment and supplies

- Camera
- Field notebook and/or necropsy forms
- Waterproof writing utensils
- Labeling tape or tags
- Measuring tape
- Sharp knife and sharpening tool
- Scalpels and razor blades
- Scissors
- Forceps
- Ax or hatchet
- Bone saw
- Sterile syringes and needles
- Blood tubes (red-tops for centrifugation and purple-top EDTA)
- Sterile polyester swabs
- Rigid plastic containers with airtight lids for samples in formalin
- Sterile airtight vials for samples (swabs, ectoparasites)
- Plastic bags (zip-lock or whirl-pack)
- Ice coolers and ice packs
- Leak-proof, break-proof containers

Fixatives and disinfectants

- 10% buffered formalin
- 95–100% ethanol for fecal and exudate swabs
- 70% ethanol for parasites
- Disinfectants (10% bleach, quaternary ammonium compounds)
- Alcohol lamp or gas burner for sterilizing instruments

when coupled with other information from the predation event. Molecular material from the feces can also provide the animal's individual identity (Chapter 4; Ernest *et al.* 2002), sex (Chapter 4; Blejwas *et al.* 2006), diet (Chapter 11), and potentially information on the predator's health (Chapter 12).

Remote cameras can be set to identify the predator of smaller carnivores that return to a carcass or cache site (Chapter 4; John Erb, Minnesota Department of Natural Resources, personal communication).

13.2 Studying disease and pathogen cycles in carnivores

Disease can lead to mortality of individuals and to population reduction or regulation (Thorne and Williams 1988; Randall *et al.* 2006; McCallum *et al.* 2007). Certain pathogens affecting carnivores, like rabies and canine distemper, are extremely virulent and have caused dramatic population decreases and local extinctions (Roelke-Parker *et al.* 1996; Laurenson *et al.* 1998; Timm *et al.* 2009). On the other hand, disease impacts may be subtle, not causing mortality or obvious clinical signs, yet still affecting reproductive success or the ability of a carnivore to secure enough food for itself and its offspring, or potentially making individuals vulnerable to other forms of mortality.

13.2.1 Detection of disease, infection, and pathogen exposure

Handling live-trapped carnivores provides a perfect opportunity to collect biological samples, including blood, exudates, feces, and parasites, for disease screening and assessment. Even if health and disease are not the primary focus of a study, archiving these samples for future use can reduce the need to resample animals to obtain health information, alleviating potential negative impacts to a population from additional handling (Chapter 6; Botzler and Armstrong-Buck 1985). Obtain technical instruction from qualified researchers or veterinarians to avoid complications that could arise from collecting and handling animals and biological samples improperly. Table 13.1 provides information about sampling techniques.

13.2.1.1 Photo documentation

Photograph a carnivore under anesthesia to create a baseline reference of visual characteristics that might vary with changes in health over time. Pelage quality can deteriorate due to mite infestations (mange) or with emaciation linked to disease. Tooth wear may indicate excessive biting due to ectoparasite infestation or a coarse diet. Wounds or other external abnormalities should be reassessed if the animal is

Table 13.1 *Biological sample collection purposes and techniques for storage and transport in the study of carnivore disease. Also consult with collaborating laboratory for lab-specific preferences on storage and shipping.*

Sample Type	Purpose	Container Type	Storage (ST=short-term; LT=long-term)	Transport
Blood: serum or plasma	Determine antibody presence and titers; biochemistry panels to detect differences in selected markers	Red-top tube or sterile container with no anticoagulant	Let clot for 20 min, centrifuge after clot forms, transfer supernatant, discard red cell clot. Avoid direct sunlight. ST: < 48 h, refrigerate. LT: > 48 h, –20 to –80°C freezer	Use coolers with ice to transport
Blood: whole blood	Assess health parameters, such as inflammation, anemia; determine pathogen presence, antibody presence and titers	Purple-top tube with an anticoagulant (e.g. EDTA)	Refrigerate, do not freeze before certain selected tests. Avoid direct sunlight. ST: < 48 h, refrigerate. LT: > 48 h, –20 to –80°C freezer	Use coolers with ice to transport
Feces (fresh or old)	Molecular detection of non-labile viruses (parvoviruses, etc.) and some endoparasites and their ova	Clearly labeled airtight plastic container; if rectal swab is used, also store in airtight plastic container	ST: Store in 95–100% ETOH and refrigerate. Avoid direct sunlight. LT: Store in 95–100% ETOH and freeze in –20 to –80°C freezer	Use coolers with ice to transport
Feces (fresh only)	Detection of endoparasites and their ova; analysis of fecal hormones	Clearly labeled airtight plastic container	ST: Refrigerate and keep cool for analysis ≤ 2 h. Avoid direct sunlight LT: If parasites must be stored, 5–10% buffered formalin to fix ova	Use coolers with ice to transport
Exudates (nasal, ocular, wounds)	Molecular detection of viruses shed in exudates; bacterial culture	Synthetic-tipped swabs placed in an airtight plastic container containing either guanidine-thyocyanate solution, 95–100% molecular grade ethanol, or a commercially-produced RNA/DNA preservative	ST and LT: freeze in –20 to –80°C freezer (or follow manufacturer's instructions). Avoid direct sunlight. For bacterial culture, refrigerate and send to lab as soon as possible	Use coolers with ice to transport

(continued)

Table 13.1 *Continued*

Sample Type	Purpose	Container Type	Storage (ST=short-term; LT=long-term)	Transport
Urine	Detection of pathogens shed in urine	Clearly labeled airtight plastic container	Neutralize urine with phosphate buffered saline solution. ST: $\leq$24 h, refrigerate LT: $>$ 24 h, freeze at –20 to –80°C	Use coolers with ice to transport
Ectoparasites (ticks, fleas, mites, etc.)	Species determination of parasites and possible pathogens they vector	Clearly labeled airtight plastic container	ST and LT: (ticks, fleas, lice): $\geq$ 70% ETOH and 5% glycerol. ST and LT: (mites): 87 parts 70% ethanol, 5 parts glycerin, 8 parts glacial acetic acid	Keep in a cool dry place; do not freeze
Endoparasites (flukes, nematodes, tapeworms, etc.)	Species determination of parasite and possible pathogens they harbor	Clearly labeled airtight plastic container	ST and LT: 92 parts 70% ethanol, 3 parts 40% formaldehyde, 5 parts glycerin	Keep in a cool dry place; do not freeze

captured again. Photographs also document conditions of animals noted in one region but not elsewhere.

13.2.1.2 Blood

Blood can provide critical clues to an animal's health status (Chapter 12). Antibodies in blood can be used to determine past pathogen exposure or potentially, active infection. Fluctuations in glucose and urea levels may indicate disease. Abnormal white blood cell counts often indicate response to viral, bacterial, and parasitic infections. Information on collection, storage and pertinent diagnostic tests is summarized in Table 13.1.

13.2.1.3 Feces

Feces can harbor bacteria, endoparasites and their ova, various environmentally resistant viruses, and some labile viruses. Collect feces using a swab within the rectum of an anesthetized animal or opportunistically collecting a scat. Feces collected for genetic or diet studies can also be used to screen for pathogens or parasites. Collection and storage methods for feces are shown in Table 13.1.

13.2.1.4 Exudates

Certain pathogens are transmitted by ocular, nasal, and oral exudates. Canine herpes virus, canine distemper virus, and influenza virus are often shed within ocular–nasal exudates, which can be tested for their presence (Williams and Barker 2001). Since many of these viruses are extremely labile, collect samples properly to maximize chances of detecting infected animals (Table 13.1).

13.2.1.5 Urine

Though uncommon, certain pathogens are shed in urine. *Leptospira* species of bacteria are shed by this route, and all carnivores can be either maintenance or accidental hosts for *Leptospira* spp. Manually express the bladder and collect urine mid-stream to avoid contaminating the sample with traces of feces or other materials near the opening of the urinary tract. If working with trained wildlife veterinarians, they can obtain a sterile urine sample by cystocentesis (collection of urine with needle and syringe).

13.2.1.6 Ectoparasites and endoparasites

Collecting ectoparasites and endoparasites (Table 13.1) provides information regarding vector-borne pathogens that may infect carnivores, such as plague (*Yersinia pestis*) and heartworm (*Dirofilaria immitis*). Many parasites themselves cause disease in carnivores, such as mange caused by mite infestation or infections with helminths or protozoans. Parasites also provide information on the life-history traits of the focal carnivore, such as habitat associations of parasites in turn providing clues to habitats the carnivore may have visited. Many endoparasites require specific intermediate hosts that are eaten by carnivores, thereby providing insight to a carnivore's diet.

13.2.1.7 Rapid diagnostic testing for disease

Several rapid diagnostic tests (RDT) are available for detecting pathogens of domestic animals, including heartworm, distemper virus, parvovirus and rabies. Although these tests are convenient, rapid diagnostic tests that have been developed for domestic animal use cannot simply be transferred for use in wildlife (Stallknecht 2007). First, validate any rapid diagnostic test to be used on wild carnivores against an acceptable gold standard test to determine its effectiveness with the focal species (Stallknecht 2007). Recent evaluation of a parvovirus RDT on fisher and gray fox (*Urocyon cinereoargenteus*) fecal samples demonstrated that the test failed to detect any parvovirus infections, while the conventional parvovirus PCR readily detected the virus in many of the same samples (Gabriel *et al.* 2010).

13.2.1.8 Disinfection considerations

Many pathogens that infect carnivores are generalists, affecting members of many sympatric carnivore species. Furthermore, many of these pathogens are stabile in the environment (e.g. parvoviruses). Steps can be taken by researchers and managers to avoid spreading disease inadvertently via indirect monitoring equipment like cameras, scent-stations, and track plates, in addition to live-traps and handling equipment. After every new animal contacts the equipment, remove all fecal material and visible exudates and then disinfect the equipment (always wear personal protective equipment). Many disinfectants are suitable for neutralizing most pathogens of concern, but safety and application methods determine which are most appropriate. Sodium hypochlorite, or bleach, is a common choice of disinfectant and is readily available and inexpensive. Use a dilution of 1:32 up to 1:10 to cover the contaminated surface and let it remain there for at least 10 minutes. Though hazardous if ingested or inhaled, it is very effective in neutralizing even highly resistant viruses (Gilman 2004). Quaternary ammonium compounds are available from veterinary and tack and feed supply stores (e.g. Roccal-D®, Parvo-sol®, Spectrasol®). This group may not be particularly effective for neutralizing resistant viruses (Kennedy *et al.* 1995; Eleraky *et al.* 2002), but should be effective for removing many pathogenic bacteria and some labile viruses. Potassium peroxymonosulfate, a relatively new class of disinfectant marketed as

Virkon-S® or Trifectant®, is effective for neutralizing a wide array of pathogens including resistant viruses. This disinfectant has low toxicity and few corrosive or irritating properties.

13.2.1.9 Indirect animal sampling

Direct sampling of live animals is the most sensitive and reliable technique for diagnosing disease, detecting infection, or documenting prior exposure to pathogens. However, constraints to capturing and processing multiple individuals, especially rare, elusive, or trap-shy carnivores, often preclude direct sampling. Two alternatives to direct animal sampling are collecting feces for pathogen detection and sampling opportunistically-found carcasses.

When collected with an a priori systematic sampling design, fecal sampling can provide prevalence data for fecally-shed pathogens that are known to infect many carnivore species, such as parvoviruses, coronaviruses, many helminthes and protozoans, and potentially canine distemper virus (Ballmann-Acton and Elaine 2009). Scrape or cut the outside of feces to ensure that the mucous, gastrointestinal epithelium, and virons lining the epithelium are included in the sample. This method also ensures that DNA from the focal animal is sampled, allowing accurate species identification (Chapter 4). Store samples in 95–100% ethanol to fix the pathogen's DNA or RNA and the focal carnivore's DNA. Sampling design should accommodate the likelihood of sampling individuals multiple times. Individual identities can often be verified using fecal DNA. Depending on research objectives, this allows the researcher to use only samples from different individuals to avoid pseudoreplication, or to track clearing of infections and shedding cycles within the same individuals over time.

Carnivore carcasses provide a wealth of information on the health status of a population. As mentioned earlier, working with agency personnel and trappers or hunters can generate a sample of carcasses greater than can be achieved by researcher-based collection alone. Collect blood directly from the right heart ventricle using a sterile syringe. Take care not to puncture the heart prior to drawing blood, as blood within the heart is the most sterile and keeping the heart intact ensures blood sterility. Collect all other samples similarly to methods described earlier in this chapter for live animals and in Table 13.1.

13.2.2 Epizootiology in carnivore populations

Epizootiology is the study of population-level patterns of disease in animals. It assesses disease risk in a population, correlating "risk factors" such as animal characteristics (e.g. sex, age, habits) and environmental variables to pathogen exposure, infection, and transmission. Knowledge of baseline information on population health is essential to assess disease risk accurately and understand whether pathogens occur at an "enzootic" rate (expected and relatively low, constant rate) or "epizootic" rate (elevated rate that is unexpected due to its temporal patterns, spatial patterns or frequency; Wobeser 2007).

13.2.2.1 Incorporating age structure

Pathogens affect some age classes more than others; thus understanding the age structure of a population in relation to disease dynamics is integral for characterizing disease risk and past epizootics. For example, a population missing a particular cohort may have experienced an epizootic of a pathogen that selectively affects juveniles when that cohort was young. When this evidence is supported by high prevalence of exposure to that particular pathogen in older cohorts, one can more conclusively infer the epizootic history of the population.

13.2.2.2 Incorporating fecundity

Pathogens affect carnivore reproductive success through many mechanisms, including abortions and reduced survival of offspring. Although challenging to obtain, data on pregnancy rates, neonatal loss, and survival of juveniles allows inferences about pathogen exposure and fecundity. Pregnancy status can be assessed in free-ranging carnivores (Chapter 12) by detecting relaxin hormone in blood, serum, plasma, or urine (Carlson and Gese 2007; De Haas van Dorsser *et al.* 2007; Bauman *et al.* 2008), detecting fecal progesterone in induced ovulators (Brown 1997), and using ultrasound (McNay *et al.* 2006; Clifford *et al.* 2007). For carnivores with known dens, young can be counted directly or less invasively by using small "peeper" cameras (e.g. Sandpiper Technologies, Manteca, CA). If examination of preweaning young is possible, collect samples for disease testing and mark young uniquely for later identification (Chapter 7) and correlation of neonate survival with pathogen exposure. Collect and test feces at dens for diseases and parasites of weaned offspring.

If pregnancy can be documented and fetuses counted via ultrasound, and pregnant females tracked via telemetry, remote cameras can be placed near dens to document emergence of the young, thereby providing an index of perinatal mortality (Clifford *et al.* 2007). Given sufficient sample sizes, perinatal mortality together with pathogen exposure histories of females allows inference regarding effects of disease on reproductive performance of females. Long-term demographic data combined with disease prevalence or pathogen exposure rates can reveal otherwise undetectable impacts of disease. Thirty years of demographic data on wolves (*Canis lupus*) in Minnesota showed that pup survival decreased dramatically after the appearance of canine parvovirus in the population, lowering annual population growth rates (Mech *et al.* 2008). Long-term datasets of placental scarring or other pregnancy stages with corresponding disease data can be collected from animals that are harvested.

13.2.2.3 Estimating contact rates

The transmission rate, or rate at which a disease is transmitted throughout a population, depends on (1) how often an animal capable of contracting the disease contacts an infected animal or infectious material (contact rate), and (2) how likely that contact is to result in disease transmission. Proximity data-loggers affixed to animal collars estimate contact rate by recording the number and specific times of contact with another collar, estimating the probability and frequency of contacts (Böhm *et al.* 2009; Hamede *et al.* 2009). Users can program the distance between collars required to log as a "contact."

From home ranges calculated as utilization distributions (use raster format), researchers can calculate the probabilities of two individuals being in their area of home-range overlap at the same time, and can test for avoidance or attraction by comparing actual use of the overlap to that predicted for random use (Chapter 9). These probabilities can be used as a proxy for contact rate and predict risk of pathogen spread throughout populations of different densities (White *et al.* 1995; Kauhala and Holmala 2006).

At natural or human-made sites where animals gather, such as watering holes, large animal carcasses, supplemental feeding stations, or latrine sites, remote cameras can provide data on interactions and log contacts between conspecifics, as well as among species (Macdonald *et al.* 2004a). These data can be used to calculate nightly contact rates (Totton *et al.* 2002). When direct contact between individuals is not necessary for pathogen transmission (as in the case of parvoviruses, coronaviruses, and many nematodes and protozoans), simple contact between an individual and an infected animal's feces may be considered an effective contact in the disease sense, and these interactions could be well-documented and quantified at latrines using cameras (Page *et al.* 1999).

Finally, another indicator of contact between individuals is multispecies latrines and areas where many animals leave scats for marking. Especially where overmarking occurs, these areas can be used to develop an index of contact rate for fecally-shed pathogens. Documenting presence of scats from domestic animals in these areas also helps assessment of spillover risk, especially from dogs and feral cats, whose scats should be collected and tested for pathogen presence.

13.2.3 Modeling techniques in disease ecology

Mathematical modeling contributes to the understanding of researchers and managers of disease transmission dynamics within and among carnivore populations, and assists in building hypotheses for the occurrence, spatial spread and population-level impacts of disease. SIR models, a form of "compartment model," are commonly used to model disease dynamics and epidemics. In these models, individuals move among compartments depending on whether they are susceptible to (*S*), infected with (*I*), or recovered from (*R*) a disease (Abbey 1952). Differential equations derived from contact rate and disease prevalence data estimate the rates at which individuals move among the *S–I–R* compartments. Population demographic parameters, density dependence, and stochasticity can be included. Spatial modeling approaches, including nearest neighbor, moving window analyses (Alexander and Boyle 1996), simulation models (Deal *et al.* 2000), and diffusion models (Moore 1999; Adjemian *et al.* 2007), can be integrated with or "added onto" an SIR framework. The result is spatially explicit predictions (hypotheses) for rates and patterns of disease spread that also can be used to evaluate the effects of different disease control actions. Stochastic simulation models combined with disease, spatial, and demographic data from field studies have been used to assess risk of disease spillover from domestic animals to wildlife (Clifford *et al.* 2009) and to examine disease dynamics in a multihost carnivore community (Craft 2008b).

Spatial scan statistics use models to perform both geographical and time surveillance on disease-occurrence data to detect and locate spatial and temporal clustering of disease. These analyses can help predict "hotspots" of disease, indicating greater pathogenic risk, and may be important for selection of the safest and most appropriate locations for carnivore reintroductions or vaccination programs. Incorporating temporal data into the model may define seasonal cycles of disease or periods of greater pathogenic risk. Examples of software for use in spatial scan statistics are highlighted in Table 13.2.

Frequency and resultant mortality rates of disease can also be incorporated into population viability analysis (PVA) models developed to assess extinction risk (Lacy 2000). This approach was used to examine the risk of quasi-extinction from canine distemper virus for endangered island foxes (Kohlmann *et al.* 2005). More complex epidemiological disease dynamic models have been embedded into population viability models to evaluate rabies vaccination strategies needed to prevent critically low post-outbreak population densities for African wild dogs (*Lycaon pictus*, Vial *et al.* 2006), Ethiopian wolves (*Canis simiensis*, Haydon *et al.* 2002), and to examine the effects of periodic canine distemper outbreaks on the persistence of the Ethiopian wolf population.

Table 13.2 *Computer software available for disease modeling, spatial analysis, and disease risk assessment. Freeware programs are marked with an asterisk (*). (All websites accessed August 21, 2011)*

Name	Application	Website and Manufacturer
@RISK	Disease risk analysis tool for Microsoft Excel using Monte Carlo simulations	http://www.palisade.com/risk/ Palisade Corp., Ithaca, NY USA
Precision Tree	Performs decision analysis in Microsoft Excel using decision tree and influence diagrams	http://www.palisade.com/precisiontree/ Palisade Corp., Ithaca, NY USA
STELLA	Icon-based and graphically oriented systems modeling and simulation software	http://www.iseesystems.com/softwares/Education/StellaSoftware.aspx Isee Systems Inc., Lebanon, NH USA
Vensim	Graphical development, analysis and packaging program for dynamic feedback models	http://www.vensim.com/software.html Ventana Systems Inc., Harvard, MA USA
AnyLogic	Simulation modeling software for systems dynamics, discrete-event and agent-based models	http://www.xjtek.com/anylogic/why_anylogic/ XJ Technologies, Hampton, NJ USA
VORTEX*	Population viability analysis	http://www.vortex9.org/vortex.html Bob Lacy, Dept. of Conservation Science, Chicago Zoological Society, Brookfield, IL, USA
Model Builder*	Graphical tool for designing, simulating and analyzing ordinary differential equation mathematical models	http://sourceforge.net/projects/model-builder/
SaTScan*	Spatial, temporal, space-time scan statistical program to detect spatial and temporal clusters of disease	http://www.satscan.org Martin Kulldorff, Harvard Medical School, 133 Brookline Avenue, 6th Floor, Boston, MA USA
FleXScan*	Spatial scan statistical program to detect spatial clusters of disease; allows users to define spatial connections among data	http://www.niph.go.jp/soshiki/gijutsu/download/flexscan (Tango and Takahashi 2005)

For PVA models to be truly useful for carnivore conservation, the hypotheses they generate (usually called "predictions") should be tested with independent data (Powell *et al.*, in press). Barring independent tests, models must be built using the best-available disease, demographic and spatial data, and incorporate uncertainty in their predictions to account for both environmental stochasticity and the limitations of the data. The extinction risk generated by these models is best used as an index. A non-exhaustive list of mathematical, spatial, and PVA modeling software and their features is provided in Table 13.2.

13.3 Prevention and control of disease

Disease is a natural component of carnivore ecology. Species coevolve with pathogens while adapting to new pathogens, developing resistance to new strains of old pathogens as they re-emerge, and undergoing population fluctuations that modify the pathogen cycles they experience. Human-induced changes may increase the frequency of disease epizootics in carnivores by introducing exotic pathogens, new strains of pathogens, and toxins to new areas; by altering disease cycles through changing ecological communities and reducing ecosystem function; and by driving wildlife populations to low numbers so that they are vulnerable to stochastic processes like disease. When humans recognize the risks posed by disease cycles to threatened species, we must decide whether intervention is warranted, feasible, and morally justified. Our decisions will be constrained by financial conditions, stakeholder opinions, odds of success, the logistical ability to implement an intervention, and often, whether the disease threatens humans or domestic animals.

13.3.1 Intervention options: removing the causative factor

In some rare cases, the agent causing disease can be removed. This option is usually available only for diseases caused by toxic agents, such as anticoagulant rodenticides (Fournier-Chambrillon *et al.* 2004; Riley *et al.* 2007) and heavy metals (Laskowski 1991). Although the ideal option is removal of a toxic agent, this management action is rife with legal, political, ethical, financial, and cultural issues, which often delay or prevent removal of the agent. This lengthy and complex process is typified by the years needed to ban lead ammunition throughout the range of the California condor (*Gymnogyps californianus*), (Title 14 California Code of Regulations, section 475). In the end, despite clear biological information, most decisions are based on sociological, political, and economic issues.

13.3.2 Intervention options: manipulating the host population

The cycle of infectious disease depends on a range of host-population characteristics that influence the intensity, infection rate, and duration of an epizootic. Demographic and population parameters, such as density, vital rates, and social systems influence how a disease behaves in a population, and therefore, changes in these parameters alter the course of the disease. Public health specialists and wildlife managers have used attributes of pathogen cycles to develop management techniques for wildlife diseases; managing the affected host population is usually more feasible than attempting to eradicate a pathogen. First, managers must understand the ecology of the pathogen and the host within the area of interest. They must know whether the pathogen is a specialist or generalist, whether it circulates through a community of many different species, and whether it has vectors and intermediate hosts.

13.3.2.1 Treating individuals

One way to reduce disease transmission is to reduce the number of susceptible or infected individuals under a threshold density below which the pathogen cannot persist. Treating infected individuals through focal animal treatment (e.g. treating bobcats, *Lynx rufus*, with ivermectin for infestations with notoedric mange; Riley *et al.* 2007) or mass distribution of medicines via food bait (e.g. antihelmintics in red foxes, *Vulpes vulpes*, against the tapeworm *Echinococcus multilocularis*; Hegglin *et al.* 2003) directly reduces the number of infected animals and inherently reduces the infection rate.

13.3.2.2 Culling

Through a somewhat controversial method to control disease, managers can reduce contact rates and disease prevalence by reducing a dense host carnivore population, or "culling." Three approaches to culling can be effective: (1) culling only infected individuals, (2) culling randomly over a large area simply to reduce population density, and (3) culling in a specific area to create a barrier to pathogen spread (e.g. a cordon sanitaire). The second approach has been attempted numerous times throughout the world with mixed success, often directed towards carnivores because of the zoonotic threat of rabies (Irsara *et al.* 1982; Rosatte 1988). Local depopulation to create a barrier to disease spread has shown some success in rabies control (Gunson *et al.* 1978). When employing any culling program for disease control, choosing the correct target species or group of species is paramount, as is understanding potential indirect effects of depopulation, such as opening territories and inducing immigration of infected individuals from adjacent areas (Woodroffe *et al.* 2006).

13.3.2.3 Vaccination

Through vaccination, managers reduce the number of susceptible hosts by switching them to the equivalent of "recovered." A number of new vaccines developed for domestic animals has been validated for wild carnivores and used in control programs, conservation programs, and carnivore reintroductions. Effective and safe vaccines are those that: (1) do not produce disease in the host, (2) provide long-term immunity, (3) protect the species of concern against all strains or varieties of the pathogen, (4) cannot revert to virulence, and (5) allow one to distinguish between individuals with vaccine-induced immunity and natural immunity (Wobeser 2007). Managers and biologists should first consult with a wildlife veterinarian to determine the safest, most effective, appropriate vaccine for the focal species.

Broad-scale vaccination using widely broadcast, vaccine-laden baits is regularly implemented to control rabies in carnivores throughout the world (Stöhr and Meslin 1996; Slate *et al.* 2005; Niin *et al.* 2008; Sterner *et al.* 2009; Capello *et al.* 2010). These programs have met with mixed, though generally positive, success after all costs are accounted (Sterner *et al.* 2009). Appropriate baits must be chosen for the target species and are typically distributed throughout a large geographic area by hand or by aircraft. Oral rabies vaccine packages contain a biomarker (often tetracycline) that allows biologists to assess whether a particular individual has consumed the bait and vaccine, thus allowing the percentage of the population that was vaccinated to be estimated. Targeting a particular host carnivore is difficult with broadcast baits, sometimes requiring that more baits be used than needed just for the target carnivore species.

Another approach to vaccination is trap–vaccinate–release (TVR), where animals are live-trapped, vaccinated typically by intramuscular or subcutaneous injection, and released at the capture site (Rosatte *et al.* 1992). TVR programs can target specific carnivore species that may not readily consume oral baits, release non-target species for which the vaccines are not suitable, and use vaccines that have not been developed in an orally administered form, such as vaccines for distemper virus and parvoviruses. TVR programs can be integrated into ongoing trapping and monitoring programs, allowing for initial and follow-up blood samples to determine pathogen exposure prior to, and post, vaccination. Survival of vaccinated individuals can be assessed through radio collaring or mark–recapture methods.

The costs of TVR programs are usually substantially higher than mass-distributions of oral vaccines, but high bait costs and unforeseen legal obstacles of oral-bait distribution may make TVR more desirable to managers. Some programs combine the two methods of mass-immunization of wildlife to achieve the greatest coverage of a focal population(s) (Sterner *et al.* 2009). Vaccination programs for free-ranging carnivores should be adaptively designed with a monitoring component to assess the effectiveness of the program, examine costs, and reassess the need to continue the program.

13.3.3 Intervention options: manipulating sympatric species including domestic animals

The objective of many mass-vaccination efforts is to minimize disease risk to humans and domestic animals. In most of North America and Europe, many domestic animals are protected from common infectious diseases by mandatory immunizations. In many other regions of the world, however, domestic dog populations are immense, largely unvaccinated, and pose a constant threat of disease spillover to nearby carnivore populations. To control a pathogen in a threatened population, one might need to manage a common, sympatric species, which might be a domestic animal population.

When domestic carnivores pose a serious disease threat to wild carnivores, as in the case for the Ethiopian wolf (Randall *et al.* 2006), domestic dog vaccination is an essential part of the disease control program. Data on a domestic dog population size and the contact rates between dogs, other wildlife, and the focal species, can be incorporated into SIR-type models to estimate the minimum proportion of the domestic dog population, or a sympatric wildlife population, that must be vaccinated in order to reduce spillover risk to focal species.

13.3.4 Intervention options: addressing human activities

Though few, if any, pathogens of humans directly threaten wild carnivore populations, human behavior and activities with domestic dogs and cats, often pose threats to carnivore populations. An effective system encouraging dog vaccination for critical pathogens can reduce disease threats to endangered carnivores (Randall *et al.* 2006), if everything goes as planned. Preventing contact between domestic animals and wild carnivores can also reduce disease threats to wildlife (Laurenson *et al.* 1997), again only when pet owners are responsible.

Humans inadvertently affect wild carnivore populations by artificially inflating the densities via supplemental feeding of wildlife with pet food left within the reaches of wild carnivores. Humans encroach on wildlife habitats and relegate diminishing carnivore populations to small, isolated habitat patches, increasing the likelihood of contact among infected individuals, and, thereby, intensifying the risk of infection throughout these populations. Elevated risks make the likelihood of catastrophic epizootics more probable in the short term, especially for urban carnivores that are notorious for perpetuating disease cycles, like raccoons (*Procyon lotor*) and striped skunks (*Mephitis mephitis*). Altering the behavior of humans is a daunting task. Developing creative solutions to these real problems must be a priority, nonetheless, in the conservation and management of wild carnivores.

14

Mitigation methods for conflicts associated with carnivore depredation on livestock

John D. C. Linnell, John Odden, and Annette Mertens

The conflict between carnivores and human livestock is as old as the history of domestication. Carnivores evolved hunting the wild ancestors of today's domestic animals, and the process of domestication has made these domestic animals especially vulnerable by stripping them of much of their antipredator behavior, alertness, and fleetness, and often placing them in landscapes to which they are not adapted. Because of this vulnerability, humans through the millennia have developed diverse strategies to protect their valuable livestock. Nonetheless, for a variety of socioeconomic, historical, and practical reasons, these techniques are often not used, resulting in many conflicts between carnivores and livestock. These conflicts fuel the bulk of the negative attitudes that some human groups hold against carnivores, and absorb large amounts of conservation resources (both financial and human). As a result carnivore–livestock conflicts present in many areas a significant barrier to carnivore conservation. From the points of views of carnivore conservation, agricultural economics, and livestock welfare, making livestock husbandry practices compatible with carnivores is imperative (Breitenmoser *et al.* 2005; Baker *et al.* 2008). The vast literature on carnivore–livestock conflicts includes both theoretical and empirically derived approaches, case studies from every continent, and several reviews (e.g. Linnell *et al.* 1996; Kaczensky 1999; Knowlton *et al.* 1999; Sillero-Zubiri and Laurenson 2001; Shivik *et al.* 2003; Breitenmoser *et al.* 2005; Graham *et al.* 2005; Bangs *et al.* 2005; Sillero-Zubiri *et al.* 2006; Baker *et al.* 2008; Inskip and Zimmermann 2009). This chapter summarizes the topic and focuses on mitigation strategies pertinent to the twenty-first century. We mainly focus on the large carnivore species (>15 kg) and domestic herbivores, excluding poultry and pigs.

14.1 Who kills whom?

Depredation occurs on every continent and in every habitat where domestic animals and carnivores occur together. The extent of depredation, however, and the species involved, vary widely. The most basic parameter leading to the potential for conflict is the size ratio between carnivore and livestock. Small stock (e.g. sheep and goats) are vulnerable to depredation by more carnivore species than are large livestock (e.g. cattle, horses, water buffalo) and juveniles are vulnerable to more carnivores than are adults. Table 14.1 summarizes the carnivore species that are most commonly associated with livestock depredation on different continents. In all, 24 species of carnivore (8 felids, 9 canids, 5 ursids, 1 mustelid, and 3 hyenas) are commonly associated with depredation on 9 species of livestock.

Being aware of the community structure of the carnivore guild in any area is a vital first step in planning mitigation strategies for the various life-cycle stages of the different livestock species. As progress is made with large carnivore recovery, old conflicts will likely return as carnivores reappear within an ecological community. Reappearance will require continual readjustments to husbandry and management strategies, as the necessary mitigation measures are tailored for the specific species combinations.

14.2 Documenting depredation

Even for a given carnivore–livestock combination, depredation is highly variable in space and time (Kaczensky 1999; Baker *et al.* 2008; Inskip and Zimmermann 2009). Landscape, pasture characteristics, age and sex of the carnivores, availability of wild prey, season, use of different husbandry methods, all affect levels of depredation. Livestock die of a wide range of other factors, including starvation, disease, and accidents, and are stolen. Dead livestock are often found some time after death by carnivores who are facultative scavengers, especially in extensive ranching operations, making cause of death hard to identify objectively. Signs of presence at a kill can be related to depredation or scavenging. Accordingly, a crucial first step in addressing depredation is to document the extent to which depredation actually occurs and to identify the species of the carnivores responsible. Documenting the extent of depredation assists in determining the costs and benefits of addressing depredation vs. other mortality factors (Moberly *et al.* 2004; Azevedo and Murray 2007). Documenting the extent of depredation is also crucial for ensuring effective operation of potential compensation systems. Identifying the species of the responsible carnivore is crucial for targeting mitigation or lethal control activities.

Table 14.1 *Vulnerability of various life-stages of livestock species to depredation from the various carnivore species most commonly implicated in conflicts.*

	Europe	North America	South America	Africa	Middle East and Asia	Australia
Cattle						
—calves	wolf/brown bear	wolf/brown bear/ black bear/ cougar/coyote	cougar/jaguar/ Andean bear	lion/spotted hyena/ leopard/cheetah/African wild dog	wolf/brown bear/Asiatic black bear/leopard/snow leopard/tiger/golden jackal/dhole	dingo
—adults	wolf/brown bear	wolf/brown bear/ black bear/cougar	cougar/jaguar/ Andean bear	lion/spotted hyena/ leopard	wolf/brown bear/Asiatic black bear/leopard/tiger	dingo
Sheep and goats						
—lambs/kids	wolf/brown bear/ Eurasian lynx/ wolverine/red fox/ golden jackal	wolf/brown bear/ black bear/ cougar/coyote/ bobcat	cougar/jaguar/ culpeo fox	lion/spotted hyena/ stripped hyena/brown hyena/leopard/cheetah/ caracal/African wild dog/ black-backed jackal/ Ethiopian wolves	wolf/brown bear/Asiatic black bear/sub bear/ leopard/snow leopard/ tiger/golden jackal/dhole	dingo/ red fox
—adults	wolf/brown bear/ Eurasian lynx/ wolverine	wolf/brown bear/ black bear/ cougar/coyote/ bobcat	cougar/jaguar/ culpeo fox	lion/spotted hyena/ leopard/cheetah/caracal/ African wild dog/black-backed jackal	wolf/brown bear/Asiatic black bear/sun bear/ leopard/snow leopard/ tiger/golden jackal/dhole	dingo
Reindeer						
—calves	wolf/brown bear/ Eurasian lynx/ wolverine				wolf/brown bear/Eurasian lynx/wolverine	
—adults	wolf/brown bear/ Eurasian lynx/ wolverine				wolf/brown bear/Eurasian lynx/wolverine	

Horses					
—unspecified age	wolf/brown bear	wolf/brown bear/ cougar	cougar/jaguar		wolf/tiger/leopard/snow leopard/Asiatic black bear
Donkeys					
				lion	leopard
—Llama/alpaca					
—juveniles			cougar		
—adults			cougar		
Camels					
—calves					wolf/snow leopard
—adults					wolf
Yaks					
—unspecified age					wolf/tiger/leopard/snow leopard/Asiatic black bear
Water buffalo					
—unspecified age					tiger

Refs: Bjärvall and Franzen 1981; Bowland *et al.* 1992; Linnell *et al.* 1996; Nowell and Jackson 1996; Ciucci and Boitani 1998; Mills and Hofer 1998; Servheen *et al.* 1999; Ernest and Boyce 2000; Greentree *et al.* 2000; Sillero-Zubiri and Laurenson 2001; Odden *et al.* 2002; Mishra *et al.* 2003; Nemtzov 2003; Ogada *et al.* 2003; Jackson and Wangchuk 2004; Sillero-Zubiri *et al.* 2004; Bradley *et al.* 2005; Graham *et al.* 2005; Marker *et al.* 2005b; Goldstein *et al.* 2006; Michalski *et al.* 2006; Holmern *et al.* 2007; Namgail *et al.* 2007; Baker *et al.* 2008; Gula 2008; Henle *et al.* 2008; Kissui 2008; Land *et al.* 2008; Sangay and Vernes 2008; Dar *et al.* 2009; Gusset *et al.* 2009; Inskip and Zimmermann 2009.
Latin names: wolf/dingo *Canis lupus*, dhole *Cuon alpinus*, coyote *Canis latrans*, golden jackal *Canis aureus*, black-backed jackal *Canis mesomelas*, tiger *Panthera tigris*, leopard *Panthera pardus*, snow leopard *Panthera unica*, jaguar *Panthera onca*, bobcat *Lynx rufus*, Eurasian lynx *Lynx lynx*, cougar *Puma concolar*, cheetah *Acinonyx jubatus*, brown bear *Ursus arctos*, black bear *Ursus americanus*, Asiatic black bear *Ursus thibetanus*, Andean bear *Tremarctos ornatus*, Sun bear *Helarctos malayanus*, spotted hyena *Crocuta crocuta*, stripped hyena *Hyaena hyaena*, brown hyena *Hyaena brunnea*, wolverine *Gulo gulo*.

The most basic step is to examine all livestock found dead. Find and examine kills as quickly after death as possible (Chapter 13); this requires frequent inspection of pastures. Depredation is always associated with physical trauma of some type, so examining a carcass carefully should reveal bites or claw marks. Some carnivores, such as large felids, kill very efficiently with one or few bites (usually to the neck or throat), so the signs may be subtle. Skinning a carcass is almost always necessary to reveal the full extent of trauma. Bite marks accompanied by subcutaneous bruising and bleeding separate depredation from scavenging (no bruising and bleeding). Most carnivores have distinctive prey killing and handling techniques, the sign of which often allows an experienced observer to identify the species of carnivore responsible in the field (Bowland *et al.* 1992; Kaczensky and Huber 1994; Molinari *et al.* 2000; Levin *et al.* 2008; http://icwdm.org/). Some taxonomically similar species, however, sometimes leave similar signs. Separation between canid species, such as coyotes (*Canis latrans*), wolves (*C. lupus*), and domestic dogs (*C. familiaris*) is difficult yet critical, as the desired response to depredation differs among these carnivores (Ciucci and Boitani 1998). Similarly, responses differ according to different large felids (jaguar *Panthera onca* vs. puma *Puma concolor*, leopard *Panthera pardus* vs. tiger *Panthera tigris*). Although experienced fieldworkers and technicians may be able to identify the carnivore species responsible for some kills in areas of sympatry, visual separation is impossible for many cases. Genetic methods that can identify species on the basis of DNA, extracted from a carnivore's saliva left in a bite wound, provides a powerful tool for identifying the responsible carnivore objectively (Chapter 13; Ernest and Boyce 2000; Williams *et al.* 2003; Williams and Johnston 2004; Blejwas *et al.* 2006; Sundqvist *et al.* 2008). Furthermore, the ability to identify sex and individual identity using salivary DNA provides a powerful tool for determining whether problem individuals exist (Linnell *et al.* 1999). Although these methods are expensive, they are rapidly becoming quicker and cheaper.

In some cases the extent of depredation has been highly controversial and hard to quantify because livestock are free-ranging. In response, depredation rates have been studied using radio-telemetry equipment that sends a signal when a sheep, reindeer, or calf dies (remains motionless for a set time; Chapter 7). This technology allows the rapid discovery and examination of the carcass, increasing the chances of accurately assessing cause of death (Bjärvall and Franzén 1981; Warren and Mysterud 2001; Oakleaf *et al.* 2003; Tveraa *et al.* 2003; Knarrum *et al.* 2006). With standard husbandry, this technology may help establish baseline levels of livestock mortality and resolve uncertainty when livestock losses suddenly increase in an area. An important issue, however, is to leave a carcass in the field following autopsy, otherwise the predator may be forced to kill again to replace the lost food, which inflates predation rates.

Enormous potential exists for widespread social conflict surrounding the uncertainty of cause of death in livestock, especially when many animals are simply "missing" (Linnell and Brøseth 2003). Therefore, documenting such losses is very important. Technicians and fieldworkers must receive adequate training in standardized methods, and field inspections and documentation must be rigorous (Wobeser 1996). Rigorous, standardized methods are especially important in cases where compensation may be paid, because the consequences of the identification have economic and legal consequences for the livestock producer.

14.3 The ecology of depredation and its mitigation

Mitigating depredation requires understanding the ecology of predation. At its most basic, depredation occurs because carnivores eat other animals and they do not differentiate between wild and domestic animals! Predation, however, is much more complicated, consisting of a set of six specific, sequential steps (Linnell *et al.* 1996): (1) searching for and locating an animal, (2) identifying this animal as potential prey, (3) approaching the animal closely enough to attack, (4) attacking the animal and establishing physical contact with it, (5) killing it, and (6) consuming it. Depredation is basically similar, with the exception that prey may not be consumed fully, due either to surplus killing (Kruuk 1972a) or to the high risk of disturbance at the kill by a livestock guardian. From the perspective of mitigation, opportunities exist at every step to interrupt the progression.

Humans have sought new ways to protect their livestock since antiquity, providing thousands of years of human experience. Table 14.2 places many mitigation measures in the context of the sequence of events that describe the predation process. Mitigation measures that hold the most promise focus on two broad categories: those focused on carnivores (e.g. lethal control or non-lethal removal), and those focused on livestock (husbandry methods). Addressing livestock depredation effectively inevitably requires use of both approaches (Bangs *et al.* 2005), though the relative use of each varies greatly with circumstances.

14.3.1 Avoiding encounters between carnivores and livestock

Throughout human history, humans have attempted, on a broad scale, to eliminate carnivores large enough to kill livestock (Boitani 1995). By the late nineteenth and early twentieth centuries, this goal had almost been achieved across many landscapes for many species (e.g. brown bears (*Ursus arctos*) and wolves in densely human populated areas of Europe and North America), and resulted in a dramatic reduction in depredation. In the context of conservation, however, this approach is clearly no longer acceptable. Zoning is a compromise approach that separates carnivores and livestock geographically (Linnell *et al.* 2005a). Zoning takes advantage of natural limits to species' distributions, habitat selection, and active regulation of carnivore distribution and density. For example, hunting and lethal control methods can be used to minimize carnivore densities in areas where livestock have priority. Well-designed zoning can increase the predictability of carnivore depredation, which allows producers to plan their future needs and to adopt appropriate husbandry techniques. Zoning also enables a geographical prioritization of economic instruments, such as those subsidizing mitigation measures (Rondinini and Boitani 2007). The long dispersal distances of many carnivores (100s of km; Linnell *et al.* 2005b) dictate that zoning will work only at very large spatial scales and will never exclude carnivores absolutely from any zone. Enough areas with sufficient connectivity need to be zoned for large carnivores to ensure that conservation goals can be achieved demographically and genetically with ecologically viable populations (Linnell *et al.* 2005c).

Another prerequisite for zoning is that managers must be able to control carnivore populations using methods that are economically and socially acceptable. Influencing the densities and distributions of abundant carnivores that have high reproductive rates (e.g. coyotes, *Canis latrans*; red foxes, *Vulpes vulpes*) often requires resorting to poison, which is not allowed in most countries (Knowlton *et al.* 1999; Greentree *et al.* 2000). For these carnivores, effective control is unlikely to be cost-effective, and when benefits are gained, they may be small and hard to document (Harris and Saunders 1993; Knowlton *et al.* 1999; Greentree *et al.* 2000; Moberly *et al.* 2004; Berger 2006). For other species that occur at lower densities and have lower reproductive rates (e.g. bears, *Ursus* spp.; wolves; pumas; lions, *Panthera leo*; Eurasian lynx, *Lynx lynx*), a wealth of examples document where human persecution has limited their densities and distributions, both through depredation-motivated lethal control and through recreational harvest (e.g. Herfindal *et al.* 2005; Treves 2009). Two challenges of using lethal control and a zoning system with these species are: (1) ensuring that control does not threaten the viability of a population (i.e. by creating a sink), and (2) gaining public acceptance for widespread population reduction of carnivores or their exclusion from certain areas (Treves and Naughton-Treves 2005). Zoning is both socially and ecologically complex and must be approached with caution (Linnell *et al.* 2005a).

A final consideration for using widespread lethal control is that reducing the densities of large carnivores in some areas allows smaller carnivores to increase in density (e.g. Blaum *et al.* 2009), which can lead to depredation or other conflicts with them (through mesopredator release, Palomares *et al.* 1995).

At a finer scale, much can be gained from locating pastures in the parts of landscapes that are associated with habitats that carnivores avoid. Pastures and farms most exposed to attack are usually close to areas of high carnivore density (e.g. wolf rendez-vous sites or dens, national park borders; Oakleaf *et al.* 2003; Bradley and Pletscher 2005; Holmern *et al.* 2007; Van Bommel *et al.* 2007; Gula 2008; Kaartinen *et al.* 2009). Avoiding areas with forest cover, riparian habitat, or good stalking terrain, and selecting areas with high densities of roads and close to human habitation, should reduce depredation (Treves *et al.* 2004; Bradley and Pletscher 2005; Michalski *et al.* 2006; Azevedo and Murray 2007; Palmeria *et al.* 2008). Simply clumping livestock into fenced pastures, rather than allowing them to spread throughout the forest, reduces encounters dramatically and, therefore, depredation (Swenson and Andrén 2005; Kissling *et al.* 2009).

Whether abundant, alternate prey reduces depredation is equivocal. Studies of wolves and Eurasian lynx, where wild prey were locally abundant, documented increases in depredation, possibly because carnivores were attracted to these habitats (Stahl *et al.* 2002; Treves *et al.* 2004; Bradley and Pletscher 2005; Moa *et al.* 2006; Odden *et al.* 2008). Other studies, in contrast, on African wild dog (*Lycaon pictus*), jaguar (*Panthera onca*), and wolves, found the opposite: where livestock are perceived as alternative prey they will be killed, when or where wild prey are not abundant (Meriggi and Lovari 1996; Polisar *et al.* 2003; Sidorovich *et al.* 2003; Woodroffe *et al.* 2005a; Kolowski and Holekamp 2006). The difference may simply be a matter of scale. In large areas with an overall low densities of wild prey, carnivores may increase their depredation pressure on livestock to compensate for having few prey (Odden *et al.* 2006), but in areas where wild prey is widely distributed and abundant, carnivores may spend most time in the most prey-rich patches, leading to high encounter rates with livestock there and more depredation (Odden *et al.* 2008).

14.3.2 Preventing the recognition of livestock as potential prey

For the last 30 years, much research has targeted aversive conditioning. The basic principle of aversive conditioning is that carnivores experiencing a negative stimulus when attacking livestock will associate the negative stimulus with livestock and not attack livestock again. Negative stimuli tried include chemicals that induce vomiting (or at least taste bad) placed on carcasses, electric shock collars placed on predators, shooting predators with rubber bullets or exploding cracker shell, and using guard dogs (Smith *et al.* 2000b; Shivik 2006; Hawley *et al.* 2009). Tests in captivity have taught individual carnivores to avoid eating carcasses but success at stopping them from killing has been minimal. Furthermore, no field trials have been successful (Land *et al.* 1998; Smith *et al.* 2000b; Shivik *et al.* 2003; Cowan *et al.* 2005; Shivik 2006). To work, aversive condition needs to be applied continually to every individual carnivore of each species that depredates livestock. The hope that an individual might teach other members of its social group to avoid livestock has no support.

A common approach to resolving depredation problems has been to selectively remove those individual carnivores that prey on livestock, the so called "problem individuals" (Linnell *et al.* 1999; Treves 2009). Although this idea is appealing, such individuals often do not exist (Odden *et al.* 2002; Herfindal *et al.* 2005), and the logistical problems associated with targeting them for removal are myriad. Only when the individual is observed making a kill and is removed immediately, where it can be tracked from the kill and killed, or when toxic collars (Livestock Protection Collars, Connolly 1993) poison the attacker directly, can the right individual be removed for certain. Often, the animals most commonly responsible for depredation are the hardest to target (Sacks *et al.* 1999). Even where individuals can be removed selectively, their territories will usually be filled rapidly, potentially by more than one animal, which can lead to even more conflicts (e.g. Robinson *et al.* 2008). The effectiveness of selective removal varies widely among situations, implying the need to test the underlying assumptions before building a response strategy on this principle. The benefits of selective removal are probably mainly social, in that livestock producers may feel appeased or empowered if they are allowed to kill the occasional, presumed, problem individual. Even this benefit is limited to particular segments of society, as other social groups find even this killing of carnivores controversial (Treves and Naughton-Treves 2005). In response to the public dislike for lethal control, selected individuals have been removed and translocated, a supposedly non-lethal alternative. Despite its widespread use and popularity, translocation is largely unsuccessful as a routine conflict-management tool (Linnell *et al.* 1997; Bradley *et al.* 2005). Translocated animals often die, roam over large areas, return to the site of capture, or depredate livestock at the release site. In specific cases, translocation may cause conflicts to escalate (Athreya *et al.* 2010).

The main mitigation strategy that does cause predators not to consider livestock as prey is choosing large livestock species or breeds (Rook *et al.* 2004). Using water buffalo or cattle instead of sheep or goats effectively excludes depredation by small carnivores (Table 14.2). Significant benefits also come from switching to breeds or selectively breeding individuals that exhibit strong antipredator behavior, that are amenable to herding, or that are amendable to other mitigation measures (May *et al.* 2008a). Much more research is needed with this strategy, and the strategy

Table 14.2 *The behavioral steps in a predation sequence with the associated mitigation measures that can interrupt the escalation of attack.*

Behavior	Mitigation measure	Mechanism (theory/assumption)
Search	Eradication of carnivores	If all carnivores are removed there will be no encounters—historically the measure of choice, but obviously not suitable within a conservation context.
↓	Zoning	On a coarse scale (measured in 1000s of km^2) it is possible to reduce depredation by avoiding livestock production in regions with highest density carnivore populations. This can be based on natural patterns of carnivore distribution or on the active regulation of their distribution and density, e.g. through hunting.
↓	Placement of livestock in the landscape	On a finer scale it is possible to take advantage of carnivore patterns of habitat selection to place flocks in parts of the landscape that carnivores use less, or to invest more heavily in mitigation measures in high risk areas.
↓		
Identify	Aversive conditioning	The principle is to provide negative experiences associated with livestock that should lead the carnivore to avoid regarding the livestock as suitable prey.
↓	Selective removal	If depredation is due to a few specific problem individuals, their selective removal should in theory reduce depredation.
↓	Different livestock species	Moving from small stock (sheep and goats) to large stock (cattle, water buffalo) production will prevent depredation by many smaller carnivores.
↓	Promote wild prey	The existence of wild alternative prey is a prerequisite for effective depredation mitigation. The greater the availability of wild prey, the less likely it is that carnivores will depend on livestock.
↓		
Approach	Avoid certain habitats	Keeping livestock in open habitats as opposed to closed habitats and away from stalking cover may discourage many species in their final approach.
↓	Carnivore-proof fencing	The use of carnivore-proof enclosures (e.g. electric fences) around whole pastures or for night-time enclosures (e.g. bomas), by definition effectively stops depredation.
↓	Lights, sirens	The principle is that these devices will scare carnivores away as they make their final approach.

(continued)

Table 14.2 *Continued*

Behavior	Mitigation measure	Mechanism (theory/assumption)
↓	Livestock-guarding dogs	These dogs will remain with the flock and either will drive the carnivores away or interfere enough with their attack sequence so that shepherds can arrive.
↓	Shepherds	Most carnivores will be deterred from their attack by the arrival of multiple human shepherds.
↓		
Attack	Livestock-guarding dogs	Dogs will interfere with the carnivore's attack, preventing it from completing the kill.
↓	Shepherds	Shepherds will interfere with the carnivore's attack, preventing it from completing the kill.
↓		
Kill	Protective collars	In principle these collars will form a physical barrier to the carnivore's bite.
↓		
Consume	Livestock-guarding dogs	Dogs will prevent the carnivore from being able to consume its kill by driving it away.
	Shepherds	Shepherds will prevent the carnivore from being able to consume its kill by driving it away.

should be compatible with agricultural conservation initiatives that focus on conserving traditional and rare breeds (Hall and Bradley 1995). Increasing protection for vulnerable juveniles of all livestock species by confining them to sheds or areas close to human habitation during and after birth provides further benefits.

14.3.3 Preventing access to livestock by carnivores

Most successful mitigation measures operate at this stage of the predation process. Devices (often high tech) that produce loud sounds and lights to scare carnivores, and flag-lines ("fladry") for wolves, deter predators from entering pastures in the short term (Musiani *et al.* 2003). Nonetheless, no real evidence supports more than a temporary respite from depredation because carnivores become habituated (Shivik *et al.* 2003; Bangs *et al.* 2005; Shivik 2006). These devices may be useful for rapid deployment in crisis situations to buy time to introduce more effective measures.

Two approaches have produced effective results: modern electric fencing and traditional shepherding systems. In many situations in Europe and North America,

Box 14.1 Electric fencing

Many different electric fences have been tested. The most thoroughly tested is the type most commonly used in Europe and North America to prevent depredation by wolves, coyotes, Eurasian lynxes, and brown bears. The basic design has 5–6 parallel strands of high-tensile wire (1.4 to 2.5 mm diameter) mounted on solid poles reaching to a height of 1.1 to 1.6 m. The lowest wires should be less than 20 cm above the ground. Voltage in the fence should be 4000 to 7000 V. Grounding is crucial, and requires at least three 1-m long grounding rods. In areas where the ground is a poor conductor (snow covered or very dry), disconnecting some of the fence wires may be necessary to use them as grounding wires. Take care with streams, ditches, or other places where carnivores can crawl under. Details of designs have been provided by Angst (2002), Levin (2002), and the Wildlife Damage Centre (2003). Power can be drawn from power mains, generators, batteries, or solar panels. Although these fences have been designed to use solid poles and to be fixed, a lightweight version can also be used to form a small night-time enclosure (Mertens *et al.* 2002).

These fences are highly effective at discouraging large carnivores from entering fields and even from entering large fenced areas of forest pasture where sheep graze 24 hours a day without supervision. Such a fenced pasture system can be combined with free-ranging, livestock-guarding dogs, or used to gather sheep into a more solid night-time enclosure. Further testing of fence designs is needed for the very large felids.

livestock are grazed on permanent pastures that are fields or open forests. Normally livestock are constrained by simple wire-netting fences or lightweight electric fences that hinder movement by livestock but are permeable to carnivores. While containing livestock in this way prevents a great deal of depredation by reducing chance encounters (Swenson and Andrén 2005), it is relatively simple to upgrade the fencing to carnivore-proof electric fencing. Upgrading requires 5–7 strands of high-tensile wire and high voltage (Box 14.1) and is effective for many species of carnivore (e.g., coyotes, wolves, bears); electric fences may be less effective at stopping large felids that jump. Even though some carnivores still enter these enclosures, losses are greatly reduced. In Norway, large (up to 20 km^2) areas of forest pasture have been enclosed. This approach has the large problems of constructing and maintaining very long fences and has side-effects on the movements of other wildlife. Initial investment costs are high but maintenance costs are

Box 14.2 Shepherding

The most important prerequisite for shepherding is that a flock must be under control, which is impossible in extensive systems. The following shepherding methods represent an increasing intensity of protection in response to increasing threats of depredation.

1. Synchronize reproduction and ensure that birthing occurs under close supervision, either indoors or in well-protected pastures. Neonates should reach a critical size before being released onto pasture.
2. Avoid grazing livestock (especially when juveniles are present) in high-risk pastures, associated with stalking cover. Graze livestock in clumped patterns.
3. Gather livestock into carnivore-proof night-time enclosures (electric fences, high wire fences, barns).
4. Have shepherds accompany herds constantly during day time.
5. Have livestock-guarding dogs remain with the herds by day and night. Shepherds should sleep nearby to react to signs of an attack during the night.

Refs: Linnell *et al.* 1996; Smith *et al.* 2000a; Rigg 2001; Ogada *et al.* 2003; Woodroffe *et al.* 2007.

relatively low. In countries with high labor costs, carnivore-proof electric fencing around permanent pastures will probably be one of the best solutions to depredation.

Traditional shepherding systems in Europe, Asia, and Africa (Box 14.2) have shepherds, often accompanied by dogs, who guard livestock while they graze during daytime, and enclose the livestock into corrals or sheds at night. Protection is provided by the presence of the shepherd and the dogs during day, and by the physical structure of the night-time enclosure and the proximity of the shepherd and dogs at night (Mertens *et al.* 2001; Ogada *et al.* 2003; Wambuguh 2007; Woodroffe *et al.* 2007). Some extensive systems, especially those associated with nomadic pastoralists, have no night-time enclosures. Instead livestock bed close to a campsite and are guarded by shepherds and dogs at night. These traditional systems have permitted livestock production on landscapes with high densities of large carnivores for millennia. Many studies have demonstrated their success (Kruuk 1980; Smith *et al.* 2000a; Ogada *et al.* 2003; Espuno *et al.* 2004; Bauer and de Iongh 2005; Woodroffe *et al.* 2007; Gusset *et al.* 2009) and the negative consequences of lax husbandry (e.g. Linnell and Brøseth 2003; Wang and Macdonald 2006).

The traditional systems are just as applicable today as in the past, with minimal changes. One potential change is in the availability of specialized livestock-guarding dogs beyond their areas of origin. The many breeds, which were developed in Europe and the Middle East (Coppinger and Schneider 1995; Rigg 2001), are currently being spread around the world (Marker *et al.* 2005a). Furthermore, far better knowledge is available now about the best techniques of bonding, integrating dogs into flocks, and correcting inappropriate behavior. Second, new alternatives exist for constructing night-time enclosures, including chain-link and electric fences. New materials exist to construct mobile lightweight electric fences suitable for nomadic systems (Mertens *et al.* 2002), as well as very solid permanent structures. Many large carnivores, especially large felids (lions; leopards, *Panthera pardus*; snow leopards, *Uncia uncia*) and spotted hyenas (*Crocuta crocuta*), can be extremely aggressive and persistent when entering night-time enclosures, either by jumping fences, squeezing through small openings, or using brute force (often aided by panicking the livestock that break down the enclosure wall from the inside). In addition, these carnivores may not be deterred by the presence of humans and may be dangerous. For these species, enclosures must be constructed from very solid materials, often with a roof. Clearly, more work is needed to identify practical, functional materials for structures, appropriate for different landscapes and socioeconomic consequences, and that will deter the different predator species (e.g. Ogada *et al.* 2003; Kolowski and Holekamp 2006; Woodroffe *et al.* 2007). Predation on guarding dogs by predators, such as leopards (Wambuguh 2007; Dar *et al.* 2009), may be decreased by integrating specialist and large Eurasian dog breeds into other husbandry systems (Marker *et al.* 2005b).

The main problem with traditional herding systems is that they are labor intensive. In systems where livestock are milked, the addition of guarding measures comes at relatively low extra costs because the livestock need to be herded for twice daily processing. In systems where meat is the main product, guarding has a high cost because production can, in theory, exist without shepherds, if carnivores are absent. Development of very solid night enclosures will eliminate the need for herders to be awake all night. The socioeconomic status of the country will determine the relative costs of labor intensive vs. technical solutions. The benefits of having large vs. small herds appear mixed (e.g. Schiess-Meier *et al.* 2007; Van Bommel *et al.* 2007; Hemson *et al.* 2009), making the impact of adopting economies of scale unclear. Because depredation can have seasonal patterns (e.g. Rasmussen 1999; Pattersonl *et al.* 2004; Kolowski and Holekamp 2006; Sangay and Vernes 2008; Dar *et al.* 2009) or be confined to certain age classes of livestock

(Table 14.2), use of mitigation measures can be limited to seasonal needs, thereby providing potential savings. A careful spatial analysis of conflict risk should help target the appropriate mitigation measures into the correct areas (e.g. Treves *et al.* 2004; Kolowski and Holekamp 2006; Inskip and Zimmermann 2009; Kissling *et al.* 2009).

Secondary impacts of changes in livestock husbandry to livestock growth and health, and the impacts of livestock on vegetation, also are important. Livestock allowed to graze freely, and those that are shepherded and confined at night, have different activity patterns and different access to forage. Livestock with access to abundant forage during the day may be able to compensate for night-time confinement (e.g. Iason *et al.* 1999), but confinement and herding will probably reduce growth rates in other circumstances. Changing the breed of livestock to one that has behavioral adaptations that are more compatible with the husbandry system (e.g. flocking behavior) might be necessary. Furthermore, changes in grazing pressure caused by fencing or herding will probably increase grazing pressure in some areas and decrease pressure in others. The resultant impacts on vegetation biomass and biodiversity may be complicated and hard to predict (Austrheim *et al.* 1999).

14.4 Compensation

The payment of compensation for livestock losses due to depredation is a widespread, but far from universal, method used to protect livestock producers from economic losses and to try to increase public acceptance of conflicts (Fourli 1999; Nyhus *et al.* 2005; Box 14.3). Where compensation is paid, it is funded by the state, by non-government organizations, or by agricultural insurance schemes. Compensation is usually paid only for depredation by carnivores of specific species, requiring identification of the responsible species. In addition, conditions may be attached to the payments, such as minimum husbandry requirements. The assumption is that receiving an economic compensation increases tolerance of carnivore depredation. This assumption is not always valid (Naughton-Treves *et al.* 2003; Gusset *et al.* 2009; Boitani *et al.* 2010) and its validity varies with the socioeconomic and cultural contexts (Maclennan *et al.* 2009). Furthermore, throughout most of the developed (and much of the developing) world, agriculture is heavily subsidized to achieve a range of strategic, social, and cultural goals beyond food production. Often, societal interests promote protecting livestock producers from economic loss.

Compensation schemes can be very expensive (transaction costs and amounts paid) and controversial (where depredation must be documented), which makes them unsustainable (Nemtzov 2003) and out of reach for many developing countries. Many compensation systems are also so poorly administrated that many people do not bother to use them (Madhusudan 2003). In many cases, finding all carnivore-killed livestock, or having them inspected rapidly to verify cause of death, is impossible (Linnell and Brøseth 2003). Furthermore, many people feel that compensation schemes reward passivity and do not motivate producers to adopt effective mitigation strategies (Nyhus *et al.* 2005; Bulte and Rondeau 2006; Boitani *et al.* 2010). Insurance programs appear to work in some

Box 14.3 Economic instruments

A fundamental premise is that the presence of large carnivores and the associated risk of depredation is natural (e.g. in the same was as weather) and, therefore, livestock producers must accept that they have a responsibility to protect their flocks. Society, however, often has a range of motives for encouraging livestock production that goes beyond the simple economics of meat, milk, or wool production. Therefore, some economic assistance to producers will be desired to insulate them from the economic costs associated with depredation or depredation avoidance. Such assistance should be provided with the following priorities:

1. Assistance (incentives) to adopt carnivore-compatible husbandry. This could take the form of providing materials (e.g. electric fences, livestock-guarding dogs, building materials for night-time enclosures and shepherds cabins), labor, advice, or guidance. This assistance is most effective if the producers are also required to co-fund purchases. Follow-up by qualified technical staff is required.
2. Cash incentives, including the payment for the presence of carnivores.
3. An insurance scheme where producers pay premiums, or parts of premiums, to insure their livestock against predation.
4. *Ex post facto* payments for losses.

For the latter two systems, the following aspects are highly desirable.

- Some minimum requirements for husbandry systems.
- Depredation cases verified by trained inspectors.
- A deductible, so that a producer carries a portion of the cost.

Refs: Naughton-Treves *et al.* 2003; Nyhus *et al.* 2005; Bulte and Rondeau 2006; Schwedtner and Gruberb 2007; Zabel and Holm-Müller 2008; Maclennan *et al.* 2009.

countries, where producers pay premiums to insure their stock against losses. Even when the system is subsidized, it induces a sense of responsibility into the system. Nonetheless, theory and experience suggest that financial mechanisms that pay incentives for carnivore presence (paying for risk), rather than paying *ex post facto* for damage, should work better (Ferraro and Kiss 2002; Schwerdtner and Gruberb 2007; Zabel and Holm-Müller 2008). Such incentive systems encourage depredation prevention, rather than documentation, and have significantly lower transaction costs than compensation and insurance systems. The major cost for incentive systems is the need to map carnivore distributions accurately and to agree on a rate of payment that is fair. In addition, spending money to monitor carnivores, rather than to document cause of death among livestock, is a far better investment from a conservation perspective.

The classical *ex post facto* system may still have a role within an incentive system. Payments may be needed when depredation occurs outside the known range of certain carnivore species, where livestock producers would expect to need mitigation measures. Payment might also be needed for extreme depredation despite the use of effective measures (Box 14.3).

14.5 Integrating mitigation into agricultural policy

Any successful mitigation system will probably adopt part of each of the three components of an integrated approach (Balme *et al.* 2009). The first component consists of economic instruments used to protect livestock producers from extreme economic loss caused by depredation or used to reduce depredation (Box 14.3). The second component consists of actions directed at the carnivores; most often this is lethal control. The third component consists of actions aimed at livestock husbandry to reduce the abilities of carnivores to attack and kill livestock (Boxes 14.1 and 14.2). Based on our experience and our review of the literature, we strongly recommend that most emphasis be placed on the third component. Emphasizing the third component focuses on the most effective mitigation strategy, encourages carnivore conservation, addresses serious animal welfare issues that are associated with allowing livestock to be exposed to depredation, and allows meat production with the associated job satisfaction for livestock producers.

Some need for carnivore removal, most often through lethal control, will undoubtedly be needed (Linnell *et al.* 1997). Although lethal control may well be controversial with the urban public (Treves and Naughton-Treves 2005; Bruskotter *et al.* 2009; Treves 2009), rural communities and livestock producers support immediate retaliatory action for livestock depredation (Woodroffe and Frank 2005; Kissisi 2008; Balme *et al.* 2009; Hemson *et al.* 2009). In some cases, retaliatory action removes genuine problem individuals. In other cases, it simply reduces social conflicts through the empowerment of producers (Bjerke *et al.* 2000). As human–wildlife conflicts have both material and social components (Dickman 2010), both real effects and perceived effects of actions must be considered, which requires a solid understanding of both the social (Lindsey *et al.* 2005; Marker *et al.* 2005b; Bagchi and Mishra 2006; Wambuguh 2007; Lagendijk and Gusset 2008; Dar *et al.* 2009; Hemson *et al.* 2009) and biological aspects of depredation conflicts. No matter what the situation, lethal control must occur within a regulated context that does not have unintended negative impacts on carnivore populations (Woodroffe and Frank 2005). Furthermore, the appropriateness of lethal control depends on the conservation status of the species in question.

Many of the effective mitigation measures involving husbandry are extremely labor-intensive and costly. In areas with recovering populations of large carnivores, the return to effective husbandry practices will require dramatic increases in costs. In areas where carnivores have been present continuously and effective, traditional husbandry practices exist, these practices must be maintained. Maintaining these practices means that producers face significant opportunity costs, because the carnivores will prevent changes to more cost-effective, extensive forms of livestock production that are usable in carnivore-free areas. In addition, adopting carnivore mitigation measures may result in reduced weight gain and increased parasite burdens caused by changes in pasture use and reduced grazing times. Extensive planning will be needed regarding how to fund the adoption and maintenance of the required husbandry practices. A central issue is how to redistribute the costs and the benefits of conservation, such that not all costs fall on a few people. Finding mechanisms for this redistribution is a central challenge within the present discourse on ecosystem services (Naidoo and Ricketts 2006; Jack *et al.* 2008; Linnell *et al.* 2010). In the modern, competitive, globalized economy, subsidizing agriculture is controversial; depredation mitigation, however, should fit within the framework of the "green" subsidies that are permitted.

Simply giving producers money or instructions is not enough. Long-term outreach systems are needed to motivate the introduction of, and to follows-up, the long-term use of effective mitigation measures. If such a system is to be up-scaled from a few case studies (e.g. Jackson and Wangchuk 2004) to widespread implementation, it must be institutionalized within existing agricultural structures and must not be left to conservation organizations. Changes of this magnitude require the adoption of an holistic view of the entire livestock production system, a view that simultaneously considers traditions, habitat (forage), disease, depredation, economics, markets, and labor at both local and global scales, and that varies with context (Mishra *et al.* 2003; Robinson and Milner-Gulland 2003; Nyariki *et al.* 2009). The actual objectives of livestock production need to be clarified. In different areas of the world, these objectives will vary from subsistence food production, to commercially viable meat, milk, and wool production, to maintenance of traditional practices, to landscape conservation (i.e. conserving modified landscapes for biodiversity or aesthetic reasons). Such an holistic view requires a close coordination between agricultural and environmental policies (Henle *et al.* 2008).

Finally, people must recognize that some systems of livestock production are so extensive (e.g. Sami semi-domestic reindeer herding in Fennoscandia, Australian and Argentine sheep ranching) that making effective changes to their form of husbandry is almost impossible. In such cases, the only adequate responses may well lie with widespread lethal control or with some form of economic compensation for losses. The former option raises the question of whether viable carnivore populations should be conserved in these areas. The second option accepts that depredation is inevitable and losses must simply be repaid. These systems, however, represent extreme cases. Everywhere else, systems of livestock production can adapt to the presence of large carnivores to the extent that livestock depredation is kept to levels that are acceptable to the range of economic, ecological, social and ethical interests that exist.

15

Carnivore restoration

Michael K. Stoskopf

The success of a species' restoration requires that, at a minimum, the problems that led to the need for the restoration have been corrected. Restoration requires finding and integrating solutions to complex challenges in human, environmental, and animal dimensions that are unique to each restoration challenge. Excellent prescriptions for successful restorations have been available for over two decades (Kleiman 1989) and guidelines for species restoration, including reintroduction, translocation, reinforcement, and conservation introduction for over a decade (IUCN 1998). Unfortunately, restorations have failed as result of ignoring this literature more than from other causes (Macdonald 2009). Whether existing guidelines are a complete map to success is less the issue than the reality that restoration projects focus, all too often, only on subsets of the challenges that must be addressed.

In this chapter, we will address briefly the factors in the human dimension, including cultural, political, jurisdictional, and economic issues that affect a carnivore restoration. We will then examine important factors in the environmental dimension, in particular, climate, topography and risk of infectious disease. Finally, we move to the most comfortable realm for restoration biologists, the factors in the animal dimension of restoration, including inter- and intraspecies' interactions, selection of founder animals, control of genetic introgression, and management of population health.

Can carnivore restorations succeed? Synthetic reviews published through the early 1990s concluded that reintroductions of carnivores, and particularly mega-carnivores (those carnivores perceived as capable of predating on humans), were not viable (Wemmer and Sunquist 1988; Mills 1991), despite the successes achieved in the 1960s and 1970s. Though many carnivore biologists consider carnivore restoration feasible, even if challenging (Hayward *et al.* 2007a, 2007b; Powell *et al.* in press), estimates of success rates range from 11 to 53% (Jule *et al.*

2008), the wide range depending, in part, on variable, arbitrary definitions of success.

15.1 Human dimension

"The biology is easy. The human issues are hard." Biologists working on carnivore restorations inevitably express some version of this statement. Many biologists lack education and experience with how to deal with people, and dealing with people is hard work. Biologists need greater knowledge and better tools for affecting human attitudes toward change and, in particular, carnivore restorations. The literature on humans and carnivores lacks prescriptive methods for affecting human attitudes toward restorations.

Human attitudes are informed by cultural, political, educational, and economic factors that shape personal experience. Planning a carnivore restoration is complicated further because human attitudes evolve (Schwartz *et al.* 2003). Fortunately, this evolution provides hope that individual and social thinking can be modified.

15.1.1 Cultural issues

To affect public opinion, restoration biologists must understand that differences in attitude are informed more by knowledge, both cultural and educational, than by a fear of the unknown. The public views mega-carnivores differently from small carnivores, and different cultures have different attitudes toward different assemblages of carnivores. Europeans have a longer history than North Americans of persecuting large carnivores and then restoring carnivore populations (Schwartz *et al.* 2003). The shift in human attitudes in Europe parallels the decline in utilitarian approaches to wildlife and the recognition of intrinsic values of all species. Nonetheless, the shift in attitude occurs at different rates in different regions. The assumption that resistance to carnivore restoration is simply lack of knowledge (Morzillo *et al.* 2007a, 2007b; Rice *et al.* 2007) is a gross oversimplification. Compelling data and attitudinal theory predict that people with strongly held opinions become more extreme in their opinions after receiving more information (Meadow *et al.* 2005). That prediction is consistent with attitudes of ranchers in Montana, USA, toward the reintroduction of black-footed ferrets (*Mustela nigripes*) and of people in Sweden (Ericsson and Heberleiln 2003) toward reestablishing wolves.

Cultural change in attitudes toward predators appears to be very slow (Chavez *et al.*2005). Rural residents tend to harbor more negative attitudes toward mega-carnivores than urbanites, and these attitudes seem not to be based on recent experience. Attitude may be based on accounts of experiences across many generations. Despite strong efforts, education regarding the status of carnivore populations and the obstacles to recovery of carnivores are unsuccessful, even in relatively urbanized and economically advantaged areas. For a restoration biologist, the take-home message might be that the time frame for change in cultural attitudes is too long to be realized practically in most restoration projects. Nevertheless, biologists must recognize the cultural foundations for public attitudes and understand the basis for resistance to efforts to modify public opinions.

15.1.2 Political and jurisdictional issues

Political and legal treatments of carnivores need to provide a balance between obligations to reestablish a species and the rights of those who may suffer losses as a result of restoration (Chapter 14; Rees 2001). How different countries, and even political regions within a country, address that balance varies tremendously and is affected by political systems (Williams *et al.* 2002a).

In democratic nations, policy and law makers weigh perceived risk by members of the public much more heavily than they weigh technical assessments made by experts (Slovic, 1987). Decision-making and the interactions of participants in stakeholders are influenced strongly by social factors, such as leadership, communication, teamwork, the presence or absence of evaluation, organizational culture, and the ideologies of participants (Wallace 2003). These political and public behaviors affect attitudes and actions regarding restoration of carnivore populations. Even favorable responses from public consultation exercises will not guarantee success of a species' reintroduction program, primarily because a few active opponents can exert a disproportionate impact on public acceptance (Rees 2001). Here the lesson for restoration biologists is clear. Though by principle, the majority may rule in democratic societies, political success depends more on the correct identification and successful cooption of key active opponents, than on the ability to persuade large numbers of citizens who lack the tools to influence decision-making at political levels.

15.1.3 Economics

An argument frequently used to support carnivore-restoration programs is that communities and nations will benefit economically through non-consumptive ecotourism and, sometimes, hunting tourism. In Africa, economic returns from ecotourism and non-consumptive, photo-safaris may exceed 10% on investment (Cotterill 1997), and the annual return on the $160,000 annual investment in lions in Kenya's Pilanesburg area is reported to be $4,160,000 (Stuart-Hill and Grossman 1993), though this may be overoptimistic. Nonetheless, the value of ecotourism is being associated with carnivore reintroduction schemes in Africa to justify relocation costs. Elsewhere, where government-led programs rather than private carnivore-restoration programs are more common, this is not the case and carnivore reintroductions involve substantial capital outlay with little opportunity of direct capital return (Hayward *et al.* 2007b). Unfortunately, indirect returns are infrequently assessed. Where data are available, evidence suggests that reintroducing large predators is financially beneficial when both direct and indirect returns are considered. One African study looking at lion reintroduction into a managed park, attributed a 31% increase in tourist occupancy and a 71% increase in revenue following predator reintroduction. (Hayward *et al.* 2007b).

In contrast to non-consumptive ecotourism, hunting-based tourism receives a quieter nod within carnivore-restoration circles as an economic engine. Both the impact of carnivores on prey species and the economics of hunting carnivores themselves are important. Carnivore reintroduction or recovery generates an interestingly divided response among hunters. In the United States for example, deer hunters opposed to reintroductions of wolves perceive a threat to deer populations. This view is most pronounced where deer-hunting seasons have been closed because of low deer numbers (Lohr *et al.* 1996). The negative economic impact of increased predator populations on popular sport-hunting species is poorly documented in the literature and better assessment is needed on the impact of the prey base on sustainability of prey species (Garrott et all 2005). Similarly, few examples exist of the potentially positive financial returns from the hunting of carnivores, especially mega-carnivores. One example is, that, although polar bear hides acquired on subsistence hunts in the Canadian arctic have commercial value, revenues from nonresident trophy hunting provide a much greater economic return to Inuit communities (Freeman and Wenzel 2006).

On the cost side of economic calculations, carnivore predation on livestock receives the most attention but the great diversity of situations (type of livestock, environmental context, husbandry methods, acceptance of carnivores, etc.) makes generalization impossible (Chapter 14; Breitenmoser 1998; Fritts *et al.* 2003; Oakleaf *et al.* 2003; Azevedo and Murray 2007). In many situations, carnivore-restoration programs require significant changes in livestock-management practices to minimize depredation. Predation management requires a partnership among producers and wildlife managers to tailor programs to specific damage situations so that the most appropriate techniques can be selected (Chapter 14; Knowlton *et al.* 1999).

Successful carnivore-restoration programs assess broadly the economic drivers that relate to the target species. They identify the key potential factors on both the negative and positive sides of the economic equations and focus resources on mitigating losses in politically and culturally acceptable ways, while optimizing gains from the most beneficial contributors to the local economy.

15.2 Environmental and habitat dimension

Selection of recovery and release sites is critical for successful reintroductions and restorations (IUCN 1998). Topography, climate, anthropogenic features, habitat, and prey base must meet the requirements of target carnivores. Other factors in the environmental dimension, such as habitat health-risk assessment, are included in restoration planning infrequently, unfortunately. In practice, until our understanding of carnivore biology and our technical abilities to gather and to interpret large amounts of time-sensitive data greatly improve, no recovery site will be completely characterized before restoration efforts begin. In addition, the environmental and habitat dimension is always changing, driven by complex natural and anthropogenic forces, and unlikely to persist unchanged over the long timescales envisioned for restoration success. Ongoing adaptive evaluation and management of factors within the environmental and habitat dimension is essential throughout a species' restoration effort, and even after success is declared.

15.2.1 Topography

Topography affects access to resources for both large and small carnivores. Home ranges of black bears (*Ursus americanus*) are oriented on major topographic features, such as watersheds and ridges, and access to resources is affected by topography (Powell and Mitchell 1998). Pine marten (*Martes martes*) home-ranges include more wood and scrub vegetation, and less arable land, than found on the entire landscape, and favor watercourses with continuous vegetation (Rondinini and Boitani 2002). Similarly, travel routes by wolves clearly favor corridors for easy travel, including roads affecting resource selection by wolves (Ciucci *et al.* 2003). Snow leopards (*Uncia uncia*) have a strong affinity for steep and rugged terrain and habitat edges (McCarthy *et al.* 2005). Finally, topologic features may not affect a population uniformly. Home ranges of female polecats (*Mustela putorius*) contain more farms and ponds than do those of adult males in the breeding season (Rondinini *et al.* 2006a).

Identifying the key topographical features that affect home range and habitat use by members of a species is critical to selecting restoration sites capable of supporting the intended population (Chapters 9 and 10). When key topographic requirements are known for a species, they need to be quantified to develop population projections for a restoration site. The more common challenge is a paucity of detailed understanding of the impact of topography on the target carnivores. Broad variations in topography across a species' historical ranges can lull restoration biologists into assuming that their species has no specific topographical requirements. The vegetation associated with a particular topography differs in different places, causing topography not to be a good predictor of habitat. This disconnect can cause biologists to select as recovery or release sites places that cannot sustain the target population.

15.2.2 Climate

Both long-term climatic conditions and the probability of short-term catastrophic weather influence recovery site selection and carnivore restoration in other ways as well. The historic range of a species includes climatic extremes that were compatible with survival in the context of past conditions. Climate variations across the historic range may not all be acceptable today. Examine climate carefully at both mesoclimate and microclimate scales to identify recovery sites most likely to succeed. Identify recovery areas that avoid historic climate extremes.

One might hypothesize that a restoration area with relatively homogeneous weather would be best and would reduce the need to vary for management efforts across the recovery area over time. When the climate requirements of a species are not well known, however, selecting a recovery area that incorporates some climatic variation across both time and space has merit.

The timing of management efforts in relation to climate and weather can affect restoration success. Release timing should always consider weather. Optimization of the timing of release or other management activities may not always be intuitive, nor necessarily correspond with the most comfortable weather conditions for biologists. For example, in a recovery area of moderate temperate conditions, winter weather conditions appear to maximize survival and reduce post-release movements of black bears (Clark *et al.* 2002). Releasing fishers *(Martes pennant)* in autumn and early winter ensures that females have time to establish home ranges before the blastocysts they carry implant, improving reproductive success (Powell, unpublished data).

Consider the probability of catastrophic weather (e.g. tornado, hurricane) at a restoration site. Restoration efforts need well-considered and updated plans for project management both pre- and post catastrophic weather. Response plans should identify animals most likely to be affected and should include a plan for post-event impact assessment that can be adapted readily to fit the actual event. Droughts can play havoc with restoration programs. Have pre-established plans for drought interventions that include criteria that trigger implementation, pre-established water sources, and transport and deployment options. Drought plans need to consider the potential impacts of congregation at scarce water sources including disease transmission.

15.2.3 Anthropogenic features

Wild carnivores tend to avoid humans (but not always), even very large carnivores. Kodiak bears (*Ursus arctos*) in Katmai National Park, Alaska, alter their temporal and spatial use of rivers to avoid humans. Coyotes (*Canis latrans*) and bobcats (*Lynx rufus*) are more nocturnal than otherwise where human activity is high (Kitchen *et al.* 2000).

Roads affect survival of many carnivores (e.g. Kramer-Schadt *et al.* 2004; Reynolds-Hogland and Mitchell 2007a). Backwoods roads provide access to hunters, increasing hunting and poaching mortality of carnivores. Wide, multi-lane, high-speed highways present formidable barriers to dispersal (e.g. Riley *et al.* 2003). Even with as much as 30% of populations crossing the highway barrier, populations on either side can become genetically differentiated at a level consistent with a much lower migration fraction. To avoid restricted gene flow, even for wide-ranging species, more wildlife under- and over-passes may be needed than predicted on the basis of migration.

15.2.4 Prey base

A diverse, abundant, and stable prey base is essential to successful restoration (Fernandez *et al.* 2003). For some small- to medium-sized carnivores, such as stone martens (*Martes foina*) and badgers (*Meles meles*), food availability can be more important than any other resource. Within certain limits, suitability of small or relatively isolated food patches can be improved by modifying the relative amounts of prey and their distribution within patches (Mortelliti and Boitani 2008).

Other factors may supersede prey availability in determining successful restoration. Population monitoring post-release must be able to distinguish if either (or both) reproduction or survival limits population growth of the target population. For either, factors other than prey availability may be limiting, such as disease, limited den sites, or stress from human disturbance. Differentiate prey base from other potentially limiting factors and assess prey base throughout a carnivore restoration effort.

15.2.5 Health-risk assessment

Health-risk assessments are generally descriptive. Mathematical approaches remain weakly explored and models generally fail to be predictive, in large part due to lack of data as the basis of quantitative work. In addition, health-risk assessment is usually unbudgeted prior to the eruption of a catastrophic problem and is routinely given the lowest priority in setting restoration management goals. This is unfortunate because systematic pathological assessment of carnivores resident at a restoration site, and assessment of non-infectious mortality such as road-kill and hunting, provide insight into the future mortality risk of released target animals (Chapter 13). Surveys of domestic animal health should be conducted for all restoration sites (Chapter 13). Health risk is dynamic. Emerging health-risks, including newly evolving infectious diseases, have caused catastrophic failures for some high-profile restoration efforts, including black-footed ferrets and African hunting dogs (*Lycaon pictus*) (Williams *et al.* 1988; Fitzjohn *et al.* 2002).

A useful program for health-risk assessment for a carnivore restoration has two parts: evaluating health of animals to be released and evaluating the health risks at the release site. Animals to be released should be screened for general health and disease exposure. Chapter 13 provides details.

Evaluation of health risks of candidate sites should start before restoration and continue throughout the restoration effort. Evaluate the epidemiology of diseases established in sympatric carnivores, prey species, and domestic animals in the restoration area, and the potential for emerging diseases. The absence of an infectious disease may signal a need to be particularly vigilant to avoid introduction of that disease and to have monitoring in place for early detection. Look not only at definitive hosts for parasites and diseases, but also known and suspected vectors and reservoirs. Base assessments on real data collected in the restoration areas, evaluated using a combination of pathological, serologic, and molecular diagnostic approaches (Chapters 12 and 13). For a health-risk assessment to be of much use, it must be quantitative and integrate the quantitative behavior and natural history of the species being assessed with epidemiologic data (Chapter 13; Wobeser 1994).

15.3 Animal dimension

15.3.1 Carnivore–carnivore interactions

Restoration areas will inevitably have resident carnivores of species other than the target species, presenting the potential for competition with, predation on, or being prey for, the target carnivores. Niche differentiation among carnivores is usually clear (reviewed by Macdonald and Norris 2001). Competition for food is strongest among carnivores of roughly similar size and build, though these species tend to forage in different habitats or have different foraging behaviors. Evaluate not just the prey base, but also the expected impact of resident carnivores on the prey base of the restored carnivore. This requires knowledge of the dynamics of both the local prey populations and of the resident carnivore populations. The same applies to other resources. Assessing the availability of resting and denning sites, water, and the connectivity and patch sizes of critical habitats, requires a solid understanding of the biology of the resident carnivores.

Intraguild predation has the potential to thwart a restoration program. Predation by coyotes can be a major source of mortality for swift foxes (*Vulpes velox*, Kitchen *et al.* 2006a) and both coyotes and bobcats prey on gray foxes (*Urocyon cinereoargenteus*) (Farias *et al.* 2005). Intraguild predation threatened the success of all reintroductions of black-footed ferrets (Biggins *et al.* in press a; Poessel *et al.* 2011), necessitating experimental releases of sterile, non-endangered Siberian polecats (*Mustela eversmanii*, an ecological surrogate) to test restoration strategies (Biggings *et al.* in press b). Resident predators were removed before any black-footed ferrets were reintroduced. Experience of the target carnivores with other carnivores, especially of the resident species, improves restoration success (Miller *et al.* 1990b). Mortality rates are higher when lion-naive cheetahs (*Acinonyx jubatus*) are reintroduced compared to cheetahs from areas with resident (Hayward *et al.* 2007a, 2007b).

Intraguild predation is not limited to mammals. Recovery of a subspecies of the island fox (*Urocyon littoralis*) on the northern Channel Islands of California, USA, required the reduction of the golden eagle (*Aquila chrysaetos*) population (Roemer and Donlan 2005). Evaluation of a potential restoration area requires knowledge of raptors, crocodilians, venomous snakes, and other potential, non-mammalian predators.

Knowing the seasonal population and predation dynamics of expected predators of the restoration species should inform release timing, location, and methods. Selecting carnivores for release based on experience with predators, and providing that experience, if possible, can improve chances of restoration success (Miller *et al.* 1990a).

Intraguild predation becomes complex when multiple predators are restored in an area. Release competitively subordinate or vulnerable carnivores prior to dominant species to allow the more sensitive species the opportunity to locate refugia and to build up populations able to sustain some level of predation before the arrival of potential competitors and predators (Durant 1998; Hayward *et al.* 2007b).

Invasive carnivores constitute a special concern because target carnivores are evolutionarily naive about these carnivores. On the other hand, large-scale restoration of native carnivores can sometimes displace or reduce populations of invasive carnivores. The recovery of Eurasian otters (*Lutra lutra*) in Great Britain, displacing invasive American minks (*Neovison vison*) in the late 1990s and early 2000s, is an example. Successful control of invasive carnivores through restoration of native carnivores requires studies of both species to inform management decisions.

15.3.2 Carnivore-prey interactions

An unanticipated drop in prey abundance can thwart restoration success. The management of key prey species to optimize prey availability for restored carnivores is a powerful tool to sustain restoration efforts during natural low prey numbers. Targeted delivery of dead prey as supplemental food or strategic released of live prey, particularly during prolonged soft release, has been used more often than efforts to manage prey populations *in situ*.

15.3.3 Selecting founder populations

In the past, source populations for reintroductions were chosen mostly on availability. In the United States, personnel in wildlife agencies in different states traded animals to release (you give me species *x* and I'll give you species *y*), often with surprising success (Bolen and Robinson 2003). Nonetheless, genetic assessments of source populations can be important, especially to avoid accidental selection of hidden linked lethal or fitness-compromising traits. As assessment of functional genomics becomes more feasible, the challenge will be to construct genetic criteria for founder populations that will increase restoration success. Despite the conservation genetics concern for matching the genetics of founder animals to the genetics of the original population, there is evidence from real reintroductions suggesting that the greater the genetic diversity of the founder population, irrespective of the genetics of the original population, the better the chances of a successful reintroduction (Powell *et al.* in press).

15.3.4 Use of captive animals for restoration

Captive animals often lack social and behavioral skills important for survival in the wild and wild carnivores do appear to have somewhat higher survival rates than captive animals in restoration programs (Jule *et al.* 2008). Nonetheless, restoration efforts using captive animals have succeeded. Sometimes, captive animals are the only option. The entire successful, reintroduction program for black-footed ferrets has been based on captive animals (Miller *et al.* 1996; Wisely *et al.* 2008).

Issues of concern about captive animals have included lacking learned skills, such as hunting and predator avoidance, being conditioned to human presence, and lacking appropriate social behaviors related to mating and dominance (Snyder *et al.* 1996; Soorae and Price 1997; Wallace 2000; Rabin 2003; Vickery and Mason 2003). The reintroduction of captive-bred, African wild dogs (*Lycaon pictus*) in Etosha in the 1990s appears to have failed primarily because the released animals could not hunt successfully (Scheepers and Venzke 1995; Hayward *et al.* 2007b). Failure to prepare captive animals for survival may cause restoration failure as much as the innate impact of captivity (Hayward *et al.* 2007b). To prepare captive black-footed ferrets for release in the wild, biologists experimented with ways to teach ferrets to forage and to avoid predation (Miller *et al.* 1990a, 1990b; Biggins *et al.* in press a, b; Poessel *et al.* 2011). In some situations captive animals can be prepared for release situations that might be difficult to manage with wild carnivores. Large, social predators can sometimes be bonded into social groups with unrelated individuals (Graf *et al.* 2006; Gusset *et al.* 2006), sometimes with the judicious use of sedatives during the initial phase of introductions in captivity (Hayward *et al.* 2007b).

Impacts of captivity also include the actual expression of genes key to basic physiology. Adult, captive red wolves (*Canis rufus*) exhibit an over-activation of pro-inflammatory and stress responses (Kennerly *et al.* 2008).

The sophistication of captive breeding programs has increased dramatically over the past two decades. Captive breeding yields research opportunities into the physiology, disease susceptibility and prevention, genetics, and behavior. Well-managed breeding efforts can largely eliminate concerns of wildlife restoration specialists regarding uncertain genetic lineage. The challenge now is for field biologists and conservationists to find innovative, effective ways to use captive animals for reintroductions and augmentations.

Capital investments in caging, labor, and food are significant. Animals slated for possible release often require complex caging to provide stimuli for proper development of young animals and to isolate animals from humans. The argument that the resources spent on captive management would be better spent on field operations is largely specious. In reality, resources available for captive breeding are largely not transferable.

Including captive bred animals in restorations has a profound positive impact on resource availability by energizing private fund-raising efforts and increasing public awareness. Because they are charismatic, carnivores benefit particularly from educational efforts and fund raising. Developing large consortia for captive breeding and management provide economy of scale and addresses the old adage of not holding all of one's eggs in one basket. Nonetheless, sophisticated and effective management is needed to maintain quality control and protocol compliance. Captive management institutions must also be proactive with educational efforts to reduce public concerns about the losses of captive reared animals in restoration efforts.

15.3.5 Genetic management

Restoration decisions are often affected by efforts to maintain genetic diversity, particularly when the founder population is small. Although the classic definition of genetic diversity "the average level of heterozygosity within a population," conservation geneticists often substitute the number of alleles per locus examined, or the percentage of polymorphic alleles in a set of alleles examined. Genetic information can contribute to restorations but have limitations and peculiarities.

Mitochondrial DNA (mtDNA) has a strict matrilineal inheritance, meaning that restriction of female gene-flow in a population can affect estimates of genetic variability and calculations of viable population sizes. For wolverines (*Gulo gulo*), significant matrilineal structuring and restricted female gene-flow suggests that dispersal of female wolverines will be critical for maintaining genetic variability in a wolverine population (Cegelski *et al.* 2006).

Nuclear DNA (nDNA), via microsatellites or simple sequence repeats (SSR), are widely used to infer levels of genetic diversity in populations. The typical practice of selecting only highly polymorphic markers, unfortunately, can lead to ascertainment bias that reduces sensitivity for judging overall genetic diversity within a population (Vali *et al.* 2008). Point estimates of nucleotide diversity can vary by an order of magnitude or more, despite very similar levels of microsatellite marker heterozygosity (Vali *et al.* 2008). Carnivore restoration scientists, therefore, need to consider other approaches to assess genetic diversity for small founder populations.

Microsatellite loci, developed for use with well-characterized species, can be extremely useful. Microsatellite loci developed for domestic cats (*Felis catus*) can be used to identify individuals of other felid species (Waits *et al.* 2007). Cross-species' divergence, however, can lead to heterozygotes being misidentified as homozygotes, leading to the false interpretation that a population is in more severe genetic straits than it actually is. Microsatellite analysis can also be used to identify extra pair mating and mate switching (Kitchen *et al.* 2006b).

Minisatellites are used to study DNA turnover and are the basis of the technique called "DNA fingerprinting," used to identify individuals. The technique estimates the probability of finding identical genotypes in a population, generally assuming random associations between alleles within and among loci. These assumptions are violated for species with complex population substructure, making interpretations inaccurate. This problem has been examined for endangered populations of wolves, brown bears, and northern hairy-nosed wombats (*Lasiorinyus krefftii*). The theoretical estimates of the probability of identifying identical genotypes ($P_{(ID)}$) for these species were consistently lower than the observed probabilities, by as much as three orders of magnitude (Waits *et al.* 2001).

Sampling for any of these analyses can be labor intensive and expensive. Although blood or skin biopsies, classically used for genetic assessments, necessitate capture and handling of the animals, using hair follicles from hair traps, and feces, has revolutionized genetic assessment of wild carnivores (Chapter 4; red wolf, Adams *et al.* 2003, 2007; Iberian lynx, Fernandez *et al.* 2006; Adams and Waits 2007).

Assessing functional genomics of restoration species has lagged behind efforts to assess diversity or lineage. Careful assessment of health and mortality records of captive populations can identify heritable diseases that could affect restoration success (Acton *et al.* 2000, 2006), and this information should be incorporated into health-screening strategies. The development of gene chip technology allows the rapid assessment of thousands of functional genes simultaneously from a single sample. Efforts to adapt these tools to fitness assessment of carnivores in restoration efforts, offers promise. The conservative nature of many key functional genes allows useful data to be obtained from a portion of genes on chips developed to assess of human health. Rapid advances in whole-genome projects provide the potential to develop similar tools for individual carnivore species, though economics will probably dictate that tools be developed first with dog, cat, and ferret (*Mustela furo*) genomes. Because habitat and diet appear to affect functional gene expression (Kennerly *et al.* 2008), advances in this area must be cautious.

As a final caution, some small, isolated, carnivore populations carry unique alleles identified using either mtDNA or nDNA. Without knowing whether these alleles are adaptive, their existence should not be used to prevent augmentation of the population, if other analyses suggest the need.

15.3.6 Hybridization and introgression management

Hybridization and genetic introgression occur when target carnivores interbreed with members of other species. The classic tool for management of introgression on a restoration site has been to remove or exclude the other species, but this approach is labor intensive and, potentially, an unending effort, as the undesired carnivores disperse to fill empty space rapidly (Palmer *et al.* 2005). Establishing sterile barriers to reduce introgression by sterilizing undesired carnivores (via vasectomy and tubal ligation to maintain territorial behavior) may be modestly less labor-intensive with longer term stability (Beck 2006; Roth *et al.* 2008). The advantages of having sterile animals hold territories, excluding fertile animals from restoration areas, include the reduction of trap and removal effort, and the opportunity to have expendable place holders that can be removed when the target population expands, either via reproduction or augmentation. A sterile barrier of coyotes has aided the restoration of red wolves in the United States. Intensive monitoring and genetic testing to reduce coyote introgression are needed to maintain the barrier (Beck 2006). Surgically sterilized coyotes maintain pair bonds, territory fidelity, and may even have a higher survival rate than reproductive animals (Bromley and Gese 2001).

15.3.7 Health management and biosecurity

Infectious diseases have dealt serious, if not program-ending, blows to carnivore restoration efforts (Chapter 13; MoehrenschlagerScheepers and Venzke 1995; Moehrenschlager and Somers 2004; Hayward *et al.* 2007b). In the late 1980s, fewer than 24% of reintroduction programs included any form of health screening (Griffith *et al.* 1989). Today, health management is a must for any well-designed restoration effort. Disease management requires identifying the diseases of potential importance to carnivores and the ecological conditions associated with their spread and severity (Murray *et al.* 1999).

Biosecurity is the prevention of disease transmission and the movement of infectious disease agents. The most common biosecurity tool employed in restoration work is prerelease screening of animals slated to be introduced (Chapter 13). Early adoption of prerelease screening was partially driven by concerns that captive-born animals might lack immunity to viruses and other infectious diseases prevalent in their wild counterparts (Bush 1994; Woodford and Rossiter 1994; Cunningham 1996). Although both wild and captive animals are now screened, prerelease screening usually focuses on a relatively small suite of diseases known to affect the target carnivores for which a validated, non-lethal test exists and, ideally, preventive measures (e.g. effective and safe vaccine) are also available. Much of the screening focuses on viral diseases, though the actual array of health risks from infectious diseases is far broader. Carnivores slated for release should be screened for key, non-viral, infectious agents and for zoonotic disease concerns including, but not limited to, parasitic diseases such as *Echinococcus* and various tick-borne agents (Nutter *et al.* 1998).

Diseases with the potential for acute impacts and rapid spread among populations through respiratory or gastrointestinal routes of infection receive more attention. The literature near the turn of the century found 52 diseases reported, 44% of which were viral and 31% bacterial, for 34 large, terrestrial, carnivore species (>20 kg; Murray *et al.* 1999). Many of those diseases were endemic in carnivores and infected multiple taxa (Murray *et al.* 1999). The broad range of diseases of carnivores emphasizes the importance of broad health screening beyond the target species, to include sympatric carnivores and prey.

Most prerelease screening uses serological data (Chapter 13). Because serology only provides information on exposure, demographics of the species being studied become very important to interpretation. Interpreting serologic data can be counter-intuitive, in that seroconversion might be a positive trait for a restoration animal destined to be inserted into a habitat where the disease is established. The advantages for such an animal are obvious when looking at diseases that cause acute mortality with high infectivity for naive individuals. On the other hand, if a disease is not present in a restoration area, then even a small risk of introducing it through accidental introduction of a carrier with a titer should be avoided. Without knowing the status of the disease in the habitat, it is impossible to optimize this important aspect of biosecurity.

The influence of habitat and associated prey assemblages on the prevalence of diseases in carnivores is understudied. The threat of epizootics in carnivores may be serious if potentially lethal infections are endemic in reservoir hosts and transmitted horizontally between taxa, as occurs with rabies and canine distemper viruses (Murray *et al.* 1999). Many factors affect disease transmission. A study of two coyote populations, in different habitats with different prey, found no difference in prevalence of canine parvovirus, canine distemper virus, canine adenovirus, or tularemia exposure between the two populations (Arjo *et al.* 2003). In contrast, the prevalence of exposure to plague (*Yersinia pestis*) differed between the populations, probably due to differences in prey species available (Arjo *et al.* 2003).

Health screening routinely looks only at diseases known to exist. Our knowledge of infectious diseases of carnivores is certainly incomplete. New molecular approaches promote the discovery of cryptic pathogens and emerging pathogens, while broadening the known hosts for many known pathogens. A genetically unique, small Babesia-like piroplasm, recently found in erythrocytes of apparently healthy, wild caught, North American otters (*Lontra canadensis*) in North Carolina, USA, has potential pathogenic implications for other species (Birkenheuer *et al.* 2007). The potential for an emerging disease to have a catastrophic impact on a population makes it important to pay close attention to potential emerging diseases in a restoration area, even if no morbidity or mortality has yet been demonstrated.

15.3.8 Health interventions

Health interventions are diagnostic or therapeutic efforts to gather information about, or to mitigate, ongoing health problems. Restoration policies and rules should anticipate the need for health interventions and provide useful guidance for how intervention decisions will be made.

A frequently employed health intervention in a restoration is vaccination for infectious diseases (Chapter 13). The practice is credited with securing or salvaging several restorations, including averting the extinction of the Catalina Island fox (*Urocyon 1ittoralis catalinae*) after an outbreak of canine distemper virus (Roemer and Donlan 2005). Canine distemper virus has a much broader host-range than its name implies (Carpenter *et al.* 1998) and it has decimated several carnivore restoration efforts, including some for black-footed ferrets (Williams *et al.* 1988) and African wild dogs (Fitzjohn *et al.* 2002).

The decision of which diseases to vaccinate a restoration population against is most frequently dictated by the availability of safe and effective vaccines. Unfortunately, useful vaccines are not available for all diseases that can challenge a restoration. Few vaccines are actually tested in non-domestic carnivores, and safety and efficacy are generally extrapolated from information on domestic carnivores. Experience with captive animals sometimes provides useful safety information, but challenge studies, which would confirm efficacy, are rarely done. Traditional vaccine types have a tradeoff between increased safety from using killed vaccines and potentially improved efficacy with modified live vaccines. The risk of a vaccine strain outbreak of a disease in the restoration population generally supports use of killed vaccines, when available. The vast majority of vaccines must be delivered by injection and relatively few vaccines are available for carnivores that can be delivered by oral or inhalation routes. Therefore, vaccinations for most diseases occur only when animals are handled (Chapter 13). Oral bait rabies vaccines are available.

A key weakness of most health assessment is the inability to obtain timely necropsy information for animals dying in the field (Chapter 13).

Assess the health of live animals whenever they are trapped and handled for whatever reason. Veterinarians involved in restorations must remember, however, that health assessment alone often cannot justify trapping and handling restoration animals. Consider noninvasive approaches to health screening (Chapters 4, 12 and 13). Disease detection through scat analysis holds potential (Acton *et al.* 2006; Whittier *et al.* 2010).

15.3.9 Adaptive management

Reintroductions have been compared to the process of biological invasions, which have divided into either two or three distinct phases (Seddon 1999; Armstrong and Seddon 2008). In the two-phase approach, establishment, when stochastic mortality and sex ratios can be most critical to success, is followed by spread, when factors influencing birth rates assume greater importance (Bright and Smithson 2001). In a three-phase approach, establishment, population growth, and population regulation are distinguished (Sarrazin and Barbault 1996). The key point from a management perspective, however, is that conditions change, and having a restoration management system that can change with changing needs is important. Evaluating outcomes can be misleading and often unhelpful in making practical recommendations to improve program function without the use of scientific investigation. Program managers have a strong tendency to measure successes in terms of outcome and problems in terms of process (Wallace 2003). This tendency often leads to poor decision-making. Positive outcomes can be the result of inefficient or even damaging processes (Wallace 2003).

Blending science into an adaptive management process offers significant advantages to a restoration effort. One viable model establishes an independent team of scientists charged with the scientific assessment of information gathered during restoration management. This team then advises restoration managers, suggesting approaches to optimize the scientific credibility of information being gathered (Stoskopf *et al.* 2005). Advice may include small-scale "evaluative" efforts and data assessment before reacting to perceived risks or opportunities. Scientific teams work best if they meet face to face frequently enough to provide timely advice. The scientific team must be large enough to accomplish the necessary tasks but small enough to be effective at generating advice and analyses. Each member of the team should have specific assignments for data analyses, and all advice provided to the restoration biologists needs to be based on the principles of adaptive management (Stoskopf *et al.* 2005). Each team member must be skilled and knowledgeable about, but not necessarily experts with, the species being restored; they must be able to identify dogma, which is often easier for outsiders (Chapter 17).

An insidious force countering the adaptive management approach for a restoration is the reluctance to change procedures and methods that are perceived as successful. This reluctance is particularly problematic for teams who have worked successfully for many years. Concern is strong that change could lead to catastrophic failure. Specific procedures and techniques, particularly those used to monitor a population, become artificially associated with the actual success of the animals being restored. Management teams must re-examine their commitment to adaptive management frequently throughout a restoration.

15.3.10 Release methods

"Soft" and "hard" releases have been debated for decades. In a soft release, animals are maintained at their release site in pens or enclosures for days to months before release, to allow them to habituate to the restoration area. Animals are released simply by opening a door and giving them access to the world beyond; food is often supplemented after release. In a classic hard release, restoration personnel release animals from their traps or travel cages upon arrival at the release site, and they receive no supplemental food. Obviously, a continuum extends from hard to soft. Different release situations require different approaches.

For soft releases, pens must restrain the new animals and protect them from harassment by resident predators (Hayward *et al.* 2007b). Preventing carnivore dependence on humans is important, as a way to prevent increased risk of released animals attacking humans (Hayward *et al.* 2007b). In Africa, where only soft release is used for carnivores, the construction details of containment pens are crucial, ensuring that animals are exposed and habituated to management devices, such as electric fencing and game-viewing vehicles (Linnell *et al.*1997). Pens can even allow safe food supplementation (Hayward *et al.* 2007b). Bonding within human-made social groups is not always successful and constructed prides or packs may fragment despite long prerelease acclimation together (Killian and Bothma 2003). Before black-footed ferrets were reintroduced, restoration personnel tested different designs for release pens and release protocol, using Siberian polecats as surrogates (Biggins *et al.* in press a, b). Martens, fishers, and sables have been reintroduced using the whole spectrum from classic hard to extremely soft without noticeable difference in reintroduction success (Powell *et al.* in press).

Time releases to an appropriate season in the annual cycle of the target carnivores. Avoid seasons when females are pregnant or about to give birth. Releasing mothers with their kits, especially with food supplements, may stimulate the mothers to settle near their release sites (Hobson *et al.* 1989). Sometimes the optimal season for release is a difficult time to obtain animals, forcing compromise.

15.3.11 Population augmentation

An augmentation is the adding of animals to an existing population, usually a small population that has habitat that can support a larger population, but that is not expanding apparently due to stochastic events or demographic limitations. Augmentation most often entails translocating animals from a source population to the restoration population. Different carnivore species respond differently to translocation, requiring different translocation and release approaches. Leopards are difficult to translocate successfully because they try to return to their original locations (Hayward *et al.* 2007b). Black bear restoration success appears correlated with translocation distance, and is greater when subadult animals are translocated, as opposed to mature bears (Clark *et al.* 2002). Female black bears tend to stay put better than translocated than males (Clark *et al.* 2002).

Litter augmentation is the placement of one or more cross-fosterlings with a dam that is judged capable of rearing more young than she has in her current litter. Complete litter replacements insert an entire litter with a dam that has lost her litter. Cross-fostering captive born young to wild parents is an augmentation technique that is frequently overlooked. When it is used, avoid high expectations for success. Remember that young carnivore mortality is naturally high in stable and growing populations. A 30 to 35% yield to maturity from a complete litter cross-fostering is an important, positive contribution to the population, considering that otherwise the mother would not have reared any progeny at all. Survival of augmented infants must be evaluated in the context of any excess loss experienced by the existing litter. Nonetheless, a 30 to 35% yield to maturity is a success for most carnivores.

One challenge to this technique is identifying dens with litters suitable for augmentation or litters whose complete litter can be detected. The matching challenge is having appropriately aged youngsters available for insertion. For coyote litters, where pups were augmented, pup survival was dependent on augmenting when litters are <1-month-old (Kitchen and Knowlton 2006). Black bear mothers readily accept augmented cubs if their noses are filled with Vick's Vaporub or a similar substance that allows a new cub or cubs to gain the litter's odor before the mother can smell again.

Concerns about den disturbance are often invoked against cross-fostering and augmentation. Test the actual impact of den disturbance before abandoning the techniques (Beck *et al.* 2009). Reasonable care to reduce scent transfer through the use of gloves, minimization of on-site personnel, and timing insertions when adults are not at the den, provide for considerable success.

15.4 Exit strategy

All restoration projects must assess their ultimate success or failure. Make this a continuous process with end points established a priori, based on scientific data. Rarely, however, is knowledge of a target species sufficient during project design to establish hard, accurate, quantitative criteria for restoration end points. The trend toward quantitative approaches in biology is good, but beware of pseudo-quantitative criteria with no data serving as a basis for the hard numbers. Too often, criteria for success are expressed as absolute numbers rather than by population performance criteria (Jule *et al.* 2008). The adoption of numbers by restoration plans, as opposed to performance criteria, is a recipe for failed decision-making.

Developing performance standards for a restoration can be challenging. For example, even choosing a number of self-sustaining populations to represent success is an arbitrary choice that balances hopeful optimism with political and economic realities. Trends over time (increasing numbers of breeding pairs, increasing numbers of young surviving to breeding age, decreasing mortality due to specific causes over several years) can seem more suitable than absolute numbers, but generally these criteria must include thresholds to account for expected stochastic fluctuations. They also need to be integrated into an assessment of the risk of extinction over a long time to allow population decreases to be detected with minimal monitoring and intervention. Such a demographic modeling exercise should be supplemented with success criteria that speak to identifying habitat, health, and environmental threats to the population being restored.

Ultimately, every restoration project will end. Choosing criteria to mark that end will allow you to exit, successful or not, satisfied that now is the right time.

16

Designing a monitoring plan

Eric M. Gese, Hilary S. Cooley, and Frederick F. Knowlton

Monitoring is the collection and analysis of repeated observations or measurements to determine whether a management action is having the desired effect of meeting management objectives and demonstrating success or failure of a management strategy (Elzinga *et al.* 2001). Monitoring is composed of a series of surveys (*sensu* Chapter 2) framed in a design aimed at answering specific management questions. There are many reasons to establish monitoring plans, such as when a carnivore species is of a high social or economic value, is rare and decreasing in numbers, is in eminent danger of extinction, or is part of a legally mandated planning process. Monitoring is commonly conducted in combination with a formal research program with ecological objectives to provide managers and policy makers with information for making informed decisions and formulating conservation plans with some level of certainty or success (Nichols and Williams 2006; Sauer and Knutson 2008; McComb *et al.* 2010). Monitoring can also be useful for adaptive management strategies by treating management as a hypothesis and incorporating learning into the process with the data collected providing feedback about the effectiveness of alternative actions (McComb *et al.* 2010).

Designing a monitoring plan involves identifying the goals of the associated management plan, developing key questions, and designing a rigorous sampling scheme. Analyses must be pertinent to management objectives and capable of assigning probabilities to observed trends. Finalizing a monitoring design is a precursor to initiating data collection. Some monitoring programs fail to provide the information needed due to unclear or unspecific objectives, flawed or poor study design, low statistical precision or power to detect change, inconsistent commitment to implement or adjust the monitoring plan, or failing to communicate results to stakeholders (Elzinga *et al.* 2001).

This chapter provides the conceptual framework for designing a monitoring program with special emphasis on carnivores, but details on surveys that are part of monitoring programs are covered in Chapter 2. Elzinga *et al.* (2001) and McComb *et al.* (2010) describe the design and implementation of monitoring programs in more detail. Details of field techniques are covered in Chapters 4, 5, 6, 7, 12, and 13.

16.1 Identifying questions and monitoring designs

To design a monitoring program, one must understand the biological system to be monitored and know the gaps in knowledge. Background knowledge is needed to articulate questions clearly; questions that ensure the data collected will be adequate to address the questions, fill knowledge gaps, test assumptions, and able to identify thresholds for altering management actions. Detail and focus are important at this stage. Use of vague or unclear terms, overly broad or ambiguous questions, and ill-defined spatial and temporal scales increase the risk data collected will not adequately address the key questions at scales that are meaningful. Questions that guide a monitoring program must be anchored to the objectives of the associated management plan. The questions must address the gaps in information about the target population that prevent managers from understanding how the target population is responding to management actions, or predicting how the target population will respond to proposed future management actions. Many monitoring programs are set within a research program, allowing the key questions to be stated as hypotheses or as a number of alternative hypotheses. If the management plan dictates that managers need to know if the target population is increasing, then this need becomes a question for the monitoring program.

The four basic monitoring designs (McComb *et al.* 2010) address monitoring questions of different complexity.

1. *Incidental observations* are opportunistic observations of animals or sign. These are usually of little use within a monitoring framework except, perhaps, to provide preliminary information to a more structured plan.
2. *Inventory designs* document the presence or absence of the target species in an area (often referred to as a survey, *sensu* Chapter 2). The rarity of the species and the level of confidence in determining presence/absence are critical.
3. *Status and trend monitoring (aka surveillance) designs* establish trends over time by monitoring populations over long time-spans. The design of a monitoring plan should consider the scope of inference. Monitoring may be needed only for a local population, or may cover a large portion of the target species' geographic range and require participation by multiple agencies. For monitoring trends, sampling intensity must be designed to detect

change, or lack of change, over time on the appropriate spatial scale. Chapter 2 provides background on sampling design.

4. *Cause and effect monitoring designs* allow evaluation of short- or long-term effects of a management action on a population and include such approaches as retrospective comparative mensurative designs or Before–After Control–Impact (BACI) designs (Stewart-Oaten *et al.* 1986; Gotelli and Ellison 2004).

16.2 Developing a monitoring program

Chapters 2, 4, and 8 outline the critical aspects of setting boundaries, selecting indicators to measure, developing sampling design, choosing sampling units and sites, calculating effect size, and choosing statistical analyses. In addition, developing a useful monitoring plan usually requires the simultaneous consideration of several major issues, each with embedded components. The monitoring plan must include techniques that are biologically appropriate and feasible, legally and socially acceptable, and must provide useful results with the resources available. Many potential problems can be avoided by careful thought during the design phase and asking advice from research and managerial personnel working in similar environments. Issues intuitive to experienced biologists may not be to a naive biologist.

Monitoring programs must stay within their budgets. If the optimal sampling design and sampling methods preclude meeting budget constraints, the monitoring questions, and perhaps the objectives of the management plan, need to be re-evaluated. Developing a management program that cannot be carried out is a waste of time and money.

Gaining the necessary permits from governments and agencies involved in the area is an important hurdle to resolve early in the planning process. Often, such entities need to be consulted and even involved in the study design. Terms and methodologies should be clearly defined in research protocols and proposals to avoid confusion. If samples are to cross international boundaries, special permits may be required for export. Depending upon the capture methodologies involved, knowledge of, and permission to, handle non-target species must also be obtained. Procedures for handling target and any non-target species should be outlined. Completion of an approved handling and immobilization course from a qualified veterinarian should be considered. Some countries or agencies require a veterinarian be present when animals are captured, immobilized, and handled.

The ability to conduct a monitoring program could be curtailed if the social, political, or cultural values prohibit either the presence of you or your equipment. Cultural and social sensitivities related to the animals should be respected, particularly where the local populace retains religious or cultural ties to the carnivore involved.

The manner in which individuals of a species distribute themselves across a landscape in both time and space is an important issue. When designing a monitoring plan, several issues should be considered: (1) Whether the animal is solitary or gregarious; for example, packs of animals are more readily sighted than solitary individuals thus influencing the probability of detection. (2) Does recognition of one individual influence recognition of others? Often for social canids, finding sign of one individual indicates the presence of others in the vicinity. (3) Is the interest focused toward assessing individuals or groups (e.g. packs or clans)? For some estimates, knowing a pack or social group is present may be sufficient for monitoring; while pack size may be necessary in other situations. (4) Whether seasonal movements, such as migrations or movements among different habitats, are apt to be involved and do they apply equally to all sex and age classes? (5) Whether the species is territorial, which may result in them being distributed in some regular fashion, and if territorial, how large are the territories and how does this relate to the size of area for which the assessment is being attempted?

Equally important in designing a program are attributes of the study area. The physical attributes, including size, topography, and nature of the environment play a role in determining what sort of activities are feasible and practical. This starts with a clear designation of the area or areas for monitoring with clearly defined boundaries. This is essential if complete enumeration is feasible or whether constraints on time or resources dictate some type of sampling. Size of area would be an important aspect but topographic and vegetative features would also be involved. Ultimately, the demarcation of the boundary of the population area to be assessed would be critical if an estimate of species' density is needed. Some carnivores occupy rough terrains, dense habitats, extreme habitats, roadless areas, or high elevations. The terrain can be used to an advantage. Placing remote cameras in situations where animals funnel down trails into a valley or through a mountain pass allows concentrating sampling efforts and increasing success of "capture." Stratifying track sampling along trails commonly traveled by the species may increase probability of detection. Prominent landscape features used by carnivores for scrapes or scent-marking can be also useful for sign surveys. Techniques that increase detection, however, may introduce bias for many sampling designs (Chapters 2, 4, and 8).

Identifying when to collect data will determine not only the merits of the information obtained, but also the inferences made from the data. Several issues dealing with the timing of sampling should be considered: (1) Seasonal changes in the activity or visibility of the animals. (2) Whether a seasonal pattern of population phenology is involved, when might sampling be best accomplished, and how does that relate to the question? (3) Among species with seasonal breeding patterns, characterizing the breeding population may be more important than making assessments at times that will include young of the year. (4) Among many species, dispersal patterns must be considered as they relate to naive animals moving across unfamiliar landscapes with consequent changes in population structure. In instances where some measure of reproductive performance is desired, conducting the assessments at the proper time of year may be required (if young animals can be discriminated from adults). The activity periods of many species are influenced by the prevailing weather or even lunar patterns.

Whenever possible, use typical behaviors or products of behaviors, to detect animals or assess species' abundance rather than elicited responses. Elicited responses may be influenced by social status or environmental conditions. For some species, especially when documenting "presence" is the primary objective, sampling can take advantage of stereotypic activities. Many felids, for example, have an affinity for traveling within narrow canyons or along specific ridge-tops. Similarly, since the distribution of black-footed ferrets (*Mustela nigripes*) appears limited primarily to prairie dog (*Cynomys* spp.) towns, it is reasonable to limit assessments to such areas. During development of inventory procedures for coyotes (*Canis latrans*), the use of elicited vocalizations to assess abundance was considered (Okoniewski and Chambers 1984). Early trials determined that a four-fold difference in response rates resulted from three different types of sirens used to elicit the vocalizations, and coyotes were likely to respond at times they were active but unlikely to respond when they were inactive. Wolfe (1974) reported that while dominant (alpha) individuals were most likely to respond, transient individuals were appreciably less likely to respond. While taking advantage of such behaviors can increase sampling efficiency, researchers need to accommodate for potential biases that unequally represent specific sex, age, and social classes.

Many carnivore species have an innate curiosity to novel situations in their environment. As the objects become familiar through repeated exposures, they elicit less interest. Sometimes simply moving the stimulus a small distance will revive interest; however, new or rearranged objects can induce neophobic responses. Sensitivity to such situations varies widely among species. While coyotes react strongly, and warily (neophobia), to novel stimuli (Windberg and Knowlton 1988; Windberg 1996; Harris and Knowlton 2001), bobcats (*Lynx rufus*) are much less reactive to novel situations and can be repeatedly trapped, even in the same locations with the same attractants. Knowledge of the repertoires of species can be important for selecting sampling methods.

Statistical hypotheses are widely used because they provide objective, standardized criteria for decision-making. However, this has received much criticism over the last decade (Johnson 1999, 2002; Anderson *et al.* 2000; Burnham and Anderson 2002; Ellison 2004; Guthery 2008). Null hypothesis testing is uninformative in some cases (Johnson 1999), and often results in conclusions that lack meaningful insights for conservation, planning, management, or further research (Guthery 2008). Additionally, the significance level (α) used in a test is often based on convention (i.e. $\alpha = 0.1, 0.05$), classifying results into biologically meaningless categories (significant and non-significant) (Anderson *et al.* 2000). There may be times when a biologist, faced with a test statistic with a p-value of >0.05 but <0.10, may decide that a result is biologically meaningful or suggestive of a relationship. Bayesian and information-theoretic approaches are often more applicable for analyses in monitoring programs than traditional parametric, or even non-parametric statistical techniques. Learn the analytical techniques most appropriate for the monitoring program.

16.3 Evaluating the monitoring plan

After data has been collected and analyzed, biologists and managers must decide: given the information, what should we do? Several alternatives can be considered: (1) continue to monitor,(2) use the information to make changes in the management programs, as well as the monitoring plan, (3) evaluate the risk of changing versus continuing with the status quo, and (4) determine if integrating the data with data from other programs will produce a broader picture of the species or system (McComb *et al.* 2010).

16.3.1 Thresholds and trigger points

Within any monitoring program there are a multitude of issues to be addressed by managers and stakeholders before making any changes in the plan. One suggested approach is to agree with the stakeholders at the outset that if a particular threshold or trigger point is reached, alternative management actions need to be implemented (McComb *et al.* 2010). Trigger points might be considered points initiating a change to a management program, whereas thresholds indicate success in a management action (Block *et al.* 2001). Preferably, stakeholders have agreed beforehand to a series of steps to be taken, if a trigger point is reached. A potential problem with thresholds is they may result from social negotiation among stakeholders, and define a socially and mutually acceptable level of progress that may not be biologically dependable (McComb *et al.* 2010).

16.3.2 Forecasting trends

With several years of data, trends may emerge providing information to guide management actions. However, one must remember that the degree of precision decreases, the further that forecasts are predicted into the future, so forecasting trends beyond the dataset should be viewed cautiously and serve as one tool in guiding management decisions. Variation associated with trends and trend analyses, especially for rare species, is often high and the power associated with detecting a significant trend is often low.

Computer simulations have been used to model carnivore populations under diverse conditions (e.g. Connolly 1978; Mowbray *et al.* 1979; Lindzey and Meslow 1980; Sterling *et al.* 1983; Pitt *et al.* 2003; Conner *et al.* 2008). These models can be used to simulate population responses when one or more demographic variables are manipulated. Trigger points can be established from the risk assessment, prompting alternative management actions. Examining the sensitivity of the model to changes in the initial conditions of the system, parameter values, and structural features of the equations is useful for model assessment (Williams *et al.* 2002b). Population viability analysis (PVA) and population and habitat viability assessment (PHVA) can be used to evaluate the outcomes of various management actions, environmental perturbations, and stochastic events on the population viability of a species over a predetermined period of time (Shaffer 1981; Boyce 1992; Reed *et al.* 1998). Biologists using such models should consider the "realism" of the models and should ensure that the models are adaptive in response to ecological, environmental, and management factors (Williams *et al.* 2002b). A PVA or PHVA is only a model and is only as valid as the assumptions and information upon which they are based. They may not reflect or predict population persistence, and should not be the primary tool for developing conservation plans. Macdonald *et al.* (1998) suggest that PVAs may be most useful to biologists for developing and guiding management actions and identifying practical monitoring methods. Always evaluate the accuracy of the data incorporated and the levels of uncertainty (Reed *et al.* 1998; Williams *et al.* 2002b). Some PVAs and PHVAs may be used to raise questions and formulate hypotheses for future testing (Macdonald *et al.* 1998; Reed *et al.* 1998; Williams *et al.* 2002b).

16.3.3 Predicting patterns over space and time

Biologists and managers like to know where on a landscape a species is likely to occur, so management actions might increase or decrease populations, or might have minimal effects on the target species (McComb *et al.* 2010). Monitoring the presence of carnivores across a landscape provides information on the spatial distribution of individuals within populations, and provides a better understanding of meta-population structure and connectivity among subpopulations. If demographic rates are known, the value of subpopulations as sources or sinks can be examined (Chapter 10), as well as the probability of subpopulations becoming locally extinct and subsequently recolonized.

16.3.4 Integrating monitoring data

Data from a monitoring plan can be integrated with other environmental data to produce an integrated view of a landscape, thereby allowing managers to evaluate individual parts, as well as the whole landscape (McComb *et al.* 2010). These approaches use data to parameterize a spatial and temporal model to increase understanding of possible future conditions on the landscape. This allows for examination of various "what if" scenarios for comparing alternative actions. In addition, the approach can identify key parameters to be monitored in the future to help stakeholders understand whether the results of a management action are being realized. The danger of using these models is that they may not have been tested with independent data and, therefore, their accuracy is completely unknown. The potential for wildly incorrect management actions is very real (Chapter 11).

16.3.5 Risk analysis

Risks from environmental stressors, disturbances, or human activities may be important when evaluating a management plan. Monitoring data can be used in risk analysis in a stepwise process to assess threats (e.g. Hull and Swanson 2006). Risk assessment is a procedure to determine threats and understand uncertainty providing an estimate of the likelihood and severity of species, population, or habitat loss or gain, and an evaluation of the potential tradeoffs associated with various management actions (McComb *et al.* 2010). Kerns and Ager (2007) proposed a quantitative and probabilistic risk assessment to provide a bridge between planning and policy.

16.4 Changing the monitoring plan

Extensive time and money are expended in executing a monitoring plan. Consequently, the design of these programs must be scientifically and statistically rigorous, and managers and stakeholders must understand exactly how the information will be used to make decisions (McComb *et al.* 2010). Decisions should be made using a sequence of steps: characterize the problem or question, identify the full range of alternatives, determine a set of criteria for selecting one, collect information about each option and evaluate it based on the criteria. Then make a decision. Data collected in an adaptive management framework should use the information gained to refine the monitoring plan and improve the quality and utility of the data (McComb *et al.* 2010).

Changing a plan has consequences that must be considered carefully. If the data collected and analyzed suggest the goals and objectives are not being adequately met, changes to the plan may be required. Adding or dropping variables to be measured may be required as information reveals new patterns or processes, or budget constraints necessitate reducing the number of variables that can be measured. Changing the location or periodicity of sampling, or attempting to increase precision in data collection, are changes that can be considered. Avoid making changes that cause some or all of the data already collected to be incompatible with data collected after the change (McComb *et al.* 2010). Changing a monitoring program should not be done lightly and necessitates as much preparation as establishing the initial plan. Gaining information and revising management approaches based on that new information is the main objective of monitoring. If changing conditions preclude managers following the sampling design and, therefore, the questions addressed by the monitoring program and the objects of the management plan cannot be met, the management plan and its objectives need to be reconsidered (Chapter 2).

Deciding when to terminate a monitoring program is equally difficult. Generally the decision to terminate monitoring should be based upon whether the questions associated with the objective of the management plan have been answered. The decision of when to end the program should be outlined in the monitoring plan itself (Chapter 15). If the data collected through the monitoring plan indicate a carnivore population has been increasing over the last 4–5 years and may be reaching carrying capacity, and this is the main objective of the program, then terminating monitoring may be a logical step. If the data indicate a declining population and the key questions have not been answered, the main objective of the program has not been attained and continuation of data collection may be necessary.

17

Assessing conservation status and units for conservation

Urs Breitenmoser, Christine Breitenmoser-Würsten, and Luigi Boitani

Two features make the assessment of carnivore populations and the definition of units for their conservation particularly challenging: (1) the spatial scale of their movements and (2) the conflicts they cause with human interests. Viable populations of (large) carnivores need living space that usually goes beyond the size of protected areas (e.g. Woodroffe and Ginsberg 1998). Consequently, most self-sustaining carnivore populations extend into cultivated and multiple-use landscapes, where they often come into conflict with human interests. Most carnivores are resilient and can live in human-modified and human-dominated environments; as a matter of fact, the carrying capacity of such landscapes is often high, especially for large predators preying on wild and domestic ungulates. Indeed, the main conflict with, and the intense persecution of, carnivores result from their ability to live in human-altered environments. In multiple-use landscapes, carnivore populations are often kept below the ecological carrying capacity by illegal killings and management interventions to reduce conflict. An extended carnivore population may, hence, be a patchwork of diverse sites and situations, where the ecological potential and the observed or tolerated population status do not always match. In this chapter, we address three questions:

1. How do we assess the risk of extinction for carnivore populations?
2. How do we identify and delineate carnivore conservation units?
3. How do we assess, monitor, and manage carnivore conservation units?

The impressive diversity of terrestrial mammalian predators extends from least weasels (*Mustela nivalis,* commonly <100 g) to polar bears (*Ursus maritimus*, up to 800 kg), with species adapted to all climatic zones and biomes across the world, and diets ranging from strictly carnivorous (e.g. most felids) and omnivorous (e.g. canids such as red foxes *Vulpes vulpes*) species, to monophagous vegetarians (giant pandas *Ailuropoda melanoleuca*). This tremendous diversity prohibits all but the simplest generalities. Nonetheless, we focus mainly on a "typical" situation and use it as the general framework for this chapter: carnivores that hunt wild or domestic prey, so that they come into conflict with human interests, and that live in a human-dominated and altered environment, with (meta-) populations spreading over distinct management units and often across international borders. A classic example for this situation is Europe, a densely settled, highly fragmented continent, where large carnivores, such as brown bears (*Ursus arctos*), wolves (*Canis lupus*), and Eurasian lynx (*Lynx lynx*), are making remarkable comebacks, and where generalist carnivores, such as red foxes and stone martens (*Martes foina*), are common inhabitants of large cities (e.g. Gloor *et al.* 2001; Herr *et al.* 2009). Many other examples exist around the world.

17.1 Assessing extinction risks for carnivore populations

Large carnivores had high extinction rates even before *Homo sapiens* played a role (Steneck 2005). Although today, human interactions, such as direct persecution and destruction of habitats and prey bases, are the main causes of population decreases, extinction risk is still determined more by biology than human population density (Cardillo *et al.* 2004). On a population level, extinction-prone species are characterized by large body size, wide-ranging movements, low densities, low recruitment rates, and limited dispersal opportunities (Woodroffe 2001). These characteristics typify large carnivores, but also many small carnivores, when compared with their main prey. In a tested six factors associated with vulnerability (small geographic ranges, low densities, high trophic level, "slow" life histories, large body size, tolerance to altered habitats) for Brazilian carnivores, large body size carried the highest extinction risk because large species are most vulnerable to human activities, such as killing, habitat destruction and fragmentation, and the small sizes of protected areas (Forero-Medina *et al.* 2009). The typical candidate for extinction is a large carnivore living at low density and eating a strictly carnivorous diet (Purvis *et al.* 2004). Populations of large carnivores have more trouble recovering from prey decreases than do populations of small ones (Carbone *et al.* 2011), largely because total numbers of large predators are small. Carnivores suffer from human-made fragmentation because isolated populations are small and face high extinction risks through demographic stochasticity, hybridization (wildcat, *Felis silvestris* in Scotland, Macdonald *et al.* 2004b), diseases (Iberian lynx, *Lynx pardinus*, López *et al.* 2009), and inbreeding and genetic drift (Allendorf and Luikart 2007), often amplified by human-induced mortality, such as poaching (Amur tiger, *Panthera tigris altaica*, Goodrich *et al.* 2008), retaliatory killings (lion, *Panthera leo*, in Kenya, Frank *et al.* 2006), and traffic losses (Eurasian lynx in the Jura Mountains; Breitenmoser-Würsten *et al.* 2007a).

Population viability assessments (PVAs) estimate the minimum viable population size (MVP) required for a population to survive over a long time period or many generations with a specified probability (Schaffer 1981). PVAs identify and quantify factors reducing the survival of a certain age or sex/social class of individuals, and assess the survival probability of a population under the given circumstances. Computations are usually done using computer programs, such as VORTEX (Miller and Lacy 2005). Input values are empirical biological and ecological values for life history and population parameters, but the software packages also accept rough estimations, guesses, and assumptions, where genuine data are incomplete or lacking. Deficient input values, sensitivity to parameter estimates, the limited suitability of demographic and ecological values from the literature, the significant differences between average population models and individual-based models (White 2000), and the difficulty of validating outcomes, sparked a debate over the practical value of MVP concepts (e.g. Akçakaya and Syögren-Gulve 2000; Beissinger and McCullogh 2002; Reed *et al.* 2002).

Even though the absolute values of MVP estimations are often controversial, no generally accepted alternatives exist (Reed *et al.* 2003), and defining conservation goals or management interventions for carnivores without discussing population numbers is almost impossible. PVAs do provide a transparent process with explicit and potentially testable assumptions (Chapron and Arlettaz 2006). And their importance lays not so much with their estimates of extinction probabilities and minimum population sizes, but with their assessments of vulnerabilities to, and relative importance of, various threats, impacts of human activities, ranking of management options (Akçakaya and Syögren-Gulve 2000), and prioritizing of research needs. The International Union for Conservation of Nature (IUCN) recommends using quantitative analyses by means of PVAs for Red List (www.iucnredlist.org) assessments wherever adequate data are available (Criterion E; IUCN/SSC 2001), but specifies that "in presenting the results of quantitative analyses, the assumptions (which must be appropriate and defensible), the data used and the uncertainty in the data or quantitative model must be documented."

Population viability depends on both demographic and genetic conditions (Beissinger and McCullogh 2002). Franklin (1980) proposed for practical conservation the 50/500 rule-of-thumb: an effective population size (N_e) of 50 individuals to avoid an inbreeding depression in the short term, and 500 to ensure long-term genetic variability. This rule-of-thumb has many problems (Allendorf and Luikart 2007), most importantly that these effective population sizes are not threshold values (though often treated as such), as the loss of genetic diversity is a continuous process. Despite the problems, the rule-of-thumb is a useful guide for managing populations. The numbers, however, should not be taken as targets but as warning lights (Allendorf and Luikart 2007).

In reality, demographic and genetic viability are interrelated. Reed *et al.* (2003) used VORTEX 8.01 (Miller and Lacy 2005) to compute MVPs for 102 vertebrate species (including 18 terrestrial carnivores) for a 99% probability of survival over 40 generations ($MVP_{corrected}$), including carrying capacities, demographic and environmental stochasticities, catastrophes, reproductive rates, and inbreeding depression. The mean $MVP_{corrected}$ across all taxa was 7316 adult animals. MVP decreased to 4700 for a survival probability of 90% and to 550 for 50%, respectively.

These results illustrate one problem for using PVAs to define conservation goals: What survival probability is appropriate? Related to that question, Reed *et al.* (2003) showed that the duration of the studies used to parameterize the model affects the simulation results. Parameters derived from long studies predict larger MVPs, because long studies experience greater variation in ecological and physical conditions. MVPs increased considerably (by a factor 4–9) when corrected for study duration. Reed *et al.* (2003) concluded that PVAs based on short-term studies result not only in less precise estimations, but that they lead to a systematic underestimation of MVPs and, hence, of the extinction risk. Long-term studies or studies from several populations under various ecological and physical conditions are available for only a few carnivore species. In spite of all the problems with PVAs, they indicate that the magnitude of viable carnivore populations is in the range of many 100s to a few 1000s.

Few case studies of bottlenecks for carnivore population are sufficiently documented to allow retrospective assessments of the applicability of PVAs or the importance of demographic and genetic threats. Some spectacular recoveries from low population sizes have occurred; for example, the Scandinavian brown bear population that increased from 130 individuals in 1930 to 3300 in 2008 (Swenson *et al.* 1995, 2010); tigers (*Panthera tigris*) in the Russian Far East that increased from 20–30 individuals in the 1940s (Kaplanov 1948) to 428–502 in 2005 (Miquelle *et al.* 2007). Many recovering populations, although they seem to function demographically, may face problems as a consequence of genetic impoverishment. The well known Isle Royal wolf population, founded by 3 individuals in 1949 (with subsequent supplements), numbered 24 wolves (apparent carrying capacity) in 2009, but more than half of the individuals examined exhibited skeletal malformations as a consequence of inbreeding (Vucetich and Peterson 2008; Räikkönen *et al.* 2009). The wolf population in Sweden was founded by a pair immigrating from the Fenno-Russian population in 1980 and supplemented by another immigrating male in 1991. The population increased to 272 wolves in 2010 but shows a heavy genetic load and signs of inbreeding depression (Wabakken *et al.* 2001; Liberg *et al.* 2005; Sand *et al.* 2010). Populations of Eurasian lynx reintroduced in the 1970s in Switzerland (Alps and Jura Mountains) and Slovenia with very few (<10) founder individuals show severe genetic drift and high loss of rare alleles (Breitenmoser-Würsten and Obexer-Ruff 2007c).

Noteworthy in this context, more than the size of MVPs, is the consistency of MVPs across taxa and trophic levels (Reed *et al.* 2003 discussed this result). Regarding MVPs, mammalian predators seem not to differ fundamentally from other taxonomic groups, including their prey. Nonetheless, corresponding to the low abundance of predators compared to their prey, the spatial requirements of viable carnivore populations are huge. For some large carnivores, the estimated global population is already smaller than 7000 mature individuals (www.iucnredlist.org). Considering the high degree of fragmentation and the complete isolation of many of the occupied habitat patches, even populations in large protected areas bear a high probability of extinction.

A third concept of viability (after demographic and genetic) is the "ecological viability," referring to the interactions between species and with their environment. Besides incorporating the conservation of existing patterns of biodiversity, this concept incorporates the preservation of ecological and evolutionary processes. For carnivores, this comprises not only all elements and resources a particular species needs to survive, but also its ecological and selective role, hence its impact on prey populations (Linnell *et al.* 2008). Ecological viability, or the functionality of ecosystems and the conservation of processes, is difficult to assess quantitatively. It entails much wider distribution ranges and larger numbers of individuals than required for demographic or genetic viability alone, and it implies the maintenance of given species in many different ecosystems (Allendorf and Luikart 2007). If predation, as an important natural process shaping an ecosystem, becomes a conservation goal in itself (Linnell *et al.* 2008), predators should be maintained across all suitable habitats.

The definition of "suitable habitat" has an ecological and an anthropogenic component. If the prey base is sufficient, and human-caused mortality sustainable, carnivores can live in almost all altered and multiple-use habitats, but they may not be tolerated there. Prey species are often very abundant on agricultural land, for instance rodents, which are important food sources for foxes, mustelids, and small cats. Though small carnivores may be tolerated in our neighborhood, large carnivores generally are not. As a consequence of conflicts and fears, large carnivores are often excluded from cultivated landscapes, even if the ecological conditions, e.g. wild prey availability, would allow their presence. Large herbivores, such as wild boar (*Sus scrofa*), moose (*Alces alces*), and nilgai (*Boselaphus tragocamelus*), reach peak densities in mixed and multiple-use landscapes where they would be perfect prey for wolves or tigers. If predation is considered a conservation goal or a part of herbivore management, then appropriate predators should be maintained on all landscapes inhabited by their prey species. If we learn to mitigate the conflicts (Chapter 14), we cannot only increase the ecological viability, but also use the cultivated landscapes (even if they would host carnivores below the ecological carrying capacity density) as corridors between protected source populations, increasing demographic and genetic viability.

Although the concept of MVP has received considerable attention over the past three decades, the practical use of PVAs for carnivore conservation is limited. At present, even in the rare cases when threatened carnivore populations receive proper attention and funding for conservation, MVP sizes in the 100s or 1000s are often already illusory. Furthermore, the willingness to conserve charismatic and problematic species, such as tigers, wolves, or brown bears, does not depend on the quantified probabilities of their survival as much as on societal and political considerations. Relying on, or insisting in, theoretical MVP numbers is not practical, and is sometimes counterproductive, when developing pragmatic conservation objectives in a participatory process with partners, or when ranking species or populations regarding their conservation importance.

In the end, nevertheless, practical benefits of using PVAs and viability concepts do exist for conservation planning:

1. Used appropriately, correcting for poor data, PVAs can be valuable for exploring which parameters and sensitivities should be used to rank the importance of threats and to define priorities for conservation actions and research.
2. MVP considerations should be used as broad references to set long-term goals for the sizes and shapes of (meta-) populations and to define the spatial scale for achieving long-term conservation goals. In general for carnivores, conservation practitioners underestimate the required population sizes and special scales. Estimating population sizes and spatial scales, however, requires reliable, long-term datasets.
3. Ecological viability should always be considered in carnivore conservation, in addition to demographic and genetic viabilities, to avoid saving a species without its key ecological functions. This concept requires maintaining predation as a selective process and also requires developing multilevel and meta-population concepts with protected source populations connected through managed (sink) populations.

17.2 Identifying and delineating carnivore conservation units

The variety of theoretical models and legislative terms for conservation units presently used is confusing. Scientific concepts partly overlap but also differ in essential points, whereas terms used in legislation or international treaties sometimes lack (biological) definitions. Many terms were born out of the necessity to define distinct, situation- and landscape-specific conservation units below the level of species or subspecies, as listed in laws and treaties. With the exception of viable populations living entirely within protected areas (a rare situation for carnivores), the definition of conservation units always includes the following components: (1) a biological entity (species, subspecies, population), (2) an assessment of cultural and political constraints and opportunities, and (3) the (spatially explicit) translation of these analyses into a geographic unit. These steps do not represent a chronological order; the geographic unit may be the first and easiest to identify. But, finally, for the implementation of conservation actions and management principles, a concrete map is needed.

17.2.1 Choosing biological entities

Biological entities addressed by legislation are typically species or subspecies. The myriad international treaties, such as the Convention on International Trade in Endangered Species of Wild Fauna and Flora (CITES) and the Convention on Biological Diversity (CBD), and important national laws, such as the US Endangered Species Act (ESA), all include lists of protected or endangered species and subspecies. National legislation often protects a specific list of species "and their living space," but some modern laws also address varieties or distinct populations (Allendorf and Luikart 2007). Identifying species is only rarely a problem for the Carnivora, e.g. the debate on the specific status of the Algonquin wolf *Canis (lupus) lyacon* (Kyle *et al.* 2006), or the golden jackal *Canis aureus lupaster* in Egypt that may be a wolf (Rueness *et al.* 2011). The identification of intraspecific units for conservation, however, is a challenge though a clear necessity for all species with wide distributions, many distinct populations and large dispersal distances. The tradition of basing subspecies on morphological differences and (often incomplete) reproductive isolation, results in an immense and confusing number of carnivore subspecies (see, for eaxmple, Wilson and Reeder 2005). Over-splitting was common in traditional carnivore taxonomy: for the puma (*Puma concolor*), molecular genetics approaches reduced subspecies from 32 to 6 (Culver *et al.* 2000), for the leopard (*Panthera pardus*) from 27 to 9 (Uphyrkina *et al.* 2001), for the wolf from more than 30 to fewer than 10 (Nowak 2003). Over-splitting can lead to wasting conservation resources (Allendorf and Luikart 2007) or impeding pragmatic conservation measures. In addition, traditional designation of subspecies can fail to recognize valuable, local populations. The decline and critical status of the distinct subspecies of Eurasian lynx (*Lynx lynx martinoi*: Mirić 1978) in the south-western Balkans, was ignored for decades because the phylogenetic distinctiveness of this local population was not recognized (Breitenmoser *et al.* 2009).

A current approach for defining intraspecific conservation entities based only on biological traits is to identify evolutionarily significant units (ESU), defined as populations or meta-populations of special conservation interest because of their genetic or ecological distinctness. Ryder (1986) first used this term but the review of proposed definitions by Allendorf and Luikart (2007) found three underlying concepts pertinent to evolutionarily significant units: (1) long-term reproductive isolation (generally hundreds of generations) and ecological and adaptive uniqueness, representing a reservoir of genetic and phenotypic variation potentially important for future evolution (Waples 1991); (2) reciprocal monophyly for mtDNA and significant divergence of allele frequencies at nuclear loci (Moritz 1994); and (3) ecological and genetic exchangeability (Crandall *et al.* 2000), meaning that individuals from different populations of the same ESU are ecologically and genetically so similar that they can be exchanged without any reduction in fitness. For carnivores, considering the concept of ecological viability, concept (3) should include that they can be exchanged without altering their selective impact on prey.

The concept of an evolutionarily significant unit sometimes collides with presently described subspecies. A proposal to augment the almost extinct leopard population in the northern Caucasus through translocation of Persian leopards (*Panthera pardus saxicolor*, Pocock 1927) from the southern edge of the range was challenged because the leopards in Russia were described as distinct subspecies *P. p. ciscaucasicus* (Satunin 1914; Lukarevsky *et al.* 2007). IUCN encourages the assessment and listing of subspecies in the Red List (www.iucnredlist.org) but over-splitting can be a considerable constraint for conservation, especially for augmentations and reintroductions. Indeed, the term "evolutionarily significant units" was coined because current mammalian taxonomy, especially subspecies, seldom represents significant adaptive variation (Ryder 1986). Recent definitions of subspecies include such criteria as that individuals belonging to a subspecies must share a geographic range, a set of "phylogenetically concordant phenotypic characters," a suite of molecular genetic similarities, and derived adaptations relative to other subspecies (O'Brien and Mayr 1991; Waples 1991; Culver 2009). These rules result in a high degree of similarity between subspecies and evolutionary stable units for carnivores. This result would be highly desirable for practical conservation work, as the subspecies concept appears simple and is broadly used in legislation, conventions, and red lists.

How do subspecies or evolutionary stable units relate to populations or metapopulations? Populations, as ecological functional units, may show significant adaptive differentiation to ecological niches and represent potential evolutionary lineages (Frankham *et al.* 2002), but genetic differences between neighboring populations do not, a priori, represent significant adaptations. Anthropogenic fragmentation of landscapes has split many closed distribution ranges of carnivores into isolated populations, and differences observed in the new populations may be the result of a segmented (genetic) cline, inbreeding depression, genetic drift or, in special cases, hybridization.

The discussion of local adaptation for these population usually lacks sufficient (molecular genetic) information. In the Trentino region of the Italian Alps, some brown bears survived and had a high symbolic value as the only surviving brown bear population in the Alps since the early twentieth century. Since the 1970s, augmenting of the declining population with individuals from the nearby Dinaric population has been discussed. Specific behavioral features, such as a pronounced shyness (H. Roth, personal communication), however, indicated some local adaptation, resulting in a lasting debate over the adaptive uniqueness of this remnant population. The dispute was indeed about the obvious risks of inbreeding depression versus the hypothetical risk of outbreeding. Finally, in 1999–2002, after many years without signs of local reproduction and only 2–3 bears surviving in the wild, 10 bears from the Dinaric population were released in the Trentino area (Mustoni *et al.* 2003). The population is now growing, but all young bears are offspring of translocated individuals (De Barba *et al.* 2010). None of the indigenous "Alpine bears" have contributed to the re-emerging population. Hence the "augmentation" was genetically a reintroduction. Earlier action might have saved part of the supposed "adapted Alpine gene pool." In this case, the discussion over adaptive uniqueness disguised the awareness of critical threats (e.g. inbreeding). The symbolic value of the last Alpine bears promoted the belief in their (genetic) uniqueness, and the controversial dispute among scientists helped to delay political decisions.

Gebremedhin *et al.* (2009a) predicted that, in the near future, adaptive genetic variation will become an increasingly important topic of conservation genetics and noted that "preserving local adaptations must be a major goal of conservation." Further, Festa-Bianchet (2009) specified that both "taxonomy and conservation require a holistic approach, considering genetics, morphology, ecology and, above all, *evidence* of local adaptation" (italics added). This last point is clearly a considerable challenge and, in the case of the Alpine brown bears, such evidence was anecdotal at best.

In reality, defining conservation units is often simpler than theory implies, because the choice is limited to a definite number of practicable options. For saving the Alpine brown bear, the true question was not *what* to do, but *when* to do it. The Florida panther (*Puma concolor coryi*) presents an example of successful, though also highly controversial, population level management where the genetic introgression from released female pumas from Texas (*P. c. stanleyana*) improved kitten survival (Hostetler *et al.* 2010).

17.2.2 Socio-political considerations

The broad agreement for defining conservation units on an intermediate level between species and populations lacks consensus on how to do it. Part of the problem is that a purely biological definition collides with socio-political constraints, with cultural and economic demands, and with requirements of the real world. Theoretical concepts mean nothing to local people who must live with carnivores impeding their way of living. Wildlife biologists assessing conservation needs in a scientific, value-free process tend to regard themselves as objective and neutral. Local people, however, often consider these theorists as part of the problem rather than part of the solution.

The US Endangered Species Act protects not only species and subspecies, but "any distinct population segment of any species." The intention was to enable the US Fish and Wildlife Service to list important populations and to "tailor management practices to unique circumstances and grant varied levels of protection in different parts of a species' range" (Pennock and Dimmick 1997). In the 1990s, distinct population segments were interpreted as evolutionary stable units. Pennock and Dimmick (1997), however, argued that such a redefinition would restrain conservation and management options because the original definition of distinct population segments also considered (in contrast to the biological ESU concept) demographic and behavioral information, and non-biological aspects such as cultural, economic, and geographic justifications. In a human-dominated world, the definition of conservation units cannot be based on a biological assessment alone. The EU Habitats Directive introduced the concept of Favourable Conservation Status (FCS), defined through the Favourable Reference Range and the Favourable Reference Population, to be applied on the level of member states. As is usual for legislation ranging from lichens to lynx, an ongoing debate ensued on how to interpret these terms. In an attempt to make this concept functional for large carnivores, Linnell *et al.* (2008) concluded that favorable conservation status includes both demographic and genetic viability and recognizes the importance of ecological viability. As a consequence, the concept of favorable conservation status is, for species such as wolf, brown bear, and Eurasian lynx, not practical to be applied on a national level because, for most European countries, both reference range and reference population would not be viable. Furthermore, large carnivores generally live in cross-border populations, because favorable habitats are found most often along international boundaries. In Europe, large carnivore conservation units meeting the favorable conservation status requirements will be (meta-) populations stretching over several countries, but managed under common objectives and a set of agreed rules (Linnell *et al.* 2008).

Integrating biological needs and socio-political considerations into a carnivore conservation unit will eventually result in a diversified management landscape, where, in different parts of the unit, varied and specific management measures are implemented (as for wolves in the Greater Yellowstone Ecosystem, Jimenez *et al.* 2010), but under the common goal of maintaining a demographically, genetically and ecologically viable and phylogenetically genuine carnivore population. Maintaining carnivores on a multiple-use landscape in coexistence with local people will ultimately lead to graded management zones (Linnell *et al.* 2005a) and, hence, a managed meta-population (Chapter 14; Figure 17.1).

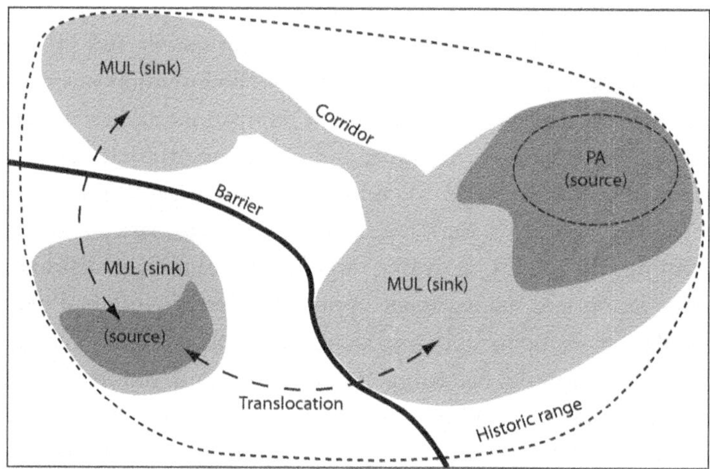

Fig. 17.1 Spatial concept for a carnivore conservation unit in the form of a managed meta-population MMP. The historic range of a taxon (dotted line) now consists of unsuitable habitats (white), suitable multiple-use landscape (MUL, light grey) and prime habitat (e.g. in protected areas PA). The protected core populations (dark grey) are demographically viable, whereas the part of the population in the MUL may be (artificial) sinks, but still grant the genetic viability and connect source populations. If barriers prevent natural migration through corridors, genetic management is done by means of translocations. As a whole, the MMP also maintains ecological viability.

17.2.3 Geographic delineation

For practical conservation and management, conservation units must be transferred to real landscapes, no matter whether the units are based on broad assessments of genetic, phylogeographic, ecological and cultural, human dimension, and political concerns. When Gebremedhin *et al.* (2009b) identified the endangered Ethiopian walia ibex as a distinct taxon (*Capra walie*) and conservation unit, they used a combination of phylogenetics, population genetics, and ecological modeling that translated genetic and taxonomic results directly into a geographic conservation concept. For carnivore populations that stretch beyond protected areas onto multiple-use landscapes and across state borders, biological and socio-political considerations must be expressed and visualized in spatially explicit maps. The three levels of viability, demographic, genetic and ecological, are for most carnivore species reflected by three different, spatial scales (Figure 17.1). Only in (endemic) species with small and compact distribution ranges, such as the Chinese mountain cat (*Felis bieti*), the Colombian weasel (*Mustela felipei*), and the Bornean ferret-badger (*Melogale everetti*), are these three levels identical. If the distribution range is fragmented, even if the total area is small, as for Iberian lynx (Simón *et al.* 2009), a spatial differentiation is inevitable. The two remnant populations of the Iberian lynx, in the Sierra Morena and in the Coto Doñana, though demographically still viable, are genetically impoverished (Godoy *et al.* 2009) and must be managed as one meta-population, even though they are geographically isolated. For species with wide (e.g. continental) or very wide distributions (circumpolar, intercontinental, e.g. wolves, brown bears leopards), geographical conservation units cannot, obviously, be the total ranges, on the one hand for practical reasons and on the other hand because global status does not justify local activities. The leopard is Near Threatened at a global scale, but two subspecies (Amur leopard *P. p. amoyensis* and Arabian leopard *P. p. nimr*) are Critically Endangered (www.iucnredlist.org). These subspecies are the logical conservation units. The geographic extensions of the two units, however, are totally different: the Amur leopard exists as a closed population in about 2500 km^2 in the Russian Far East (Pikunov *et al.* 2000), while the Arabian leopard is fragmented into probably 13 small patches in four different countries (Spalton and Al Hikmani 2006), distributed over a distance of nearly 4000 km along the mountain chains of the Arabian Peninsula. Many carnivore conservation projects are not justified by the threat to a species or subspecies but, rather, by the ecological significance of a local population. This is the case for many reintroduction projects, such as for Eurasian lynx in Western and Central Europe (Breitenmoser and Breitenmoser-Würsten 2008; Linnell *et al.* 2009) or wolves in Yellowstone

National Park (Smith and Bangs 2009). But, if a reintroduction succeeds, the demographic and genetic viability of the population becomes a concern and calls for the geographic delineation of a conservation unit. Even a protected area of almost 9000 km^2, like Yellowstone National Park, cannot host a genetically viable wolf population at length, and conservation and management planning must, consequently, go beyond the park boundaries to the Greater Yellowstone Ecosystem. In 2009, 96 wolves lived within the Park in Wyoming and 224 outside the Park (Jimenez *et al.* 2010).

The Bengal tiger (*Panthera tigris bengalensis*), the most numerous of all remnant tiger subspecies, is estimated to number 1920–2570 animals in India, Nepal, Bhutan, Bangladesh, and Myanmar (www.globaltigerinitiative.org). The entire population, however, is fragmented into many small patches across South Asia. The single largest population in India is in the Western Ghats with 290 tigers. Some of the isolated populations are so small (20 tigers in Simlipal, 24 in Panna; Jhala *et al.* 2008) that their demographic survival is at stake. This situation is primarily a consequence of poaching in tiger reserves in recent times and, therefore, protecting core populations is the first priority (Walston *et al.* 2010). Nonetheless, even maintaining Bengal tigers at their carrying capacities in the designated protected areas will not secure their genetic viability in the long run, and tigers will no longer play their ecological role as top predators. A forward-looking strategy is the concept of "tiger conservation landscapes" (Dinerstein *et al.* 2006; Sanderson *et al.* 2006), where core areas and buffer zones are linked with habitat corridors, allowing conservation of the ecological requirements of tigers as well. In all, 76 tiger conservation landscapes have been identified (40 in the historic distribution range of the Bengal tiger) using a thorough GIS-based analysis, resulting in a spatially explicit conservation concept (Sanderson *et al.* 2006). This concept is similar to the managed meta-populations presented in Figure 17.1.

The combination of the three concepts of viability allows a meaningful and practical definition of geographic conservation units. Demographically viable core subpopulations act as sources for the surrounding, multiple-use landscapes and for the areas and corridors connecting the subpopulations. Core zones may be protected areas but, for large carnivores living at low densities, demographically meaningful source populations will often live across larger areas. Multiple-use landscapes, where a variety of economic activities, such as livestock husbandry, logging, hunting, etc., take place, may offer favorable ecological conditions for carnivores, but coexistence with people may require that carnivore population density is kept below the ecological potential. These parts of the population might even be (temporary) sinks, but will still be important for maintaining the genetic viability and to support the corridors. Carnivore densities in the multiple-use landscape may not be at their ecological carrying capacities but should be sufficiently high to grant ecological viability. Local land-users, such as livestock breeders and hunters, are often willing to accept the presence of carnivores, if their abundances are limited to agreed levels (Chapter 14). In situations where maintaining or recreating corridors that allow natural migration is impossible, the genetic viability of isolated subpopulations must be managed through regular translocation of individuals. Maintaining such subpopulations is justified for the sake of the ecological viability and to preserve "backup populations."

To identify and delineate a carnivore conservation unit, follow these considerations:

1. A taxonomic and phylogenetic assessment is the first step to identifying carnivore conservation units. A subspecies is often a good choice but sometimes the distribution range is too large to form one commonly managed unit and, occasionally, over-splitting results in a too restrictive view that obstructs the exchange of individuals for genetic management. In these cases, a different evolutionarily significant unit (ESU) must be identified.
2. A purely biological assessment (as in the concept of an evolutionarily significant unit) does not adequately reflect the complexity of carnivore conservation, especially regarding the ecological roles of predators and their conflicts with human land use. A socio-political assessment must reveal where members of a carnivore species can live and at what abundances they are accepted by local people (Chapter 14).
3. Biological and socio-political considerations must be translated into a spatially explicit geographic unit (Chapter 14). The resulting map outlines a managed meta-population with (protected) core areas (source populations), multiple-use areas where a set of management rules are applied, possibly leading to a lower density (sink populations), and several subpopulations inter-connected or jointly managed through translocations. The whole metapopulation allows maintaining the genetic and ecological viability of the identified evolutionary significant unit.

17.3 Designating and establishing carnivore conservation units

Phylogenetic analyses allow biologists to identify the biological unit of a carnivore species to be conserved, and habitat modeling allows its visualization in form of a hypothetical but spatially explicit map. Such academic exercises are useful only if they are transformed into a realistic conservation plan accepted by local people and endorsed by the relevant governmental institutions. Political considerations go beyond managing populations across borders between states with differing wildlife management traditions and laws. It includes foremost mitigating conflicts between human interests and the conservation of carnivores. The contemporary, relatively strong legal protection of carnivores provides a top-down framework for their conservation. But even where protective laws are a result of a democratic process, they may not represent local opinion or have the support of all interest groups, particularly local land-users. Law enforcement in carnivore protection is difficult and, worldwide, generally weak; effective and sustainable conservation of conflict species requires the support of local people and implies considering their opinions and needs in a bottom-up approach. Public involvement is needed at a very early stage and the final designation of a carnivore conservation unit should be done in a participatory process.

A number of conservation planning approaches, both for species and areas, have been developed and used (IUCN/SSC 2008 provides an overview). The Species Survival Commission (SSC) of IUCN emphasizes that strategic planning for species' conservation requires a multi-stakeholder participation, including species' specialists (scientists), range state governments, regional politicians, and members of local communities (IUCN/SSC 2008). After a scientific-technical review of the species' or population's status and of the human dimension situation (threats, conflicts, policy, etc.), goals, objectives, and actions are defined in a participatory process, including all groups that might be involved in, or affected by, the implementation of the strategy. The guidelines for population-level management plans for large carnivores in Europe recommend a similar approach. Linnell *et al.* (2008) emphasized that the participatory process is an integral part of the product (a conservation or management plan in the form of a document and a map), and participants in the process should have some real influence on that product. The room to maneuver in management plans is tight, however, as national and international legislation set clear preconditions. Public involvement is usually organized in the form of workshops with participation of all important interest groups. As an input to these workshops, documents or maps summarizing the biological analyses are helpful, but they must be presented as baseline information or as proposals. Local people should be involved at all stages of the planning process, but the large geographic scale of some carnivore conservation units sometimes requires splitting the process into several steps. For the conservation of leopards in the Caucasus ecoregion, for example, a group of experts first produced a status review and situation analyses including a map with the potential distribution (IUCN/SSC Cat Specialist Group and WWF 2007). This report was used in a workshop where representatives from all six Caucasian countries developed a conservation strategy (Breitenmoser-Würsten *et al.* 2007b). This strategy again provided a framework for the development of national actions plans, together with local stakeholders. The freedom of the participants in the national workshops was somewhat restricted through the instructions given by the Caucasian strategy but, for practical reasons (e.g. language), organizing an international workshop with local participation would have been impossible.

A proposal for a carnivore conservation unit, based on thorough scientific analyses and translated into a pragmatic, spatially explicit plan with the agreement of local interest groups, has a higher chance to be endorsed and implemented than one lacking such participation.

Establishing a carnivore conservation unit able to host several hundred to a few thousand individuals, especially of large carnivores, can have considerable consequences far beyond the conservation community. Implementation might require adopting national legislation and changing land-use plans (e.g. designating new protected areas, changing exploitation habits) or landscape-management measures (creating corridors, improving habitat, recovering prey populations). The creation and maintenance of the designated conservation unit must be monitored. Monitoring a managed meta-population is demanding and includes monitoring the distribution and abundance of the target species in all subpopulations, monitoring the genetic structure of the entire population and its subunits, and monitoring ecological processes, this is, predation and its impact on prey. Furthermore, the effectiveness of management measures must be monitored. In tiger conservation areas, not only tiger numbers and the abundance of important prey species are monitored, but also the efficiency of law enforcement (all supported by a software tool called MIST, Management Information System; www.ecostats.com). Understanding why a carnivore population changes is critical for assessing the correctness of the management plan.

Monitoring is a prerequisite for an adaptive process. No carnivore conservation unit can (or should) be designed to last forever. Given the complexity of designating and conserving a managed meta-population of carnivores, and the assumed conflict potential, both the conservation unit and the management plan must be explicitly changeable through an adaptive approach. Revision and the revision process must be part of the plan. Achieving the conservation targets may require reshaping the conservation unit but, more importantly, a consensus with local people is easier to reach if they know that they will be involved continuously.

To put a carnivore conservation unit into practice and to make it operational for the long-term maintenance of a carnivore meta-population, respect the following points:

1. After a thorough analysis of the situation, relevant governmental authorities and local people must be involved in the designation of the carnivore

conservation unit. A participatory process is required to gain the support of local stakeholders, the political endorsement, and the subsequent establishment of the conservation unit.

2. Creating and maintaining a managed meta-population of several hundred to a few thousand carnivores requires monitoring the population (distribution, abundance, dynamic) and its genetics, and monitoring ecological processes and the public attitude (conflicts).
3. Designing a carnivore conservation unit is an adaptive process. Changes of the unit or the management plan may be needed to respond to results from monitoring or changes in local conditions. The option of revision is important for continuous support from local communities.

17.4 Final thoughts

Biodiversity conservation is more than protecting species or sites, although these two approaches are still mainly represented in our legal frameworks. Holistic conservation includes protecting existing patterns of diversity and the processes that generate diversity, which are the prerequisites for future adaptation and evolution. The necessity to go beyond protecting species and areas has given rise to a number of theoretical concepts and their legislative expressions that are difficult to interpret. If we narrow the scope to a certain group of species (the carnivores) or to a particular problem, the situation is less complicated because we face a reduced complexity and a limited number of options. Nonetheless, carnivore conservation is complex because it requires preserving habitat and prey, solving conflicts with human land-use, and extending areas for viable population. We fail to conserve carnivores almost never because we lack scientific understanding or concepts, but because we are not able to implement these concepts in the real world, where scientific and biological models need the support of politicians and local people. We need to make complex scientific ideas and cryptic legal language functional and easily understood.

The concept of a managed meta-population allows the integration of most of the requirements for a carnivore conservation unit. It is relatively simple and clear because it can be demonstrated on a real map, and it is negotiable and adaptable. It requires involving all interest groups when designing and changing the zones or the management rules in a given zone. Under a managed meta-population concept, we can secure all levels of viability (demographic, genetic and ecological) of an identified carnivore conservation unit. We can also maintain a continuous and concrete dialogue among scientists, politicians, conservationist, managers, and local land-users.

References

Aars, J., Marques, T. A., Buckland, S. T. *et al.* (2009). Estimating the Barents Sea polar bear subpopulation size. *Marine Mammal Science*, **25**, 35–52.

Aarts, G., MacKenzie, M., McConnell, B., Fedak, M., and Matthiopoulos, J, (2008). Estimating space-use and habitat preference from wildlife telemetry data. *Ecography*, **31**, 140–160.

Aastrup, P. (2009). Caribou movements in West Greenland. Studies in relation to proposed industrial development. In R. E. Haugerud, B. Åhman, and O. Danell (eds). *Abstracts of the 15th Nordic Conference on Reindeer and Reindeer Husbandry Research*, pp.88–89. Rangifer Report No. 13. (Luleå, Sweden, 26th to 29th January 2009. Luleå). Nordic Council for Reindeer Husbandry Research (NOR).

Abbey, H. (1952). An examination of the Reed-Frost theory of epidemics. *Human Biology; An International Record of Research*, **24**, 201–233.

Abrams, P. A. and Ginzburg, L. R. (2000). The nature of predation: prey dependent, ratio dependent or neither? *Trends in Ecology and Evolution*, **15**, 337–341.

Acton, A. E., Munson, L. and Waddell, W. T. (2000). Survey of necropsy results in captive red wolves (*Canis rufus*), 1992–1996. *Journal of Zoo and Wildlife Medicine*, **31**, 2–8.

Acton, A. E., Beale, A. B., Gilger, B. C. and Stoskopf, M. K. (2006). Sustained release cyclosporine therapy for bilateral *keratoconjunctivitis sicca* in a red wolf (*Canis rufus*). *Journal of Zoo and Wildlife Medicine*, **37**, 562–564.

Adams, J. R. and Waits, L. P. (2007). An efficient method for screening faecal DNA genotypes and detecting new individuals and hybrids in the red wolf (*Canis rufus*) experimental population area. *Conservation Genetics*, **8**, 123–131.

Adams, J. R., Kelly, B. T. and Waits, L. P. (2003). Using faecal DNA sampling and GIS to monitor hybridization between red wolves (*Canis rufus*) and coyotes (*Canis latrans*). *Molecular Ecology*, **12**, 2175–2186.

Adams, J. R., Lucash, C., Schutte, L. and Waits, L. P. (2007). Locating hybrid individuals in the red wolf (*Canis rufus*) experimental population area using a spatially targeted sampling strategy and faecal DNA genotyping. *Molecular Ecology*, **16**, 1823–1834.

Adjemian, J. Z., Foley, P., Gage, K. L. and Foley, J. E. (2007). Initiation and spread of traveling waves of plague, *Yersinia pestis*, in the western United States. *The American Journal of Tropical Medicine and Hygiene*, **76**, 365–375.

Admasu, E., Thirgood, S. J., Bekele, A. and Laurenson, M. K. (2004). Spatial ecology of golden jackal in farmland in the Ethiopian Highlands. *African Journal of Ecology*, **42**, 144–152.

Ågren, E. O., Nordenberg, L. and Mörner, T. (2000). Surgical implantation of radiotelemetry transmitters in European badgers (*Meles meles*). *Journal of Zoo and Wildlife Medicine*, **31**, 52–55.

Aitchinson, J. (1986) *The Statistical Analysis of Compositional Data.* Chapman and Hall, London.

Akçakaya, H. R. and Sjögren-Gulve, P. (2000). Population viability analysis in conservation planning: an overview. *Ecological Bulletins*, **48**, 9–21.

Alexander, F. E. and Boyle, P. (1996). *Methods for investigating localized clustering of disease.* International Agency for Research on Cancer. World Health Organization, Geneva, Switzerland.

Alho, J. M. (1990). Logistic regression in capture-recapture models. *Biometric*, **46**, 623–635.

Allen, D.L. (1939). Winter habits of Michigan skunks. *Journal of Wildlife Management*, **3**, 212–228.

Allen, T. F. H. and Starr, T. B. (1982). *Hierarchy, perspectives for ecological complexity.* University of Chicago Press, Chicago, IL.

Allendorf, F. W. and Luikart, G. (2007). *Conservation and the genetics of populations.* Blackwell Publishing Ltd., Oxford.

Alpers, D. L., Taylor, A. C., Sunnucks, P., Bellman, S. A. and Sherwin, W. B. (2003). Pooling hair samples to increase DNA yield for PCR. *Conservation Genetics*, **4**, 779–788.

American Association of Zoo Veterinarians. (2006). Guidelines for euthanasia of non-domestic animals. American Association of Zoo Veterinarians (AAZV), Lawrence, KS.

American Society of Mammalogists Animal Care and Use Committee. (1998). Guidelines for the capture, handling, and care of mammals as approved by the American Society of Mammalogists (ASM). *Journal of Mammalogy*, **79**, 1416–1431.

American Veterinary Medical Association. (2007). *AVMA guidelines on euthanasia* (Formerly the Report of the AVMA Panel on Euthanasia). American Veterinary Medical Association (AVMA). Schaumburg, IL.

Amlaner, C. J., Jr. (ed.) (1989). *Biotelemetry X.* University of Arkansas Press, Fayetteville, AK.

Amlaner, C. J., Jr. and Macdonald, D. W. (eds) (1980). *A handbook on biotelemetry and radio tracking.* Pergamon Press, Oxford.

Amstrup, S. C., McDonald, T. L. and Durner, G. M. (2004). Using satellite radiotelemetry data to delineate and manage wildlife populations. *Wildlife Society Bulletin*, **32**, 661–679.

Amstrup, S. C., McDonald, T. L. and Manly, B. F. (eds). (2005). *Handbook of capture–recapture analysis.* Princeton University Press, Princeton, NJ.

Amstrup, S. C., Durner, G. M., McDonald, T. L., Mulcahy, D. M. and Garner, G. W. (2001). Comparing movement patterns of satellite-tagged male and female polar bears. *Canadian Journal of Zoology*, **79**, 2147–2158.

Amstrup, S. C., Durner, G. M., Stirling, I., Lunn, N. J., Messler, F. and Messier, F. (2000). Movements and distribution of polar bears in the Beaufort Sea. *Canadian Journal of Zoology*, **78**, 948–966.

Andelt, W. F. and Gipson, P. S. (1980). Toe-clipping coyotes for individual identification. *Journal of Wildlife Management*, **44**, 293–294.

Anderka, F. W. (1987). Radiotelemetry techniques for furbearers. In M. Novak, J. A. Baker, M. E. Obbard, and B. Malloch (eds). *Wild furbearer management and conservation in North America*, pp. 216–227. Ontario Ministry of Natural Resources, Toronto.

Anderson, J. L. (1986). Age determination of the nyala *Tragelaphus angasi. South African Journal of Wildlife Research*, **16**, 82–90.

Andersen, R. (1991). Habitat deterioration and the migratory behaviour of moose (*Alces alces*) in Norway. *Journal of Applied Ecology*, **28**, 102–108.

Anderson, D. R. (2001). The need to get the basics right in wildlife field studies. *Wildlife Society Bulletin*, **29**, 1294–1297.

Anderson, C. R. and Lindzey, F. G. (2003). Estimating cougar predation rates from GPS location clusters. *Journal of Wildlife Management*, **67**, 307–316.

Anderson, E. M. and Lovallo, M. J. (2003). Bobcat and Lynx. In G A. Feldhamer, B. C. Thompson and J. A. Chapman (eds). *Wild Mammals of North America: Biology, Management and Conservation*, 2nd edn, pp. 758–786. The John Hopkins University Press, Baltimore, MD.

Andersen, D. E. and Rongstad, O. J. (1989). Home-range estimates of red-tailed hawks based on random and systematic relocations. *Journal of Wildlife Management*, **53**, 802–807.

Anderson, D. R., Burnham, K. P., White, G. C. and Otis, D. L. (1983). Density estimation of small-mammal populations using a trapping web and distance sampling methods. *Ecology*, **64**, 674–680.

Anderson, D. R., Burnham, K. P. and Thompson, W. L. (2000). Null hypothesis testing, problems, prevalence, and an alternative. *Journal of Wildlife Management*, **64**, 912–923.

Andersen, M., Derocher, A. E., Wiig, O. and Aars, J. (2008). Movements of two Svalbard polar bears recorded using geographical positioning system satellite transmitters. *Polar Biology*, **31**, 905–911.

Andrabi, S. M. and Maxwell, W. M. (2007). A review on reproductive biotechnologies for conservation of endangered mammalian species. *Animal Reproduction Science*, **99**, 223–243.

Andruskiw, M., Fryxell, J. M., Thompson, I. D. and Baker, J. A. (2008). Habitat-mediated variation in predation risk by the American marten. *Ecology*, **89**, 2273–2280.

Anel, L., Martinez, F., Alvarez, M., Anel, E., Boxio, J. C., Kaabi, M. *et al.* (1999). Post-mortem spermatozoa recovery and freezing in a Cantabric brown bear. *Theriogenology*, **51**, 277.

Angerbjörn, A., Tannerfeldt, M. and Erlinge, S. (1999). Predator-prey relationships: arctic foxes and lemmings. *Journal of Animal Ecology*, **69**, 34–49.

Angst, C. (2002). Fence systems and grounding systems: some practical advice. *Carnivore Damage Prevention News*, **5**, 14–16.

Araújo, M., Martins, E., Cruz, L. *et al.* (2010). Nested diets: a novel pattern of individual-level resource use. *Oikos*, **119**, 81–88.

Arjo, W. M., Gese, E. M., Bromley, C., Kozlowski, A. and Williams, E. S. (2003). Serologic survey for diseases in free-ranging coyotes (*Canis latrans*) from two ecologically distinct areas of Utah. *Journal of Wildlife Diseases*, **39**, 449–455.

Armstrong, D. P. and Seddon, P. J. (2008). Directions in reintroduction biology. *Trends in Ecology and Evolution*, **23**, 20–25.

Arnason, A. N. (1972). Parameter estimates from mark-recapture experiments on two populations subject to migration and death. *Researches on Population Ecology*, **13**, 97–113.

Arnason, A. N. (1973). The estimation of population size, migration rates, and survival in a stratified population. *Researches on Population Ecology*, **15**, 1–8.

Arnemo, J. M. and Caulkett, N. (2007). Stress. In G. West, D. Heard, and N. Caulkett (eds). *Zoo animal and wildlife immobilization and anesthesia*, pp. 103–109. Blackwell Publishing Professional, Ames, Iowa.

Arnemo, J. M., Ahlqvist, P., Andersen, R., Berntsen, F., Ericsson, G., Odden, J., Brunberg, S., Segerström, P. and Swenson, J. E. (2006). Risk of capture-related mortality in large free-ranging mammals: experiences from Scandinavia. *Wildlife Biology*, **12**, 109–113.

Arthur, S. M. (1988). An evaluation of techniques for capturing and radiocollaring fishers. *Wildlife Society Bulletin*, **16**, 417–421.

Asa, C. S. (1996). Physiological and social aspects of reproduction of the wolf and their implications for contraception. In L. Carbyn (ed.). *Ecology and conservation of wolves in a changing world*, pp. 283–286. Canadian Circumpolar Institute, Edmonton.

Asa, C. S. (1999). Dogs (Canidae). In E. Knobil and J. D. Neill (eds). *Encyclopedia of reproduction*, pp. 902–909. Academic Press, New York.

Asa, C. S. (2005). Types of contraception: the choices. In C. S. Asa and I. J. Porton (eds). *Wildlife contraception: issues, methods and application*, pp. 29–52. Johns Hopkins University Press, Baltimore, MD.

Asa, C. S. (2010). Reproductive physiology. In D. G. Kleiman, K. V. Thompson and C. K. Baer (eds). *Wild mammals in captivity: principles and techniques for zoo management*, pp. 411–428. University of Chicago Press, Chicago, IL.

Asa, C. S. and Porton, I. J. (2005). The need for wildlife contraception: Problems related to unrestricted population growth. In C. S. Asa and I. J. Porton (eds). *Wildlife contraception: issues, methods and application*, pp. xxv–xxxii, Johns Hopkins University Press, Baltimore, MD.

Asa, C. S. and Porton, I. J. (2010). Contraception as a management tool for controlling surplus animals. In D. G. Kleiman, K. V. Thompson and C. K. Baer (eds). *Wild mammals in captivity: principles and techniques for zoo management*, pp. 469–482. University of Chicago Press, Chicago, IL.

Asa, C. S. and Valdespino, C. (2003). A review of small canid reproduction. In M. A. Sovada and L. Carbyn (eds). *The swift fox: ecology and conservation of swift foxes in a changing world*, pp. 117–123. Canadian Plains Research Center, Regina, Saskatchewan.

Asa, C. S., Seal, U. S., Plotka, E. D., Letellier, M. A. and Mech, L. D. (1986). Effect of anosmia on reproduction in male and female wolves (*Canis lupus*). *Behaviour and Neural Biology*, **46**, 272–284.

Asa, C. S., Mech, L. D., Seal, U. S. and Plotka, E. D. (1990). The influence of social and endocrine factors in scent-marking by wolves (*Canis lupus*). *Hormones and Behavior*, **24**, 497–509.

Asa, C. S., Junge, R. E., Bircher, J. S., Noble, G. and Plotka, E. D. (1992). Assessing reproductive cycles and pregnancy in cheetahs (*Acninonyx jubatus*) by vaginal cytology. *Zoo Biology*, **11**, 139–151.

Asa, C. S., Bauman, K., Callahan, P., Bauman, J., Volkmann, D. H. and Jochle, W. (2006). GnRH-agonist induction of fertile estrus with either natural mating or artificial insemination, followed by birth of pups in gray wolves (*Canis lupus*). *Theriogenology*, **66**, 1778–1782.

Asa, C. S., Bauman, J. E., Coonan, T. J., and Gray, M. M. (2007). Evidence for induced estrus or ovulation in a canid, the island fox (*Urocyon littoralis*). *Journal of Mammalogy*, **88**, 436–440.

Association for the Study of Animal Behaviour/Animal Behavior Society (ASAB/ABS). (2000). Guidelines for the treatment of animals in behavioural research and teaching. *Animal Behavior*, **59**, 253–257.

Athreya, V., Odden, M., Linnell, J. D. C. and Karanth, K. U. (2010). Translocation as a tool for mitigating conflict with leopards in human-dominated landscapes of India. *Conservation Biology*, **25**, 133–141.

Atkinson, R. P. D., Rhodes, C. J., Macdonald, D. W. and Anderson, R. M. (2002). Scale-free dynamics in the movement patterns of jackals. *Oikos*, **98**, 134–140.

Aubry, K. B. and Jagger, L. A. (2006). The importance of obtaining verifiable occurrence data on forest carnivores and an interactive website for archiving results from standardized surveys. In M. Santos-Reis, J. D. S. Birks, E. C. O'Doherty and G. Proulx (eds). *Martes in Carnivore Communities*, pp. 159–176. Alpha Wildlife Publications, Sherwood Park, Alberta.

Augusteyn, R. C. (2007). On the relationship between rabbit age and lens dry weight: improved determination of the age of rabbits in the wild. *Molecular Vision*, **13**, 2030–2034.

Austin, J., Chamberlain, M. J., Leopold, B. D. and Burger, L. W., Jr. (2004). An evaluation of EGG™ and wire cage traps for capturing raccoons. *Wildlife Society Bulletin*, **32**, 351–356.

Austrheim, G., Olsson, E. G. A. and Grøntvedt, E. (1999). Land-use impact on plant communities in semi-natural sub-alpine grasslands of Budalen, central Norway. *Biological Conservation*, **87**, 369.

Azevedo, de, F. C. C. and Murray, D. L. (2007a). Evaluation of potential factors predisposing livestock to predation by jaguars. *Journal of Wildlife Management*, **71**, 2379–2386.

Azevedo, de, F. C. C., Murray, D. L. (2007b). Spatial organization and food habits of jaguars (*Panthera onca*) in a floodplain forest. *Biological Conservation*, **137**, 391–402.

Bachmann, P., Bramwell, R. N., Fraser S. J. *et al.* (1990). Wild carnivore acceptance of baits for delivery of liquid rabies vaccine. *Journal of Wildlife Diseases*, **26**, 486–501.

Badyaev, A. V. (1998). Environmental stress and developmental stability in dentition of the Yellowstone grizzly bears. *Behavioral Ecology*, **9**, 339–344.

Badyaev, A. V. and Foresman, K. R. (2000). Extreme environmental change and evolution: stress-induced morphological variation is strongly concordant with patterns of evolutionary divergence in shrew mandibles. *Proceedings of the Royal Society of London B*, **267**, 371–377.

Badyaev, A. V., Foresman, K. R. and Fernandes, M. V. (2000). Stress and developmental stability: vegetation removal causes increased fluctuating asymmetry in shrews. *Ecology*, **81**, 336–345.

Baevsky, U. B. (1963). The effect of embryonic diapause on the nuclei and mitotic activity of mink and rat blastocysts. In A. C. Enders (ed.) *Delayed implantation*. Chicago, IL: University of Chicago Press.

Bagchi, S. and Mishra, C. (2006). Living with large carnivores: predation on livestock by the snow leopard (*Uncia uncia*). *Journal of Zoology*, **268**, 217–224.

Bailey, L. L., Hines, J. E., Nichols, J. D., and MacKenzie, D. I. (2007). Sampling design trade-offs in occupancy studies with imperfect detection, examples and software. *Ecological Applications*, **17**, 281–290.

Bailey, L. L., Reid J. A., Forsman, E. D. and Nichols, J. D. (2009). Modeling co-occurrence of northern spotted and barred owls, accounting for detection probability differences. *Biological Conservation*, **142**, 2983–2989.

Baker, J. R. (1938). The evolution of breeding seasons. In G. R. De Beer (ed.). *Evolution. Essays presented to E.S. Goodrich*. Oxford University Press, Oxford.

Baker, P. J., Boitani, L., Harris, S., Saunders, G. and White, P. C. L. (2008). Terrestrial carnivores and human food production: impact and management. *Mammal Review*, **38**, 123–166.

Balkenhol, N., Waits, L. P. and Dezzani, R. J. (2009). Statistical approaches in landscape genetics: an evaluation of methods for linking landscape and genetic data. *Ecography*, **32**, 818–830.

Ballard, W. R., Franzmann A. W. and Gardner, C. I. (1982). Comparison and assessment of drugs used to immobilize Alaskan gray wolves (*Canis lupus*) and wolverines (*Gulo gulo*) from a helicopter. *Journal of Wildlife Diseases*, **18**, 339–342.

Ballard, W. B., Krausman, P. R., Boe, S., Cunningham, S. and Whitlaw, H. A. (2000). Short-term response of gray wolves, *Canis lupus*, to wildfire in Northwestern Alaska. *Canadian Field-Naturalist*, **114**, 241–247.

Ballesteros, C., Perez de la Lastra, J. M. and de la Fuente J. (2007). Recent developments in oral bait vaccines for wildlife. *Recent Patents on Drug Delivery and Formulation*, **1**, 230–235.

Ballmann-Acton, B. and Elaine, A. (2009). *Evaluation of non-invasive molecular monitoring for fecal pathogens among free-ranging carnivores*. North Carolina State University, Raleigh, NC.

Balme, G. A., Hunter, L. T. B. and Slotow, R. (2009a). Evaluating methods for counting cryptic carnivores. *Journal of Wildlife Management*, **73**, 433–441.

Balme, G. A., Slotow, R. and Hunter, L. T. B. (2009b). Impact of conservation interventions on the dynamics and persistence of a persecuted leopard (*Panthera pardus*) population. *Biological Conservation*, **142**, 2681–2690.

Balser, D. S. (1965). Tranquilizer tabs for capturing wild carnivores. *Journal of Wildlife Management*, **29**, 438–442.

Banci, V. A. (1987). *Ecology and behavior of wolverine in the Yukon*. M.Sc. Thesis, Simon Fraser University, Burnaby, British Columbia.

Banci, V. and Proulx, G. (1999). Resiliency of furbearers to trapping in Canada. In G. Proulx (ed.). *Mammal trapping*, pp. 175–203. Alpha Wildlife Research and Management Ltd., Sherwood Park, Alberta.

Bangs E. E., Fontaine J. A., Jimenez M. D. *et al.* (2005). Managing wolf-human conflict in the northwestern United States. In R. Woodroffe, S. Thirgood, and A. Rabinowitz (eds), pp. 340–356. *People and wildlife: conflict or co-existence*. Cambridge University Press Cambridge.

Banks, P. B., Hume, I. D. and Crowe, O. (1999). Behavioural, morphological and dietary response of rabbits to predation risk from foxes. *Oikos*, **85**, 247–256.

Barea-Azcon, J. M., Virgos, E., Ballesteros-Duperon, E., Moleon, M. and Chirosa, M. (2007). Surveying carnivores at large spatial scales: a comparison of four broad-applied methods. *Biodiversity and Conservation*, **16**, 1213–1230.

Barja, I., de Miguel, F. J. and Barcena, F. (2005). Faecal marking behaviour of Iberian wolf in different zones of their territory. *Folia Zoologica*, **54**, 21–29.

Barja, I., Silvan, G. and Illera, J. (2008). Relationships between sex and stress hormone levels in feces and marking behavior in a wild population of Iberian wolves (*Canis lupus signatus*). *Journal of Chemical Ecology*, **34**, 697–701.

Barja, I., Silvan, G., Rosellini, S. *et al.* (2007). Stress physiological responses to tourist pressure in a wild population of European pine marten. *Journal of Steroid Biochemistry and Molecular Biology*, **104**, 136–142.

Barrett, M. W., Nolan J. W. and Roy, L. D. (1982). Evaluation of a hand-held net-gun to capture large mammals. *Wildlife Society Bulletin*, **10**, 108–114.

Barrett, M. W., Proulx, G., Hobson, D., Nelson, D. and Nolan, J. W. (1989). Field evaluation of the C120 Magnum trap for marten. *Wildlife Society Bulletin*, **17**, 299–306.

Bascompte, J. and Vilà, C. (1997). Fractals and search paths in mammals. *Landscape Ecology*, **12**, 213–221.

Basille, M., Herfindal, I., Santin-Janin, H. *et al.* (2009). What shapes Eurasian lynx distribution in human dominated landscapes, selecting prey or avoiding people? *Ecography*, **32**, 1–9.

Bassett, C. F. and Leakley, J. R. (1942). Determination of estrum in the fox vixen. *North American Veterinarian*, **23**, 454–457.

Bates, K., Steenhof, K. and Fuller, M. R. (2003). *Recommendations for finding PTTs on the ground without VHF telemetry*. Proceedings of the Argos Animal Tracking Symposium. Annapolis, Maryland, 24–26 March 2003. Available on a CD from Service Argos, Inc., Largo MD 20774.

Bauer, H. and de Iongh, H. H. (2005). Lion (*Panthera leo*) home ranges and livestock conflicts in Waza National Park, Cameroon. *African Journal of Ecology*, **43**, 208–214.

Bauman, J. E., Clifford, D. L. and Asa, C. S. (2008). Pregnancy diagnosis in wild canids using a commercially available relaxin assay. *Zoo Biology*, **27**, 406–413.

Beach, F. A. (1976). Sexual attractivity, proceptivity and receptivity in female mammals. *Hormones and Behavior*, **7**, 105–138.

Bearhop, S., Adams, C. E., Waldron, S., Fuller, R. A. and MacLeod, H. (2004). Determining trophic niche width: a novel approach using stable isotope analysis. *Journal of Animal Ecology*, **73**, 1007–1012.

Beasley, J. C. and Rhodes, O. E., Jr. (2007). Effect of tooth removal on recaptures of raccoons. *Journal of Wildlife Management*, **71**, 266–270.

Beasom, S. L., Evans, W. and Temple, L. (1980). The drive net for capturing big game. *Journal of Wildlife Management*, **44**, 478–480.

Beck, K. (2006). *Epidemiology of coyote introgression into the red wolf genome*. PhD Dissertion, North Carolina State University, Raleigh, NC.

Beck, K., Lucash, C. and Stoskopf, M. K. (2009). Lack of impact of den disturbance on neonatal red wolves (*Canis rufus*). *Southeastern Naturalist*, **8**, 631–638.

Becker, E.F., Spindler, M.A. and Osborne, T.O. (1998). A population estimator based on network sampling of tracks in the snow. *Journal of Wildlife Management*, **62**, 968–977.

Becker, E. F., Golden, H. N., and Gardner, C. L. (2004). Using probability sampling of animal tracks in snow to estimate population size. In W. L. Thompson (ed.). *Sampling rare or elusive species*, pp. 248–270. Island Press, Washington DC.

Beckmann, J. P. and Berger, J. (2003). Using black bears to test ideal-free distribution models experimentally. *Journal of Mammalogy*, **84**, 594–606.

Begg, C. M., Begg, K. S., Du Toit, J. T. and Mills, M. G. L. (2005). Spatial organization of the honey badger *Mellivora capensis* in the southern Kalahari: home-range size and movement patterns. *Journal of Zoology*, **265**, 23–35.

Beier, L. R., Lewis, S. B., Flynn, R. W., Pendleton, G. and Schumacher, T. V. (2005). From the field: a single-catch snare to collect brown bear hair for genetic mark-recapture studies. *Wildlife Society Bulletin*, **33**, 766–773.

Beissinger, S. R. and McCullogh, D. R. (eds) (2002). *Population viability analysis*. University of Chicago Press, Chicago.

Beja-Pereira, A., Oliveira, R., Alves, P. C., Schwartz, M. K. and Luikart, G. (2009). Advancing ecological understandings through technological transformations in noninvasive genetics. *Molecular Ecology Resources*, **9**, 1279–1301.

Bekoff, M. and Wells, M. C. (1981). Behavioural budgeting by wild coyotes; the influence of food resources and social organization. *Animal Behaviour*, **29**, 794–801.

Belant, J. L. (1992). Efficacy of three types of live traps for capturing weasels *Mustela* spp. *Canadian Field-Naturalist*, **106**, 394–397.

Belant, J. L. and Follmann, E. H. (2002). Sampling considerations for American black and brown bear home range and habitat use. *Ursus*, **13**, 299–315.

Bell, W. J. and Kramer, E. (1979). Search and anemotactic orientation of cockroaches. *Journal of Insect Physiology*, **25**, 631–640.

Beltran, J. F. and Tewes, M. E. (1995). Immobilization of ocelots and bobcats with ketamine hydrochloride and xylazine hydrochloride. *Journal of Wildlife Diseases*, **31**, 43–48.

Ben-David, M., Flynn, R. W. and Schell, D. M. (1997). Annual and seasonal changes in diets of martens: evidence from stable isotope analysis. *Oecologia*, **111**, 280–291.

Ben-David, M., Blundell, G. M., Kern, J. W., Maier, J. A. K., Brown, E. D. and Jewett, S. C. (2005). Communication in river otters: creation of variable resource sheds for terrestrial communities. *Ecology*, **86**, 1331–1345.

Benhamou, S. (2004). How to reliably estimate the tortuosity of an animal's path, straightness, sinuosity, or fractal dimension. *Journal of Theoretical Biology*, **229**, 209–220.

Bennett, A. and Hayssen, V. (2010). Measuring cortisol in hair and saliva from dogs: coat color and pigment differences. *Domestic Animal Endocrinology*, **39**, 171–180.

Berchielli, L. T., Jr. and Tullar, B. F., Jr. (1980). Comparison of a leg snare with a standard leg-gripping trap. *New York Fish and Game Journal*, **27**, 63–71.

Berger, K. M. (2006). Carnivore-livestock conflicts: effects of subsidized predator control and economic correlates on the sheep industry. *Conservation Biology*, **20**, 751–761.

Berger, K. M. and Gese E. M. (2007). Does interference competition with wolves limit the distribution and abundance of coyotes? *Journal of Animal Ecology*, **76**, 1075–1085.

Berger, J., Testa, J. W., Roffe, T., and Monfort, S. L. (1999). Conservation endocrinology: a non-invasive tool to understand relationships between carnivore colonization and ecological carrying capacity. *Conservation Biology*, **13**, 980–989.

Berland, A., Nelson, T., Stenhouse, G., Graham, K. and Cranston, J. (2008). The impact of landscape disturbance on grizzly bear habitat use in the Foothills Model Forest, Alberta, Canada. *Forest Ecology and Management*, **256**, 1875–1883.

Berry, O., Sarre, S. D., Farrington, L. and Aitken, N. (2007). Faecal DNA detection of invasive species: the case of feral foxes in Tasmania. *Wildlife Research*, **34**, 1–7.

Bertschinger, H. J., Asa, C. S., Calle, P. P. *et al.* (2001). Control of reproduction and sex related behaviour in exotic wild carnivores with the GnRH analogue deslorelin: preliminary observations. *Journal of Reproduction and Fertility*, Suppl., **57**, 275–283.

Beyer, H. L., Haydon, D. T., Morales, J. M. *et al.* (2010). The interpretation of habitat preference metrics under use-availability designs. *Philosophical Transactions of the Royal Society B*, **365**, 2245–2254.

Bibby, C. J. (2004). Bird diversity survey methods. In W. J. Sutherland, I. Newton and R. E. Green (eds). *Bird ecology and conservation*, pp.1–16. Oxford University Press, Oxford.

Bidlack, A. L., Reed, S. E., Palsboll, P. J. and Getz, W. M. (2007). Characterization of a western North American carnivore community using PCR-RFLP of cytochrome b obtained from fecal samples. *Conservation Genetics*, **8**, 1511–1513.

Biggins, D. E., Godbey, J. L., Miller, B. J. and Hanebury, L. R. (2006). *Radio-telemetry for black-footed ferret research and monitoring.* Proceedings of the Symposium on the Status of the Black-footed Ferret and its Habitat. Ft. Collins, Colorado, 28–29 January 2004. Scientific Investigations Report 2005–5293: U.S. Geological Survey.

Biggins, D., Hanesbury, L., Miller, B. and Powell, R. A. (2011a). Mortality of Siberian polecats and black-footed ferrets released onto prairie dog colonies. *Journal of Mammalogy*, **92**, 721–731.

Biggins, D., Hanesbury, L., Miller, B. and Powell, R. A. (2011b). Black-footed ferrets and Siberian polecats as ecological surrogates and ecological equivalents. *Journal of Mammalogy*, **92**, 710–720.

Bingham, B. B. and Noon, B. R. (1997). Mitigation of habitat "take": application to habitat conservation planning. *Conservation Biology*, **11**, 127–139.

Birkenheuer, A. J., Harms, C. A., Neel, J. *et al.* (2007). The identification of a genetically unique piroplasma in North American river otters (*Lontra canadensis*). *Parasitology*, **134**, 631–635.

Birks, J. D. S., Hanson P., Jermyn D. L., Schofield, H. W. and Walton, K. C. (1994). Development of a method for monitoring polecats. In J. D. S. Birks and A. C. Kitchener (eds). *The distribution and status of the polecat Mustela putorius in Britain in the 1990s*, pp. 55–83. The Vincent Wildlife Trust, UK.

Bjärvall, A. and Franzen, R. (1981). Mortality transmitters - an important tool for studying reindeer calf mortality. *Ambio*, **10**, 126–128.

Bjerke, T., Vittersø, J. and Kaltenborn, B. P. (2000). Locus of control and attitudes toward large carnivores. *Psychological Reports*, **86**, 37–46.

Bjorge, R. R. and Gunson, J. R. (1989). Wolf, *Canis lupus*, population characteristics and prey relationships near Simonette River, Alberta. *Canadian Field-Naturalist*, **103**, 327–334.

Blackwell, G. L., Bassett, S. M. and Dickman, C. R. (2006). Measurement error associated with external measurements commonly used in small-mammal studies. *Journal of Mammalogy*, **87**, 216–223.

Blanco, J. C. and Cortés, Y. (2007). Dispersal patterns, social structure and mortality of wolves living in agricultural habitats in Spain. *Journal of Zoology*, **273**, 114–124.

Blaum, N., Tietjen, B. and Rossmanith, E. (2009). Impact of livestock husbandry on small- and medium-sized carnivores in Kalahari savannah rangelands. *Journal of Wildlife Management,* **73**, 60–67.

Blejwas, K. M., Williams, C. L., Shin, G. T., McCullough, D. R. and Jaeger, M. M. (2006). Salivary DNA evidence convicts breeding male coyotes of killing sheep. *Journal of Wildlife Management,* **70**, 1087–1093.

Block, W. M., Franklin, A. B., Ward, J. P., Jr., Ganey, J. L. and White, G. C. (2001). Design and implementation of monitoring studies to elucidate the success of ecological restoration on wildlife. *Restoration Ecology*, **9**, 293–303.

Bloemendal, H. (1977). The vertebrate eye lens. *Science*, **197**, 127–138.

Blundell, G. M., Ben-David, M. and Bowyer, R. T. (2002). Sociality in river otters: cooperative foraging or reproductive strategies? *Behavioral Ecology*, **13**, 134–141.

Blundell, G. M., Kern J. W., Bowyer R. T. and Duffy, L. K. (1999). Capturing river otters, a comparison of Hancock and leg-hold traps. *Wildlife Society Bulletin*, **27**, 184–192.

Boggess, E. and Loegering, P. (1985). *Minnesota Trapper Education Manual.* Department of Natural Resources, St. Paul and Minnesota Trappers Association, Monticello, MN.

Boggess, E. K., Batcheller G. R., Linscombe, R. G. *et al.* (1990). Traps, trapping and furbearer management, a review. *Wildlife Society Technical Review*, **90/91**, 1–30.

Böhm, M., Hutchings, M. R. and White, P. C. L. (2009). Contact networks in a wildlife-livestock host community: identifying high-risk individuals in the transmission of bovine TB among badgers and cattle. *PloS ONE,* **4**, 1–12.

Böhm, M., Palphramand, K. L., Newton-Cross, G., Hutchings, M. R. and White, P. C. L. (2008). Dynamic interactions among badgers: implications for sociality and disease transmission. *Journal of Animal Ecology*, 77, 735–745.

Boitani, L. (1995). Ecological and cultural diversities in the evolution of wolves-humans relationships. In L. N. Carbyn, S. H. Fritts and D. R. Seip (eds). *Ecology and conservation of wolves in a changing world*, pp. 3–12. Canadian Circumpolar Institute, Occasional Publ. N. 35, Edmonton, Alberta.

Boitani, L., Ciucci, P. and Raganella-Pelliccioni, E. (2010). Ex-post compensation payments for wolf predation on livestock in Italy: a tool for conservation ? *Wildlife Research*, **37**, 722–730.

Boitani, L., Sinibaldi, I., Corsi, F. *et al.* (2008). Distribution of medium- to large-sized african mammals based on habitat suitability models. *Biodiversity and Conservation*, **17**, 605–621.

Bolen, E. G. and Robinson, W. L. (2003). *Wildlife ecology and management.* Prentice Hall, Upper Saddle River, NJ.

Bolton, M., Butcher, N., Sharpe, N., Stevens, D. and Fisher, G. (2007). Remote monitoring of nests using digital camera technology. *Journal of Field Ornithology*, **78**, 213–220.

Bonney, R. C., Moore, H. D. M. and Jones, J. M. (1981). Plasma concentrations of oestradiol-17β and progesterone and laparoscopic observations of the ovary in the puma (*Felis concolor*) during oestrus, pseudo-pregnancy and pregnancy. *Journal of Reproduction and Fertility*, **63**, 523–531.

Bonsnes, R. W. and Taussky, H. H. (1945). On the colorimetric determination of creatinine by the Jaffe reaction. *Journal Biological Chemistry*, **158**, 581–591.

Boone, W. R., Keck, B. B., Catlin, J. C. *et al.* (2004). Evidence that bears are induced ovulators. *Theriogenology*, **61**, 1163–1169.

Boonstra, R. (2005). Equipped for life: the adaptive role of the stress axis in male mammals. *Journal of Mammalogy*, **86**, 236–247.

Boonstra, R., Hik, D., Singleton, G. R. and Tinnikov, A. (1998). The impact of predator-induced stress on the snowshoe hare cycle. *Ecological Monographs,* **68,** 371–394.

Borchers, D. L. and Efford, M. G. (2008). Spatially explicit maximum likelihood methods for capture-recapture studies. *Biometrics,* **64,** 377–385.

Bothma, J., Nel, J. A. J. and Macdonald, A. (1984). Food niche separation between four sympatric Namib Desert carnivores. *Journal of Zoology,* **202,** 327–340.

Botzler, R. G. and Armstrong-Buck, S. B. (1985). Ethical considerations in research on wildlife diseases. *Journal of Wildlife Diseases,* **21,** 341–345.

Boue, F., Delhomme, A. and Chaffaux, S. (2000). Reproductive management of silver foxes (*Vulpes vulpes*) in captivity. *Theriogenology,* **53,** 1717–1728.

Boulanger, J. B., McLellan, B. N., Woods, J. G., Proctor, M. and Strobeck, C. (2004). Sampling design and bias in DNA-based capture-mark-recapture population and density estimates of grizzly bears. *Journal of Wildlife Management,* **68,** 457–469.

Boulanger, J., Kendall, K. C., Stetz, J. D., Roon, D. A., Waits, L. P and Paetkau, D. (2008). Multiple data sources improve DNA-based mark–recapture population estimates of grizzly bears. *Ecological Applications,* **18,** 577–589.

Boulanger, J. B., White, G. C., McLellan, B. N., Woods, J. G., Proctor, M. and Himmer, S. (2002). A meta-analysis of grizzly bear DNA mark-recapture projects in British Columbia, Canada. *Ursus* **13,** 137–152.

Boutelle, S. and Bertschinger, H. (2010). Reproductive management in captive wild canids: contraception challenges. *International Zoo Yearbook,* **44,** 109–120.

Boutelle, S., Lenahan, K., Krisher, R., Bauman, K. L., Asa, C. S. and Silber, S. (2011). Vitrification of oocytes from endangered Mexican gray wolves (*Canis lupus baileyi*). *Theriogenology,* **75,** 647–654.

Bovet, P. and Benhamou, S. (1988). Spatial analysis of animals' movements using a correlated random walk model. *Journal of Theoretical Biology,* **131,** 419–433.

Bowland, A. E., Mills, M. G. L. and Lawson, D. (1992). *Predators and farmers.* Endangered Wildlife Trust, Pretoria, South Africa.

Boyce, M. S. (1992). Population viability analysis. *Annual Review of Ecology and Systematics,* **23,** 481–506.

Boyce, M. S. (2006). Scale for resource selection functions. *Diversity and Distributions,* **12,** 269–276.

Boyce, M. S. and McDonald, L. L. (1999). Relating populations to habitats using resource selection functions. *Trends in Ecology and Evolution,* **14,** 268–272.

Boyce, M. S. and Waller, J. S. (2003). Grizzly bears for the Bitterroot, predicting potential abundance and distribution. *Wildlife Society Bulletin,* **31,** 670–683.

Boyce, M. S., McDonald, L. L. and Manly, B. F. J. (1999). Relating populations to habitats, Reply. *Trends in Ecology and Evolution,* **14,** 490.

Boyce, M. S., Blanchard, B. M. Knight, R. R. and Servheen, C. (eds) (2001). *Population viability for grizzly bears, a critical review.* Monograph Series Number 4. International Association of Bear Research and Management.

Boyce, M. S., Vernier, P. R., Nielsen, S. E and Schmiegelow, F. K. A. (2002). Evaluating resource selection functions. *Ecological Modelling,* **157,** 281–300.

Boyce, M. S., Pitt, J., Northrup, J. M. *et al.* (2010). Temporal autocorrelation functions for movement rates from global positioning system radiotelemetry data. *Philosophical Transactions of The Royal Society B*, **365**, 2213–2219.

Boydston, E. E., Kapheim, K. M., Van Horn, R. C., Smale, L. and Holekamp, K. E. (2005). Sexually dimorphic patterns of space use throughout ontogeny in the spotted hyena (*Crocuta crocuta*). *Journal of Zoology*, **267**, 271–281.

Bradley, E. H. and Pletscher, D. H. (2005). Assessing factors related to wolf depredation of cattle in fenced pastures in Montana and Idaho. *Wildlife Society Bulletin*, **33**, 1256–1265.

Bradley, E. H., Pletscher, D. H., Bangs, E. E. *et al.* (2005). Evaluating wolf translocation as a nonlethal method to reduce livestock conflicts in the northwestern United States. *Conservation Biology*, **19**, 1498–1508.

Brand, W. A. (1996). High precision isotope ratio monitoring techniques in mass spectrometry. *Journal of Mass Spectrometry*, **31**, 225–235.

Braun, C. E. (ed.) (2005). *Techniques for wildlife investigations and management.* The Wildlife Society, Bethesda, MD.

Braun, B. C., Frank, A., Dehnhard, M., Voigt, C. C., Vargas, A., Goritz F. *et al.* (2009). Pregnancy diagnosis in urine of Iberian lynx (*Lynx pardinus*). *Theriogenology*, **71**, 754–761.

Breitenmoser, U. (1998). Large predators in the Alps, the rise and fall of man's competitors. *Biological Conservation*, **83**, 279–290.

Breitenmoser, U. and Breitenmoser-Würsten, Ch. (2008). *Der Luchs—Ein Grossraubtier in der Kulturlandschaft.* Salm Verlag, Bern.

Breitenmoser, U., Angst, C., Landry, J. M., Breitenmoser-Würsten, C., Linnell, J. D. C. and Weber, J. M. (2005). Non-lethal techniques for reducing depredation. In R. Woodroffe, S. Thirgood and A. Rabinowitz (eds). *People and wildlife: conflict or coexistence?*, pp. 49–71. Cambridge University Press, Cambridge.

Breitenmoser, U., von Arx, M., Bego, F. *et al.* (2009). Strategic planning for conservation of the Balkan lynx. Proceedings of the III Congress of Ecologists of Macedonia, Skopije, Macedonia, pp. 242–248.

Breitenmoser-Würsten, Ch. and Obexer-Ruff, G. (2007c). Reintroduction of lynx in Switzerland—a molecular evaluation 30 years after translocation. Presentation at the Conference "*Biology and Conservation of Wild Felids*," September 2007, Department of Zoology, Oxford. Abstract book p. 42.

Breitenmoser-Würsten, Ch., Vandel, J.-M., Zimmermann, F. and Breitenmoser, U. (2007a). Demography of lynx *Lynx lynx* in the Jura Mountains. *Wildlife Biology*, **13**, 381–392.

Breitenmoser-Würsten, Ch., Breitenmoser, U., Mallon, D. and Zazanashvili, N. (eds) (2007b). *Strategy for the conservation of the leopard in the Caucasus ecoregion.* IUCN/SSC Cat Specialist Group and WWF Caucasus Programme Office. IUCN, Gland, Switzerland.

Bremner-Harrison, S., Harrison, S. W. R., Cypher, B. L., Murdoch, J. D., Maldonado, J. and Darden, S. K. (2006). Development of a single-sampling non-invasive hair snare. *Wildlife Society Bulletin*, **34**, 456–461.

Bridges, A. S., Olfenbuttel, C. and Vaughan, M. R. (2002). A mixed regression model to estimate neonatal black bear cub age. *Wildlife Society Bulletin*, **30**, 1253–1258.

Bridges, A. S., Vaughan, M. R. and Klenzendorf, S. (2004a). Seasonal variation in American black bear *Ursus americanus* activity patterns: quantification via remote photography. *Wildlife Biology*, **10**, 277–284.

Bridges, A. S., Fox, J. A., Olfenbuttel, C. and Vaughan, M. R. (2004b). American black bear denning behavior: observations and applications using remote photography. *Wildlife Society Bulletin*, **32**, 188–193.

Bright, P. W. and Smithson, T. J. (2001). Biological invasions provide a framework for reintroductions, selecting areas in England for pine marten releases. *Biodiversity and Conservation*, **10**, 1247–1265.

Brockelman, W. Y. and Kobayashi, N.K. (1971). Live capture of free-ranging primates with a blowgun. *Journal of Wildlife Management*, **35**, 852–855.

Brogliatti, G. M. and Adams, G. P. (1996). Ultrasound-guided transvaginal oocyte collection in prepuberal calves. *Theriogenology*, **46**, 1163–1176.

Bromley, C. and Gese, E. M. (2001). Effects of sterilization on territory fidelity and maintenance, pair bonds, and survival rates of free-ranging coyotes. *Canadian Journal of Zoology*, **79**, 386–392.

Bronson, F. H. (1989). *Mammalian reproductive biology*. University of Chicago Press, Chicago, IL.

Brook, B. W., O'Grady, J. J., Chapman, A. P., Burgman, M. A., Akçakaya, H. R. and Frankham, R. (2000). Predictive accuracy of population viability analysis in conservation biology. *Nature*, **404**, 385–387.

Brooks, R. P. (1997). Improving habitat suitability index models. *Wildlife Society Bulletin*, **25**, 163–167.

Brøseth, H., Knutsen, K. and Bevanger, K. (1997). Spatial organization and habitat utilization of badgers *Meles meles*: effects of food patch dispersion in the boreal forest of central Norway. *Mammalian Biology*, **62**, 12–22.

Brosi, B. J. and Biber, E. G. (2009). Statistical inference, Type II error, and decision making under the US Endangered Species Act. *Frontiers in Ecology and the Environment*, 7, 487–494.

Brown, J. L. (1969). Territorial behavior and population regulation in birds: a review and re-evaluation. *Wilson Bulletin*, **81**, 293–329.

Brown, J. L. (1997). Fecal steroid profiles in black-footed ferrets exposed to natural photoperiod. *Journal of Wildlife Management*, **61**, 1428–1436.

Brown, J. L. and Wildt, D. E. (1997). Assessing reproductive status in wild felids by noninvasive faecal steroid monitoring. *International Zoo Yearbook*, **35**, 173–191.

Brown, J. L., Wildt, D. E., Phillips, L. G. *et al.* (1989). Adrenal-pituitary-gonadal relationships and ejaculate characteristics in captive leopards (*Panthera pardus kotiya*) isolated on the island of Sri Lanka. *Journal of Reproduction and Fertility*, **85**, 605–613.

Brown, J. L., Wasser, S. K., Wildt, D. E. and Graham, L. H. (1994). Comparative aspects of steroid hormone metabolism and ovarian activity in felids, measured noninvasively in feces. *Biology of Reproduction*, **51**, 776–786.

Brown, J. L., Wildt, D. E., Graham, L. H. *et al.* (1995). Natural versus chorionic gonadotropin-induced ovarian responses in the clouded leopard (*Neofelis nebulosa*) assessed by fecal steroid analysis. *Biology of Reproduction*, **53**, 93–102.

Brown, J. S., Laundre J. W. and Gurung, M. (1999). The ecology of fear, optimal foraging, game theory and trophic interactions. *Journal of Mammalogy*, **80**, 385–399.

Browne, C., Anderson, D. R., Burnham, K. P. and Robson, D. S. (1985). *Statistical inference from band-recovery data – a handbook.* US Fish and Wildlife Service Resource Publication 156.

Brownie, C., Hines, J. E., Nichols, J. D., Pollock, K. H. and Hestbeck, J. B. (1993). Capture-recapture studies for multiple strata including non-Markovian transition probabilities. *Biometrics*, **49**, 1173–1187.

Bruskotter, J. T., Vaske, J. J. and Schmidt, R. H. (2009). Social and cognitive correlates of Utah residents' acceptance of the lethal control of wolves. *Human Dimensions of Wildlife*, **14**, 119–132.

Buck, S. (1982). *Habitat utilization by fisher (Martes pennant) near Big Bar, California.* M.Sc. Thesis, Humboldt State University, Arcata, California.

Buckland, S. T., Anderson, D. R., Burnham, K. P., Laake, J. L., Borchers, D. L. and Thomas, L. (2001). *Introduction to distance sampling.* Oxford University Press, Oxford.

Bull, E. L., Heater, T. W. and Culver, F.G. (1996). Live-trapping and immobilizing American martens. *Wildlife Society Bulletin*, **24**, 555–558.

Bullard, F. (1999). *Estimating the home range of an animal: the Brownian bridge approach.* MS thesis. University of North Carolina, Chapel Hill, NC.

Bulte, E. H. and Rondeau, D. (2006). Why compensating wildlife damages may be bad for conservation. *Journal of Wildlife Management*, **69**, 14–19.

Burnham, K. P. (1972). *Estimation of population size in multiple capture-recapture studies when capture probabilities vary among animals.* PhD Thesis, Oregon State University, Corvallis, Oregon.

Burnham, K. P. (1993). A theory for combined analysis of ring recovery and recapture data. In J. D. Lebreton and P. M. North (eds). *Marked individuals in the study of bird population*, pp. 199–213. Birkhauser Verlag, Berlin.

Burnham, K. P. and Anderson, D. R. (2002). *Model selection and multimodal inference, a practical information-theoretic approach.* (2nd edn). Springer, New York.

Burnham, K. P. and Overton, W. S. (1978). Estimation of the size of a closed population when capture probabilities vary among animals. *Biometrika*, **65**, 625–633.

Burnham, K. P., Anderson, D. R., White, G. C., Brownie, C. and Pollock, K. H. (1987). *Design and analysis methods for fish survival experiments based on release-recapture.* American Fisheries Society, Bethesda, MD.

Burt, W. H. (1943). Territoriality and home range concepts as applied to mammals. *Journal of Mammalogy*, **24**, 346–352.

Bush, M. (1994). Reintroduction Medicine. *Zoos' Print*, **Jan/Feb**, 36–37.

Byrom, A. E. (2002). Dispersal and survival of juvenile feral ferrets *Mustela furo* in New Zealand. *The Journal of Applied Ecology*, **39**, 67–78.

Cagnacci, F. and F. Urbano. (2008). Managing wildlife: A spatial information system for GPS collars data. *Environmental Modelling and Software*, **23**, 957–959.

Cagnacci, F., Boitani, L., Powell, R. A. and Boyce, M. S. (2010). Animal ecology meets GPS-based radiotelemetry: a perfect storm of opportunities and challenges. *Philosophical Transactions of The Royal Society B*, **365**, 2157–2162.

Cain, J. W. III, Krausman, P. R., Jansen, B. D. and Morgart, J. R. (2005). Influence of topography and GPS fix interval on GPS collar performance. *Wildlife Society Bulletin*, **33**, 926–934.

Calenge, C., Dray S. and Royer-Carenzi, M. (2009). The concept of animals' trajectories from a data analysis perspective. *Ecological Informatics*, 4, 34–41.

Calhoun, J. B. and Casby, J. U. (1958). *Public Health Monograph No. 55*. Government Printing Office, Washington DC.

Calvert, W. and Ramsay, M. A. (1998). Evaluation of age determination of polar bears by counts of cementum growth layer groups. *Ursus*, **10**, 449–453.

Campbell, L. A., Long, R. A. and Zielinski, W. J. (2008). Integrating multiple methods to achieve survey objectives. In R. A. Long, P. MacKay, W. J. Zielinski and J. C. Ray (eds). *Noninvasive survey methods for carnivores*, pp. 223–237. Island Press, Washington DC.

Can, O. E. and Togan, I. (2009). Camera trapping of large mammals in Yenice Forest, Turkey: local information versus camera traps. *Oryx*, **43**, 427–430.

Canadian Cooperative Wildlife Health Centre. (2010). *Wildlife disease investigation manual*, third edn. Canadian Cooperative Wildlife Health Centre (CCWHC), Saskatoon, Saskatchewan, Canada.

Canadian Council on Animal Care. (2003). *CCAC guidelines on, the care and use of wildlife*. Canadian Council on Animal Care (CCAC), Ottawa, Canada.

Canadian Council on Animal Care. (2008). *Wildlife research and three Rs*. Canadian Council on Animal Care, Ottawa, Canada. Available at, www.ccac.ca/en/alternatives/wildlife_faune.html [Accessed 25 March 2011].

Canadian General Standards Board. (1996). *Animal (mammal) trap-mechanically-powered, trigger-activated killing traps for use on land*. National Standards of Canada/CGSB-144.1-96, Ottawa, Ontario, Canada.

Caniglia, R., Fabbri, E., Greco, C., Galaverni, M. and Randi, E. (2009). Forensic DNA against wildlife poaching: identification of a serial wolf killing in Italy. *Forensic Science International: Genetics*, 4, 334–338.

Canivenc, R. (1966). A study of progestation in the European badger (*Meles meles* L.). *Symposium Zoological Society London*, **15**, 15–26.

Canivenc, R., Bonnin, M., and Ribes, C. (1975). Ecological regulation of luteal function in mammals with delayed implantation. In Klotz, H.-P. (ed.) *Problemes actuels d'endocrinologie et de nutrition*. Expansion Scientifique Francaise, Paris.

Canon, S. K., Bryant, F. C., Bretzlaff, K. N. and Hellman, J. M. (1997). Pronghorn pregnancy diagnosis using trans-rectal ultrasound. *Wildlife Society Bulletin*, **25**, 832–834.

Capello, K., Mulatti, P., Comin, A. *et al.* (2010). Impact of emergency oral rabies vaccination of foxes in northeastern Italy, 28 December 2009–20 January 2010: preliminary evaluation. *Euro Surveillance*, **15**, 1–4.

Carbone, C. and Gittleman, J. L. (2002). A common rule for the scaling of carnivore density. *Science*, **295**, 2273–2276.

Carbone, C., Pettorelli, N. and Stephens, P. A. (2011). The bigger they come, the harder they fall: body size and prey abundance influence predator-prey ratios. *Biology Letters*, 7, 312–315.

Carde, R. T. and Elkinton, J. S. (1984). Field trapping with attractants, methods and interpretation. In H. E. Hummel and T. A. Miller (eds). *Techniques in pheromone research*, pp. 111–129. Spring-Verlag, New York.

Cardillo, M., Purvis, A., Sechrest, W., Gittleman, J. L., Bielby, J. and Mace, G. M. (2004). Human population density and extinction risk in the world's carnivores. *PLoS Biology*, **2**, 909–914.

Carlson, D. A. and Gese, E. M. (2007). Relaxin as a diagnostic tool for pregnancy in the coyote (*Canis latrans*). *Animal Reproduction Science*, **101**, 304–312.

Carlson, D. A. and Gese, E. M. (2008). Reproductive biology of the coyote (*Canis latrans*): integration of mating behavior, reproductive hormones, and vaginal cytology. *Journal of Mammalogy*, **89**, 654–664.

Caro, T. (1976). Observations on ranging behavior and daily activity of lone silverback mountain gorilla (*Gorilla gorilla beringei*). *Animal Behaviour*, **24**, 889–897.

Carpenter, F. L. and MacMillen, R. E. (1976). Threshold model of feeding territoriality and test with a Hawaiian honeycreeper. *Science*, **194**, 634–642.

Carpenter, M. A., Appel, M. J. G., Roelke-Parker, M. E. *et al.* (1998). Genetic characterization of canine distemper virus in Serengeti carnivores. *Veterinary Immunology and Immunopathology*, **65**, 259–266.

Carrel, W. K. (1994). Reproductive history in dental cementum of female black bears from dental cementum. *International Conference on Bear Research and Management*, **9**, 205–212.

Carroll, C. (2007). Interacting effects of climate change, landscape conversion, and harvest on carnivore populations at the range margin, marten and lynx in the northern Appalachians. *Conservation Biology*, **21**, 1092–1104.

Carroll, C. and Miquelle, D. G. (2006). Spatial viability analysis of Amur tiger *Panthera tigris altaica* in the Russian Far East, the role of protected areas and landscape matrix in population persistence. *Journal of Applied Ecology*, **43**, 1056–1068.

Carroll, C., Noss, R. F., Paquet, P. C. and Schumaker, N. H. (2003a). Use of population viability analysis and reserve selection algorithms in regional conservation plans. *Ecological Applications*, **13**, 1773–1789.

Carroll, C., Phillips, M. K., Schumaker, N. H. and Smith, D. W. (2003b). Impacts of landscape change on wolf restoration success, planning a reintroduction program based on static and dynamic spatial models. *Conservation Biology*, **17**, 536–548.

Castro-Arellano, I., Madrid-Luna, C., Lacher, T. E. and Leon-Paniagua, L. (2008). Hair-trap efficacy for detecting mammalian carnivores in the tropics. *Journal of Wildlife Management*, **72**, 1405–1412.

Catling, P. C., Corbett, L. K. and Westcott, M. (1991). Age determination in the dingo and crossbreeds. *Wildlife Research*, **18**, 75–84.

Cattet, M. R. L. and Obbard, M. E. (2005). To weigh or not to weigh: conditions for the estimation of body mass by morphometry. *Ursus*, **16**, 102–107.

Cattet, M., Shury, T. and Patenaude, R. (eds). (2005). *The Chemical Immobilization of Wildlife*, second edn. Canadian Association of Zoo and Wildlife Veterinarians.

Cattet, M. R., Christison, K., Caulkett, N. A. and Stenhouse, G. B. (2003). Physiological responses of grizzly bears to different methods of capture. *Journal of Wildlife Diseases*, **39**, 649–654.

Cattet, M., Caulkett, N. A., Wilson, C., Vandenbrink, T. and Brook, R. K. (2004). Intranasal administration of xylazine to reduce stress in elk captured by net gun. *Journal of Wildlife Diseases*, **40**, 562–565.

Cattet, M. R. L., Bourque, A., Elkin, B. T., Powley K. D., Dahlstrom D. B. and Caulkett, N. A. (2006). Evaluation of the potential for injury with remote drug delivery systems. *Wildlife Society Bulletin*, **34**, 741–749.

Cattet, M., Boulanger, J., Stenhouse, G., Powell, R. A. and Reynolds-Hogland, M. J. (2008a). An evaluation of long-term capture effects in ursids, implications for wildlife welfare and research. *Journal of Mammalogy*, **89**, 973–990.

Cattet, M., Stenhouse G. and Bollinger, T. (2008b). Exertional myopathy in a grizzly bear (*Ursus arctos*) captured by leg-hold snare. *Journal of Wildlife Diseases*, **44**, 973–978.

Catullo, G., Masi, M., Falcucci, A., Maiorano, L., Rondinini, C. and Boitani, L. (2008). A gap analysis of southeast asian mammals based on habitat suitability models. *Biological Conservation*, **141**, 2730–2744.

Caughley, G. (1994). Directions in conservation biology. *Journal of Animal Ecology*, **63**, 215–244.

Cavalcanti, S. M. C. and Gese, E. M. (2010). Kill rates and predation patterns of jaguars (Panthera onca) in the southern Pantanal, Brazil. *Journal of Mammalogy*, **91**, 722–736.

Cavallini, P. and Volpi, T. (1995). Biases in the analysis of the diet of the red fox *Vulpes vulpes*. *Wildlife Biology*, **1**, 243–248.

Ceballos, G., Erlich, P. R., Soberon, J., Salazar, I. and Fay, J. P. (2005). Global mammal conservation, What must we manage? *Science*, **309**, 603–607.

Cegelski, C. C., Waits, L. P., Anderson, N. J., Flagstad, O., Strobeck, C. and Kyle, C. J. (2006). Genetic diversity and population structure of wolverine (*Gulo gulo*) populations at the southern edge of their current distribution in North America with implications for genetic viability. *Conservation Genetics*, 7, 197–211.

Chadwick, J., Fazio, B. and Karlin, M. (2010). Effectiveness of GPS-based telemetry to determine temporal changes in habitat use and home-range sizes of red wolves. *Southeastern Naturalist*, **9**, 303–316.

Chao, A. (1989). Estimating population size from sparse data in capture-recapture experiments. *Biometrics*, **45**, 427–438.

Chapron, G. and Arlettaz, R. (2006). Using Models to manage carnivores. *Science*, **314**, 1682–1683.

Chapron, G., Miquelle, D. G., Lambert, A., Goodrich, J. M., Legendre, S. and Clobert, J. (2008). The impact on tigers of poaching versus prey depletion. *Journal of Applied Ecology*, **45**, 1667–1674.

Charlton, K. G., Hird, D. W. and Fitzhugh, E. L. (1998). Physical condition, morphometrics, and growth characteristics of mountain lions. *California Fish and Game*, **84**, 104–111.

Charnov, E. L. (1976a). Optimal foraging, attack strategy of a mantid. *American Naturalist*, **110**, 141–151.

Charnov, E., L. (1976b). Optimal foraging, the marginal value theorem. *Theoretical Population Biology*, **9**, 129–136.

Charnov, E. L., Orians G. H. and Hyatt, K. (1976). Ecological implications of resource depression. *American Naturalist*, **110**, 247–259.

Chavez, A. S. and Gese, E. M. (2006). Landscape use and movements of wolves in relation to livestock in a wildland-agriculture matrix. *Journal of Wildlife Management*, **70**, 1079–1086.

Chavez, A. S., Gese, E. M., Krannich, R. S., Ghavez, A. S. S. and Kranich, F. S. (2005). Attitudes of rural landowners toward wolves in northwestern Minnesota. *Wildlife Society Bulletin*, **33**, 517–527.

Cheeseman, C. L. and Mallinson, P. J. (1980). Radio tracking in the study of bovine tuberculosis in badgers. In C. J. Amlaner, Jr. and D. W. Macdonald (eds). *A Handbook on Biotelemetry and Radio Tracking*, pp. 649–656. Pergamon Press, Oxford.

Chen, M. T., Tewes, M. E., Pei, K. J. and Grassman, L. I. (2009). Activity patterns and habitat use of sympatric small carnivores in southern Taiwan. *Mammalia*, **73**, 20–26.

Chesson, J. (1978). Measuring preference in selective predation. *Ecology*, **59**, 211–215.

Christ, T. O., Guertin, D. S., Weins, J. A. and Milne, B. T. (1992). Animal movement in heterogeneous landscapes, an experiment with *Elodes* beetles in shortgrass prairie. *Functional Ecology*, **6**, 536–544.

Christ, A., Ver Hoef, J. and Zimmerman, D. L. (2008). An animal movement model incorporating home range and habitat selection. *Environmental and Ecological Statistics*, **15**, 27–38.

Christie, D. W. and Bell, E. T. (1971). Changes in rectal temperature during the normal oestrous cycle in the beagle bitch. *British Veterinary Journal*, **127**, 93–98.

Christman, M. C. (2004). Sequential sampling for rare and geographically clustered populations. In W. L. Thompson (ed.). *Sampling rare or elusive species: concepts, designs, and techniques for estimating population parameters*, pp. 134–148. Island Press, Washington DC.

Ciarniello, L. M., Boyce, M. S., Heard, D. C. and Seip, D. R. (2007). Components of grizzly bear habitat selection, density, habitats, roads, and mortality risk. *Journal of Wildlife Management*, **71**, 1446–1457.

Ciucci, P. and Boitani, L. (1998). Wolf and dog depredation on livestock in central Italy. *Wildlife Society Bulletin*, **26**, 504–514.

Ciucci, P., Boitani, L., Francisci, F. and Andreoli, G. (1997). Home range, activity and movements of a wolf pack in central Italy. *Journal of Zoology*, **243**, 803–819.

Ciucci, P., Masi, M. and Boitani, L. (2003). Winter habitat and travel route selection by wolves in the northern Apennines, Italy. *Ecography*, **26**, 223–235.

Ciucci, P., Tosoni, E. and Boitani, L. (2004). Assessment of the point-frame method to quantify wolf *Canis lupus* diet by scat analysis. *Wildlife Biology*, **10**, 149–153.

Ciucci, P., Reggioni, W., Maiorano, L. and Boitani, L. (2009). Long-distance dispersal of a rescued wolf from the northern Apennines to the western Alps. *Journal of Wildlife Management*, **73**, 1300–1306.

Ciucci, P., Boitani, L., Pelliccioni, E. R., Rocco, M. and Guy, I. (1996). A comparison of scat-analysis methods to assess the diet of the wolf *Canis lupus*. *Wildlife Biology*, **2**, 37–48.

Clark, T. W. and Wallace, R. L. (2002). Understanding the human factor in endangered species recovery, an introduction to human social process. *Endangered Species Update*, **19**, 87–94.

Clark, J. D., Dunn, J. E. and Smith, K. G. (1993). A multivariate model of female black bear habitat use for a geographic information system. *Journal of Wildlife Management*, **57**, 519–526.

Clark, J. D., Huber, D. and Servheen, C. (2002). Bear reintroductions, lessons and challenges. *Ursus*, **13**, 335–345.

Clarke, G. P., White, P. C. L. and Harris, S. (1998). Effects of roads on badger *Meles meles* populations in south-west England. *Biological Conservation*, **86**, 117–124.

Clayton, N. S., Reboreda J. C. and Kacelnik, A. (1997). Seasonal changes of hippocampus volume in parasitic cowbirds. *Behavioral Processes*, **41**, 237–243.

Clevenger, A. P., and Waltho, N. (2000). Factors influencing the effectiveness of wildlife underpasses in Banff National Park, Alberta, Canada. *Conservation Biology*, **14**, 47–56.

Clifford, D. L., Woodroffe, R., Garcelon, D. K., Timm, S. F. and Mazet, J. A. K. (2007). Using pregnancy rates and perinatal mortality to evaluate the success of recovery strategies for endangered island foxes. *Animal Conservation*, **10**, 442–451.

Clifford, D. L., Schumaker, B. A., Stephenson, T. R. *et al.* (2009). Assessing disease risk at the wildlife-livestock interface: a study of Sierra Nevada bighorn sheep. *Biological Conservation*, **142**, 2559–2568.

Cochran, W. G. (1977). *Sampling techniques*, 3rd edn. John Wiley and Sons, New York.

Cochran, W. W. (1980). Wildlife telemetry. In S. D. Schemnitz (ed.). *Wildlife management techniques manual*, 4th edn, pp. 507–520. The Wildlife Society. Washington DC.

Cochran, W. W. and Lord, Jr., R. D. (1963). A radio-tracking system for wild animals. *Journal of Wildlife Management*, **27**, 9–24.

Cockrem, J. F. (2005). Conservation and behavioral neuroendocrinology. *Hormones and Behavior*, **48**, 492–501.

Coelho, C. M., Bandeira de Melo, L. F., Sábato, M. A. L., Rizel, D. N. and Young, R. J. (2007). A note on the use of GPS collars to monitor wild maned wolves *Chrysocyon brachyurus* (Mammalia, Canidae). *Applied Animal Behaviour Science*, **105**, 259–264.

Colli, L., Cannas, R., Deiana, A. M., Gandolfi, G. and Tagliavini, J. (2005). Identification of mustelids (Carnivora: Mustelidae) by mitochondrial DNA markers. *Mammalian Biology*, **70**, 384–389.

Colwell, R. K. and Futuyma, D. J. (1971). On the measurement of niche breadth and overlap. *Ecology*, **52**, 567–576.

Compton, B. W., Rhymer, J. M. and McCollough, M. (2002). Habitat selection by wood turtles (*Clemmys insculpta*), an application of paired logistic regression. *Ecology*, **83**, 833–843.

Conaway, C. H. (1971). Ecological adaptation and mammalian reproduction. *Biology of Reproduction*, **4**, 239–247.

Concannon, P., Hansel, W. and Mcentree, E. (1977a). Changes in LH, progesterone and sexual behavior associated with preovulatory luteininzation in the bitch. *Biology of Reproduction*, **17**, 604–613.

Concannon, P. W., Powers, M. E., Holder, W. and Hansel, W. (1977b). Pregnancy and parturition in the bitch. *Biology of Reproduction*, **16**, 517–526.

Conley, R. H. and Jenkins, J. H. (1969). An evaluation of several techniques for determining the age of bobcat (*Lynx rufus*) in the Southeast. *Proceedings of the Annual Conference of the Southeast Association of Fish and Wildlife Agencies*, **23**, 104–109.

Conner, M. C. and Labisky, R. F. (1985). Evaluation of radioisotope tagging for estimating abundance of raccoon populations. *Journal of Wildlife Management*, **49**, 326–332.

Conner, M. M., Ebinger, M. R. and Knowlton, F. F. (2008). Evaluating coyote management strategies using a spatially explicit, individually-based, socially structured model. *Ecological Modelling*, **219**, 234–247.

Connolly, G. E. (1978). Predator control and coyote populations, a review of simulation models. In M. Bekoff (ed.). *Coyotes, biology, behavior, and management*, pp. 327–345. Academic Press, New York.

Connolly, G. E. (1993). Livestock protection collars in the United States, 1988–1993. *Proceedings of the Great Plains Wildlife Damage Control Workshop*, **11**, 25–33.

Conover, M. R. (2002). *Resolving human-wildlife conflicts, the science of wildlife damage management.* CRC Press, Boca Raton, Florida.

Cooch, E. G. and White, G. C. (2009). *Using MARK- a gentle introduction.* http//www.phidot.org/software/(accessed January 2011)

Cook, C. J. (2002). Rapid non-invasive measurement of hormones in transdermal exudate and saliva. *Physiology and Behavior*, **75**, 169–181.

Cooke, S. J., Hinch, S. G., Wikelski, M., Andrews, R. D., Kuchel, L. J., Wolcott, T. G. and Butler, P. J. (2004). Biotelemetry: a mechanistic approach to ecology. *Trends in Ecology and Evolution*, **19**, 334–343.

Copeland, J. P., Cesar, E., Peek, J. M., Harris, C. E., Long, C. D. and Hunter, D. L. (1995). A live trap for wolverine and other forest carnivores. *Wildlife Society Bulletin*, **23**, 535–538.

Coppinger, R. and Schneider, R. (1995). Evolution of working dogs. In J. Serpell (ed.). *The domestic dog*, pp. 21–47. Cambridge University Press, Cambridge.

Cormack, R. M. (1964). Estimates of survival from the sighting of marked animals. *Biometrika*, **51**, 429–438.

Corn, J. L. and Conroy, M. J. (1998). Estimation of density of mongooses with capture-recapture and distance sampling. *Journal of Mammalogy*, **79**, 1009–1015.

Corsi, F., de Leeuw, J. and Skidmore, A. K. (2000). Modeling species distribution with GIS. In L. Boitani and T. K. Fuller (eds). *Research techniques in animal ecology*, pp. 389–434. Columbia University Press, New York.

Costa, G. M., Chiarini-Garcia, H., Morato, R. G., Alvarenga, R. L. and Franca, L. R. (2008). Duration of spermatogenesis and daily sperm production in the jaguar (*Panthera onca*). *Theriogenology*, **70**, 1136–1146.

Costello, C. M., Inman, K. H., Jones, D. E., Inman, R. M., Thompson, B. C. and Quigley, H. B. (2004). Reliability of the cementum annuli technique for estimating age of black bears in New Mexico. *Wildlife Society Bulletin*, **32**, 169–176.

Cotterill, A. (1997). The economic viability of lions (*Panthera leo*) on a commercial wildlife ranch, examples and management implications for a Zimbabwean case study. In B. L. Penzhorn (ed.). *Lions and Leopards as Game Ranch Animals*, pp. 189–197. SAVWA, Onderstepoort, South Africa.

Coulter, M. W. (1966). *Ecology and management of fishers in Maine.* PhD Thesis, State University New York-ESF , NY.

Coy, P. L. and Garshelis, D. L. (1992). Reconstructing reproductive histories of black bears from incremental layering in dental cementum. *Canadian Journal of Zoology*, **70**, 2150–2160.

Coyne, M. S. and Godley, B. J. (2005). Satellite tracking and analysis tool (STAT): An integrated system for archiving, analyzing and mapping animal tracking data. *Marine Ecology Progress Series*, **301**, 1–7.

Crabb, W. D. (1941). A technique for trapping and tagging spotted skunks. *Journal of Wildlife Management*, **5**, 371–374.

Craft, M. E. (2008a). Capture and rapid handling of jackals (*Canis mesomelas* and *Canis adustus*) without chemical immobilization. *African Journal of Ecology*, **46**, 214–216.

Craft, M. E. (2008b). *Predicting disease dynamics in african lion populations*. PhD Thesis, University of Minnesota, Minneapolis, MN.

Craighead, J. J. and Craighead, F. C. (1956). *Hawks, owls and wildlife*. Stackpole, Harrisburg and Wildlife Management Institute, Washington DC.

Craighead, J. J. and Craighead, F. C. (1971). Grizzly bear - man relationships in Yellowstone National Park. *Bioscience*, **21**, 845–857.

Craighead, F. C., Craighead, J. J. and Davies, R. S. (1963). Radiotracking grizzly bears. In L. E. Slater (ed.). *Biotelemetry: the use of telemetry in animal behavior and physiology in relation to ecological problems*, pp. 133–148. Macmillan, New York.

Craighead, J. J., Hornocker, M. G. and Craighead Jr, F. C. (1969). Reproductive biology of young grizzly bears. *Journal of Reproduction and Fertility*, (Supplement) **6**, 447–475.

Crandall, K. A., Bininda-Emonds, O. R. P., Mace, G. M. and Wayne, R. K. (2000). Considering evolutionary processes in conservation biology. *Trends in Ecology and Evolution*, **15**, 290–295.

Crawford, K., McDonald, R. A. and Bearhop, S. (2008). Applications of stable isotope techniques to the ecology of mammals. *Mammal Review*, **38**, 87–107.

Crawshaw, G. J., Mills, K. J., Mosley, C. and Patterson, B. R. (2007). Field implantation of intraperitoneal radiotransmitters in eastern wolf (*Canis lycaon*) pups using inhalation anesthesia with sevoflurane. *Journal of Wildlife Diseases*, **43**, 711–718.

Creed, R. F. S. (1960). Observations on reproduction in the wild fox (*Vulpes vulpes*). An account with special reference to the occurrence of fox-dog crosses. *British Veterinary Journal*, **166**, 419–426.

Creel, S. (2005). Dominance, aggression, and glucocorticoid levels in social carnivores. *Journal of Mammalogy*, **86**, 255–264.

Creel, S., Creel, N., Wildt, D. E. and Monfort, S. L. (1992). Behavioural and endocrine mechanisms of reproductive suppression in Serengeti dwarf mongoose. *Animal Behaviour*, **43**, 231–245.

Creel, S., Fox, J. E., Hardy, A., Sands, J., Garrot, B. and Peterson, R. (2002). Snowmobile activity in glucocoricoid stress responses in wolves and elk. *Conservation Biology*, **16**, 809–814.

Creel, S., Spong, G., Sands, J. L. *et al.* (2003). Population size estimation in Yellowstone wolves with error-prone noninvasive microsatellite genotypes. *Molecular Ecology*, **12**, 2003–2009.

Creel, S., Christianson, D., Liley, S. and Winnie, J. A. (2007). Predation risk affects reproductive physiology and demography of elk. *Science*, **315**, 960–960.

Crockford, J. A., Hayes, F. A., Jenkins, J. H. and Feurt, S. D. (1957). Field application of nicotine salicylate for capturing deer. *Transactions North American Wildlife Conference*, **22**, 579–583.

Crompton, A. E., Obbard, M. E., Petersen, S. D. and Wilson, P. J. (2008). Population genetic structure in polar bears (*Ursus maritimus*) from Hudson Bay, Canada: Implications of future climate change. *Biological Conservation*, **141**, 2528–2539.

Crosbie, S. F. and Manly, B. F. J. (1985). Parsimonious modelling of capture-mark-recapture studies. *Biometrics*, **41**, 385–398.

Crowe, D. M. (1975). Aspects of aging, growth, and reproduction in bobcats from Wyoming. *Journal of Mammalogy*, **56**, 177–198.

Culver, M. (2009). Lessons and insights from evolution, taxonomy, and conservation genetics. In M. Hornocker and S. Negri (eds). *Cougar—ecology and conservation*, pp. 27–40. The University of Chicago Press, Chicago.

Culver, M., Johnson, W. E., Pecon-Slattery, J. and O'Brien, S. J. (2000). Genomic ancestry of the American puma (*Puma concolor*). *Journal of Heredity*, **91**, 186–197.

Cunningham, A. A. (1996). Disease risks of wildlife translocations. *Conservation Biology*, **10**, 349–353.

Curi, N. H. A. and Talamoni, S. A. (2006). Trapping, restraint and clinical-morphological traits of wild canids (Carnivora, Mammalia) from the Brazilian Cerrado. *Revista Brasileira de Zoologia*, **23**, 1–7.

Curio, E. (1976). *The ethology of predation*. Springer-Verlag, Berlin.

Currie, D. and Robertson, E. (1992). *Alberta wild fur management study guide*. Alberta Forestry, Lands and Wildlife, Fish and Wildlife, Edmonton, Alberta.

Cushman, S. A., McKelvey, K. S. and Schwartz, M. K. (2009). Use of empirically derived source-destination models to map regional conservation corridors. *Conservation Biology*, **23**, 368–376.

Czekala, N. M., Durrant, B. S., Callison, I., Williams, M. H. and Millard, S. (1994). Fecal steroid hormone analysis as an indicator of reproductive function in the cheetah. *Zoo Biology*, **13**, 119–128.

Dalerum, F. and Angerbjörn, A. (2005). Resolving temporal variation in vertebrate diets using naturally occurring stable isotopes. *Oecologia*, **144**, 647–658.

Dalziel, B. D., Morales J. M. and Fryxell, J. M. (2008). Fitting probability distributions to animal movement trajectories, using artificial neural networks to link distance, resources, and memory. *American Naturalist*, **172**, 248–258.

Dar, N. I., Minhas, R. A., Zaman, Q. and Linkie, M. (2009). Predicting the patterns, perceptions and causes of human-carnivore conflict in and around Machiara National Park, Pakistan. *Biological Conservation*, **142**, 2076–2082.

Darr, R. L. and Hewitt, D. G. (2008). Stable isotope trophic shifts in white-tailed deer. *Journal of Wildlife Management*, **72**, 1525–1531.

Darrow, P. A., Skirpstunas, R. T., Carlson, S. W. and Shivik, J. A. (2009). Comparison of injuries to coyote from 3 types of cable foot-restraints. *Journal of Wildlife Management*, **73**, 1441–1444.

Darwin, C. (1861). *On the origin of species*, 3rd edn. Murray, London.

Datta, A., Anand, M. O. and Naniwadekar, R. (2008). Empty forests: Large carnivore and prey abundance in Namdapha National Park, north-east India. *Biological Conservation*, **141**, 1429–1435.

Davenport, M. D., Tiefenbacher, S., Lutz, C. K., Novak, M. A. and Meyer, J. S. (2006). Analysis of endogenous cortisol concentrations in the hair of rhesus macaques. *General and Comparative Endocrinology*, **147**, 255–261.

Davis, J. L., Chestkiewicz, C.-L. B., Bleich, V. C. *et al.* (1996). A device to safely remove immobilized mountain lions from trees and cliffs. *Wildlife Society Bulletin*, **24**, 537–539.

Davis, M. L., Kelly, M. J. and Stauffer, D. S. (2011). Carnivore coexistence and habitat use in the Mountain Pine Ridge Forest Reserve, Belize. *Animal Conservation*, **14**, 56–65.

Davison, A., Birks, J. D. S., Brookes, R. C., Braithwaite, T. C. and Messenger, J. E. (2002). On the origin of faeces: morphological versus molecular methods for surveying rare carnivores from their scats. *Journal of Zoology*, **257**, 141–143.

Dawkins, M. S. (1998). Evolution and animal welfare. *Quarterly Review of Biology*, **73**, 305–328.

Day, G. I., Schemnitz, S. D. and Taber, R. D. (1980). Capturing and marking mammals. In S. D. Schemnitz (ed.). *Wildlife management techniques manual*, pp. 61–80. The Wildlife Society, Washington.

Day, M. G. (1966) Identification of hair and feather remains in the gut and faeces of stoats and weasels. *Journal of Zoology*, **148**, 201–217.

De Barba, M., Waits, L. P., Garton, O. E., Genovesi, P., Randi, E, Mustoni, A. and Groff, C. (2010). The power of genetic monitoring for studying demography, ecology, and genetics of a reintroduced brown bear population. *Molecular Ecology*, **19**, 3938–3951.

De Haas Van Dorsser, F. J., Swanson, W. F., Lasano, S. and Steinetz, B. G. (2006). Development, validation, and application of a urinary relaxin radioimmunoassay for the diagnosis and monitoring of pregnancy in felids. *Biology of Reproduction*, 74, 1090–1095.

De Haas van Dorsser, F. J., Lasano, S., and Steinetz, B. G. (2007). Pregnancy diagnosis in cats using a rapid, bench top kit to detect relaxin in urine. *Reproduction in Domestic Animals*, **42**, 111–112.

De Luca, D. W. and Ginsberg, J. R. (2001). Dominance, reproduction and survival in banded mongooses, towards an egalitarian social system? *Animal Behaviour*, **61**, 17–30.

de Waal, H. O., Combrinck, W. J. and Borstlap, D. G. (2004). A comprehensive procedure to measure the body dimensions of large African predators with comments on the repeatability of measurements taken from an immobilized African lion (*Panthera leo*). *Journal of Zoology*, **262**, 393–398.

Deal, B., Farello, C., Lancaster, M., Kompare, T. and Hannon, B. (2000). A dynamic model of the spatial spread of an infectious disease: the case of fox rabies in Illinois. *Environmental Modeling and Assessment*, **5**, 47–62.

Deanesly, R. (1944). The reproductive cycle of the female weasel (*Mustela nivalis*). *Proceedings of the Zoological Society of London*, **114**, 339–349.

Debrot, S. (1984). Dynamique du renouvellement et structure d'âge d'une population d'hermines (*Mustela herminea*). *La Terre et la Vie – Revue d'Écologie*, **39**, 77–88.

DeCesare, N. J., Squires, J. R. and Kolbe, J. A. (2005). Effect of forest canopy on GPS-based movement data. *Wildlife Society Bulletin*, **33**, 935–941.

Dehnhard, M., Hildebrandt, T. B., Knauf, T., Jewgenow, K., Kolter L. and Goritz, F. (2006). Comparative endocrine investigations in three bear species based on urinary steroid metabolites and volatiles. *Theriogenology*, **66**, 1755–1761.

Dehnhard, M., Naidenko, S., Frank, A., Braun, B., Goritz, F. and Jewgenow, K. (2008). Noninvasive monitoring of hormones: a tool to improve reproduction in captive breeding of the Eurasian lynx. *Reproduction of Domestic Animals*, **43** (Suppl. 2), 74–82.

DeMatteo, K. E., Porton, I. J., Kleiman, D. G. and Asa, C. S. (2006a). The effect of the male bush dog (*Speothos venaticus*) on the female reproductive cycle. *Journal of Mammalogy*, **87**, 723–732.

DeMatteo, K. E., Porton, I. J., Kleiman, D. G., and Asa, C. S. (2006b). The effect of the male bush dog (*Speothos venaticus*) on the female reproductive cycle. *Journal of Mammalogy*, **87**, 723–732.

DeMatteo, K. E., Rinas, M. A., Sede, M. M. *et al.* (2009). Detection dogs: an effective technique for bush dog surveys. *Journal of Wildlife Management*, **73**, 1436–1440.

Demma, D. J. and Mech, L. D. (2009). Wolf use of summer territory in Northeastern Minnesota. *Journal of Wildlife Management*, **73**, 380–384.

Demma, D. J., Barber-Meyer, S. M. and Mech, L. D. (2007). Testing global positioning system telemetry to study wolf predation on deer fawns. *Journal of Wildlife Management*, **71**, 2767–2775.

DeNiro, M. J. and Epstein, S. (1978). Influence of diet on the distribution of carbon isotopes in animals. *Geochimica et Cosmochimica Acta*, **42**, 495–506.

D'Eon, R. G. (2003). Effects of a stationary GPS fix-rate bias on habitat-selection analyses. *Journal of Wildlife Management*, **67**, 858–863.

D'Eon, R. G. and Delparte, D. (2005). Effects of radio-collar position and orientation on GPS radio-collar performance, and the implications of PDOP in data screening. *Journal of Applied Ecology*, **42**, 383–388.

D'Eon, R. G., Serrouya, R., Smith, G. and Kochanny, C. O. (2002). GPS radiotelemetry error and bias in mountainous terrain. *Wildlife Society Bulletin*, **30**, 430–439.

Derocher, A. E. and Stirling, I. (1998). Maternal investment and factors affecting offspring size in polar bears (*Ursus maritimus*). *Journal of Zoology*, **245**, 253–260.

DeVan, R. (1982). *The ecology and life history of the long-tailed weasel* (Mustela frenata). PhD Thesis, University of Cincinnati, Cincinnati, Ohio.

Devery, S. (2010). *A retrospective study of pyometra and endometrial hyperplasia in captive African wild dogs* (Lycaon pictus), *bush dogs* (Speothos venaticus) *and fennec foxes* (Vulpes zerda) *in North America*. Master of Science, University of London.

Di Bitetti, M. S., Paviolo, A. and De Angelo, C. (2006). Density, habitat use and activity patterns of ocelots (*Leopardus pardalis*) in the Atlantic Forest of Misiones, Argentina. *Journal of Zoology*, **270**, 153–163.

Di Bitetti, M. S., Di Blanco, Y. E., Pereira, J. A., Paviolo, A. and Perez, I. J. (2009). Time partitioning favors the coexistence of sympatric crab-eating foxes (*Cerdocyon thous*) and pampas foxes (*Lycalopex gymnocercus*). *Journal of Mammalogy*, **90**, 479–490.

Di Orio, A. P., Callas, R. and Schaefer, R. J. (2003). Performance of two GPS telemetry collars under different habitat conditions. *Wildlife Society Bulletin*, **31**, 372–379.

Dickman, A. J. (2010). Complexities of conflict: the importance of considering social factors for effectively resolving human-wildlife conflict. *Animal Conservation*, **13**, 458–466.

Dickson, B. G., Jenness, J. S. and Beier, P. (2005). Influence of vegetation, topography, and roads on cougar movement in Southern California. *Journal of Wildlife Management*, **69**, 264–276.

Diefenbach, D. R. and Alt, G. L. (1998). Modelling and evaluation of ear tag loss in black bears. *Journal of Wildlife Management*, **62**, 1292–1300.

Dillon, A., and Kelly, M. J. (2007). Ocelot (*Leopardus pardalis*) in Belize: the impact of trap spacing and distance moved on density estimates. *Oryx*, **41**, 469–477.

Dillon, A., and Kelly, M. J. (2008). Ocelot home range, overlap and density: comparing radio telemetry with camera trapping. *Journal of Zoology*, **275**, 391–398.

Dinerstein, E., Loucks, C., Heydlauff, A. *et al.* (2006). *Setting priorities for the conservation and recovery of wild tigers: 2005–2015. A user's guide.* WWF, WCS, Smithsonian, and NFWF-STF, Washington DC. and New York.

Dische, Z., Borenfreund, E. and Zelmenis, G. (1956). Changes in lens proteins in rats during aging. *Archives of Ophthalmology*, **55**, 471–483.

Doncaster, C. P. (1990). Non-parametric estimates of interaction from radio-tracking data. *Journal of Theoretical Biology*, **143**, 431–443.

Doncaster, C. P. and Macdonald, D. W. (1991). Drifting territoriality in the red fox *Vulpes vulpes*. *Journal of Animal Ecology*, **60**, 423–439.

Doncaster, C. P. and Macdonald, D. W. (1992). Optimum group size for defending heterogeneous distributions of resources; a model applied to red foxes, *Vulpes vulpes*, in Oxford City. *Journal of Theoretical Biology*, **159**, 189–198.

Doncaster, C. P. and Macdonald, D. W. (1996). Intra-specific variation in movement behaviour of foxes (*Vulpes vulpes*). A reply to White, Saunders and Harris. *Journal of Animal Ecology*, **65**, 126–127.

Doncaster, C. P. and Woodroffe, R. (1993). Den site can determine shape and size of badger territories, implications for group-living. *Oikos*, **66**, 88–93.

Dorazio, R. M. and Royle, J. A. (2003). Mixture models for estimating the size of a closed population when capture rates vary between individuals. *Biometrics*, **59**, 351–364.

Drew, M. L. (2006). Wildlife issues. In C. K. Baer (ed), *Guidelines for euthanasia of nondomestic animals*, pp. 19–22. American Association of Zoo Veterinarians,Yulee, FL.

Drewe, J. A. (2010). Who infects whom? Social networks and tuberculosis transmission in wild meerkats. *Proceedings of the Royal Society B*, **277**, 633–642.

Drewe, J. A., Madden, J. R. and Pearce, G. P. (2009). The social network structure of a wild meerkat population, 1. Inter-group interactions. *Behavioral Ecology and Sociobiology*, **63**, 1295–1306.

Durant, S. M. (1998). Competition refuges and coexistence, an example from Serengeti carnivores. *Journal of Animal Ecology*, **67**, 370–386.

Durner, G. M. and Amstrup, S. C. (1996). Mass and body dimension relations of polar bears in northern Alaska. *Wildlife Society Bulletin*, **24**, 480–484.

Durrant, B. S., Olson, A. M., Amodeo, D. *et al.* (2003). Vaginal cytology and vulvar swelling as indicators of impending estrus and ovulation in the giant panda (*Ailuropoda melanoleuca*). *Zoo Biology*, **22**, 313–321.

Durrant, B. S., Ravida, N., Spady, T. and Cheng, A. (2006). New technologies for the study of carnivore reproduction. *Theriogenology*, **66**, 1729–1736.

Dussault, C., Courtois, R., Ouellet, J.-P. and Huot, J. (1999). Evaluation of GPS telemetry collar performance for habitat studies in the boreal forest. *Wildlife Society Bulletin*, **27**, 965–972.

Eagle, T. C., Choromanski-Norris, J. and Kuechle, V. B. (1984). Implanting radio transmitters in mink and Franklin's ground squirrels. *Wildlife Society Bulletin*, **12**, 180–184.

Earle, R. D., Lining, D. M., Tuvalu, V. R. and Shivik, J. A. (2003). Evaluating injury mitigation and performance of # 3 Soft Catch® trap to restrain bobcats. *Wildlife Society Bulletin*, **31**, 617–629.

Eason, T. H., Smith, B. H. and Pelton, M. R. (1996). Research variation in collection of morphometrics on black bears. *Wildlife Society Bulletin*, **24**, 485–489.

Ebling, F. J. P. and Foster, D. L. (1989). Seasonal breeding- A model for puberty. In H. A. Delemarre-Van de Waal, T. M. Plant, G. P. Van Rees and J. Shoemaker (eds). *Control of the onset of puberty III*, pp. 253–264. Elsevier, Amsterdam.

Edwards M. A., Nagy, J. A. and Derocher, A. E. (2009). Low site fidelity and home range drift in a wide-ranging, large Arctic omnivore. *Animal Behaviour*, 77, 23–28.

Efford, M. (2004). Density estimation in live-trapping studies. *Oikos*, **106**, 598–610.

Eiler, H. and Oliver, J. (1980). Combined dexamethasone suppression and Cosyntropin (synthetic ACTH) stimulation test in the dog: new approach to testing of adrenal gland function. *American Journal of Veterinary Research*, **41**, 1243–1246.

Eilts, B. E., Williams, D. B. and Moser, E. B. (1993). Ultrasonic measurement of canine testes. *Theriogenology*, **40**, 819–828.

Eleraky, N. Z., Potgieter, L. N. D. and Kennedy, M. A. (2002). Virucidal efficacy of four new disinfectants. *Journal of the American Animal Hospital Association*, **38**, 231–234.

Elith, J. and Leathwick, J.R. (2009). Species distribution models: ecological explanation and prediction across space and time. *Annual Review of Evolution Ecology and Systematics*, **40**, 677–697.

Elith, J., Graham, C. H, Anderson, R. P. *et al.* (2006). Novel methods improve prediction of species' distributions from occurrence data. *Ecography*, **29**, 129–151.

Elkaim, G. H., Decker, E. B., Oliver G. and Wright, B. (2006). Go deep marine mammal marker for at-sea monitoring. *GPS World*, **17**, 30–33.

Ellison, A.M. (2004). Bayesian inference in ecology. *Ecology Letters*, 7, 509–520.

Ellner, S. and Real, L. A. (1989). Optimal foraging models for stochastic environments; are we missing the point? *Comments Theoretical Biology*, **1**, 129–158.

Elmeros, M. and Hammershoj, M. (2006). Experimental evaluation of the reliability of placental scar counts in American mink (*Mustela vison*). *European Journal of Wildlife Research*, **52**, 132–135.

Elzinga, C. L., Salzer, D. W., Willoughby, J. W. and Gibbs, J. P. (2001). *Monitoring plant and animal populations*. Blackwell Science, Malden, Massachusetts.

Enders, R. K. and Enders, A. C. (1963). Morphology of the female reproductive tract during delayed implantation in the mink. In A. C. Enders (ed.). *Delayed implantation*, pp. 129–140. University of Chicago Press, Chicago, IL.

Enders, R. K. and Leekley, J. R. (1941). Cyclic changes in the vulva of the marten (*Martes americana*). *Anatomical Record*, **79**, 1–5.

Englund, J. (1970). Some aspects of reproduction and mortality rates in Swedish foxes (*Vulpes vulpes*), 1961–1963 and 1966–1969. *Viltrevy*, **8**, 1–82.

Englund, J. (1982). A comparison of injuries to leghold trapped and foot snared red foxes. *Journal of Wildlife Management*, **46**, 1113–1120.

Ensley, P. K., Wing, A. E., Gosenk, B. B., Lasley, B. L. and Durrant, B. (1982). Application of noninvasive techniques to monitor reproductive function in a brown hyena (*Hyena brunnea*). *Zoo Biology*, **1**, 333–343.

Ericsson, G. and Heberlein, T. A. (2003). Attitudes of hunters, locals, and the general public in Sweden now that the wolves are back. *Biological Conservation*, **111**, 149–159.

Erlinge, S. and Sandell, M. (1986). Seasonal changes in the social organization of male stoats, *Mustela erminea*. An effect of shifts between two decisive resources. *Oikos*, 47, 57–62.

Ernest, H. B. and Boyce, W. M. (2000). DNA identification of mountain lions involved in livestock predation and public safety incidents and investigations. *Proceedings of the Vertebrate Pest Conference*, **19**, 290–294.

Ernest, H. B., Rubin, E. S. and Boyce, W. M. (2002). Fecal DNA analysis and risk assessment of mountain lion predation of bighorn sheep. *Journal of Wildlife Management*, **66**, 75–85.

Errington, P. L. (1943). *An analysis of mink predation on muskrats in North-Central United States*. Iowa Agricultural Experiment Station, Research Bulletin 320. Ames, IA.

Espuno, N., Lequette, B., Poulle, M. L., Migot, P. and Lebreton, J. D. (2004). Heterogeneous response to preventive sheep husbandry during wolf recolonization of the French Alps. *Wildlife Society Bulletin*, **32**, 1195–1208.

European Community, Government of Canada, and Government of the Russian Federation. (1997). Agreement on international humane trapping standards. *Official Journal of the European Communities*, **42**, 43–57.

European Union. (1986). European Directive 8/609/EEC of 24 November 1886 on the approximation of laws, regulations and administrative provisions of the member states regarding the protection of animals used for experimental and other scientific purposes. *Official Journal of European Communities L*, **358**, 18/12/1986.

Ewer, R. F. (1968). *Ethology of mammals*. Logos Press, London.

Ewer, R. F. (1973). *The carnivores*. Cornell University Press, Ithaca, NY.

Fagen, R. M. and Young, D. Y. (1978). Temporal patterns of behavior: durations, intervals, latencies, and sequences. In P. Colgen (ed.). *Quantitative ethology*, pp. 79–114. John Wiley and Sons, New York.

Fagerstone, K. A. and Jones, B. E. (1987). Transponders as permanent identification markers for domestic ferrets, black-footed ferrets, and other wildlife. *Journal of Wildlife Management*, **51**, 294–297.

Fahlman, Å., Arnemo, J.M., Persson, J., Segerström, P. and Nyman, G. (2008). Capture and medetomidine-ketamine anesthesia of free-ranging wolverines (*Gulo gulo*). *Journal of Wildlife Diseases*, **44**, 133–142.

Falcucci, A., Ciucci, P., Maiorano, L., Gentile L. and Boitani, L. (2009). Assessing habitat quality for conservation using an integrated occurrence-mortality model. *Journal of Applied Ecology*, **46**, 600–609.

Farias, V., Fuller, T. K., Wayne, R. K. and Sauvajot, R. M. (2005). Survival and cause-specific mortality of gray foxes (*Urocyon cinereoargenteus*) in southern California. *Journal of Zoology*, **266**, 249–254.

Farstad, W. (1996). Semen cryopreservation in dogs and foxes. *Animal Reproduction Scince*, **42**, 251–260.

Farstad, W. (2000). Current state in biotechnology in canine and feline reproduction. *Animal Reproduction Science*, **60–61**, 375–388.

Favreau, J. M. (2006). *The effects of food abundance, foraging rules and cognitive abilities on local animal movements*. PhD thesis, North Carolina State University, Raleigh, NC.

Federal Provincial Committee for Humane Trapping. (1981). *Final report*. Committee of the Federal Provincial Committee for Humane Trapping (FPCHT), Ottawa, Ontario.

Federoff, N. E. (2001). Antibody response to rabies vaccination in captive and free- ranging wolves (*Canis lupus*). *Journal of Zoo and Wildlife Medicine*, **32**, 127–129.

Fernández, N., Palomares, F. and Delibes, M. (2002). The use of breeding dens and kitten development in the Iberian lynx (*Lynx pardinus*). *Journal of Zoology*, **258**, 1–5.

Fernández, N., Delibes, M., Palomares, F. and Mladenoff, D. J. (2003). Identifying breeding habitat for the Iberian Lynx, Inferences from a fine-scale spatial analysis. *Ecological Applications*, **13**, 1310–1324.

Fernández, N., Delibes, M. and Palomares, F. (2006). Landscape evaluation in conservation, molecular sampling and habitat modelling for the Iberian lynx. *Ecological Applications*, **16**, 1037–1049.

Fernández-Morán, J., Saavadra, D. and Manteca-Vilanova, X. (2002). Reintroduction of the Eurasian otter (*Lutra lutra*) in northeastern Spain, trapping, handling, and medical management. *Journal of Zoo and Wildlife Medicine*, **33**, 222–227.

Ferrari, P. J. and Kiss, A. (2002). Direct payments to conserve biodiversity. *Science*, **298**, 1718–1719.

Festa-Bianchet, M. (2009). The walia ibex is a valuable and distinct conservation unit. *Animal Conservation*, **12**, 101–102.

Fieberg, J. (2007a). Utilization distribution estimation with weighted kernel density estimators. *Journal of Wildlife Management*, **71**, 1669–1675.

Fieberg, J. (2007b). Kernel density estimators of home range, smoothing and the autocorrelation red herring. *Ecology*, **88**, 1059–1066.

Fieberg, J. and Börger, L. (In press). *Journal of Mammalogy*.

Field, S. A., Tyre, A. J. and Possingham, H. P. (2005). Optimizing allocation of monitoring effort under economic and observational constraints. *Journal of Wildlife Management*, **69**, 473–482.

Fielding, A. H. and Bell, J. F. (1997). A review of methods for the assessment of prediction errors in conservation presence/absence models. *Environmental Conservation*, **24**, 38–49.

Fiorello, C. V., Cunningham, M. W., Cantwell, S. L., Levy, J. K., Neer, E. M., Conley, K. and Rist, P. M. (2007). Diagnosis and treatment of presumptive postobstructive pulmonary edema in a Florida panther (*Puma concolor coryi*). *Journal of Zoo and Wildlife Medicine*, **38**, 317–322.

Fitzjohn, T. R., Kuiken, T., Lema, S. *et al.* (2002). Distemper outbreak and its effect on African wild dog conservation. *Emerging Infectious Diseases*, **8**, 211–213.

Fleming, P. J. S., Allen, L. R., Berghout, M. J. *et al.* (1998). The performance of wild-canid traps in Australia, efficiency, selectivity and trap-related injuries. *Wildlife Research*, **25**, 327–338.

Fletcher, T. J. (1974). The timing of reproduction in red deer (*Cervus elaphus*) in relation to altitude. *Journal of Zoology*, **172**, 363–367.

Flick L., Mitchell, D. and Fuller, A. (2007). Long-acting neuroleptics used in wildlife management do not impair thermoregulation or physical activity in goats (*Capra hircus*). *Comparative Biochemistry and Physiology, Part A*, **147**, 445–452.

Foley, C. A. H., Papageorge, S. and Wasser, S. K. (2001). Non-invasive stress and reproductive measures of social and ecological pressures in free-ranging African elephants. *Conservation Biology*, **15**, 1134–1142.

Follmann, E. H., Manning, A. E. and Stuart, J. L. (1982). A long-range implantable heart rate transmitter for free-ranging animals. *Biotelemetry and Patient Monitoring*, **9**, 205–212.

Follman, E. H., Savarie, P. J., Ritter, D. G. and Baer, G. M. (1987). Plasma marking of arctic foxes with iophenoxic acid. *Journal of Wildlife Diseases*, **23**, 709–712.

Folse L. J., Packard, J. M. and Grant, W. E. (1989). AI modelling of animal movements in a heterogeneous habitat. *Ecological Modelling*, **46**, 57–72.

Foote, R. H., Swierstra, E. E. and Hunt, W. L. (1972). Spermatogenesis in the dog. *Anatomical Record*, **173**, 341–352.

Forcada, J. and Aguilar, A. (2000). Use of photographic identification in capture-recapture studies of Mediterranean monk seals. *Marine Mammal Science*, **16**, 767–793.

Forero-Medina, G., Viniciu Vieira, M., de Viveiros Grelle, C. E. and Almeidal, P. J. (2009). Body size and extinction risk in Brazilian carnivores. *Biota Neotropioca*, **9**, 45–49.

Foresman, K. R.(2001). *Key to the mammals of Montana*. University of Montana Bookstore, Missoula, Montana.

Forester, J. D., Im, H. K. and Rathouz, P. J. (2009). Accounting for animal movement in estimation of resource selection functions, sampling and data analysis. *Ecology*, **90**, 3554–3565.

Foreyt, W. J. and Rubenser, A. (1980). A live trap for multiple captures of coyote pups from dens. *Journal of Wildlife Management*, **44**, 487–488.

Forman, D. W. and Williamson, K. (2005). An alternative method for handling and marking small carnivores without the need of sedation. *Wildlife Society Bulletin*, **33**, 313–316.

Fortin, D., Beyer, H. L., Boyce, M. S., Smith, D. W., Duchesne, T. and Mao, J. S. (2005) Wolves influence elk movements: behavior shapes a throphic cascade in Yellowstone National Park. *Ecology*, **86**, 1320–1330.

Fourli, M. (1999). *Compensation for damage caused by bears and wolves in the European Union*. European Commission (D. G. Environment), Bruxelles.

Fournier-Chambrillon, C., Berny, P. J., Coiffier, O. *et al.* (2004). Evidence of secondary poisoning of free-ranging riparian mustelids by anticoagulant rodenticides in France: implications for conservation of European mink (*Mustela lutreola*). *Journal of Wildlife Diseases*, **40**, 688–695.

Fowler, M. E. (2008). *Restraint and Handling of Wild and Domestic Animals*, 3rd ed. Blackwell Publishing, Williston, Vermont.

Frafjord, K. and Prestrud, P. (1992). Home range and movements of arctic foxes *Alopex lagopus* in Svalbard. *Polar Biology*, **12**, 519–526.

Frair, J. L., Nielsen, S. E., Merrill, E. H. *et al.* (2004). Removing GPS collar bias in habitat selection studies. *Journal of Applied Ecology*, **41**, 201–212.

Frair, J. L., Fieberg, J., Hebblewhite, M., Cagnacci, F., DeCesare, N. J. and Pedrotti, L. (2010). Resolving issues of imprecise and habitat-biased locations in ecological analyses using GPS telemetry data. *Philosophical Transactions of The Royal Society B*, **365**, 2187–2200.

Frame, P. F. and Meier, T. J. (2007). Field-assessed injury to wolves captured in rubber-padded traps. *Journal of Wildlife Management*, **71**, 2074–2076.

Franceschini, M. D., Rubenstein, D. I. Low, B. and Romero, L. M. (2008). Fecal glucocorticoid metabolite analysis as an indicator of stress during translocation and acclimation in an endangered large mammal, the Grevy's zebra. *Animal Conservation*, **11**, 263–269.

Frank, L., Simpson, D. and Woodroffe, R. (2003). Foot snares, an effective method for capturing African lions. *Wildlife Society Bulletin*, **31**, 309–314.

Frank, L., Maclennan, S., Hazzah, L. *et al.* (2006). *Lion killing in the Amboseli-Tsavo ecosystem, 2001–2006, and its implications for Kenya's lion population.* Project report, Living with lions www.lionconservation.org. 9 pp.

Franke, A., Caelli, T., Kuzyk, G. and Hudson, R. J. (2006) Prediction of wolf (*Canis lupus*) kill-sites using hidden Markov models. *Ecological Modelling,* **197**, 237–246.

Frankham, R. (2005). Genetics and extinction. *Biological Conservation,* **126**, 131–140.

Frankham, R., Ballou, J. D. and Briscoe, D. A. (2002). *Introduction to conservation genetics.* Cambridge University Press, Cambridge.

Franklin, I. R. (1980). Evolutionary changes in small populations. In M. E. Soulé and B. A. Wilcox (eds). *Conservation Biology: an Evolutionary-Ecological Perspective,* pp. 135–149. Sinauer, Sunderland MA.

Franklin, S. E., Stenhouse, G. B., Hansen, M. J., Popplewell, C. C., Dechka, J. A. and Peddle, D. R. (2001). An integrated decision tree approach (IDTA) to mapping land-coverusing satellite remote sensing in support of grizzly bear habitat analysis in the Alberta Yellowhead Ecosystem. *Canadian Journal of Remote Sensing,* **27**, 579–592.

Franklin, S. E., Peddle, D. R., Dechka, J. A. and Stenhouse, G. B. (2002). Evidential reasoning with Landsat TM, DEM and GISdata for land cover classification in support of grizzly bear habitat mapping. *International Journal of Remote Sensing,* **23**, 4633–4652.

Frederick, C., Kyes, R., Hunt, K., Collins, D., Durrant, B. and Wasser, S. K. (2010). Methods of estrus detection and correlates of the reproductive cycle in the sun bear (*Helarctos malayanus*). *Theriogenology,* **74**, 1121–1135.

Fredrickson, L. F. (1983). Use of radiographs to age badger and striped skunk. *Wildlife Society Bulletin,* **11**, 297–299.

Freeman, M. M. R. and Wenzel, G. W. (2006). The nature and significance of polar bear conservation hunting in the Canadian Arctic. *Arctic,* **59**, 21–30.

Freitag, S., Hobson, C., Biggs, H. C. and Van Jaarsveld, A. S. (1998). Testing for potential survey bias: the effect of roads, urban areas and nature reserves on a southern african mammal data set. *Animal Conservation,* **1**, 119–127.

Freitas, C., Kovacs, K. M., Lydersen, C. and Ims, R. A. (2008). A novel method for quantifying habitat selection and predicting habitat use. *Journal of Applied Ecology,* **45**, 1213–1220.

Fretwell, S. D. (ed.). (1972). *Populations in a seasonal environment.* Princeton University Press, Princeton.

Fretwell, S. D. and Lucas, J. H. J. (1970). On territorial behavior and other factors influencing habitat distribution in birds. *Acta Biotheoretica,* **19**, 16–36.

Frid, A. L. E. J. (1997). Vigilance by female Dall's sheep: interactions between predation risk factors. *Animal Behaviour,* **53**, 799–808.

Friend, M. (1967). A review of research concerning eye-lens weight as a criterion of age in mammals. *New York Fish and Game Journal,* **14**, 152–165.

Fritts, S. H., Stephenson, R. O., Hayes R. D. and Boitani, L. (2003). Wolves and humans. In L. D. Mech and L. Boitani (eds). *Wolves. Behaviour, ecology and conservation,* pp. 289–316. University of Chicago Press, Chicago, IL.

Frost, H. C. and Krohn, W. B. (1994). *Capture, care, and handling of fishers (Martes pennanti).* Technical Bulletin, Maine Agriculture and Forest Experiment Station, University of Maine, Orono, ME.

Fryxell, J. M. and Lundberg, P. (eds). (1997). *Individual behavior and community dynamics.* Chapman and Hall, New York.

Fuglei, E., Mercer, J. B. and Arnemo, J. M. (2002). Surgical implantation of radio transmitters in arctic foxes (*Alopex lagopus*) on Svalbard, Norway. *Journal of Zoo and Wildlife Medicine*, **33**, 342–349.

Fujiwara, M. and H., Caswell. (2002). A general approach to temporary emigration in mark-recapture analysis. *Ecology*, **83**, 3266–3275.

Fuller, A. K. and Harrison, D. J. (2010). Movement paths reveal scale-dependent habitat decisions by Canada lynx. *Journal of Mammalogy*, **91**, 1269–1279.

Fuller, T. K. and Sievert, P. R. (2001). Carnivore demography and the consequences of changes in prey availability. In J. L. Gittleman, S. M. Funk, D. MacDonald, and R. K. Wayne (eds). *Carnivore Conservation*, pp. 163–178. Cambridge University Press, Cambridge.

Fuller, T. K., Biknevicius, A. R., Kat, P. W., Van Valkenburgh, B. and Wayne, R. K. (1989). The ecology of three sympatric jackal species in the Rift Valley of Kenya. *African Journal of Ecology*, **27**, 313–323.

Fuller, T. K., Kat, P. W., Bulger, J.B. *et al.* (1992). Population dynamics of African wild dogs. In D. R. McCullough and R. H. Barrett (eds). *Wildlife 2001, populations*, pp. 1125–1139. Elsevier Applied Science, London.

Fuller, T. K., Mech, L. D. and Cochrane, J. F. (2003). Wolf population dynamics. In L. D. Mech and L. Boitani (eds). *Wolves, behavior, ecology and conservation*, pp. 161–190. University of Chicago Press, Chicago, IL.

Fuller, M. R, Millspaugh, J. J., Church, K. E. and Kenward, R. E. (2005). Wildlife radio telemetry. In C. E. Braun (ed.). *Techniques for wildlife investigations and management*, 6th edn, pp. 377–417. The Wildlife Society, Bethesda, Maryland.

Funk, V. A. and Richardson, K.S. (2002). Systematic data in biodiversity studies: use it or lose it. *Systematic Biology*, **51**, 303–316.

Fyhn, M., Molden, S., Witter, M. P., Moser, E. I. and Moser, M-B. (2004). Spatial representation in the entorhinal cortex. *Science*, **305**, 1258–1264.

Gabriel, M. W., Wengert, G. M., Matthews, S. M. *et al.* (2010). Effectiveness of rapid diagnostic tests to assess pathogens of fishers (*Martes pennanti*) and gray foxes (*Urocyon cinereoargenteus*). *Journal of Wildlife Diseases*, **46**, 966–970.

Galea L. A. M., Kavaliers, M. and Ossenkopp, K-P. (1996). Sexually dimorphic spatial learning in meadow voles *Microtus pennsylvanicus* and deer mice *Peromyscus maniculatus*. *The Journal of Experimental Biology*, **199**, 195–200.

Ganan, N., Gomendio, M. and Roldan, E. R. (2009). Effect of storage of domestic cat (*Felis catus*) epididymides at 5 degrees C on sperm quality and cryopreservation. *Theriogenology*, **72**, 1268–1277.

Garcelon, D. K. (1977). An expandable drop-off transmitter collar for young mountain lions. *California Fish and Game*, **63**, 185–189.

Gardner, B., Royle, J. A. and Wegan, M. T. (2009). Hierarchical models for estimating density from DNA mark-recapture studies. *Ecology*, **90**, 1106–1115.

Gardner, B., Reppucci, J., Lucherini, M. and Royle, J. A. (2010). Spatially explicit inference for open populations, estimating demographic parameters for camera-trap studies. *Ecology*, **91**, 3376–3383.

Garnier, J. N., Holt, W. V. and Watson, P. F. (2002). Non-invasive assessment of oestrous cycles and evaluation of reproductive seasonality in the female wild black rhinoceros (*Diceros bicornis minor*). *Reproduction*, **123**, 877–889.

Garrott, R. A., Gude, J. A., Bergman, E. J., Gower, C., White, P. J. and Hamlin, K. L. (2005) Generalizing wolf effects across the Greater Yellowstone Area, a cautionary note. *Wildlife Society Bulletin*, **33**, 1245–1255.

Garshelis, D. L. (2000). Delusions in habitat evaluation: measuring use, selection and importance. In L. Boitani and T. K. Fuller (eds). *Research techniques in animal ecology*, pp. 111–164. Columbia University Press, New York.

Garshelis, D. L. and McLaughlin, C. R. (1998). Review and evaluation of breakaway devices for bear radiocollars. *Ursus*, **10**, 459–465.

Garshelis, D. L. and Pelton, M. R. (1980). Activity of black bears in the Great Smoky Mountains National Park. *Journal of Mammalogy*, **61**, 8–19.

Garshelis, D. L. and Pelton, M. R. (1981). Movements of black bears in the Great Smoky Mountains National Park. *Journal of Wildlife Management*, **45**, 912–925.

Garshelis, D. L. and Visxser, L. G. (1997). Enumerating megapopulations of wild bears with an ingested biomarker. *Journal of Wildlife Management*, **61**, 466–480.

Garshelis, D. L., Quigley, H. B., Villarrubia, C. R. and Pelton, M. R. (1982). Assessment of telemetric motion sensors for studies of activity. *Canadian Journal of Zoology*, **60**, 1800–1805.

Gaston, K. J. (1991). How large is a species geographic range? *Oikos*, **61**, 434–438.

Gaston, K. J. (1993). *Rarity*. Chapman and Hall, London.

Gaston, K. J. (2003). *The structure and dynamics of geographic ranges*. Oxford University Press, Oxford.

Gau, R. J., Mulders, R., Ciarniello, L. J. *et al.* (2004). Uncontrolled field performance of Televilt GPS-Simplex (TM) collars on grizzly bears in western and northern Canada. *Wildlife Society Bulletin*, **32**, 693–701.

Gautestad, A. O. and Mysterud, I. (1993). Physical and biological mechanisms in animal movement processes. *Journal of Applied Ecology*, **30**, 523–535.

Gautestad, A. O. and Mysterud, I. (1995). The home range ghost. *Oikos*, **74**, 195–204.

Gautestad, A. O., Mysterud I. and Pelton, M. R. (1998). Complex movement and scale-free habitat use. Testing the multiscale home range model on black bear telemetry data. *International Conference on Bear Research and Management*, **10**, 219–234.

Gavin, M. (1991). The importance of asking why? evolutionary biology in wildlife management. *Journal of Wildlife Management*, **55**, 760–766.

Gazey, W. J., and Staley, M. J. (1986). Population estimation from mark-recapture experiments using a sequential bayes algorithm. *Ecology*, **67**, 941–951.

Gazit, I., and Terkel, J. (2003). Explosives detection by sniffer dogs following strenuous physical activity. *Applied Animal Behaviour Science*, **81**, 149–161.

Gebremedhin, B., Ficetola, G. F., Naderi, S. *et al.* (2009a). Frontiers in identifying conservation units: from neutral markers to adaptive genetic variation. *Animal Conservation*, **12**, 107–109.

Gebremedhin, B., Ficetola, G. F., Naderi, S. *et al.* (2009b). Combining genetic and ecological data to assess the conservation status of the endangered Ethiopian walia ibex. *Animal Conservation*, **12**, 89–100.

Gehring, T. M. and Swihart, R. K. (2000). Field immobilization and use of radiocollars on long-tailed weasels. *Wildlife Society Bulletin*, **28**, 579–585.

Gehrt, S. D. and Fritzell, E. K. (1996). Sex-biased response of raccoons (*Procyon lotor*) to live traps. *American Midland Naturalist*, **135**, 23–32.

Geiger, G., Bromel, J. and Habermehl, K. H. (1977). Concordance of various methods of determining the age of the red fox (*Vulpes vulpes* L., 1758). *Zeitshrift fur Jagdwissenschaft*, **23**, 57–64.

Genovesi, P. and Boitani, L. (1997). Social ecology of the stone marten in central Italy. In G. Proulx, H. N. Bryant and P. M. Woodard (eds). *Martes, taxonomy, ecology, techniques, and management*, pp. 110–120. Provincial Museum of Alberta, Edmonton, Alberta.

Gerall, A. A., Napoli, A. M. and Cooper, U. C. (1973). Daily and hourly estrous running in intact, spayed, and estrone implanted rats. *Physiology and Behaviour*, **10**, 225–229.

Gerber, B., Karpanty, S. M. and Kelly, M. J. (2011). Evaluating the potential biases in carnivore capture-recapture studies associated with the use of lure and varying density estimation techniques using photographic-sampling data of the Malagasy civet. *Population Ecology*, DOI: 10.1007/s10144-011-0276-3.

Gervasi, V., Brunberg, S. and Swenson, J. E. (2006). An individual-based method to measure animal activity levels: A test on brown bears. *Wildlife Society Bulletin*, **34**, 1314–1319.

Gese, E. M. (2006). Depredation management techniques for coyotes and wolves in North America, lessons learned and possible application to Brazilian carnivores. In *Manejo e Conservação de Carnivoros Neotropicais*, Ch. 12. Ministério do Meio Ambiente, Instituto Brasiliero do Meio Ambiente e dos Recursos Naturals Renováveis, São Paulo, Brazil.

Gese, E. M., Andersen, D. E. and Rongstad, O. J. (1990). Determining home-range size of resident coyotes from point and sequential locations. *Journal of Wildlife Management*, **54**, 501–506.

Gese, E. M., Rongstad, O. J. and Mytton, W. R. (1987). Manual and net-gun capture of coyotes from helicopters. *Wildlife Society Bulletin*, **15**, 444–445.

Getz, W. M. and Wilmers, C. C. (2004). A local nearest-neighborhood convex-hull construction of home ranges and utilization distributions. *Ecography*, **27**, 489–505.

Getz, W. M., Fortmann-Roe, S., Cross, P. C., Lyons, A. J., Ryan, S. J. and Williams, C. C. (2007). LoCoH, nonparametric kernel methods for constructing home ranges and utilization distributions. *PloS ONE*, **2**(2), e207.

Gibbs, J. P. (2000). Monitoring populations. In L. Boitani and T. K. Fuller (eds). *Research techniques in animal ecology*, pp. 213–252. Columbia University Press, New York.

Gibeau, M. L., Clevenger, A. P., Herrero, S. and Wierzchowski, J. (2002). Grizzly bear response to human development and activities in the Bow River Watershed, Alberta, Canada. *Biological Conservation*, **103**, 227–236.

Gilbert, F. F. (1981). Maximizing the humane potential of traps—the Vital and the Conibear 120. In J. A. Chapman and D. Pursley (eds). *Proceedings of the Worldwide Furbearer Conference*, pp. 1630–1646. University of Maryland, Frostburg, MD.

Gilbert, F. F. and Gofton, N. (1982). Terminal dives in mink, muskrat and beaver. *Physiology and Behavior*, **8**, 835–840.

Gilg, O., Hanski, I. and Sittler, B. (2003). Cyclic dynamics in a simple vertebrate predator-prey community. *Science*, **302**, 866–868.

Gilg, O., Sittler, B., Sabard, B. *et al.* (2006). Functional and numerical responses of four lemming predators in high arctic Greenland. *Oikos,* **113**, 193–216.

Gillies, C., Hebblewhite, M., Nielsen, S. E. *et al.* (2006). Application of random effects to the study of resource selection by animals. *Journal of Animal Ecology*, **75**, 887–898.

Gilman, N. (2004). Sanitation in the animal shelter. In L. Miller and S. Zawistowski (eds). *Shelter medicine for veterinarians and staff,* pp. 67–68. Blackwell Publishing, Oxford.

Gilsdorf, J. M., Vercauteren, K. C., Hygnstrom, S. E., Walter, W. D., Boner, J. R. and Clements, G. M. (2008). An integrated vehicle-mounted telemetry system for VHF telemetry applications. *Journal of Wildlife Management*, **72**, 1241–1246.

Gipson, P. S., Ballard, W. B., Nowak, R. M. and Mech, L. D. (2000). Accuracy and precision of estimating age of gray wolves by tooth wear. *Journal of Wildlife Management*, **64**, 752–758.

Girard, I., Ouellet, J.-P., Courtois, R., Dussault, C. and Breton, L. (2002). Effects of sampling effort based on GPS telemetry on home-range size estimations. *Journal of Wildlife Management*, **66**, 1290–1300.

Gloor, S., Bontadi, F., Hegglin, D., Deplazes, P. and Breitenmoser, U. (2001). The rise of urban fox populations in Switzerland. *Mammalian Biology*, **66**, 155–164.

Godoy, J. A. Casas-Marce, M. and Fernández, J. (2009). Genetic issues in the implementation of the Iberian Lynx Ex situ Conservation Programme. In A. Vargas, Ch. Breitenmoser and U. Breitenmoser U. (eds). *Iberian lynx ex situ conservation: a multidisciplinary approach*, pp. 86–99. Fundación Biodiversidad, Madrid.

Goldstein, I., Paisley, S., Willea, R., Jorgenson, J. P., Cuesta, F. and Castellanos, A. (2006). Andean bear - livestock conflicts: a review. *Ursus*, **17**, 8–15.

Gompper, M. E., Kays, R. K., Ray, J. C., LaPoint, S. D., Bogan, D. A. and Cryan, J. R. (2006). A comparison of non invasive techniques to survey carnivores communities in Northeastern North America. *Wildlife Society Bulletin*, **34**, 1142–1151.

Gonzalez-Esteban, J., Villate, I. and Irizar, I. (2006). Differentiating hair samples of the European mink (*Mustela lutreola*), the American mink (*Mustela vison*) and the European polecat (*Mustela putorius*) using light microscopy. *Journal of Zoology*, **270**, 458–461.

Goodrich, J. M., Kerley, L. L., Schleyer, B. O. *et al.* (2001). Capture and chemical anaesthesia of Amur (Siberian) tigers. *Wildlife Society Bulletin*, **29**, 533–542.

Goodrich, J. M., Kerley, L. L., Smirnov, E. N. *et al.* (2008). Survival rates and causes of mortality of Amur tigers on and near the Sikhote-Alin Biosphere Zapovednik. *Journal of Zoology*, **276**, 323–329.

Gorman, M. L. and Trowbridge, B. J. (1989). Role of odor in the social life of Carnivores. In J. L. Gittleman (ed.). *Carnivore Behavior, Ecology, and Evolution*, pp. 57–88. Cornell University Press, New York.

Gorman, T. A., Erb, J. D., McMillan, B. R. and Martin, D. J. (2006). Space use and sociality of river otters (*Lontra canadensis*) in Minnesota. *Journal of Mammalogy*, **87**, 740–747.

Goss-Custard, J. D. (1970). Feeding dispersion in some overwintering wading birds. In J. H. Crook (ed.). *Social behavior in birds and mammals*, pp. 1–35. Academic Press, London.

Goss-Custard, J. D. (1977). Optimal foraging and the size selection of worms by redshank *Tringa totanus. Animal Behaviour*, **25**, 10–29.

Gosselink, T. E., Van Deelen, T. R., Warner, R. E. and Mankin, P. C. (2007). Survival and cause-specific mortality of red foxes in agricultural and urban areas of Illinois. *Journal of Wildlife Management*, 71, 1862–1873.

Gotelli, N. J. and Ellison, A. M. (2004). *A primer of ecological statistics*. Sinauer Associates Publishers, Sunderland, MA.

Gotelli, D., Wang, J. L., Bashir, S. and Durant, S. M. (2007). Genetic analysis reveals promiscuity among female cheetahs. *Proceedings of the Royal Society B-Biological Sciences*, **274**, 1993–2001.

Gould, W. R. and Pollock, K. H. (1997a). Catch-effort maximum likelihood estimation of population parameters. *Canadian Journal of Fisheries and Aquatic Science*, **54**, 890–897.

Gould, W. R. and Pollock, K. H. (1997b). Catch-effort estimation of population parameters under the robust design. *Biometrics*, **53**, 207–216.

Graf, J. A., Gusset, M., Reid, C, Janse van Rensburg, S., Slotow, R. and Somers, M.J. (2006). Evolutionary ecology meets wildlife management, artificial group augmentation in the re-introduction of endangered African wild dogs (*Lycaon pictus*). *Animal Conservation*, **9**, 398–403.

Graham, K., Beckerman, A. P. and Thirgood, S. (2005). Human-predator-prey conflicts: ecological correlates, prey losses and patterns of management. *Biological Conservation*, **122**, 159–171

Graham, L. H., Byers, A. P., Armstrong, D. L., Loskutoff, N. M., Swanson, W. F., Wildt, D. E., and Brown, J. L. (2006). Natural and gonadotropin-induced ovarian activity in tigers (*Panthera tigris*) assessed by fecal steroid analyses. *General and Comparative Endocrinology*, **147**, 362–370.

Grassman, L. I., Haines, A. M., Janecka, J. E. and Tewes, M. E. (2006a). Activity periods of photo-captured mammals in north central Thailand. *Mammalia*, **70**, 306–309.

Grassman, L. I., Jr., Janecka, J. E., Austin, S. C., Tewes, M. E. and Silvy, N. J. (2006). Chemical immobilization of free-ranging dhole (*Cuon alpinus*), binturong (*Arctictis binturong*), and yellow-throated marten (*Martes flavigula*) in Thailand. *European Journal of Wildlife Research*, **52**, 297–300.

Grau, G. A., Sanderson, G. C. and Rogers, J. P. (1970). Age determination in raccoons. *Journal of Wildlife Management*, **34**, 364–372.

Graves, T. A. and Waller, J. S. (2006). Understanding the causes of missed global positioning system telemetry fixes. *Journal of Wildlife Management*, **70**, 844–851.

Green, J. S., Adair, R. A., Woodruff, R. A. and Stellflug, J. N. (1984). Seasonal variation in semen production by captive coyotes. *Journal of Mammalogy*, **65**, 506–509.

Green, J. S., Knowlton, F. F. and Pitt, W. C. (2002). Reproduction in captive wild-caught coyotes (*Canis latrans*). *Journal of Mammalogy*, **83**, 501–506.

Greentree, C., Saunders, G., McLeod, L. and Hone, J. (2000). Lamb predation and fox control in south-eastern Australia. *Journal of Applied Ecology*, **37**, 935–943.

Greenwood, J. J. D. and Elton, R. A. (1979). Analyzing experiments on frequency-dependent selection by predators. *Journal of Animal Ecology*, **48**, 721–737.

Greenwood, J. J. D. And Elton, R. A. (1987). Frequency-dependent selection by predators—comparison of parameter estimates. *Oikos* **48**, 268–272.

Greenwood, J. J. D. and Robinson, R. A. (2006). Principles of sampling. In W. J. Sutherland (ed.). *Ecological census techniques*, pp.11–86. Cambridge University Press, Cambridge.

Griffen, B. D. (2009). Consumers that are not "ideal" or "free" can still approach the ideal free distribution using simple patch-leaving rules. *Journal of Animal Ecology*, **78**, 919–927.

Griffith, B., Scott, J. M., Carpenter, J. W. and Reed, C. (1989). Translocation as a species conservation tool, status and strategy. *Science*, **245**, 477–480.

Griffin, P. C., Bienen, L., Gillin, C. M. and Mills, L. S. (2003). Estimating pregnancy rates and litter size in snowshoe hares using ultrasound. *Wildlife Society Bulletin*, **32**, 1006–1072.

Grimm, K. A. and Lamont, L. A. (2007). Clinical pharmacology. In G. West, D. Heard and N. Caulkett (eds). *Zoo Animal and Wildlife Immobilization and Anesthesia*, pp. 3–36. Blackwell Publishing Professional, Ames, Iowa.

Gruver, K. S., Philips, R. L. and Williams, E. S. (1996). Leg injuries to coyotes captured in standard and modified Soft Catch traps. *Proceedings Vertebrate Pest Conference*, 17, 91–93.

Gu, W. and Swihart, R. K. (2004). Absent or undetected? Effects of non-detection of species occurrence on wildlife-habitat models. *Biological Conservation*, **116**, 195–203.

Guggisberg, C. A. W. (1977). *Early wildlife photographers*. Taplinger, New York.

Guisan, A. and Zimmermann, N. E. (2000). Predictive habitat distribution models in ecology. *Ecological Modelling*, **135**, 147–186.

Guisan, A., Edwards, T. C. and Hastie, T. (2002). Generalized Linear and Generalized Additive Models in studies of species distributions, setting the scene. *Ecological Modelling*, **157**, 89–100.

Gula, R. (2008). Wolf depredation on domestic animals in the Polish Carpathian mountains. *Journal of Wildlife Management*, 72, 283–289.

Gulamhusein, A. P. and Thawley, A. R. (1972). Ovarian cycle and plasma progesterone levels in the stoat, *Mustela erminea*. *Journal of Reproduction and Fertility*, **31**, 492–493.

Gunson, J. R., Dorward, W. J. and Schowalter, D. B. (1978). An evaluation of rabies control in skunks in Alberta. *The Canadian Veterinary Journal*, **19**, 214–220.

Gusset, M., Slotow, R. and Somers, M. J. (2006). Divided we fail, the importance of social integration for the re-introduction of endangered wild dogs (*Lycaon pictus*). *Journal of Zoology*, **270**, 502–511.

Gusset, M., Swarner, M. J., Mponwane, L., Keletile, K. and McNutt, J. W. (2009). Human-wildlife conflict in northern Botswana: livestock predation by endangered African wild dog *Lycaon pictus* and other carnivores. *Oryx*, **43**, 67–72.

Guthery, F. S. (2008). Statistical ritual versus knowledge accrual in wildlife science. *Journal of Wildlife Management*, 72, 1872–1875.

Guthery, F. S. and Beasom, S. L. (1978). Effectiveness and selectivity of neck snares in predator control. *Journal of Wildlife Management*, **42**, 457–459.

Hadidian, J., Jenkins, S. R., Johnston, D. H., Savarie, P. J., Nettles, V. F., Manski, D. and Baer, G. M. (1989). Acceptance of simulated oral rabies vaccine baits by urban raccoons. *Journal of Wildlife Diseases*, **25**, 1–9.

Hadley, M. E. and Levine, J. E. (2007). *Endocrinology*. Pearson Prentice Hall, Upper Saddle River, NJ.

Hadow, H. H. (1972). Freeze-branding, a permanent marking technique for pigmented mammals. *Journal of Wildlife Management*, **36**, 645–649.

Haigh, J. C. and Hopf, H. C. (1976). The blowgun in veterinary practice, its uses and preparation. *Journal of the American Veterinary Medical Association*, **169**, 881–883.

Hall, S. J. G. and Bradley, D. G. (1995). Conserving livestock breed biodiversity. *Trends in Ecology and Evolution*, **10**, 267–270.

Hall, L. S., Krausman, P. R. and Morrison, M. L. (1997). The habitat concept and a plea for standard terminology. *Wildlife Society Bulletin*, **25**, 173–182.

Hallett, J. G., O'Connell, M. A., Sanders, J. D. and Seidensticker, J. (1991). Comparison of population estimators for medium-sized mammals. *Journal of Wildlife Management*, **55**, 81–93.

Hamede, R. K., Bashford, J., McCallum, H. and Jones, M. (2009). Contact networks in a wild Tasmanian devil (*Sarcophilus harrisii*) population: using social network analysis to reveal seasonal variability in social behaviour and its implications for transmission of devil facial tumour disease. *Ecology Letters*, **12**, 1147–1157.

Hamilton, W. J., Jr. and Eadie, W. R. (1964). Reproduction in the river otter, *Lutra canadensis*. *Journal of Mammalogy*, **45**, 242–252.

Hamlett, G. W. D. (1935). Delayed implantation and discontinuous development in the mammals. *Quarterly Review of Biology*, **10**, 432–447.

Hammerstedt, R. H., Graham, J. K. and Nolan, J. P. (1990). Cryopreservation of mammalian sperm: What we ask them to survive. *Journal of Andrology*, **11**, 73–88.

Hammond, J. and Marshall, F. H. A. (1930). Oestrus and pseudopregnancy in the ferret. *Proceedings of the Royal Society, B*, **105**, 607–630.

Hansen, M. C. and R. A. Riggs. (2008). Accuracy, precision, and observation rates of global positioning system telemetry collars. *Journal of Wildlife Management*, **72**, 518–526.

Hanski, I. (1999). *Metapopulation ecology*. Oxford University Press, Oxford.

Hanski, I. and Gaggiotti, O. E. (eds) (2004). *Ecology, genetics and evolution of metapopulations*. Academic Press, Amsterdam.

Hansson, A. (1947). The physiology of reproduction in mink (*Mustela vison*). *Acta Zoologica*, **28**, 1–136.

Harmsen, B. J., Foster, R. J., Silver, S. C., Ostro, L. E. T. and Doncaster, C. P. (2009). Spatial and temporal interactions of sympatric jaguars (*Panthera onca*) and pumas (*Puma concolor*) in a neotropical forest. *Journal of Mammalogy*, **90**, 612–620.

Harmsen, B. J., Foster, R. J., Silver, S. C., Ostro, L. E. T. and Doncaster, C. P. (2010). Differential use of trails by forest mammals and the implications for camera-trap studies: a case study from Belize. *Biotropica*, **42**, 126–133.

Harrington, F. H. and Mech, L. D. (1982). An analysis of howling response parameters useful for wolf pack censusing. *Journal Wildlife Management*, **46**, 686–693.

Harrison, R. L. (2006). A comparison of survey methods for detecting bobcats. *Wildlife Society Bulletin*, **34**, 548–552.

Harris, C. E. and Knowlton, F. F. (2001). Differential responses of coyotes to novel stimuli in familiar and unfamiliar settings. *Canadian Journal of Zoology*, **79**, 2005–2013.

Harris, R. B., Fancy, S. G., Douglas, D. C. *et al.* (1990a). *Tracking wildlife by satellite: current systems and performance*. Technical Report **30**. U. S. Fish and Wildlife Service, Patuxtent, MD.

Harris, S. and Saunders, G. (1993). The control of canid populations. *Symposia of the Zoological Society of London*, **65**, 441–464.

Harris, S., Cresswell, W. J., Forde, P. G., Trewhella, W. J., Wollard, T. and Wray, S. (1990b). Home-range analysis using radio-telemetry data—A review of problems and techniques particularly as applied to the study of mammals. *Mammal Review*, **20**, 97–123.

Harris, S., Cresswell, W. J. and Cheeseman, C. L. (1992). Age determination of badgers (*Meles meles*) from tooth wear: the need for a pragmatic approach. *Journal of Zoology*, **184**, 91–117.

Hass, C. C. and Valenzuela, D. (2002). Anti-predator benefits of group living in white-nosed coatis (*Nasua narica*). *Behavioral Ecology and Sociobiology*, **51**, 570–578.

Haugen, T. O., Winfield, I. J., Vollestad, L. A., Fletcher, J. M., James, J. B. and Stenseth, N. C. (2006). The ideal free pike, 50 years of fitness-maximizing dispersal in Windermere. *Proceedings of the Royal Society B-Biological Sciences*, **273**, 2917–2924.

Hawkins, R. E., Autry, D. C. and Klimstra, W. D. (1967). Comparison of methods used to capture white-tailed deer. *Journal of Wildlife Management*, **31**, 460–464.

Hawley, J. E., Gehring, T. M., Schultz, R. N., Rossler, S. T. and Wydeven, A. P. (2009). Assessment of shock collars as nonlethal management for wolves in Wisconsin. *Journal of Wildlife Management*, **73**, 518–525.

Haydon, D. T., Laurenson, M. K. and Sillero Zubiri, C. (2002). Integrating epidemiology into population viability analysis: managing the risk posed by rabies and canine distemper to the Ethiopian wolf. *Conservation Biology*, **16**, 1372–1385.

Hayne, D. W. (1949). Calculation of size of home range. *Journal of Mammalogy*, **30**, 1–18.

Hayne, D. W. (1978). Experimental designs and statistical analyses. In D. P. Snyder (ed.). *Populations of small mammals under natural conditions*, pp 3–10. Pymatuning Symposia in Ecology Number 5. Pymatuning Laboratory of Ecology, University of Pittsburgh, Pittsburgh, PA.

Hayssen, V., Van Tienhoven, A. and Van Tienhoven, A. (1993). *Asdell's Patterns of Mammalian Reproduction*. Comstock Publishing Associates, Ithaca, NY.

Hayward, M. W., Adendorff, J., O'Brien, J. *et al.* (2007a). The reintroduction of large predators to the Eastern Cape Province, South Africa, an assessment. *Oryx*, **42**, 205–214.

Hayward, M. W., Adendorff, J., O'Brien, J. *et al.* (2007b). Practical considerations for the reintroduction of large, terrestrial, mammalian predators based on reintroductions to South Africa's Eastern Cape Province. *The Open Conservation Biology Journal*, **1**, 1–11

Heard, D. C., Ciarniello, L. M. and Seip, D. R. (2008). Grizzly bear behavior and global positioning system collar fix rates. *Journal of Wildlife Management*, **72**, 596–602.

Heath, R. B., Calkins, D. B. and Spraker, T. (1996). Telazol and isoflurane anesthesia in free-ranging stellar sea lions (*Eumetopias jubatus*). *Journal of Zoo and Wildlife Medicine*, **27**, 35–43.

Hebblewhite, M. (2005). Predation interacts with the North Pacific Oscillation (NPO) to influence western North American elk population dynamics. *Journal of Animal Ecology*, **74**, 226–233.

Hebblewhite, M. (2009). Linking wildlife populations with ecosystem change, state-of-the-art satellite technology for National park science. *Park Science*, **26**, 1–14.

Hebblewhite, M. and Merrill, E. (2008). Modelling wildlife-human relationships for social species with mixed-effects resource selection models. *Journal of Applied Ecology*, **45**, 834–844.

Hebblewhite, M., Pletscher, D. H. and Paquet, P. C. (2002). Elk population dynamics in areas with and without predation by recolonizing wolves in Banff National Park, Alberta. *Canadian Journal of Zoology*, **80**, 789–799.

Hebblewhite, M., Paquet, P. C., Pletscher, D. H., Lessard, R. B., and Callaghan, C. J. (2004). Development and application of a ratio-estimator to estimate wolf killing rates and variance in a multiple prey system. *Wildlife Society Bulletin*, **31**, 933–946.

Hebblewhite, M., Merrill, E. H. and McDonald, T. E. (2005). Spatial decomposition of predation risk using resource selection functions, an example in a wolf-elk system. *Oikos*, **111**, 101–111.

Hebblewhite, M., Percy, M. and Merrill, E. H. (2007). Are all global positioning system collars created equal? Correcting habitat-induced bias using three brands in the Central Canadian Rockies. *Journal of Wildlife Management*, 71, 2026–2033.

Hebblewhite, M., Merrill, E. H. and McDermid, G. (2008). A multi-scale test of the forage maturation hypothesis for a partially migratory montane elk population. *Ecological Monographs*, **78**, 141–166.

Hebblewhite, M., Miquelle, D. G., Aramilev, V. V. and Pikonov, D. G. (2011). Predicting potential habitat and population size for reintroduction of the Far Eastern leopard in the Russian Far East. *Biological Conservation*, **144**, 2403–2413.

Hedmark, E., Flagstad, O., Segerstrom, P., Persson, J., Landa, A. and Ellegren, H. (2004). DNA-based individual and sex identification from wolverine (*Gulo gulo*) faeces and urine. *Conservation Genetics*, **5**, 405–410.

Hedmark, E., Persson, J., Segerstrom, P., Landa, A. and Ellegren, H. (2007). Paternity and mating system in wolverines *Gulo gulo*. *Wildlife Biology*, **13**, 13–30.

Hegglin, D., Ward, P. I. and Deplazes, P. (2003). Anthelmintic baiting of foxes against urban contamination with *Echinococcus multilocularis*. *Emerging infectious diseases*, **9**, 1266–1272.

Heikkinen, R. K., Luoto, M., Virkkala, R., Pearson, R. G. and Korber, J. H. (2007). Biotic interactions improve prediction of boreal bird distributions at macro-scales. *Global Ecology and Biogeography*, **16**, 754–763.

Heilburn, R. D., Silvy, N. J., Peterson, M. J. and Tewes, M. E. (2006). Estimating bobcat abundance using automatically triggered cameras. *Wildlife Society Bulletin*, **34**, 69–73.

Heinemeyer, K. S. and Copeland, J. P. (1999). *Wolverine denning habitat and surveys on the Targhee National Forest, 1998–1999 Annual report*. Unpubl. Report, GIS/ISC Laboratory, Dept. of Environmental Studies, University of California, Santa Cruz, CA.

Heinemeyer, K. S., Ulizio, T. J. and Harrison, R. L. (2008). Natural signs: tracks and scats. In R. A. Long, P. MacKay, W. J. Zielinski and J. C. Ray (eds). *Noninvasive survey methods for carnivores*, pp. 45–74. Island Press, Washington DC.

Hellstedt, P. and Henttonen, H. (2006). Home range, habitat choice and activity of stoats (*Mustela erminea*) in a subarctic area. *Journal of Zoology*, **269**, 205–212.

Hellstedt, P., Sundell, J., Helle, P. and Henttonen, H. (2006). Large-scale spatial and temporal patterns in population dynamics of the stoat, *Mustela erminea*, and the least weasel, *M. nivalis*, in Finland. *Oikos*, **115**, 286–298.

Hemson, G., Maclennan, S., Mills, M. G. L., Johnson, P. and Macdonald, D. W. (2009). Community, lions, livestock and money: a spatial and social analysis of attitudes to wildlife and the conservation value of tourism in a human-carnivore conflict in Botswana. *Biological Conservation*, **142**, 2718–2725.

Henle, K., Alard, D., Clitherow, J. *et al.* (2008). Identifying and managing the conflicts between agriculture and biodiversity conservation in Europe - a review. *Agriculture, Ecosystems and Environment*, **124**, 60–71.

Henry, C. (2004). *Organisation socio-spatiale d'une population de renards roux (Vulpes vulpes) en milieu rural, nature des relations et degrés de parenté entre individus de mêmes groupes spatiaux.* D.Sc. Thesis, Université Louis Pasteur, Strasbourg I, France.

Hensel, R. J. and Sorenson, F. E., Jr. (1980). Age determination of live polar bears. *International Conference on Bear Research and Management*, **4**, 93–100.

Henshaw, R. E. (1981). Toe-clipping coyotes for individual identification, a critique. *Journal of Wildlife Management*, **45**, 1005–1007.

Her Majesty's Stationary Office (1986). *Animals (scientific procedures) Act 1986.* HMSO, London.

Herfindal, I., Linnell, J. D. C., Moa, P. F., Odden, J., Austmo, L. B. and Andersen, R. (2005). Does recreational hunting of lynx reduce depredation losses of domestic sheep. *Journal of Wildlife Management*, **69**, 1034–1042.

Herr, J., Schley, L. and Roper, R. J. (2009). Socio-spatial organization of urban stone martens. *Journal of Zoology*, **277**, 54–62.

Herrick, J. R., Bond, J. B., Campbell, M. *et al.* (2010). Fecal endocrine profiles and ejaculate traits in black-footed cats (*Felis nigripes*) and sand cats (*Felis margarita*). *General and Comparative Endocrinology*, **165**, 204–214.

Herzog, C. J., Kays, R. W., Ray, J. C., Gompper, M. E. and Zielinski, W. J. (2007). Using patterns in track-plate footprints to identify individual fishers. *Journal of Wildlife Management*, **71**, 955–963.

Hestbeck, J. B., Nichols, J. D., and Malecki, R. A. (1991). Estimates of movement and site fidelity using mark-resight data of wintering Canada geese. *Ecology*, **72**, p.

Hiby, L., Lovell, P., Patil, N., Kumar, N. S., Gopalaswamy, A. M. and Karanth, K. U. (2009). A tiger cannot change its stripes, using a three-dimensional model to match images of living tigers and tiger skins. *Biology Letters*, **5**, 383–386.

Hillman, C. N. and Carpenter, J. W. (1983). Breeding biology and behavior of captive black-footed ferrets. *International Zoo Yearbook*, **23**, 186–191.

Hines, J. E. (2006). *PRESENCE*, version 2.2. Available at <http://www.mbr-pwrc.usgs.gov/software.html>

Hines, J. E., Nichols, J. D, Royle, J. A. *et al.* (2010). Tigers on trails, occupancy modeling for cluster sampling. *Ecological Applications*, **20**, 1456–1466.

Hirzel, A. H. and Le Lay, G. (2008). Habitat suitability modelling and niche theory. *Journal of Applied Ecology*, **45**, 1372–1381.

Hirzel, A. H., Hausser, J., Chessel, D. and Perrin, N. (2002). Ecological-niche factor analysis: how to compute habitat-suitability maps without absence data? *Ecology*, **83**, 2027–2036.

Hobson, D. F., Proulx, G. and Dew, B. I. (1989). Initial post-release behaviour of marten, *Martes americana*, introduced to Cypress Hills Provincial Park, Saskatchewan. *Canadian Field-Naturalist*, **103**, 398–400.

Hodges, K. E. and Sinclair, A. R. E. (2003) Does predation risk cause snowshoe hares to modify their diets? *Canadian Journal of Zoology*, **81**, 1973–1985.

Hodges, K., Brown, J. L. and Heistermann, M. (2010). Endocrine monitoring of reproduction and stress. In D. G. Kleiman, K. V. Thompson and C. K. Baer (eds). *Wild mammals in captivity: principles and techniques for zoo management*, pp. 447–468. Chicago University Press, Chicago, IL.

Hofer, H. and East, M. L. (1998). Biological conservation and stress. *Advances in the Study of Behavior*, 27, 405–525.

Hoffmann, J. C. and Sehgal, A. (1976). Effects of exogenous administration of hormones on reproductive-tract of female Hawaiian mongoose *Herpestes auropunctatus auropunctatus* (Hodgson). *Indian Journal of Experimental Biology*, **14**, 480–482.

Holekamp, K. E. and Sisk, C. L. (2003). Effects of dispersal status on pituitary and gonadal function in the male spotted hyena. *Hormones and Behavior*, **44**, 385–394.

Holling, C. S. (1959). The functional response of predators to prey density and its role in mimicry and population regulation. *Memoirs of the Entomological Society of Canada*, **45**, 1–60.

Holmern, T., Nyahongo, J. and Røskaft, E. (2007). Livestock losses caused by predators outside the Serengeti National Park, Tanzania. *Biological Conservation*, **135**, 518–526.

Holmes, E. E., Sabo, J. L., Viscido, S. V. and Fagan, W. F. (2007). A statistical approach to quasi-extinction forecasting. *Ecology Letters*, **10**, 1182–1198.

Holroyd, G. L., Van Tighem, K. J. (eds) (1983). *Ecological (biophysical) land classification of Banff and Jasper National Parks*, Vol 3. The wildlife inventory. Canadian Wildlife Service, Edmonton.

Holt, W. V. (2000). Fundamental aspects of sperm cryobiology: The importance of species and individual differences. *Theriogenology*, **53**, 47–58.

Holt, W. V. and Pickard, A. R. (1999). Role of reproductive technologies and genetic resource banks in animal conservation. *Reviews of Reproduction*, **4**, 143–150.

Hooven, E. F., Black, H. C. and Lowrie, J. C. (1979). Disturbance of small mammal live traps by spotted skunks. *Northwest Science*, **53**, 79–81.

Horne, J. S. and Garton, E. O. (2006a). Likelihood cross-validation versus least squares cross-validation for choosing the smoothing parameter in kernel home-range analysis. *Journal of Wildlife Management*, **70**, 641–648.

Horne, J. S. and Garton, E. O. (2006b). Selecting the best home range model, An information-theoretic approach. *Ecology*, **87**, 1146–1152.

Horne, J. S., Garton, E. O. and Sager-Fradkin, K. A. (2007). Correcting home-range models for observation bias. *Journal of Wildlife Management*, **71**, 996–1001.

Horne, J. S., Garton, E. O., Krone, S. M. and Lewis, J. S. (2007a). Analyzing animal movements using Brownian bridges. *Ecology*, **88**, 2354–2363.

Horne, J. S., Garton, E. O. and Rachlow, J. L. (2008). A synoptic model of animal space use; simultaneous estimation of home range, habitat selection, and inter/intra-specific relationships. *Ecological Modelling*, **214**, 338–348.

Horne, J. S., Tewes, M. E., Haines, A. M. and Laack, L. L. (2009). Habitat partitioning of sympatric ocelots and bobcats, implications for ocelot recovery in southern Texas. *Southwestern Naturalist*, **54**, 119–126.

Horner, M. A. (1986). *Internal structure of home ranges of black bears and analyses of home range overlap*. M.S. thesis, North Carolina State University, Raleigh, NC.

Horner, M. A. and Powell, R. A. (1990). Internal structure of home ranges of black bears and analyses of home-range overlap. *Journal of Mammalogy*, **71**, 402–410.

Hornocker, M. G. and Hash, H. S. (1981). Ecology of the wolverine in northwestern Montana. *Canadian Journal of Zoology*, **59**, 1286–1301.

Hosmer, D. W. and Lemeshow, S. (eds). (2000). *Applied logistic regression*. John Wiley and Sons, New York.

Hostetler, J. A., Onorato, D. P., Nichols, J. D. *et al.* (2010). Genetic introgression and the survival of Florida panther kittens. *Biological Conservation,* **143**, 2789–2796.

Houle, M., Fortin, D., Dussault, C., Courtois, R. and Ouellet, J. P. (2010). Cumulative effects of forestry on habitat use by gray wolf (*Canis lupus*) in the boreal forest. *Landscape Ecology*, **25**, 419–433.

Hounsome, T. D., Young, R. P., Davison, J. *et al.* (2005). An evaluation of distance sampling to estimate badger abundance. *Journal of Zoology*, **266**, 81–87.

Howard, W. E. (1949). A means to distinguish skull of coyotes and domestic dogs. *Journal of Mammalogy*, **30**, 169–171.

Howell-Skalla, L. A., Cattet, M. R. L., Ramsay, M. A. and Bahr, J. M. (2002). Seasonal changes in testicular size and serum LH, prolactin, and testosterone concentrations in male polar bears (*Ursus maritimus*). *Reproduction*, **123**, 729–733.

Huber, D., Kusak, J. and Radišić, B. (1996). Analysis of efficiency in live-capture of European brown bears. *Journal of Wildlife Research*, **1**, 158–162.

Hubert, G. F., Jr., Storm, G. L., Phillips, R. L. and Andrews, R. D. (1976). Ear tag loss in red fox. *Journal of Wildlife Management*, **40**, 164–167.

Hubert, G. F., Jr., Hungerford, L. L., Proulx, G., Bluett, R. D. and Bowman, L. (1996). Evaluation of two restraining traps to capture raccoons in non-drowning water sets. *Wildlife Society Bulletin*, **24**, 699–708.

Hubert, G. F., Jr., Hungeford, L. L. and Bluett, R. D. (1997). Injuries to coyotes captured in modified foothold traps. *Wildlife Society Bulletin*, **25**, 858–863.

Hubert, G. F., Jr., Wollenberg, G. K., Hungerford, L. L. and Bluett, R. D. (1999). Evaluation of injuries to Virginia opossums (*Didelphis virginiana*) captured in the EGG trap. *Wildlife Society Bulletin*, **27**, 301–305.

Huete, A., Didan, K., Miura, T., Rodriguez, E. P., Gao, X. and Ferreira, L. G. (2002). Overview of the radiometric and biophysical performance of the Modis vegetation indices. *Remote Sensing of Environment*, **83**, 195–213.

Huggard, D. J. (1993). Prey selectivity of wolves in Banff National Park. I. Prey species. *Canadian Journal of Zoology*, **71**, 130–139.

Huggins, R. M. (1989). On the statistical analysis of capture experiments. *Biometrika*, **76**, 133–140

Huggins, R. M. (1991). Some practical aspects of a conditional likelihood approach to capture experiments. *Biometrics*, **47**, 725–732.

Hull, R. N. and Swanson, S. (2006). Sequential analysis of lines of evidence—an advanced weight-of-evidence approach for ecological risk assessment. *Integrated Environmental Assessment and Management*, **2**, 302–311.

Humphrey, S. R. and Zinn, T. L. (1982). Seasonal habitat use by river otters and Everglades mink in Florida. *Journal of Wildlife Management*, **46**, 375–381.

Hunt, K. E. and Wasser, S. K. (2003). Effect of long-term preservation methods on fecal glucocorticoid concentrations of grizzly bear and African elephant. *Physiological and Biochemical Zoology*, **76**, 918–928.

Hunter, A. (2007). *Sensor-based animal tracking*. PhD Thesis, University of Calgary, Calgary, Alberta.

Hunter, J. S. (2009). Familiarity breeds contempt: effects of striped skunk color, shape, and abundance on wild carnivore behavior. *Behavioral Ecology*, **20**, 1315–1322.

Hunter, A., El-Seimy, N. and Stenhouse, G. (2005). Up close and grizzly – GPS/camera collar captures bear doings. *GPS World*, **16**, 24–31.

Huot, J. M., Poule, M. and Crete, M. (1995). Evaluation of several indices for assessment of coyote (*Canis latrans*) body composition. *Canadian Journal of Zoology*, **73**, 1620–1624.

Hupp, J. W. and Ratti, J. T. (1983). A test of radiotelemetry triangulation accuracy in heterogeneous environments. *Proceedings of the Fourth International Wildlife Biotelemetry Conference*, **4**, 31–46.

Hurlbert, S. H. (1978). The measurement of niche overlap and some relatives. *Ecology*, **59**, 67–77.

Hurlbert, S. H. (1984). Pseudoreplication and the design of field ecology experiments. *Ecological Monographs*, **54**, 187–211.

Hurlbert, I. A. R. and French, J. (2001). The accuracy of GPS for wildlife telemetry and habitat mapping. *Journal of Applied Ecology*, **38**, 869–878.

Hutchinson, G. E. (1953). The concept of pattern in ecology. *Proceedings of the American Academy of Science,* **105**, 1–12.

Hutchinson, G. E. (1957). Concluding remarks, Cold Spring Harbor Symposium. *Quantitative Biology*, **22**, 415–427.

Hwang, M. H. and Garshelis, D. L. (2007). Activity patterns of Asiatic black bears (*Ursus thibetanus*) in the Central Mountains of Taiwan. *Journal of Zoology*, **271**, 203–209.

Hwang, Y. T., Larivière, S. and Messier, F. (2005). Evaluating body condition of striped skunks using non-invasive morphometric indices and bioelectrical impedance analysis. *Wildlife Society Bulletin*, **33**, 195–203.

Hygnstrom, S. E. (1994). Black bears. Damage prevention and control methods. In S. E. Hygnstrom, R. M. Tim and G. E. Larson (eds). *Prevention and control of wildlife damage—1994*, pp. c5–c15. Cooperative Extension Division, Institute of Agriculture and Natural Resources, University of Nebraska, Lincoln, NE.

Iason, G. R., Mantecon, A. R., Sim, D. A. *et al.* (1999). Can grazing sheep compensate for a daily foraging time constraint? *Journal of Animal Ecology*, **68**, 87–93.

Inskip, C. and Zimmermann, A. (2009). Human-felid conflict: a review of patterns and priorities worldwide. *Oryx*, **43**, 18–34.

International Association of Fish and Wildlife Agencies. (1997). *Improving animal welfare in U.S. trapping programs, process recommendations and summaries of existing data.* International Association of Fish and Wildlife Agencies, Washington DC.

Irsara, A., Ruatti, A., Gagliardi, G., Orfei, Z., Bellani, L. and Mantovani, A. (1982). Control of wildlife rabies in North-Eastern Italy. *Comparative Immunology, Microbiology and Infectious Diseases,* **5**, 327–335.

IUCN (1998). *Guidelines for re-introductions.* IUCN/SSC Re-introduction Specialist Group, IUCN, Gland, Switzerland.

IUCN/SSC (2001). *IUCN red list categories: version 3.1.* IUCN Species Survival Commission, Gland, Switzerland.

IUCN/SSC (2008). *Strategic planning for species conservation: a handbook: version 1.0.* IUCN Species Survival Commission. Gland, Switzerland.

IUCN/SSC Cat Specialist Group and WWF (2007). *Status of the leopard in the Caucasus. Cat news special issue*, **2**. IUCN Species Survival Commission, Gland, Switzerland.

Iverson, S. J., Field, C., Bowen, W. D. and Blanchard, W. (2004). Quantitative fatty acid signature analysis: a new method of estimating predator diets. *Ecological Monographs*, **74**, 211–235.

Iverson, S. J., Stirling, I. and Lang, S. L. C. (2006). Spatial and temporal variation in the diets of polar bears across the Canadian Arctic: indicators of changes in prey populations and environment. In I. L. Boyd, S. Wanless and C. J. Camphuysen (eds). *Top predators in marine ecosystems*, pp. 98–117. Cambridge University Press, New York.

Jack, B.K., Kousky, C. and Sims, K.R.E. (2008). Designing payments for ecosystem services: lessons from previous experience with incentive-based mechanisms. *Proceedings of the National Academy of Sciences of the USA*, **105**, 9465–9470.

Jackson, R. M. and Wangchuk, R. (2004). A community based approach to mitigating livestock depredation by snow leopards. *Human Dimensions of Wildlife*, **9**, 307–315.

Jackson, D. H., Jackson, L. S. and Seitz, W. K. (1985). An expandable drop-off transmitter harness for young bobcats. *Journal of Wildlife Management*, **49**, 46–49.

Jackson, D. L., Gluesing, E. A. and Jacobson, H. A. (1988). Dental eruption in bobcats. *Journal of Wildlife Management*, **52**, 515–517.

Jackson, R. M., Roe, J. D., Wangchuk, R. and Hunter, D. O. (2006). Estimating snow leopard population abundance using photography and capture-recapture techniques. *Wildlife Society Bulletin*, **34**, 772–781.

Jacobs, L.F. and Spencer, W. D. (1994). Natural space-use patterns and hippocampal size in kangaroo rats. *Brain, behavior and evolution*, **44**, 125–132.

Jacques, C. N., Jenks, J. A. and Klaver, R. W. (2009). Seasonal movements and home-range use by female pronghorns in sagebrush-steppe communities of western South Dakota. *Journal of Mammalogy*, **90**, 433–441.

Janis, M. W., Clark, J. D. and Johnson, C. S. (1999). Predicting mountain lion activity using radiocollars equipped with mercury tip-sensors. *Wildlife Society Bulletin*, **27**, 19–24.

Jędrzejewski, W. and Jędrzejewska, B. (1992) Foraging and diet of the red fox Vulpes vulpes in relation to variable food resources in Bialowieźa National Park, Poland. *Ecograghy*, **15**, 212–220.

Jędrzejewska, B. and Jędrzejewski, W. (1998) *Predation in vertebrates' communities: the Bialowieza Primeval Forest as a case of study*. Springer-Verlag, New York.

Jędrzejewski, W., Schmidt, K., Theuerkauf, J., Jędrzejewski, B. and Okarma, H. (2001). Daily movements and territory use by radiocollared wolves (*Canis lupus*) in Bialowieza Primeval Forest in Poland. *Canadian Journal of Zoology*, **79**, 1993–2004.

Jenkins, S. H. and Busher, P. E. (1979). *Castor canadensis*. *Mammalian Species*, **120**, 1–8.

Jenkins, C. N. and Giri, C. (2008). Protection of mammal diversity in Central America. *Conservation Biology*, **22**, 1037–1044.

Jennings, A.P., Seymour, A.S. and Dunstone, N. (2006). Ranging behaviour, spatial organization and activity of the Malay civet (*Viverra tangalunga*) on Buton Island, Sulawesi. *Journal of Zoology*, **268**, 63–71.

Jennrich, R. I. and Turner, F. B. (1969). Measurement of non-circular home range. *Journal of Theoretical Biology*, **22**, 227–237.

Jessup, D. A. (1993). Remote treatment and monitoring of wildlife. In M. E. Fowler (ed.). *Zoo and wildlife medicine, current therapy*, 3rd edn, pp. 499–504. W. B. Saunders Co., Philadelphia, Pennsylvania.

Jhala, Y. V, Gopal, R., and Quereshi, Q. (2008). *Status of tigers, co-predators and prey in India.* National Tiger Conservation Authority and Wildlife Institute of India, New Delhi and Dehradun, India.

Jimenez, J. E. (2007). Ecology of a coastal population of the critically endangered Darwin's fox (*Pseudalopex fulvipes*) on Chiloe Island, southern Chile. *Journal of Zoology*, **271**, 63–77.

Jimenez, M. D., Smith D. W., Stahler, D. R., Albers, E. and Krischke, R. F. (2010). Wyoming Wolf Recovery 2009 Annual Report. In U.S. Fish and Wildlife Service, *Rocky Mountain Wolf Recovery 2009 Annual Report*, pp. WY-1 to WY-28. U. S. Fish and Wildlife Service, Ecological Services, Helena, MO.

Jobin, A., Molinari, P., and Breitenmoser, U. (2000). Prey spectrum, prey preference and consumption rates of Eurasian lynx in the Swiss Jura Mountains. *Acta Theriologica*, **45**, 243–252.

Jochle, W. (1975). Current research in coitus-induced ovulation: A review. *Journal of Reproduction and Fertility*, **22**, 165–205.

Jochle, W. and Andersen, A. C. (1977). The estrous cycle in the dog: a review. *Theriogenology*, 7, 113–140.

Johns, B. E. and Pan, H. P. (1981). Analytical techniques for fluorescent chemicals used as systemic or external wildlife markers. In E. W. Schafer, Jr. and C. R. Walker (eds). *Third conference, vertebrate pest control and management materials*, pp. 86–93. American Society for Testing Materials, West Conshohocken, PA.

Johnson, D. D. P. (1980) The comparison of usage and availability measurements for evaluating resource preferences. *Ecology*, **61**, 65–71.

Johnson, D. H. (1999). The insignificance of statistical significance testing. *Journal of Wildlife Management*, **63**, 763–772.

Johnson, D. H. (2002). The importance of replication in wildlife research. *Journal of Wildlife Management*, **66**, 919–932.

Johnson, C. J. and Gillingham, M. P. (2005). An evaluation of mapped species distribution models used for conservation planning. *Environmental Conservation*, **32**, 117–128.

Johnson, K. and Pelton, M. R. (1980). Prebaiting and snaring techniques for black bears. *Wildlife Society Bulletin*, **8**, 46–54.

Johnson, C. J. and Seip, D. R. (2008). Relationship between resource selection, distribution, and abundance, a test with implications to theory and conservation. *Population Ecology*, **50**, 145–157.

Johnson, D. D. P., Macdonald, D. W. and Dickman, A. J. (2000). Analysis and review of models of the sociobiology of the Mustelidae. *Mammal Review*, **30**, 171–196.

Johnson, C. J., Heard, D. C. and Parker, K. L. (2002). Expectations and realities of GPS animal location collars: results of three years in the field. *Wildlife Biology*, **8**, 153–159.

Johnson, C. J., Boyce, M. S., Schwartz, C. C. and Haroldson, M. A. (2004). Modeling survival, application of the multiplicative hazards model to Yellowstone grizzly bear. *Journal of Wildlife Management*, **68**, 966–974.

Johnson, C. J., Nielsen, S. E., Merrill, E. H., McDonald, T. L. and Boyce, M. S. (2006). Resource selection functions based on use-availability data, theoretical motivations and evaluations methods. *Journal of Wildlife Management*, **70**, 347–357.

Johnson, D. S., Thomas, D. L., Hoef, J. M. V. and Christ, A. (2008). A general framework for the analysis of animal resource selection from telemetry data. Biometrics, **64**, 968–976.

Johnson, A., Vongkhamheng, C. and Saithongdam, T. (2009). The diversity, status and conservation of small carnivores in a montane tropical forest in northern Laos. *Oryx*, **43**, 626–633.

Johnston, S. D., Ward, D., Lemon, J. *et al.* (2007). Studies of male reproduction in captive African wild dogs (*Lycaon pictus*). *Animal Reproduction Science*, **100**, 338–355.

Jolly, G. M. (1965). Explicit estimates from capture-recapture data with both death and immigration-stochastic model. *Biometrika*, **52**, 225–247.

Joly, S. E. and Joly, L. M. (1992). Pen and field tests of odor attractants for dingo. *Journal of Wildlife Management*, **56**, 452–456.

Jones, C., McShea, W. J., Conroy, M. J. and Kunz, T. H. (1996). Capturing mammals. In D. E. Wilson, F. R. Cole, J. D. Nichols, R. Rudran and M.S. Foster (eds). *Measuring and monitoring biological diversity. Standard methods for mammals*, pp. 115–155. Smithsonian Institution Press, Washington DC.

Jones, C., Moller, H. and Hamilton, W. (2004). A review of potential techniques for identifying individual stoats (*Mustela erminea*) visiting control or monitoring stations. *New Zealand Journal of Zoology*, **31**, 193–203.

Jones, E. S., Heard, D. C. and Gillingham, M. P. (2006). Temporal variation in stable carbon and nitrogen isotopes of grizzly bear guard hair and under fur. *Wildlife Society Bulletin*, **34**, 1320–1325.

Jones, M. E. (2000). Road upgrade, road mortality and remedial measures, impacts on a population of eastern quolls and Tasmanian devils. *Wildlife Research*, **27**, 289–296.

Jonkel, J. J. (1993). *A manual for handling bears for managers and researchers*. Office of Grizzly Bear Recovery, U.S. Fish and Wildlife Service, University of Montana, Missoula.

Jost, C., Devulder, G., Vucetich, J. A., Peterson, R. O. and Arditi, R. (2005). The wolves of Isle Royale display scale-invariant satiation and ratio-dependent predation on moose. *Journal of Animal Ecology*, **74**, 809–816.

Jozwiak, E. A., Andelt, W. F. and Bailey, T. N. (2006). Accuracy of GPS and map-plotted aerial telemetry locations on the Kenai Peninsula, Alaska. *Northwest Science*, **80**, 239–245.

Juang, P., Oki, H., Wang, Y., Martonosi, M. and Rubenstein, D. (2002). Energy-efficient computing for wildlife tracking: design tradeoffs and early experiences with ZebraNet. *ACM Sigplan Notices*, **37**, 96–107.

Jule, K. R., Leaver, L. A. and Lea, S. E.G. (2008). The effects of captive experience on reintroduction survival in carnivores, a review and analysis. *Biological Conservation*, **141**, 355–363.

Kaartinen, S., Luoto, M. and Kojola, I. (2009). Carnivore-livestock conflicts: determinants of wolf (*Canis lupus*) depredation on sheep farms in Finland. *Biodiversity and Conservation*, **18**, 3503–3517.

Kaczensky, P. (1999). Large carnivore depredation on livestock in Europe. *Ursus*, **11**, 59–72.

Kaczensky, P. and Huber, T. (1994). Wer war es?: Dokumentation und Identifikation von Raubtierrissen. Zentralstelle Osterreichischer Landesjagdverbaende, Wien.

Kaczensky, P., Knauer, F., Joozovic, M., Walzer, C. and Huber, T. (2002). Experience with trapping, chemical immobilization, and radiotagging of brown bears in Slovenia. *Ursus*, **13**, 347–356.

Kaczensky, P., Wagner, A. and Walzer, C. (2004). Activity monitoring of a brown bear – a model approach to test field methods. *Mammalian Biology*, **69**, 444–448.

Kadmon, R., Farber, O. and Danin, A. (2004). Effect of roadside bias on the accuracy of predictive maps produced by bioclimatic models. *Ecological Applications*, **14**, 401–413.

Kalinowski, S.T., Taper, M.L. and Marshall, T. C. (2007). Revising how the computer program CERVUS accommodates genotyping error increases success in paternity assignment. *Molecular Ecology*, **16**, 1099–1106.

Kamler, J. F. and Gipson, P. S. (2004). Survival and cause-specific mortality among furbearers in a protected area. *American Midland Naturalist*, **151**, 27–34.

Kamler, J. F., Ballard, W. B., Gilliland, R. L. and Mote, K. (2002). Improved trapping methods for swift foxes and sympatric coyotes. *Wildlife Society Bulletin*, **30**, 1262–1266.

Kaplanov, L. G. (1948). Tigers in Sikhote-Alin. In K. Lofdahl and A. Shevlakov, *Tiger, red deer, and moose*, pp. 18–49. (English translation of the original Kaplanov's book in Russian, 2005). http://www.tigers.ru/books/kaplanov/tigr_en.html (accessed March 2011).

Kappeler, P. M. and Erkert, H. G. (2003). On the move around the clock: correlates and determinants of cathemeral activity in wild red-fronted lemurs (*Eulemur fulvus rufus*). *Behavioral Ecology and Sociobiology*, **54**, 359–369.

Karanth, K. U. (1995). Estimating tiger *Panthera tigris* populations from camera-trap data using capture-recapture models. *Biological Conservation*, **71**, 333–338.

Karanth, K. U. and Nichols, J. D. (1998). Estimating tiger densities in India using photographic captures and recaptures. *Ecology*, **79**, 2852–2862.

Karanth, K. U. and Nichols, J. D. (eds). (2002). *Monitoring tigers and their prey*. Centre for Wildlife Studies, Bangalore.

Karanth, K. U. and Nichols J. D. (2010). Non-invasive survey methods for assessing tiger populations. In R. Tilson and P. Nyhus (eds). *Tigers of the world, science, politics and conservation of Panthera tigris*. Elsevier, New York.

Karanth, K. U., Nichols, J. D., Seidensticker, J. *et al.* (2003). Science deficiency in conservation practice, The monitoring of tiger populations in India. *Animal Conservation*, **6**, 141–146.

Karanth, K. U., Nichols, J. D., Kumar, N. S., Link, W. A. and Hines, J. E. (2004a). Tigers and their prey, predicting carnivore densities from prey abundance. *Proceedings of the National Academy of Sciences, USA*, **101**, 4854–4858.

Karanth, K. U., Nichols, J. D. and Kumar, N. S. (2004b). Photographic sampling of elusive mammals in tropical forests. In W. L. Thompson (ed.). *Sampling rare or elusive species: concepts, designs, and techniques for estimating population parameters*, pp. 229–247. Island Press, Washington DC.

Karanth, K. U., Nichols, J. D., Kumar, N. S. and Hines, J. E. (2006). Assessing tiger population dynamics using photographic capture-recapture sampling. *Ecology*, **87**, 2925–2937.

Karanth K. K., Nichols, J. D., Hines, J. E., Karanth, K. U. and Christensen, N. L. (2009). Patterns and determinants of mammal species occurrence in India. *Journal of Applied Ecology*, **46**, 1189–1200.

Karanth, K. U., Funston, P. J. and Sanderson E. W. (2010). Many ways of skinning a cat, tools and techniques for studying wild felids. In D. W. Macdonald and A. L. Loveridge (eds). *Biology and Conservation of Wild Felids*, pp. 197–216. Oxford University Press, Oxford.

Karanth, K. U., Gopalaswamy, A. M., Kumar, N.S., Vaidyanathan, S., Nichols, J. D. and MacKenzie, D. I. (2011). Monitoring carnivore populations at the lands cape scale, occupancy modeling of tigers from sign surveys. *Journal of Applied Ecology*, **48**, 1048–1056.

Karki, S. M., Gese, E. M. and Klavetter, M. L. (2007). Effects of coyote population reduction on swift fox demographics in southeastern Colorado. *Journal of Wildlife Management*, **71**, 2707–2718.

Katajisto, J. and Moilanen, A. (2006). Kernel-based home range method for data with irregular sampling intervals. *Ecological Modelling*, **194**, 405–413.

Kaufmann, J. H. (1962). Ecology and social behavior of the coati, *Nasua narica*, on Barro Colorado Island, Panama. *University of California Publications in Zoology*, **60**, 95–222.

Kauffman, M. J., Varley, N., Smith, D. W., Stahler, D. R., Macnulty, D. R. and Boyce, M. S. (2007). Landscape heterogeneity shapes predation in a newly restored predator-prey system. *Ecology Letters*, **10**, 690–700.

Kauhala, K. and Holmala, K. (2006). Contact rate and risk of rabies spread between medium-sized carnivores in southeast Finland. *Annales Zoologici Fennici*, **43**, 348–357.

Kays, R. W. and Slauson, K. M. (2008). Remote cameras. In R. A. Long, P. MacKay, W. J. Zielinski and J. C. Ray (eds). *Noninvasive survey methods for carnivores*, pp. 110–140. Island Press, Washington DC.

Keating, K. A. and Cherry, S. (2004). Use and interpretation of logistic regression in habitat selection studies. *Journal of Wildlife Management*, **68**, 774–789.

Keating K. A., Swartz, C. C., Haroldson, M. A. and Moody, D. (2002). Estimating numbers of females with cubs-of-the-year in the Yellowstone grizzly bear population. *Ursus* **13**, 16–174.

Keay, J. M., Singh, J., Gaunt, M. C. and Kaur, T. (2006). Fecal glucocorticoids and their metabolites as indicators of stress in various mammalian species: a literature review. *Journal of Zoo and Wildlife Medicine*, **37**, 234–244.

Keller, C. M. E. and Scallan, J. T. (1999). Potential roadside biases due to habitat changes along breeding bird survey routes. *Condor*, **101**, 50–57.

Kelly, J. F. (2000). Stable isotopes of carbon and nitrogen in the study of avian and mammalian trophic ecology. *Canadian Journal of Zoology*, **78**, 1–27.

Kelly, M. J. (2001). Computer-aided photograph matching in studies using individual identification: an example from Serengeti cheetahs. *Journal of Mammalogy*, **82**, 440–449.

Kelly, M. J. (2008). Design, evaluate, refine: camera trap studies for elusive species. *Animal Conservation*, **11**, 182–184.

Kelly, M. J. and Holub, E. L. (2008). Camera trapping of carnivores: Trap success among camera types and across species, and habitat selection by species, on Salt Pond Mountain, Giles County, Virginia. *Northeastern Naturalist*, **15**, 249–262.

Kelly, M. J., Noss, A. J., Di Bitetti *et al.* (2008). Estimating puma densities from camera trapping across three study sites: Bolivia, Argentina, and Belize. *Journal of Mammalogy*, **89**, 408–418.

Kendall, W. L. and Bjorkland, R. (2001). Using open robust design models to estimate temporary emigration from capture-recapture data. *Biometrics*, **57**, 1113–1122.

Kendall, B. E. and Fox, G. A. (2002). Variation among individuals and reduced demographic stochasticity. *Conservation Biology*, **16**, 109–116.

Kendall, K. C. and McKelvey, K. S. (2008). Hair collection. In R. A. Long, P. MacKay, W. J. Zielinski and J. C. Ray (eds). *Noninvasive survey methods for carnivores*, pp. 141–182. Island Press, Washington DC.

Kendall, W. L. and Nichols, J. D. (1995). On the use of secondary capture-recapture samples to estimate temporary emigration and breeding proportions. *Journal of Applied Statistics*, **22**, 751–762.

Kendall, W. L. and White, G. C. (2009). A cautionary note on substituting spatial subunits for repeated temporal sampling in studies of site occupancy. *Journal of Applied Ecology*, **46**, 1182–1188.

Kendall, W. L., Nichols, J. D. and Hines, J. E. (1997). Estimating temporary emigration using capture-recapture data with Pollock's robust design. *Ecology*, **78**, 563–578.

Kendall, K. C., Stetz, J. B., Roon, D. A., Waits, L. P., Boulanger, J. B. and Paetkau, D. (2008). Grizzly bear density in Glacier National Park, Montana. *Journal of Wildlife Management*, **72**, 1693–1705.

Kendall, K. C., Stetz, J. B., Boulanger, J. B., McLeod, A. C., Paetkau, D. and White, G.C. (2009). Demography and genetic structure of a recovering grizzly bear population. *Journal of Wildlife Management*, **73**, 3–17.

Kennedy, M. A., Mellon, V. S., Caldwell, G. and Potgieter, L. N. (1995). Virucidal efficacy of the newer quaternary ammonium compounds. *Journal of the American Animal Hospital Association*, **31**, 254–258.

Kennelly, J. J. (1969). The effect of Mestranol on canine reproduction. *Biology of Reproduction*, **1**, 282–288.

Kennelly, J. J. and Johns, B. E. (1976). The estrous cycle of coyotes. *Journal of Wildlife Management*, **40**, 272–277.

Kennerly, E., Ballmann, A., Martin, S. *et al.* (2008). A gene expression signature of confinement in peripheral blood of red wolves (*Canis rufus*). *Molecular Ecology*, **17**, 2782–2791.

Kenward, R. E. (2001). *A manual of wildlife radio tagging*. Academic Press, London.

Kerley, L. L. and Salkina, G. P. (2007). Using scent-matching dogs to identify individual amur tigers from scats. *Wildlife Society Bulletin*, **71**, 1341–1356.

Kern, J. W., McDonald, L. L., Strickland, M. D. and Williams, E. (1994). *Field evaluation and comparison of four foothold traps for terrestrial furbearers in Wyoming*. Technical Research Work Order for Furbearers Unlimited, Bloomington, IL.

Kerns, B. K. and Ager, A. (2007). Risk assessment for biodiversity conservation planning in Pacific Northwest forests. *Forest Ecology and Management*, **246**, 38–44.

Khan, M. Z., Altmann, J., Isani, S. S. and Yu, J. (2002). A matter of time: evaluating the storage of fecal samples for steroid analysis. *General and Comparative Endocrinology*, **128**, 57–64.

Kie, J. G., Bowyer, R. T., Boroski, B. B., Nicholson, M. C. and Loft, E. R. (2002). Landscape heterogeneity at differing scales. effects on spatial distribution of mule deer. *Ecology*, **83**, 530–544.

Kie, J. G., Matthiopoulos, J., Fieberg, J. *et al.* (2010). The home-range concept, are traditional estimators still relevant with modern telemetry technology? *Philosophical Transactions of the Royal Society B*, **365**, 2221–2231.

Killian, P. J. and Bothma, J. d P. (2003). Notes on the social dynamics and behaviour of reintroduced lions in the Welgevonden Private Game Reserve. *South African Journal of Wildlife Research*, **33**, 119–124.

Kilpatrick, H. J., DeNicola, A. J. and Ellingwood, M. R. (1996). Comparison of standard and transmitter-equipped darts for capturing white-tailed deer. *Wildlife Society Bulletin*, **24**, 306–310.

King, A. W. (1997). Hierarchy theory, a guide to system structure for wildlife biologists. In J. A. Bissonette (ed.). *Wildlife and landscape ecology, effects of pattern and scale*, pp. 185–212. Springer-Verlag, New York.

King, C. M. (1981). The effects of two types of steel traps upon captured stoats (*Mustela erminea*). *Journal Zoology*, **195**, 553–554.

King, C. M. and Powell, R. A. (2007). *The Natural History of Weasels and Stoats,* 2nd edn. Oxford University Press, New York.

King, C. M., McDonald, R. M., Martin, R. D., Tempero, G. W. and Holmes, S. J. (2007). Long-term automated monitoring of the distribution of small carnivores. *Wildlife Research*, **34**, 140–148.

Kinley, T. A. and Newhouse, N. J. (2008). Ecology and translocation-aided recovery of an endangered badger population. *Journal of Wildlife Management*, **72**, 113–122.

Kirkpatrick, J. and Frank, K. M. (2005). Contraception in free-ranging wildlife. In C. S. Asa and I. J. Porton (eds). *Wildlife contraception: issues, methods and application*, pp. 195–221. Johns Hopkins University Press, Baltimore, MD.

Kissling, W. D., Fernández, N. and Paruelo, J. M. (2009). Spatial risk assessment of livestock exposure to pumas in Patagonia, Argentina. *Ecography*, **32**, 807–817.

Kissui, B. M. (2008). Livestock predation by lions, leopards, spotted hyenas, and their vulnerability to retaliatory killing in the Maasai steppe, Tanzania. *Animal Conservation*, **11**, 422–432.

Kitchen, A. M. and Knowlton, F. F. (2006). Cross-fostering in coyotes, evaluation of a potential conservation and research tool for canids. *Biological Conservation*, **129**, 221–225.

Kitchen, A. M., Gese, E. M. and Schauster, E. R. (1999). Resource partitioning between coyotes and swift foxes, space, time, and diet. *Canadian Journal of Zoology*, **77**, 1645–1656.

Kitchen, A. M., Gese, E. M. and Schauster, E. R. (2000). Changes in coyote activity patterns due to reduced exposure to human persecution. *Canadian Journal of Zoology*, **78**, 853–857.

Kitchen, A. M., Gese, E. M., Waits, L. P., Karki, S. M. and Schauster, E. R. (2005). Genetic and spatial structure within a swift fox population. *Journal of Animal Ecology*, **74**, 1173–1181.

Kitchen, A. M., Gese, E. M. and Lupis, S. G. (2006a). Multiple scale den site selection by swift foxes, *Vulpes velox*, in southeastern Colorado. *Canadian Field-Naturalist*, **120**, 31–38.

Kitchen, A. M., Gese, E. M., Waits, L. P., Karki, S. M. and Schauster, E. R. (2006b). Multiple breeding strategies in the swift fox, *Vulpes velox*. *Animal Behaviour*, **71**, 1029–1038.

Kjelstrup, K. B., Solstad, T., Brun, V. H. *et al.* 2008. Finite scale of spatial representation in the hippocampus. *Science*, **321**, 140–143.

Kleiman, D. G. (1968). Reproduction in the Canidae. *International Zoo Yearbook*, **8**, 3–8.

Kleiman, D. G. (1989). Reintroduction of captive mammals for conservation. *BioScience*, **39**, 152–161.

Knapp, S. M., Craig, B. A. and Waits, L.P. (2009). Incorporating genotyping error into non-invasive DNA-based mark–recapture population estimates. *Journal of Wildlife Management*, 73, 598–604.

Knarrum, V., Sørensen, O. J., Eggen, T. *et al.* (2006). Brown bear predation on domestic sheep in central Norway. *Ursus*, 17, 67–74.

Knight, J. E. (1983). Skunks. In R. M. Timm (ed.). *Prevention and control of wildlife damage*, pp c81–c86. Institute of Agriculture and Natural Resources, University of Nebraska, Lincoln, NE.

Knight, R. R., Blanchard, B. M. and Eberhardt, L. L. (1988). Mortality patterns and population sinks for Yellowstone grizzly bears, 1973–1985. *Wildlife Society Bulletin*, **16**, 121–125.

Knight, R. R., Blanchard, B. M. and Eberhardt, L. L. (1995). Appraising status of the Yellowstone grizzly bear population by counting females with cubs-of-the-year. *Wildlife Society Bulletin*, **23**, 245–248.

Knopff, K. H., Knopff, A. A., Warren, M. B. and Boyce, M. S. (2009). Evaluating global positioning system telemetry techniques for estimating cougar predation parameters. *Journal of Wildlife Management*, 73, 586–597.

Knowlton, F. F. and Whittemore, S. L. (2001). Pulp cavity-tooth width ratios from known-age and wild-caught coyotes determined by radiography. *Wildlife Society Bulletin*, **29**, 239–244.

Knowlton, F. F., Gese, E. M. and Jaeger, M. M. (1999). Coyote depredation control, an interface between biology and management. *Journal of Range Management*, **52**, 398–412.

Kochanny, C. O., DelGiudice, G. D. and Fieberg, J. (2009). Comparing global positioning system and very high frequency telemetry home ranges of white-tailed deer. *Journal of Wildlife Management*, 73, 779–787.

Koen, E. L., Ray, J. C., Bowman, J., Dawson, F. N. and Magoun, A. J. (2008). *Surveying and monitoring wolverines in Ontario and other lowland, boreal forest habitats: recommendations and protocols*. Ontario NWSI Field Guide FG-06, Ontario Ministry Natural Resources, Northwest Science and Information, Thunder Bay, Ontario.

Kohlmann, S. G., Schmidt, G. A. and Garcelon, D. K. (2005). A population viability analysis for the island fox on Santa Catalina Island, California. *Ecological Modelling*, **183**, 77–94.

Kohn, M. H. and Wayne, R. K. (1997). Facts from feces revisited. *Trends in Ecology and Evolution*, **12**, 223–227.

Kojima, E., Tsuruga, H., Komatsu, T., Murase, T., Tsubota, T. and Kita, I. (2001). Characterization of semen collected from beagles and captive Japanese black bears (*Ursus thibetanus japonicus*). *Theriogenology*, **55**, 717–732.

Kojola, I., Kaartinen, S., Hakala, A., Heikkinen, S. and Voipio, H. M. (2009). Dispersal behavior and the connectivity between wolf populations in northern Europe. *Journal of Wildlife Management*, 73, 309–313.

Kolbe, J. A., Squires, J. R. and Parker, T. W. (2003). An effective box trap for capturing lynx. *Wildlife Society Bulletin*, **31**, 980–985.

Kolenosky, G. B. and Johnston, D. H. (1967). Radio-tracking timber wolves in Ontario. *American Zoologist*, 7, 289–303.

Kolowski, J. M. and Holekamp, K. E. (2006). Spatial, temporal, and physical characteristics of livestock depredations by large carnivores along a Kenyan reserve border. *Biological Conservation*, **128**, 529–541.

Koren, L., Mokady, O., Karaskov, T., Klein, J., Koren, G. and Geffen, E. (2002). A novel method using hair for determining hormonal levels in wildlife. *Animal Behaviour*, **63**, 403–406.

Kovach, A. I. and Powell, R. A. (2003). Effects of body size on male mating tactics and paternity in black bears, *Ursus americanus*. *Canadian Journal of Zoology*, **81**, 1257–1268.

Kowalczyk, R., Jędrzejewska, B. and Zalewski, A. (2003). Annual and circadian activity patterns of badgers (*Meles meles*) in Białowieża Primeval Forest (eastern Poland) compared with other Palaearctic populations. *Journal of Biogeography*, **30**, 463–472.

Kowalczyk, R., Jędrzejewska, B., Zalewski, A. and Jędrzejewski, W. (2008). Facilitative interactions between the Eurasian badger (*Meles meles*), the red fox (*Vulpes vulpes*), and the invasive raccoon dog (*Nyctereutes procyonoides*) in Bialowieza Primeval Forest, Poland. *Canadian Journal of Zoology*, **86**, 1389–1396.

Kramer-Schadt, S., Revilla, E., Wiegand, T. and Breitenmoser, U. (2004). Fragmented landscapes, road mortality and patch connectivity, modeling influences on the dispersal of Eurasian lynx. *Journal of Applied Ecology*, **41**,711–723.

Krause, T. (1989). *NTA trapping handbook. A guide for better trapping.* National Trappers Association, Bloomington, Illinois.

Krebs, C. J. (1966). Demographic changes in fluctuating populations of *Microtus californicus. Ecological Monographs*, **36**, 239–273.

Krebs, C. J. (1999) *Ecological methodology*. Addison Wesley Longman, Menlo Park, CA.

Krebs, C. J. (2009). *Ecology: the experimental analysis of distribution and abundance*, 6th ed. Benjamin Cummings, San Francisco, CA.

Krebs, J. R. and Davies, N. B. (1993). *Introduction to Behavioural Ecology.* Blackwell Publishing, Oxford.

Krebs, J. R., Sherry, D. F., Healy, S. D., Perry, V. H. and Vaccarino, A. L. (1989). Hippocampal specializations of food-storing birds. *Proceedings of the National Academy of Sciences*, **86**, 1388–1392.

Krebs, J., Lofroth, E., Copeland, J. *et al.* (2004). Synthesis of survival rates and causes of mortality in North American wolverines. *Journal of Wildlife Management*, **68**, 493–507.

Kreeger, T. J. (1999). Chemical restraint and immobilization of wild canids. In M. E. Fowler and R. E. Miller (eds). *Zoo and wildlife medicine, current therapy 4*, pp. 429–435. W. B. Saunders Co., Philadelphia, PE.

Kreeger, T. J. (2002). Analyses of immobilizing dart characteristics. *Wildlife Society Bulletin*, **30**, 968–970.

Kreeger, T. J. (2003). The internal wolf: physiology, pathology, and pharmacology. In L. D. Mech and L. Boitani (eds). *Wolves: behavior, ecology and conservation*, pp. 192–217. University of Chicago Press, Chicago, IL.

Kreeger, T. J. and Arnemo, J. M. (2007). *Handbook of Wildlife Chemical Immobilization, 3rd ed.* Sunquest Publishing, Inc., Vancouver, British Columbia.

Kreeger, T. J., Seal, U. S., Callahan, M. and Beckel, M. (1990a). Physiological and behavioral responses of gray wolves (*Canis lupus*) to immobilization with tiletamine and zolazepam. *Journal of Wildlife Diseases*, **26**, 90–94.

Kreeger, T. J., Seal, U. S., Plotka, E. D. and Asa, C. S. (1991). Characterization of prolactin secretion in gray wolves (*Canis lupus*). *Canadian Journal of Zoology*, **69**, 1366–1374.

Kreeger, T. J., Kuechle, V. B., Mech, L. D., Tester, J. R. and Seal U. S. (1990a). Physiological monitoring of gray wolves (*Canis lupus*) by radio telemetry. *Journal of Mammalogy*, **71**, 259–261.

Kreeger, T. J., White, P. J., Seal, U. S. and Tester, J. R. (1990b). Pathological responses of red foxes to foothold traps. *Journal of Wildlife Management*, **54**, 147–160.

Kreeger, T. J., Monson, D., Kuechle, V. B., Seal, U. S. and Tester, J. R. (1989). Monitoring heart rate and body temperature in red foxes (*Vulpes vulpes*). *Canadian Journal of Zoology*, **67**, 2455–2458.

Kruuk, H. (1972a). Surplus killing by carnivores. *Journal of Zoology*, **166**, 233–244.

Kruuk, H. (1972b). *The spotted hyaena: a study of predation and social behavior*. University of Chicago Press, Chicago, IL.

Kruuk, H. (1980). The effects of large carnivores on livestock and animal husbandry in Marsabit district, Kenya. *UNEP-MAB Integrated Project in Arid Lands Technical Report*, **E-4**, 1–52.

Kruuk, H. (1989). *The social badger*. Oxford University Press. Oxford.

Kruuk, H. and Parish, T. (1981) Feeding specialization of the European badger *Meles meles* in Scotland. *Journal of Animal Ecology*, **50**, 773–788.

Kruuk, H. and Parish, T. (1982). Factors affecting population density, group size and territory size of the European badger, *Meles meles*. *Journal of Zoology*, **196**, 31–39.

Kuehn, D. W., Fuller, T. K., Mech, L. D., Fritts, S. H. and Berg, W. E. (1986). Trap-related injuries to gray wolves in Minnesota. *Journal of Wildlife Management*, **50**, 90–91.

Kunkel, K. E. and Pletscher, D. H. (2000). Habitat factors affecting vulnerability of moose to predation by wolves in southeastern British Columbia. *Canadian Journal of Zoology*, **78**, 150–157.

Kunkel, K. E., Pletscher, D. H., Boyd, D. K., Ream, R. R. and Fairchild, M. W. (2004). Factors correlated with foraging behavior of wolves in and near Glacier National Park, Montana. *Journal of Wildlife Management*, **68**, 167–178.

Kurth, C., Loftin, C. and Zydlewski, Z. (2007). PIT tags increase effectiveness of freshwater mussel recaptures. *Journal of North American Benthological Society*, **26**, 253–260.

Kyle, C. J., Johnson, A. R., Patterson, B. R., Wilson *et al.* (2006). Genetic nature of eastern wolves: past, present and future. *Conservation Genetics*, 7, 272–283.

Lacy, R. C. (2000). Structure of the VORTEX simulation model for population viability analysis. *Ecological Bulletins*, **48**, 191–203.

Lagendijk, D. D. G. and Gusset, M. (2008). Human-carnivore coexistence on communal land bordering the greater Kruger Area, South Africa. *Environmental Management*, **42**, 971–976.

Lair, H. (1987). Estimating the location of the focal center in red squirrel home ranges. *Ecology*, **68**, 1092–1101.

Lancia, R. A., Nichols, J. D. and Pollock, K. H. (1994). Estimating the number of animals in wildlife populations. In T. A. Bookhout (ed.). *Research and management techniques for wildlife and habitats*, pp. 215–253. The Wildlife Society, Bethesda, MD.

Land, E. D., Shindle, D. B., Kawula, R. J., Benson, J. F., Lotz, M. A. and Onorato, D. P. (2008). Florida panther habitat selection analysis of concurrent GPS and VHF telemetry data. *Journal of Wildlife Management*, 72, 633–639.

Landa, A., Krogstad, S., Tømmerås, B. Å. and Tufto, J. (1998). Do volatile repellents reduce wolverine *Gulo gulo* predation on sheep? Results of a large scale experiment. *Wildlife Biology*, 4, 111–118.

Larivière, S. and Messier, F. (1999). Review and perspective of methods used to capture and handle skunks. In G. Proulx (ed.). *Mammal trapping*, pp. 141–154. Alpha Wildlife publications, Sherwood Park, Alberta.

Larkin, R. P., VanDeelen, T. R., Sabick, R. M., Gosselink, T. E. and Warner, R. E. (2003). Electronic signaling for prompt removal of an animal from a trap. *Wildlife Society Bulletin*, **31**, 392–398.

Larrucea, E. S., Brussard, P. F., Jaeger, M. M. and Barrett, R. H. (2007). Cameras, coyotes, and the assumption of equal detectability. *Journal of Wildlife Management*, **71**, 1682–1689.

Larson, G. E., Savarie, P. J. and Okuno, I. (1981). Iophenoxic acid and mirex for marking wild, bait consuming animals. *Journal of Wildlife Management*, **45**, 1073–1077.

Laskowski, R. (1991). Are the top carnivores endangered by heavy metal biomagnification? *Oikos*, **60**, 387–390.

Laundré, J. W. (2008). Summer predation rates on ungulate prey by a large keystone predator: how many ungulates does a large predator kill? *Journal of Zoology*, **275**, 341–348.

Laundré, J. W. and Hernández, L. (2008). The amount of time female pumas *Puma concolor* spend with their kittens. *Wildlife Biology*, **14**, 221–227.

Laurenson, K., Shiferaw, F. and Sillero-Zubiri, C. (1997). Disease, domestic dogs and the Ethiopian wolf: the current situation. In C. Sillero-Zubiri and D. W. MacDonald (eds). *The Ethiopian wolf—status survey and conservation action plan*, pp. 32–42. IUCN, Gland, Switzerland.

Laurenson, K., Sillero-Zubiri, C., Thompson, H., Shiferaw, F., Thirgood, S. and Malcolm, J. (1998). Disease as a threat to endangered species: Ethiopian wolves, domestic dogs and canine pathogens. *Animal Conservation*, **1**, 273–280.

Laver, P. N. and Kelly, M. J. (2005). A critical review of home range studies. *Journal of Wildlife Management*, **72**, 290–298.

Lawes, M. J. and Piper, S. E. (1998). There is less to binary maps than meets the eye: the use of species distribution data in the southern African sub-region. *South African Journal of Science*, **97**, 207–210.

Lawrence, B. and Bossert, W. H. (1967). Multiple character analysis of *Canis lupus, latrans,* and *familiaris*, with a discussion of the relationships of *Canis niger*. *American Zoologist*, 7, 223–232.

Lebreton, J. D., Burnham, K. P., Clobert, J. and Anderson, D. R. (1992). Modeling survival and testing biological hypotheses using marked animals: a unified approach with case studies. *Ecological Monographs*, **62**, 67–118.

Leckie, F. M., Thirgood, S. J., May, R., and Redpath, S. M. (1998). Variation in the diet of red foxes on Scottish moorland in relation to prey abundance. *Ecography*, **21**, 599–604.

Lehman, N., Eisenhawer, A., Hansen, K. *et al.* (1991). Introgression of coyote mitochondrial-DNA into sympatric North-American gray wolf populations. *Evolution*, **45**, 104–119.

Leigh, J. (2007). *Effects of aversive conditioning on behavior of nuisance Louisiana black bears*. MSc Thesis, Louisiana State University and Agricultural and Mechanical College, Louisiana State University, Baton Rouge, LA.

Lele, S. and Keim, J. (2006). Weighted distributions and estimation of resource selection probability functions. *Ecology*, **87**, 3021–3028.

Lemieux, R. and Czetwertynski, S. (2006). Tube traps and rubber padded snares for capturing American black bears. *Ursus*, **17**, 81–91.

Lemons, P. R., Sedinger, J. S., Herzog, M. P., Gipson, P. S. and Gilliland, R. L. (2010). Landscape effects on diets of two canids in Northwestern Texas: a multinomial modeling approach. *Journal of Mammalogy*, **91**, 66–78.

LeMunyan, C. D., White, W., Nyberg, E. and Christian, J. J. (1959). Design of a miniature radio transmitter for use in animal studies. *Journal of Wildlife Management*, **23**, 107–110.

Lentfer, J. W. (1968). A technique for immobilizing and marking polar bears. *Journal of Wildlife Management*, **32**, 317–321.

Leopold, A. (1949). *A sand county almanac and sketches here and there*. Oxford University Press, New York. The quote at the beginning of chapter 9 comes from the first 2 paragraphs in the entry under "December."

Leung, B., Forbes, M. R. and Houle, D. (2000). Fluctuating asymmetry as a bio indicator of stress: comparing efficacy of analyses involving multiple traits. *American Naturalist*, **155**, 101–115.

Leutgeb, J. K., Leutgeb, S., Moser, M.-B. and Moser, E. I. (2007). Pattern separation in the dentate gyrus and CA3 of the hippocampus. *Science*. **315**, 961–966.

Levins, R. (1968) *Evolution in changing environments*. Princeton University Press, Princeton.

Lewis, J. C. M. (2004). Field use of isoflurane and air anesthetic equipment in wildlife. *Journal of Zoo and Wildlife Medicine*, **35**, 303–311.

Lewis, M. A. and Murray, J. D. (1993). Modelling territoriality and wolf-deer interactions. *Nature*, **366**, 738.

Lewis, J. S., Rachlow, J. L., Garton, E. O. andVierling, L. A. (2007). Effects of habitat on GPS collar performance: using data screening to reduce location error. *Journal of Applied Ecology*, **44**, 663–671.

Li, D., Chiang, A. Y., Clawson, C. A., Main, B. W. and Leishman, D. J. (2008). Heartbeat dynamics in adrenergic blocker treated conscious beagle dogs. *Journal of Pharmacological and Toxicological Methods*, **58**, 118–128.

Li, Z., Zimmerman, F., Hebblewhite, M. *et al.* (2009). *Technical report on the identification of potential tiger habitat in the Changbaishan Ecosystem, Northeast China. A collaborative work conducted by World Wildlife Fund, Wildlife Conservation Society, Northeast Normal University, KORA, and the University of Montana*. Muri, Switzerland. [http//www.wwfchina.org/english/downloads/Changbai/Habitat/WWF_Changbai_Tiger_Habitat_Report031110.pdf] (accessed 12 May 2010)

Liberg, O. (1982). Correction factors for important prey categories in the diet of domestic cats. *Acta Theriologica*, **27**, 115–122.

Liberg, O., Andrén, H., Pedersen, H. *et al.* (2005). Severe inbreeding depression in a wild wolf (*Canis lupus*) population. *Biology letters*, **1**, 17–20.

Liche, H. and Wodzicki, K. (1939). Vaginal smears and the oestrous cycle of the cat and lioness. *Nature*, **144**, 245–246.

Lichstein, J. W., Simons, T. R., Shriner, S. A. and Franzreb, K. E. (2002). Spatial autocorrelation and autoregressive models in ecology. *Ecological Monographs*, 72, 445–463.

Lichti, N. I. and Swihart, R. K. (2011). Estimating utilization distributions with kernel versus local convex hull methods. *Journal of Wildlife Management*, 75, 413–422.

Liley, S. and Creel, S. (2008). What best explains vigilance in elk: characteristics of prey, predators, or the environment? *Behavioral Ecology*, **19**, 245–254.

Lima, S. L. (1998). Non-lethal effects in the ecology of predator-prey interactions- What are the ecological effects of anti-predator decision –making? *Bioscience*, **48**, 5–34.

Lima, S. L. and Bednekoff, P. A. (1999). Temporal variation in danger drives antipredator behavior: The predation risk allocation hypothesis. *American Naturalist*, **153**, 649–659.

Lincoln, G. A. and Short, R. V. (1980). Seasonal breeding: nature's contraceptive. *Recent Progress in Hormone Research*, **36**, 1–52.

Lincoln, G. A., Andersson, H. and Hazlerigg, D. G. (2003). Clock genes and the long-term regulation of prolactin secretion: evidence for a photoperiod/circannual timer in the *pars tuberalis*. *Journal of Neuroendocrinology*, **15**, 390–397.

Lindberg, M. S. and Walker, J. (2007). Satellite telemetry in avian research and management: sample size considerations. *Journal of Wildlife Management*, 71, 1002–1009.

Lindsey, P., Du Toit, J. and Mills, M. G. L. (2005). Attitudes of ranchers towards African wild dogs Lycaon pictus: conservation implications on private land. *Biological Conservation*, **125**, 113–121.

Lindström, E. R. (1981). Placental scar counts in the red fox (*Vulpes vulpes*) with special reference to fading of the scars. *Mammalian Review*, **11**, 137–149.

Lindström, E. R. (1994). Large prey for small cubs - on crucial resources of a boreal red fox population. *Ecography*, **17**, 17–22.

Lindzey, F. G. and Meslow, E. C. (1980). Harvest and population characteristics of black bears in Oregon (1971–74). *International Conference on Bear Research and Management*, **4**, 213–219.

Lindzey, F., Sawyer, H., Anderson, C. and Banulis, B. (2001). Performance of store-on-board GPS collars on elk, mule deer and mountain lions in Wyoming, USA. I. J. Gordon (ed), *Tracking animals with GPS*. pp. 29–31. Macaulay Land Use Research Institute, Aberdeen, UK.

Linhart, S. B. (1968). Dentition and pelage in the juvenile red fox (*Vulpes vulpes*). *Journal of Mammalogy*, **49**, 526–528.

Linhart, S. B. and Kennelly, J. J. (1967). Fluorescent bone labeling of coyotes with demethylchlortetracycline. *Journal of Wildlife Management*, **31**, 317–321.

Linhart, S. B. and Knowlton, F. F. (1975). Determining the relative abundance of coyotes by scent station lines. *Wildlife Society Bulletin*, **3**, 119–124.

Linhart, S. B., Dasch, G. J., Male, C. B. and Engeman, R. M. (1986). Efficiency of unpadded and padded steel foothold traps for capturing coyotes. *Wildlife Society Bulletin*, **20**, 63–66.

Link, W. A. (2003). Non identifiability of population size from capture-recapture data with heterogeneous detection probabilities. *Biometrics*, **59**, 1123–1130.

Link, W. A., Yoshizaki, J., Pollock, K. H., and Bailey, L. L. (2010). Uncovering a latent multinomial, analysis of mark-recapture data with misidentification. *Biometrics*, **66**, 178–185.

Linkie, M., Chapron, G., Martyr, D. J., Holden, J. and Leader-Williams, N. (2006). Assessing the viability of tiger subpopulations in a fragmented landscape. *Journal of Applied Ecology*, **43**, 576–586.

Linkie, M., Dinata, Y., Nugroho, A. and Haidir, I. A. (2007). Estimating occupancy of a data deficient mammalian species living in tropical rainforests: sun bears in the Kerinci Seblat region, Sumatra. *Biological Conservation*, **137**, 20–27.

Linn, I. J. (1978). Radioactive techniques for small mammal marking. In B. Stonehouse (ed.). *Animal marking, recognition marking of animals in research*, pp. 177–191. The MacMillan Press, London.

Linnell, J. D. C. and Brøseth, H. (2003). Compensation for large carnivore depredation on domestic sheep in Norway, 1996–2002. *Carnivore Damage Prevention News*, **6**, 11–13.

Linnell, J. D. C. and Strand, O. (2000). Interference interactions, co-existence and conservation of mammalian carnivores. *Diversity and Distributions*, **6**, 169–176.

Linnell, J. D. C., Smith, M. E., Odden, J., Kaczensky, P. and Swenson, J. E. (1996). Strategies for the reduction of carnivore - livestock conflicts: a review. *Norwegian Institute for Nature Research Oppdragsmelding*, **443**, 1–118.

Linnell, J. D. C., Aanes, R., Swenson, J. E., Odden, J. and Smith, M. E. (1997). Translocation of carnivores as a method for managing problem animals: a review. *Biodiversity and Conservation*, **6**, 1245–1257.

Linnell, J. D. C., Odden, J., Smith, M. E., Aanes, R., and Swenson, J. E. (1999). Large carnivores that kill livestock: do "problem individuals" really exist? *Wildlife Society Bulletin*, **27**, 698–705.

Linnell, J. D. C, Swenson, J. E., Andersen, R. and Barnes, B. (2000). How vulnerable are denning bears to disturbance? *Wildlife Society Bulletin*, **28**, 400–413.

Linnell, J. D. C., Nilsen, B. E., Lande, U. S. *et al.* (2005a). Zoning as a means of mitigating conflicts with large carnivores: principles and reality. In R. Woodroffe., S. Thirgood and A. Rabinowitz (eds). *People and wildlife: conflict or co-existence?* pp. 162–175. Cambridge University Press, Cambridge

Linnell, J. D. C., Brøseth, H., Solberg, E. J. and Brainerd, S. (2005b). The origins of the southern Scandinavian wolf *Canis lupus* population: potential for natural immigration in relation to dispersal distances, geography and Baltic ice. *Wildlife Biology*, **11**, 383–391.

Linnell, J. D. C., Prombxerger, C., Boitani, L., Swenson, J. E., Breitenmoser, U. and Andersen, R. (2005c). The linkage between conservation strategies for large carnivores and biodiversity: the view from the "half-full"forests of Europe. In J. C. Ray, K. H. Redford, R. S. Steneck and J. Berger (eds). *Carnivorous animals and biodiversity: does conserving one save the other?*, pp. 381–398. Island Press, Washington DC.

Linnell, J., Salvatori, V. and Boitani, L. (2008). *Guidelines for population level management plans for large carnivores in Europe*. A Large Carnivore Initiative for Europe report prepared for the European Commission. European Commission, DG-ENV, Bruxells and Istituto di Ecologia Applicata, Rome (Contract 070501/2005/424162/MAR/B2).

Linnell, J. D. C., Breitenmoser U., Breitenmoser-Würsten Ch. and von Arx M. (2009). Recovery of Eurasian lynx in Europe: what part has reintroduction played? In M. W. Hayward and M. J. Somers (eds). *Reintroduction of top-order predators*, pp.72–91. Wiley-Blackwell, Oxford.

Linnell, J. D. C., Rondeau, D., Reed, D. H. *et al.* (2010). Confronting the costs and conflicts associated with biodiversity. *Animal Conservation*, **13**, 429–431.

Linscombe, R. G. and Wright, V. L. (1988). Efficiency of padded foothold traps for capturing terrestrial furbearers. *Wildlife Society Bulletin*, **16**, 307–309.

Liu, S. S., Fung, H. P. and Liu, B. T. (2007). Ovarian cycle of the captive Formosan gem-faced civets (*Paguma larvata taivana*). *Zoo Biology*, **26**, 1–12.

Lloyd, M. (1967). Mean crowing. *Journal of Animal Ecology*, **36**, 1–30.

Lloyd, H. G. and Englund, J. (1973). The reproductive cycle of the red fox in Europe. *Journal of Reproduction and Fertility*, Suppl. **19**, 119–130.

Lockie, J. D. (1958). The estimation of the food of foxes. *Journal of Wildlife Management*, **23**, 224–227.

Lockie, J. D. (1966). Territory in small carnivores. *Symposium of the Zoological Society of London*, **18**, 143–165.

Loehle, C. (1990). Home range, a fractal approach. *Landscape Ecology*, **5**, 39–52.

Lofroth, E. C., Klafki, R., Krebs, J. A. and Lewis, D. (2008). Evaluation of live-capture techniques for free-ranging wolverines. *Journal of Wildlife Management*, **72**, 1253–1261.

Logan, K. A. and Sweanor, L. (2000). Puma. In S. Demarais and P.R. Krausman (eds). *Ecology and management of large mammals in North America*, pp. 347–377. Prentice-Hall, Englewood Cliffs, NJ.

Logan, K. A., Sweanor, L. L., Smith, J. F. and Hornocker, M. G. (1999). Capturing pumas with foot-hold snares. *Wildlife Society Bulletin*, **27**, 201–208.

Lohr, C., Ballard, W. B. and Bath, A. (1996). Attitudes toward gray wolf reintroduction to New Brunswick. *Wildlife Society Bulletin*, **24**, 414–420.

Long, R. A. and Zielinski, W. J. (2008). Designing effective noninvasive carnivore surveys. In R. A. Long, P. MacKay, W. J. Zielinski and J. C. Ray (eds). *Noninvasive survey methods for carnivores*, pp. 8–44. Island Press, Washington DC.

Long, R. A., Donovan, T. M., Mackay, P., Zielinski, W. J. and Buzas, J. S. (2007). Effectiveness of scat detection dogs for detecting forest carnivores. *Journal of Wildlife Management*, **71**, 2007–2017.

Long, R. A., MacKay, P., Zielinski, W. J. and Ray, J. C. (eds) (2008a). *Noninvasive survey methods for carnivores.* Island Press, Washington DC.

Long, R. A., MacKay, P., Ray, J. C. and Zielinski W. J. (2008b). Synthesis and future research needs. In R. A. Long, P. MacKay, W. J. Zielinski and J. C. Ray (eds). *Noninvasive survey methods for carnivores*, pp. 313–325. Island Press, Washington DC.

Lopes, F. L., Desmarais, J. A. and Murphy, B. D. (2004). Embryonic diapause and its regulation. *Reproduction*, **128**, 669–678.

López, G., Martinez, F., Meili, M. L. *et al.* (2009). A feline leukemia virus (FeLV) outbreak in the Doñana Iberian lynx population. In A. Vargas, Ch. Breitenmoser and U. Breitenmoser (eds). *Iberian lynx ex situ conservation: a multidisciplinary approach*, pp. 234–247. Fundación Biodiversidad, Madrid.

Loureiro, F., Rosallino, L., Macdonald, D. W. and Santos-Reis, M. (2007). Path tortuosity of Eurasian badgers (*Meles meles*) in a heterogeneous Mediterranean landscape. *Ecological Research*, **22**, 837–844.

Loveridge, A. J. and Macdonald, D. W. (2002). Habitat ecology of two sympatric species of jackals in Zimbabwe. *Journal of Mammalogy*, **83**, 599–607.

Lucentini, L., Vercillo, F., Palomba, A., Panara, F., and Ragni, B. (2007). A PCR-RFLP method on faecal samples to distinguish *Martes martes, Martes foina, Mustela putorius* and *Vulpes vulpes*. *Conservation Genetics*, **8**, 757–759.

Lucherini, M., Reppucci, J. I., Walker, R. S. *et al.* (2009). Activity pattern segregation of carnivores in the high Andes. *Journal of Mammalogy*, **90**, 1404–1409.

Ludders, J. W., Schmidt, R. H., Dein, F. J. and Klein, P. N. (1999). Drowning is not euthanasia. *Wildlife Society Bulletin*, **27**, 666–670.

Ludders, J. W., Schmidt, R. H., Dein, F. J. and Klein, P.N. (2001). Drowning can no longer be considered euthanasia, reply to Bluett. *Wildlife Society Bulletin*, **29**, 744–750.

Ludwig, J. R. and Dapson, R. W. (1977). Use of insoluble lens proteins to estimate age in white-tailed deer. *Journal of Wildlife Management*, **41**, 327–329.

Lukacs, P. M. and Burnham, K. P. (2005). Review of capture-recapture methods applicable to noninvasive genetic sampling. *Molecular Ecology*, **14**, 3909–3919.

Lukarevsky, V., Malkasyan, A. and Askerov, E. 2007. Biology and ecology of the leopard in the Caucasus. *Cat News Special Issue*, **2**, 9–14.

Lynch, J. W., Khan, M.Z., Altmann, J., Njahira, M. N. and Rubenstein, N. (2003). Concentrations of four fecal steroids in wild baboons: short-term storage conditions and consequences for data interpretation. *General and Comparative Endocrinology*, **132**, 264–271.

Lyver, P. O. B. (2000). Identifying mammalian predators from bite marks: a tool for focusing wildlife protection. *Mammal Review*, **30**, 31–44.

MacArthur, R. H. (1967). The theory of the niche. In R. C. Lewontin (ed.). *Population biology and evolution*, pp. 159–176. Syracuse University Press, Syracuse, NY.

MacArthur, R. H. and Levins, R. (1967) The limiting similarity, convergence, and divergence of coexisting species. *American Naturalist*, **101**, 377–385.

Macdonald, D. W. (1979a). "Helpers" in fox society. *Nature*, **282**, 69–71.

Macdonald, D. W. (1979b). Some observations and field experiments on the urine marking behaviour of the red fox, *Vulpes vulpes* L. *Zeitschrift für Tierpsychologie*, **51**, 1–22.

Macdonald, D. W. (1981). Resource dispersion and the social organization of the red fox, *Vulpes vulpes*. In J. A. Chapman and D. Pursley (eds). *Proceedings of the Worldwide Furbearer Conference*, pp. 918–949. Worldwide Furbearer Conference, Baltimore, MD.

Macdonald, D. W. (1983). The ecology of carnivore social behavior. *Nature*, **301**, 379–384.

Macdonald, D. W. (2009). Lessons learnt and plans laid, seven awkward questions for the future of reintroductions. In M. W. Hayward and M. J. Somers (eds). *Reintroductions of top-order predators*, pp. 411–448. Wiley- Blackwell, Hoboken, NJ.

Macdonald, D. W. and Carr, G. M. (1989). Food security and the rewards of tolerance. In V. Standen and R. A. Foley (eds). *Comparative socioecology; the behavioural ecology of humans and other mammals*, pp. 75–99. Blackwell, Oxford.

Macdonald, D. W. and Norris, S. (2001). *The new encyclopedia of mammals*. Oxford University Press, Oxford.

Macdonald, D. W., Mace, G. and Rushton, S. (1998). *Proposals for future monitoring of British mammals*. Department of Environment, Transport, and the Regions, London.

Macdonald, D. W., Buesching, C. D., Stopka, P., Henderson, J., Ellwood, S. A. and Baker, S. E. (2004a). Encounters between two sympatric carnivores: red foxes (*Vulpes vulpes*) and European badgers (*Meles meles*). *Journal of Zoology*, **263**, 385–392.

Macdonald, D. W., Daniels, M. J., Driscoll, C. *et al.* (2004b). *The Scottish wildcat. Analyses for conservation and action plan*. Wildlife Conservation Research Unit, Oxford, UK.

Mace, G. M. and Baillie, J. E. M. (2007). The 2010 biodiversity indicators: Challenges for science and policy. *Conservation Biology*, **21**, 1406–1413.

Mace, D. M., Waller, J. S., Manley, T. M., Lyon, J. and Zuuring, H. (1996). Relationships among grizzly bears, roads, and humans in the Swan mountains, Montana. *Journal of Applied Ecology*, **33**, 1395–1404.

Machin, K. L. (2007). Wildlife analgesia. In G. West, D. Heard and N. Caulkett (eds). *Zoo animal and wildlife immobilization and anesthesia*, pp. 43–60. Blackwell Publishing Professional, Ames, Iowa.

MacKay, P., Smith, D. S., Long, R. A. and Parker, M. (2008). Scat detection dogs. In R. A. Long, P. MacKay, W. J. Zielinski and J. C. Ray (eds). *Noninvasive survey methods for carnivores*, pp. 183–222. Island Press, Washington DC.

MacKenzie, D. I. (2005). What are the issues with presence-absence data for wildlife managers? *Journal of Wildlife Management*, **69**, 849–860.

MacKenzie, D. I. and Bailey, L. L. (2004). Assessing the fit of site-occupancy models. *Journal of Agricultural, Biological, and Environmental Statistics*, **9**, 300–318.

MacKenzie, D. I. and Nichols, J. D. (2004). Occupancy as a surrogate for abundance estimation. *Animal Biodiversity and Conservation*, **27**, 461–467.

MacKenzie, D. I. and Royle, J. A. (2005). Designing occupancy studies: general advice and allocating survey effort. *Journal of Applied Ecology*, **42**, 1105–1114.

MacKenzie, D. I., Nichols, J. D., Lachman, G. B., Droege, S., Royle, J. A. and Langtimm, C. A. (2002). Estimating site occupancy rates when detection probabilities are less than one. *Ecology*, **83**, 2248–2255.

MacKenzie, D. I., Nichols, J. D., Hines, J. E., Knutson, M. G. and Franklin, A. B. (2003). Estimating site occupancy, colonization, and local extinction when a species is detected imperfectly. *Ecology*, **84**, 2200–2207.

MacKenzie, D. I., Royle, J. A., Brown, J. A. and Nichols, J. D. (2004a). Occupancy estimation and modeling for rare and elusive populations. In W. L. Thompson (ed.). *Sampling rare or elusive populations*, pp. 149–172. Island Press, Washington DC.

MacKenzie, D. I., Bailey, L. L. and Nichols, J. D. (2004). Investigating species co-occurrence patterns when species are detected imperfectly. *Journal of Animal Ecology*, **73**, 546–555.

MacKenzie, D. I., Nichols, J. D., Sutton, N., Kawanishi. K., and Bailey, L. L. (2005). Improving inferences in popoulation studies of rare species that are detected imperfectly. *Ecology*, **86**, 1101–1113.

MacKenzie, D. I., Nichols, J. D., Royle, J. A., Pollock, K. H., Bailey, L. L. and Hines, J. E. (2006). *Occupancy estimation and modeling: inferring patterns and dynamics of species occurrence*. Academic press, New York.

MacKenzie, D. I., Nichols, J. D., Seamans, M. E. and Gutierrez, R. J. (2009). Modeling species occurrence dynamics with multiple states and imperfect detection. *Ecology*, **90**, 823–835.

Maclennan, S. D., Groom, R. J., Macdonald, D. W. and Frank, L. G. (2009). Evaluation of a compensation scheme to bring about pastoralist tolerance of lions. *Biological Conservation*, **142**, 2419–2427.

MacNulty, D. R., Plumb, G. E. and Smith, D. W. (2008). Validation of a new video and telemetry system for remotely monitoring wildlife. *Journal of Wildlife Management*, **72**, 1834–1844.

Madden, J. R., Drewe, J. A., Pearce, G. P. and Clutton-Brock, T. (2009). The social network structure of a wild meerkat population, 2. Intragroup interactions. *Behavioral Ecology and Sociobiology*, **64**, 81–95.

Madhusudan, M. D. (2003). Living amidst large wildlife: livestock and crop depredation by large mammals in the interior villages of Bhadra tiger reserve, south India. *Environmental Management*, **31**, 466–475.

Maejima, M., Inoue-Murayama, M., Tonosaki, K. *et al.* (2007). Traits and genotypes may predict the successful training of drug detection dogs. *Applied Animal Behaviour Science* **107**, 287–298.

Maffei, L. and Noss, A. J. (2008). How small is too small? Camera trap survey areas and density estimates for ocelots in the Bolivian chaco. *Biotropica*, **40**, 71–75.

Maffei, L., Cuellar, E. and Noss, A. (2004). One thousand jaguars (*Panthera onca*) in Bolivia's Chaco? Camera trapping in the Kaa-Iya National Park. *Journal of Zoology*, **262**, 295–304.

Maffei, L., Noss, A. J., Cuellar, E. and Rumiz, D. I. (2005). Ocelot (*Felis pardalis*) population densities, activity, and ranging behaviour in the dry forests of eastern Bolivia: data from camera trapping. *Journal of Tropical Ecology*, **21**, 349–353.

Magoun, A. J. (1985). *Population characteristics, ecology and management of wolverines in northwestern Alaska*. PhD Thesis. University of Alaska, Fairbanks, AK.

Malcolm, J. R. (1992). Use of tooth impressions to identify and age live *Proechimys guyannensis* and *Proechimys cuvieri* Rodentia Echimyidae. *Journal of Zoology*, **227**, 537–546.

Manly, B. F. J. (2004). Two-phase adaptive stratified sampling. In W. L. Thompson (ed.). *Sampling rare or elusive species: concepts, designs, and techniques for estimating population parameters*, pp. 123–133. Island Press, Washington DC.

Manly, B. F. J., McDonald, L. L., Thomas, D. L., McDonald, T. L. and Erickson, W. P. (eds). (2002). *Resource selection by animals, statistical analysis and design for field studies*. Second edn. Kluwer, Boston.

Manning, J. T. and Chamberlain, A. T. (1993). Fluctuating asymmetry, sexual selection and canine teeth in primates. *Proceedings of the Royal Society of London, B*, **251**, 83–87.

Manning, J. T. and Chamberlain, A. T. (1994). Fluctuating asymmetry in gorilla canines: a sensitive indicator of environmental stress. *Proceedings of the Royal Society of London, B*, **255**, 189–193.

Mansfield, K. G., Verstraete, F. J. M. and Pascoe, P. J. (2006). Mitigating pain during tooth extraction from conscious deer. *Wildlife Society Bulletin*, **34**, 201–202.

Mao, J. S., Boyce, M. S., Smith, D. W. *et al.* (2005). Habitat selection by elk before and after wolf reintroduction in Yellowstone National Park. *Journal of Wildlife Management*, **69**, 1691–1707.

Mares, R., Moreno, R. S., Kays, R. W. and Wikelski, M. (2008). Predispersal home range shift of an ocelot *Leopardus pardalis* (Carnivora, Felidae) on Barro Colorado Island, Panama. *Revista de Biología Tropical*, **56**, 779–787.

Marker, L. L. and Dickman, A. J. (2003). Morphology, physical condition, and growth of the cheetah (*Acinonyx jubatus jubatus*). *Journal of Mammalogy*, **83**, 840–850.

Marker, L. L. and Dickman A. J. (2005). Factors affecting leopard (*Panthera pardus*) spatial ecology, with particular reference to Namibian farmlands. *South African Journal of Wildlife Research*, **35**, 105–115.

Marker, L. L., Dickman, A. J., Jeo, R. M., Mills, M. G. L. and Macdonald, D. W. (2003). Demography of the Namibian cheetah, *Acinonyx jubatus jubatus*. *Biological Conservation*, **114**, 413–425.

Marker, L. L., Dickman, A. and Schumann, M. (2005a). Using livestock guarding dogs as a conflict resolution strategy on Namibian farms. *Carnivore Damage Prevention News*, **8**, 28–32. KORA, Muri, Swizerland.

Marker, L. L., Dickman, A. J. and Macdonald, D. W. (2005b). Perceived effectiveness of livestock guarding dogs placed on Namibian farms. *Rangeland Ecology and Management*, **58**, 3299–336.

Marks, C. A. (1996). A radiotelemetry system for monitoring the treadle snare in programmes for control of wild canids. *Wildlife Research*, **23**, 381–386.

Marks, S. A. and Erickson, A. W. (1966). Age determination in the black bear. *Journal of Wildlife Management*, **30**, 389–410.

Marks, C. A., Allen, L., Gigliotti, F. *et al.* (2004). Evaluation of the tranquiliser tap device (TD) for improving the humaneness of dingo trapping. *Animal Welfare*, **13**, 393–399.

Marnewick, K. and Cilliers, D. (2006). Range use of two coalitions of male cheetahs, *Acinonyx jubatus*, in the Thabazimbi district of the Limpopo Province, South Africa. *South African Journal of Wildlife Research*, **36**, 147–151.

Marshal, J. P. and Boutin, S. (1999). Power analysis of wolf-moose functional responses. *Journal of Wildlife Management*, **63**, 396–402.

Marshall, W. H., Gullion, G. W. and Schwab, R. G. (1962). Early summer activities of porcupines as determined by radio-positioning techniques. *Journal of Wildlife Management*, **26**, 75–79.

Martin, R. E., Pine, R. H. and DeBlase, A. F. (2001). *A Manual of Mammalogy—with keys to families of the world*, 3rd edn. McGraw-Hill, New York.

Martin, J., Calengea, C., Quenettec, P.-Y. and Allainé, D. (2008). Importance of movement constraints in habitat selection studies. *Ecological Modelling*, **213**, 257–262.

Martonosi, M. (2006). Embedded systems in the wild: ZebraNet software, hardware, and deployment experiences. *ACM Sigplan Notices*, **41**, 1–1.

Marucco, F. and McIntire, E. J. B. (2010). Predicting spatio-temporal recolonization of large carnivore populations and livestock depredation risk: wolves in the Italian Alps. *Journal of Applied Ecology*, 47, 789–798.

Marucco, F., Pletscher, D. H. and Boitani, L. (2008) Accuracy of scat sampling for carnivore diet analysis: Wolves in the Alps as a case study. *Journal of Mammalogy*, **89**, 665–673.

Mather, K. (1953). Genetical control of stability in development. *Heredity*, 7, 297–336.

Mathews, F., Honess, R. and Wolfensohn, S. (2002). Use of inhalation anaesthesia for wild mammals in the field. *The Veterinary Record*, **150**, 785–787.

Matson, G. M. (1981). *Workbook for cementum analysis.* Matson's Laboratory, Milltown, Montana. 30pp (refer to http://www.matsonslab.com/html/Miscellaneous/Publications/Publications.htm, accessed August 2011).

Matson, G. M., Casquilho-Gray, H. E., Paynich, J. D., Barnes, V. G., Jr., Reynolds, H. V., III, and Swenson, J. E. (1999). Cementum annuli are unreliable reproductive indicators in female brown bears. *Ursus*, **11**, 275–280.

Matthiopoulos, J. (2003). Model-supervised kernel smoothing for the estimation of spatial usage. *Oikos*, **102**, 367–377.

Matthiopoulos, J., Hebblewhite, M., Aarts, G. and Fieberg, J. (2011). Generalised functional responses for species distributions. *Ecology*, **92**, 583–589.

Mattson, D. J., Blanchard, B. M. and Knight, R. R. (1991) Food habits of Yellowstone grizzly bears, 1977–1987. *Canadian Journal of Zoology*, **69**, 1619–1629.

Mattisson, J., Andren, H., Persson, J. and Segerström, P. (2010). Effects of species behavior on global positioning system collar fix rates. *Journal of Wildlife Management*, 74, 557–563.

Mauritzen, M., Derocher, A. E., Pavlova, O. and Wiig, Ø. (2003a). Female polar bears, *Ursus maritumus*, on the Barents Sea drift ice, walking the treadmill. *Animal Behaviour*, **66**, 107–113.

Mauritzen, M., Belikov, S. E., Boltunov, A. N. *et al.* (2003b). Functional responses in polar bear habitat selection. *Oikos*, **100**, 112–124.

Maxfield, B., Bonzo, T., McLaughlin, C. and Bunnell, K. (2005). *Northern River Otter Management Plan*. Utah Division of Wildlife Resources, DWR Publication 04–30, Salt Lake City, Utah.

May, R., van Dijk, J., Forland, J. M., Andersen, R. and Landa, A. (2008a). Behavioural patterns in ewe-lamb pairs and vulnerability to predation by wolverines. *Applied Animal Behaviour Science*, **112**, 58–67.

May, R., van Dijk, J., Wabakken, P. *et al.* (2008b). Habitat differentiation within the large-carnivore community of Norway's multiple-use landscapes. *Journal of Applied Ecology*, **45**, 1382–1391.

Maynard Smith, J. (1976). Evolution and the theory of games. *American Scientist*, **64**, 41–45.

Mayor, S. J., Schneider, D. C., Schaefer, J. A., and Mahoney, S. P. (2009). Habitat selection at multiple scales. *Ecoscience*, **16**, 238–247.

McCallum, H., Tompkins, D. M., Jones, M. *et al.* (2007). Distribution and impacts of Tasmanian devil facial tumor disease. *EcoHealth*, 4, 318–325.

McCarthy, T. M., Fuller, T. K. and Munkhtsog, B. (2005). Movements and activities of snow leopards in Southwestern Mongolia. *Biological Conservation*, **124**, 527–537.

McCarthy, K. P., Fuller, T. K., Ming, M., McCarthy, T. M., Waits, L. and Jumabaev, K. (2008). Assessing estimators of snow leopard abundance. *Journal of Wildlife Management*, 72, 1826–1833.

McComb, B., Zuckerberg, B., Vesely, D. and Jordan, C. (2010). *Monitoring animal populations and their habitat: a practitioner's guide*. CRC Press, Boca Raton, FL.

McConway, K. (1992). The number of subjects in animal behaviour experiments, is still right? In M. S. Dawkins and L. M. Gosling (eds). *Ethics in research and animal behaviour*, pp. 35–38. Academic Press, London.

McCord, C. M. and Cardoza, J. E. (1982). Bobcat and lynx. In J. A. Chapman and G. A. Feldhamer (eds). *Wild mammals of North America*, pp. 728–766. The John Hopkins University Press, Baltimore, MD.

McCrown, J. W., Maehr, D. S. and Roboski, J. (1990). A portable cushion as a wildlife capture aid. *Wildlife Society Bulletin*, **18**, 34–36.

McCullough, C. R., Heggemann, L. D. and Caldwell, C. H. (1986). A device to restrain river otters. *Wildlife Society Bulletin*, **14**, 177–180.

McCusker, J. S. (1974). Breeding Malayan sun bears *Helarctos malayanus* at Fort Worth Zoo. *International Zoo Yearbook*, **14**, 118–119.

McDaniel, G. W., McKelvey, K. S., Squires, J. R. and Ruggiero, L. F. (2000). Efficacy of lures and hair snares to detect lynx. *Wildlife Society Bulletin*, **28**, 119–123.

McDermid, G. J., Franklin, S. E. and LeDrew, E. F. (2005). Remote sensing for large-area habitat mapping. *Progress in Physical Geography*, **29**, 449–474.

McDermid, G. J., Hall, R. J., Sanchez-Azofeifa, G. A. *et al.* (2009). Remote sensing and forest inventory for wildlife habitat assessment. *Forest Ecology and Management*, **257**, 2262–2269.

McDonald, J. E. and Fuller, T. K. (2001). Prediction of litter size in American black bears. *Ursus*, **12**, 93–102.

McDonald, L. L. (2004). Sampling rare populations. In W. L. Thompson (ed.). *Sampling rare or elusive species: concepts, designs, and techniques for estimating population parameters*, pp. 11–42. Island Press, Washington DC.

McFadden, K. W., Sambrotto, R. N., Medellin, R. A. and Gompper, M. E. (2006). Feeding habits of endangered pygmy raccoons (*Procyon pygmaeus*) based on stable isotope and fecal analyses. *Journal of Mammalogy*, **87**, 501–509.

McKelvey, K. S. and Schwartz, M. K. (2004) Genetic errors associated with population estimation using non-invasive molecular tagging: problems and new solutions. *Journal of Wildlife Management*, **68**, 439–448.

McKelvey, K. S., Von Kienast, J., Aubry, K. B. *et al.* (2006). DNA analysis of hair and scat collected along snow tracks to document the presence of Canada lynx. *Wildlife Society Bulletin*, **34**, 451–455.

McKelvey, K. S., Aubry, K. B. and Schwartz, M. K. (2008). Using anecdotal occurrence data for rare or elusive species: the illusion of reality and a call for evidentiary standards. *BioScience* **58**, 549–555.

McLean, P. K. and Pelton, M. R. (1994). Estimates of population density and growth of black bears in the Smoky Mountains. *International Conference of Bear Research and Management*, **9**, 253–261.

McLoughlin, P. D., Cluff, H. D. and Messier, F. (2002a). Denning ecology of barren-ground grizzly bears in the central Artic. *Journal of Mammalogy*, **83**, 188–198.

McLoughlin, P. D., Cluff, H. D., Gau, R. J., Mulders, R., Case, R. L. and Messier, F. (2002b). Population delineation of barren-ground grizzly bears in the central Canadian Artic. *Wildlife Society Bulletin*, **30**, 728–737.

McLoughlin, P. D., Walton, L. R., Cluff, H. D., Paquet, P. C. and Ramsay, M. A. (2004). Hierarchical habitat selection by tundra wolves. *Journal of Mammalogy*, **85**, 576–580.

McLoughlin, P. D., Morris, D., Fortin, D., Van der Wal, F. and Contasti, A. (2010). Considering ecological dynamics in resource selection functions. *Journal of Animal Ecology*, **79**, 4–12.

McNamara, J. M. and Houston, A. I. (1987). Starvation and predation as factors limiting population size. *Ecology*, **68**, 1515–1519.

McNay, M. E., Stephenson, T. R. and Dale, B. W. (2006). Diagnosing pregnancy, in utero litter size, and fetal growth with ultrasound in wild, free-ranging wolves. *Journal of Mammalogy*, **87**, 85–92.

McNay, M. E., Stephenson, T. R. and Dale, B. W. (2006). Diagnosing pregnancy, in utero litter size, and fetal growth with ultrasound in wild, free-ranging wolves. *Journal of Mammalogy*, **87**, 85–92.

Mead, R. A. (1968). Reproduction in western forms of the spotted skunk (genus *Spilogale*). *Journal of Mammalogy*, **49**, 373–390.

Mead, R. A. (1989). The physiology and evolution of delayed implantation in carnivores. In J. L. Gittleman (ed.). *Carnivores: behavior, ecology, and evolution*, pp. 437–464. Cornell University Press, Ithaca, NY.

Mead, R. A. (1993). Embryonic diapause in vertebrates. *Journal of Experimental Zoology A: Comparative Experimental Biology*, **266**, 629–641.

Mead, R. A., Neirinckx, S. and Czekala, N. M. (1990). Reproductive cycle of the steppe polecat. *Journal of Reproduction and Fertility*, **88**, 353–360.

Mead, R. A., Rector, M., Starypan, G., Neirinckx, S., Jones, M. and DonCarlos, M. N. (1991). Reproductive biology of captive wolverines. *Journal of Mammalogy*, **72**, 807–814.

Mead, R. A., Bowles, M., Starypan, G., and Jones, M. (1993a). Evidence for pseudopregnancy and induced ovulation in captive wolverines (*Gulo gulo*). *Zoo Biology*, **12**, 353–358.

Meadow, R., Reading, R. P., Phillips, M., Mehringer, M. and Miller, B. J. (2005). The influence of persuasive arguments on public attitudes toward a proposed wolf restoration in the southern Rockies. *Wildlife Society Bulletin*, **33**, 154–163.

Mech, L. D. (1966). *The wolves of Isle Royale*. Fauna of the National Parks of the United States, Fauna Series No. 7.

Mech, L. D. (1970). *The wolf, ecology and behavior of an endangered species*. Natural History Press, Garden City, NJ.

Mech, L.D. (1983). *Handbook of animal radio-tracking*. Minneapolis: University Minnesota Press.

Mech, L. D. (2002). Breeding season of wolves in relation to latitude. *Canadian Field-Naturalist*, **116**, 139–140.

Mech, L.D. (2006). Age-related body mass and reproductive measurements of gray wolves in Minnesota. *Journal of Mammalogy*, **87**, 80–84.

Mech, L. D. (2007). Prediction failure of a wolf landscape model. (2006). *Journal of Wildlife Management*, **71**, 1021.

Mech, L. D. and Barber, S. M. (2002). *A critique of wildlife radio-tracking and its use in national parks. A report to the U.S. National Park Service*. [Online] Available from: http://www.npwrc.usgs.gov/resource/wildlife/radiotrk/radiotrk.pdf, Accessed 31 March 2011].

Mech, L. D. and Boitani, L. (eds). (2003). *Wolves, behaviour, ecology, and conservation*. University of Chicago Press, Chicago, IL.

Mech, L. D. and Boitani, L. (2003). Wolf social ecology. In L. D. Mech and L. Boitani (eds). *Wolves, behavior, ecology and conservation*, pp. 1–34. University of Chicago Press. Chicago, IL.

Mech, L. D. and Gese, E. M. (1992). Field testing the Wildlink capture collar on wolves. *Wildlife Society Bulletin*, **20**, 221–223.

Mech, L. D., Fritts, S. H., Radde, G. L. and Paul, W. J. (1988). Wolf distribution and road density in Minnesota. *Wildlife Society Bulletin*, **16**, 85–87.

Mech, L. D., Chapman, R. C., Cochran, W. W., Simmons, L. and Seal, U. S. (1984). Radio-triggered anaesthetic-dart collar for recapturing large mammals. *Wildlife Society Bulletin*, **12**, 69–74.

Mech, L. D., Goyal, S. M., Paul, W. J. and Newton, W. E. (2008). Demographic effects of canine parvovirus on a free-ranging wolf population over 30 years. *Journal of Wildlife Diseases*, **44**, 824–836.

Meek, P. D., Jenkins, D. J., Morris, B., Ardler, A. J. and Hawksby, R. J. (1995). Use of two humane leg-hold traps for catching pet species. *Wildlife Research*, **22**, 733–739.

Melquist, W. E. and Hornocker, M. G. (1979). Development and use of a telemetry technique for studying river otter. *Proceedings of the Fourth International Wildlife Biotelemetry Conference*, **2**, 104–114.

Menard, S. (ed.). (2002). *Applied logistic regression*. Second edn. Sage Publications, London.

Meriggi, A. and Lovari, S. (1996). A review of wolf predation in southern Europe: does the wolf prefer wild prey to livestock? *Journal of Applied Ecology*, **33**, 1561–1571.

Merrill, S. B. (2000). Road densities and gray wolf (*Canis lupus*) habitat suitability, an exception. *Canadian Field-Naturalist*, **114**, 312–313.

Merrill, S. B. and Erickson, C. R. (2003). A GPS-based method to examine wolf response to loud noise. *Wildlife Society Bulletin*, **31**, 769–773.

Merrill, S. B. and Mech, L. D. (2003). The usefulness of GPS telemetry to study wolf circadian and social activity. *Wildlife Society Bulletin*, **31**, 947–960.

Merrill, S. B., Adams, L. G., Nelson, M. E. and Mech, L. D. (1998). Testing releasable GPS radiocollars on wolves and white-tailed deer. *Wildlife Society Bulletin*, **26**, 830–835.

Merrill, E., Sand, H., Zimmermann, B. *et al.* (2010). Building a mechanistic understanding of predation with GPS-based movement data. *Philosophical Transactions of the Royal Society B-Biological Sciences*, **365**, 2279–2288.

Mertens, A., Gheorghe, P. and Promberger, C. (2001). Carnivore damage to livestock in Romania. *Carnivore Damage Prevention News*, **4**, 9.

Mertens, A., Promberger, C. and Gheorghe, P. (2002). Testing and implementing the use of electric fences for night corrals in Romania. *Carnivore Damage Prevention News*, **5**, 2–4.

Metsers, E. M., Seddon, P. J. and van Heezik, Y. M. (2010). Cat-exclusion zones in rural and urban-fringe landscapes: how large would they have to be? *Wildlife Research*, **37**, 47–56.

Michalski, F., Boulhosa, R. L. P., Faria, A. and Peres, C. A. (2006). Human-wildlife conflicts in a fragmented Amazonian forest landscape: determinants of large felid depredation on livestock. *Animal Conservation*, **9**, 179–188.

Miller, P. S. and Lacy, R. C. (2005). *VORTEX: a stochastic simulation of the extinction process. Version 9.50 user's manual.* Conservation Breeding Specialist Group (IUCN/SSC), Apple Valley, MN.

Miller, B, Biggins, D., Wemmer, C., Powell, R. A., Calvo, L. and Wharton, T. (1990a). Development of survival skills in captive-raised Siberian ferrets (*Mustela eversmanni*). II. Predator avoidance. *Ethology*, **8**, 95–104.

Miller, B., Biggins, D., Wemmer, C., Powell, R. A., Calvo, L. and Wharton, T. (1990b). Development of survival skills in captive-raised Siberian ferrets (*Mustela eversmanni*). I. Locating prey. *Ethology*, **8**, 89–94.

Miller, B., Reading, R. P. and Forrest, S. (1996). *Prairie night*. Smithsonian Institution Press, Washington DC.

Miller, S. D., White, G. C., Sellers, R. A. *et al.* (1997). Brown and black bear density estimation in Alaska using radiotelemetry and replicated mark-resight techniques. *Wildlife Monographs*, **133**, 1–55.

Miller, C. R., Joyce, P. and Waits, L. P. (2005). A new method for estimating the size of small populations from genetic mark-recapture data. *Molecular Ecology*, **14**, 1991–2005.

Miller, D. A., Grand, J. B., Fondell, T. E. and Anthony, M. (2006). Predator functional response and prey survival: direct and indirect interactions affecting a marked prey population. *Journal of Animal Ecology*, **75**, 101–110.

Miller, D. A., Nichols, J. D., McClintock, B. T., Grant, E. H. C. Bailey, L. L. and Weir, L. (2011). Improving occupancy estimation when two types of observational error occur: nondetection and species misidentification. *Ecology*, **92**, 1422–1428.

Millions, D. G. and Swanson, B. J. (2006). An application of Manel's model: detecting bobcat poaching in Michigan. *Wildlife Society Bulletin*, **34**, 150–155.

Millon, A. and Bretagnolle, V. (2008). Predator population dynamics under a cyclic prey regime: numerical responses, demographic parameters and growth rates. *Oikos*, **117**, 1500–1510.

Mills, M. G. L. (1991). Conservation management of large carnivores in Africa. *Koedoe*, **34**, 81–90.

Mills, S. (2002). False samples are not the same as blind controls. *Nature*, **415**, 732–732.

Mills, L. S. (2007). *Conservation of wildlife populations: demography, genetics, and management.* Wiley-Blackwell Publishing, Oxford.

Mills, M. G. L. and Gorman, M. L. (1997). Factors affecting the density and distribution of wild dogs in the Kruger National Park. *Conservation Biology*, **11**, 1397–1406.

Mills, M. G. L. and Hofer, H. (1998). *Hyaenas. Status survey and conservation action plan.* IUCN, Gland, Swizterland.

Mills, L. S. and Knowlton, F. F. (1989). Observer performance in known and blind radio-telemetry accuracy tests. *Journal of Wildlife Management*, **53**, 340–342.

Mills, L. S. and Smouse, P. E. (1994). Demographic consequences of inbreeding in remnant populations. *American Naturalist*, **144**, 412–431.

Mills, L. S., Pilgrin, K. L., Schwartz, M. K. and McKelvey, K. (2000a). Identifying lynx and other North American felids based on MtDNA analysis. *Conservation Genetics*, **1**, 285–288.

Mills, L. S., Citta, J. J., Lair, K. P., Schwartz, M. K. and Tallmon, D. A. (2000b). Estimating animal abundance using noninvasive DNA sampling: promise and pitfalls. *Ecological Applications*, **10**, 283–294.

Mills, K. J., Patterson, B. R. and Murray, D. L. (2006). Effects of variable sampling frequencies on GPS transmitter efficiency and estimated wolf home range size and movement distance. *Wildlife Society Bulletin*, **34**, 1463–1469.

Mills, K. J., Patterson, B. R. and Murray, D. L. (2008). Direct estimation of early survival and movements in eastern wolf pups. *Journal of Wildlife Management*, **72**, 949–954.

Millspaugh, J. J. and Marzluff, J. M. (eds) (2001). *Radio tracking and animal populations.* Academic Press, San Diego, California.

Millspaugh, J. J. and Washburn, B. E. (2003). Within-sample variation of fecal glucocorticoid measurements. *General and Comparative Endocrinology*, **132**, 21–26.

Millspaugh, J. J. and Washburn, B. E. (2004). Use of fecal glucocorticoid metabolite measures in conservation biology research: considerations for application and interpretation. *General and Comparative Endocrinology*, **138**, 189–199.

Millspaugh, J. J., Nielson, R. M., McDonald, L. *et al.* (2006). Analysis of resource selection using utilization distributions. *Journal of Wildlife Management*, **70**, 384–395.

Ministry of Agriculture, Fisheries and Food. (1983). *Bovine tuberculosis in badger—seventh report by the Ministry of Agriculture, Fisheries and Food (MAFF)*. Her Majesty's Stationery Office, London.

Minta, S. C. (1992). Tests of spatial and temporal interaction among animals. *Ecological Applications*, **2**, 178–188.

Miquelle, D. G., Smirnov, E. N., Quigley, H., Hornocker, M., Nikolaev, I. and Matyushkin, E. N. (1996). Food habits of Amur tigers in Sikhote-Alin Zapovednik and the Russian Far East, and implications for conservation. *Journal of Wildlife Research*, **1**, 138–147.

Miquelle, D. G., Smirnov, E. N., Merrill, T. W. *et al.* (1999). Hierarchical spatial analysis of Amur tiger relationships to habitat and prey. In P. J. Seidensticker (ed.). *Riding the tiger, tiger conservation in human-dominated landscapes*, pp. 71–99. Cambridge University Press, Cambridge.

Miquelle, D. G., Pikunov, D. G., Dunishenko, Y. M. *et al.* (2006). A survey of Amur (Siberian) tigers in the Russian Far East, 2004–2005. Wildlife Conservation Society, New York, and World Wildlife Fund, Washington DC.

Miquelle, D. G., Pikunov, D. G., Dunishenko, Y. M. *et al.* (2007). 2005 Amur tiger census. *Cat News*, **46**, 14–16.

Mirić, D. (1978). *Lynx lynx martinoi* ssp. nova (Carnivora, Mammalia). Neue Luchsunterart von der Balkanhalbinsel. *Bulletin de museum d'histoire naturelle, Belgrade, Série B*, Livre **33**, 29–36.

Mishra, C., Prins, H. H. T. and van Wieren, S. (2003). Diversity, risk mediation, and change in a Trans-Himalayan agropastoral system. *Human Ecology*, **31**, 595–609.

Mitchell, M. S. and Powell, R. A. (2003a). Linking fitness landscapes with the behavior and distribution of animals. In J. A. Bissonette and I. Storch (eds). *Landscape ecology and resource management, Linking theory with practice*, pp. 93–123. Island Press, Washington DC.

Mitchell, M. S. and Powell, R. A. (2003b). Response of black bears to forest management in the southern Appalachian Mountains. *Journal of Wildlife Management*, **67**, 692–705.

Mitchell, M. S. and Powell R. A., (2004). A mechanistic home range model for optimal use of spatially distributed resources. *Ecological Modelling*, **177**, 209–232.

Mitchell, M. S. and Powell, R. A. (2007). Optimal use of resources structures home ranges and spatial distribution of black bears. *Animal Behaviour*, **74**, 219–230.

Mitchell, M. S. and Powell, R. A. (In press). Economic models of home ranges. *Journal of Mammalogy*.

Mitchell, M. S., Zimmerman, J. W. and Powell, R. A. (2002). Test of a habitat suitability index for black bears in the Southern Appalachians. *Wildlife Society Bulletin*, **30**, 794–808.

Mladenoff, D. J. and Sickley, T. A. (1998). Assessing potential gray wolf restoration in the northeastern United States, a spatial prediction of favorable habitat and potential population levels. *Journal of Wildlife Management*, **62**, 1–10.

Mladenoff, D. J., Sickley, T. A., Haight, R. G. and Wydeven, A. P. (1995). A regional landscape analysis and prediction of favorable gray wolf habitat in the northern Great-Lakes region. *Conservation Biology*, **9**, 279–294.

Mladenoff, D. J., Sickley, T. A. and Wydeven, A. P. (1999). Predicting gray wolf landscape recolonization, logistic regression models vs. new field data. *Ecological Applications*, **9**, 37–44.

Moa, P. F., Herfindal, I., Linnell, J. D. C., Overskaug, K., Kvam, T. and Andersen, R. (2006). Does the spatiotemporal distribution of livestock influence forage patch selection in Eurasian lynx? *Wildlife Biology*, **12**, 63–70.

Moberg, G. P. (1985). *Animal stress*. University Of Michigan: American Physiological Society.

Moberg, G. P. and Mench, J. A. (eds). (2000). *The biology of animal stress, basic principles and implications for animal welfare*. CABI Publishing, Wallingford, Oxon.

Moberly, R. L., White, P. C. L., Webbon, C. C., Baker, P. J. and Harris, S. (2004). Modelling the costs of fox predation and preventive measures on sheep farms in Britain. *Journal of Environmental Management*, **70**, 129–143.

Moehlman, P. D. (1983). Socioecology of silverbacked and golden jackals (*Canis mesomelas* and *Canis aureus*). In J. F. Eisenberg and D. G. Kleiman (eds). *Recent advances in the Study of Mammalian Behavior*, pp. 423–453. American Society of Mammalogists, Lawrence, KS.

Moehrenschlager, A. and Somers, M. J. (2004). Canid reintroductions and metapopulation management. In C. Sillero-Zubiri, M. Hoffman, D. W. Macdonald (eds). *Canids, foxes, wolves, jackals and dogs. status survey and conservation action plan*, pp. 289–298. IUCN/SSC Canid Specialist Group, IUCN, Gland, Switzerland.

Moehrenschlager, A., List, R. and Macdonald, D. W. (2007). Escaping intraguild predation: Mexican kit foxes survive while coyotes and golden eagles kill Canadian swift foxes. *Journal of Mammalogy*, **88**, 1029–1039.

Moehrenschlager, A., Macdonald, D. W. and Moehrenschlager, C. (2003). Reducing capture-related injuries and radio-collaring effects on swift foxes. In M. A. Sovada and L. Carbyn (eds). *The swift fox. ecology and conservation of swift foxes in a changing world*, pp. 107–113. Canadian Plains Research Center, University of Regina, Saskatchewan.

Moen, R., Burdett, C. and Niemi, G. J. (2008). Movement and habitat use of Canada lynx during denning in Minnesota. *Journal of Wildlife Management*, **72**, 1507–1513.

Molinari-Jobin, A., Molinari, P., Loison, A., Gaillard, J. M. and Breitenmoser, U. (2004). Life cycle period and activity of prey influence their susceptibility to predators. *Ecography*, **27**, 323–329.

Moll, R. J., Millspaugh, J. J., Beringer, J., Sartwell, J. and He, Z. (2007). A new "view" of ecology and conservation through animal-borne video systems. *Trends in Ecology and Evolution*, **22**, 660–668.

Møller, A. P. and Pomianlpwski, A. (1993). Fluctuating asymmetry and sexual selection. *Genetica*, **89**, 267–269.

Møller, A. P. and Swaddle, J. P. (1997). *Developmental stability and evolutionary biology*. Oxford University Press, Oxford.

Molsher, R. L. (2001). Trapping and demographics of feral cats (*Felis catus*) in central New South Wales. *Wildlife Research*, **28**, 631–636.

Mondain-Monval, M., Dutourne, B., Bonnin-laffargue, M., Canivenc, R., and Scholler, R. (1977). Ovarian activity during the anestrus and the reproductive season of the red fox (*Vulpes vulpes* L.). *Journal of Steroid Biochemistry*, **8**, 761–669.

Mondol, S., Karanth, K. U., Kumar, N. S., Gopalaswamy, A. M., Andheria, A. and Ramakrishnan, U. (2009). Evaluation of noninvasive genetic sampling methods for estimating tiger population size. *Biological Conservation*, **142**, 2350–2360.

Moorcroft, P. R. and Barnett, A. (2008). Mechanistic home range models and resource selection analysis, a reconciliation and unification. *Ecology*, **89**, 1112–1119.

Moorcroft, P. R. and Lewis, M. A. (2006). Mechanistic home range analysis. *Monographs in Population Biology*, **43**, 1–172. Princeton University Press, Princeton, NJ.

Moorcroft, P. R., Lewis, M. A. and Crabtree, R. L. (2006). Mechanistic home range models capture spatial patterns and dynamics of coyote territories in Yellowstone. *Proceedings of the Royal Society B*, **273**, 1651–1659.

Moore, D. A. (1999). Spatial diffusion of raccoon rabies in Pennsylvania, USA. *Preventive Veterinary Medicine*, **40**, 19–32.

Moore, J. E. and Swihart, R. K. (2005). Modeling patch occupancy by forest rodents: Incorporating detectability and spatial autocorrelation with hierarchically structured data. *Journal of Wildlife Management*, **69**, 933–949.

Morais, R. N., Mucciolo, R. G., Gomes, M. L. F. *et al.* (2002). Seasonal analysis of semen characteristics, serum testosterone and fecal androgens in the ocelot (*Leopardus pardalis*), margay (*L. wiedii*) and tigrina (*L. tigrinus*). *Theriogenology*, **57**, 2027–2041.

Morato, R., Verreschi, I., Guimaraes, M., Cassaro, K., Pessuti, C. and Barnabe, R. (2004). Seasonal variation in the endocrine–testicular function of captive jaguars (*Panthera onca*). *Theriogenology*, **61**, 1273–1282.

Moreira, N., Monteiro-Filho, E. L., Moraes, W. *et al.* (2001). Reproductive steroid hormones and ovarian activity in felids of the Leopardus genus. *Zoo Biology*, **20**, 103–116.

Morin, P. A., and McCarthy, M. (2007). Highly accurate SNP genotyping from historical and low-quality samples. *Molecular Ecology Notes*, 7, 937–946.

Morin, P. A., Luikart, G., Wayne, R. K. and Grp, S. N. P. W. (2004). SNPs in ecology, evolution and conservation. *Trends in Ecology and Evolution*, **19**, 208–216.

Moritz, C. (1994). Defining "evolutionary significant units." *Trends in Ecology and Evolution*, **9**, 373–375.

Morling, N. (2009). PCR in forensic genetics. *Biochemical Society Transactions*, **37**, 438–440.

Morris, J. (1975). Ovulation in raccoons and note on reproductive physiology of free-ranging raccoons. *Transactions of the Missouri Academy of Sciences*, 7, 261–262.

Morris, D. W. (2003a). How can we apply theories of habitat selection to wildlife conservation and management? *Wildlife Research*, **30**, 303–319.

Morris, D. W. (2003b). Toward an ecological synthesis, a case for habitat selection. *Oecologia*, **136**, 1–13.

Morris, W. F. and Doak, D. F. (eds). (2002). *Quantitative conservation biology, theory and practice of population viability analysis*. Sinauer Associates, Inc., Sunderland, MA.

Morrison, M. L. (2001). A proposed research emphasis to overcome the limits of wildlife-habitat relationship studies. *Journal of Wildlife Management*, **65**, 613–623.

Morrison, M. L., Marcot, B. G. and Mannan, R. W. (1992). *Wildlife-habitat relationships, concepts and applications*. University of Wisconsin Press, Madison, WI.

Mortelliti, A. and Boitani, L. (2008). Interaction of food resources and landscape structure in determining the probability of patch use by carnivores in fragmented landscapes. *Landscape Ecology*, **23**, 285–298.

Mortelliti, A., Cervone, C., Amori, G. and Boitani, L. (2010). The effect of non-target species in presence-absence distribution surveys: a case study with hair-tubes. *Italian Journal of Zoology*, 77, 211–215.

Mortenson, J. and Bechert, U. (2001). Carfentanil citrate used as an oral anesthetic agent for brown bears (*Ursus arctos*). *Journal of Zoo and Wildlife Medicine*, **32**, 217–221.

Morzillo, A. T., Mertig, A. G., Garner, N. and Liu, J. (2007a). Spatial distribution of attitudes toward proposed management strategies for a wildlife recovery. *Human Dimensions of Wildlife*, **12**, 15–29.

Morzillo, A. T., Mertig, A. G., Garner, N. and Liu, J. (2007b). Resident attitudes toward black bears and population recovery in East Texas. *Human Dimensions of Wildlife*, **12**, 417–428.

Mosby, H. S. (1960). *Manual of game investigational techniques*. The Wildlife Society, Ann Arbor, Michigan.

Mosser, A., Fryxell, J. M., Eberly, L. and Packer, C. (2009). Serengeti real estate, density vs. fitness-based indicators of lion habitat quality. *Ecology Letters*, **12**, 1050–1060.

Möstl, E. and Palme, R. (2002). Hormones as indicators of stress. *Domestic Animal Endocrinology*, **23**, 67–74.

Mowat, G. and Strobeck, C. (2000). Estimating population size of grizzly bears using hair capture, DNA profiling, and mark-recapture analysis. *Journal of Wildlife Management*, **64**, 183–193.

Mowat, G., Slough, B. G. and Rivard, R. (1994). A comparison of three live capturing devices for lynx, capture efficiency and injuries. *Wildlife Society Bulletin*, **22**, 267–269.

Mowat, G., Boutin, S. and Slough, B. G. (1996). Using placental scar counts to estimate litter size and pregnancy rate in lynx. *Journal of Wildlife Management*, **60**, 430–440.

Mowbray, E. E., Pursley, D. and Chapman, J. A. (1979). *The status, population characteristics, and harvest of the river otter in Maryland*. Publication in Wildlife Ecology 2, Wildlife Administration, Maryland Department of Natural Resources, Annapolis, MD.

Mudappa, D. and Chellam, R. (2001). Capture and immobilization of wild brown civets in Western Ghats. *Journal of Wildlife Diseases*, **37**, 383–386.

Mulcahy, D. M. and Garner, G. (1999). Subcutaneous implantation of satellite transmitters with percutaneous antennae into male polar bears (*Ursus maritimus*). *Journal of Zoo and Wildlife Medicine*, **30**, 510–515.

Müller, G., Liebsch, N. and Wilson, R. P. (2005). A new shot at a release mechanism for devices attached to free-living animals. *Wildlife Society Bulletin*, **33**, 337–342.

Muñoz-Igualada, J., Shivik, J. A., Domínguez, F. G., Lara, J. and González, L. M. (2002). Evaluation of cage-traps and cable restrain devices to capture red foxes in Spain. *Journal of Wildlife Management*, 72, 830–836.

Munson, L. (1999). *Necropsy of wild animals*. Wildlife Health Center, School of Veterinary Medicine, University of California, Davis, CA. http//www.vetmed.ucdavis.edu/whc/pdfs/munsonnecropsy.pdf [Accessed 26 March 2011]

Munson, L., Moresco, A. and Calle, P. P. (2005). Adverse effects of contraceptives. In C. S. Asa and I. J. Porton (eds). *Wildlife Contraception: Issues, Methods, and Application*, pp. 66–82. Johns Hopkins University Press, Baltimore, MD.

Muntifering, J. R., Dickman, A. J., Perlow, L. M. *et al.* (2006). Managing the matrix for large carnivores: a novel approach and perspective from cheetah (*Acinonyx jubatus*) habitat suitability modeling. *Animal Conservation*, **9**, 103–112.

Murie, A. (1940). *Ecology of the coyote in the Yellowstone*. US Government Printing Office, Washington, DC.

Murie, A. (1944). *The wolves of Mount McKinley*. Fauna of the National Parks of the United States, Fauna Series No. 5. Washington DC.

Murphy, B. D., Concannon, P. W., Travis, H. F. and Hansel, W. (1981). Prolactin: The hypophyseal factor that terminates embryonic diapause in mink. *Biology of Reproduction*, **25**, 487–491.

Murray, D. L. and Fuller, M. R. (2000). Effects of marking on the life history patterns of vertebrates. In L. Boitani and T. K. Fuller (eds). *Research techniques in ethology and animal ecology*, pp. 15–64. Columbia University Press, New York.

Murray, D. L., Kapke, C. A., Evermann, J. F. and Fuller, T. K. (1999). Infectious disease and the conservation of free-ranging large carnivores. *Animal Conservation*, **2**, 241–254.

Musiani, M., Mamo, C., Boitani, L. *et al.* (2003). Wolf depredation trends and the use of fladry barriers to protect livestock in western North America. *Conservation Biology*, **17**, 1538–1547.

Mustoni, A., Carlini, E., Chiarenzi, B. *et al.* (2003). The role of reintroduction in establishing a Brown bear *Ursus arctos* meta-population in the central Italian Alps. *Hystrix, Ital Journal of Mammology*, 14, 3–27.

Mysterud, A. and Ims, R. A. (1998). Functional responses in habitat use, availability influences relative use in trade-off situations. *Ecology*, **79**, 1435–1441.

Mysterud, A. and Ims, R. A. (1999). Relating populations to habitats. *Trends in Ecology and Evolution*, **14**, 489–490.

Naidenko, S. V. (2006). Body mass dynamic in Eurasian lynx *Lynx lynx* kittens during lactation. *Acta Theriologica*, **51**, 91–98.

Naidoo, R. and Ricketts, T.H. (2006). Mapping the economic costs and benefits of conservation. *PLoS Biology*, **4**, e360.

Namgail, T., Fox, J. L. and Bhatnagar, Y. V. (2007). Carnivore-caused livestock mortality in Trans-Himalaya. *Environmental Management*, **39**, 490–496.

Nams, V. O. (2006). Animal movements as behavioural bouts. *Journal of Animal Ecology*, **75**, 298–302.

Nams, V. O. and Bourgeois, M. (2004). Fractal analysis measures habitat use at different spatial scales, an example with American marten. *Canadian Journal of Zoology*, **82**, 1738–1747.

Nams, V. O. and Boutin, S. (1991). What is wrong with error polygons? *Journal of Wildlife Management*, **55**, 172–176.

Nasution, M. D., Brownie, C. and Pollock, K. H. (2002). Optimal allocation of sample sizes between regular banding and radio-tagging for estimating annual survival and emigration rates. *Journal of Applied Statistics*, **29**, 443–457.

Nasution, M. D., Brownie, C., Pollock, K. H. and Bennetts, R. E. (2001). Estimating survival from joint analysis of resighting and radiotelemetry capture-recapture data for wild animals. *Journal of Agricultural, Biological and Environmental Statistics*, **6**, 461–478.

National Biological Information Infrastructure (2010). The Gap Analysis Program. Available from http://gapanalysis.nbii.gov (accessed November 2010).

National Research Council (1982). Nutrient Requirements in Mink and Foxes. Second revised edn. Committee on Animal Nutrition. Pp. 72.

Naughton-Treves, L., Grossberg, R. and Treves, A. (2003). Paying for tolerance: rural citizens' attitudes toward wolf depredation and compensation. *Conservation Biology*, **17**, 1500–1511.

Naves, J., Wiegand, T., Revilla, E. and Delibes, M. (2003). Endangered species constrained by natural and human factors: the case of brown bears in northern spain. *Conservation Biology*, **17**, 1276–1289.

Naylor, B. J. and Novak, M. (1994). Catch efficiency and selectivity of various traps and sets used for capturing American martens. *Wildlife Society Bulletin*, **22**, 489–496.

Neill, L. O., Wilson, P., de Jongh, A., de Jong, T. and Rochford, J. (2008). Field techniques for handling, anaesthetising and fitting radio-transmitters to Eurasian otters (*Lutra lutra*). *European Journal of Wildlife Research*, **54**, 681–687.

Neiswenter, S. A. and Dowler, R. C. (2007). Habitat use of western spotted skunks and striped skunks in Texas. *Journal of Wildlife Management*, **71**, 583–586.

Nellis, C. H. (1968). Some methods for capturing coyotes. *Journal of Wildlife Management*, **32**, 402–405.

Nellis, C. H. and Keith, L. B. (1976). Population dynamics of coyotes in central Alberta, 1964–68. *Journal of Wildlife Management*, **10**, 389–39.

Nellis, D., Jenkins, J. H. and Marshall, A. D. (1967). Radio-active zinc as a faeces tag in rabbits, foxes, and bobcats. *Proceedings Annual Conference Southeastern Association Game and Fish Commissioners*, **21**, 205–207.

Nelson, R. J. (2005). *An introduction to behavioral endocrinology*, 3rd edn. Sinauer Associates, Inc. Publishers, Sunderland, MA.

Nelson, R. L. and Linder, R. L. (1972). Percentage of raccoons and skunks reached by egg baits. *Journal of Wildlife Management*, **36**, 1327–1329.

Nemtzov, S. C. (2003). A short-lived wolf depredation compensation program in Israel. *Carnivore Damage Prevention News*, **6**, 16.

New Zealand Department of Conservation. (2008). *Predator traps. The official site of the DOC series trapping systems*. Available at: http,//www.predatortraps.com/about.htm (accessed 26 March 2011).

Newsome, S. D., del Rio, C. M., Bearhop, S. and Phillips, D. L. (2007). A niche for isotopic ecology. *Frontiers in Ecology and the Environment*, **5**, 429–436.

Nichols, J. D. and Karanth, K. U. (2002). Statistical concepts, assessing spatial distributions. In K. U. Karanth and J. D. Nichols (eds). *Monitoring tigers and their prey. A manual for wildlife managers, researchers, and conservationists*, pp. 29–38. Centre for Wildlife Studies, Bangalore, India.

Nichols, J. D. and Pollock, K. H. (1990). Estimation of recruitment from immigration versus in situ reproduction using Pollock's robust design. *Ecology*, **71**, 21–26.

Nichols, J. D. and Williams, B. K. (2006). Monitoring for conservation. *Trends in Ecology and Evolution*, 21, 668–673.

Nichols, J. D., Sauer, J. R., Pollock, K. H. and Hestbeck, J. B. (1992). Estimating transition probabilities for stage-based population projection matrices using capture-recapture data. *Ecology*, 73, 306–312.

Nichols, J. D., Hines, J. E., Lebreton, J. D. and Pradel, R. (2000). Estimation of contributions to population growth, a reverse-time capture-recapture approach. *Ecology*, **81**, 3362–3376.

Nichols, J. D., Hines, J. E., MacKenzie, D. I., Seamans, M. E. and Gutierrez, R. J. (2007). Occupancy estimation with multiple states and state uncertainty. *Ecology*, **88**, 1395–1400.

Nichols, J. D., Bailey, L. L., O'Connell, J. *et al.* (2008). Multi-scale occupancy estimation and modelling using multiple detection methods. *Journal of Applied Ecology*, **45**, 1321–1329.

Nicholson, J. M., and Van Manen, F. T. (2009). Using occupancy models to determine mammalian responses to landscape changes. *Integrative Zoology*, 4, 232–239.

Nielsen, L. (1999). *Chemical immobilization of wild and exotic animals*. Iowa State University Press, Ames, IO.

Nielsen, S. E., Herrero, S., Boyce, M. S. *et al.* (2004). Modelling the spatial distribution of human-caused grizzly bear mortalities in the Central Rockies Ecosystem of Canada. *Biological Conservation*, **120**, 101–113.

Nielsen, S. E., Johnson, C. E., Heard, D. C. and Boyce, M. S. (2005). Can models of presence-absence be used to scale abundance? Two case studies considering extremes in life history. *Ecography*, **28**, 197–208.

Nielsen, S. E., Stenhouse, J. B. and Boyce, M. S. (2006). A habitat-based framework for grizzly bear conservation in Alberta. *Biological Conservation*, **130**, 217–229.

Nietfeld, M. T., Barrett, M. W. and Silvy, N. (1994). Wildlife marking techniques. In T. K. Bookout (ed.). *Research and management techniques for wildlife and habitats*, pp. 140–168. The Wildlife Society, Bethesda, MD.

Niin, E., Laine, M., Guiot, A. L., Demerson, J. M. and Cliquet, F. (2008). Rabies in Estonia: situation before and after the first campaigns of oral vaccination of wildlife with SAG2 vaccine bait. *Vaccine*, **26**, 3556–3565.

Nilsen, E. B., Linnell, J. D. C., Odden, J. and Andersen, R. (2009a) Climate, season, and social status modulate the functional response of an efficient stalking predator: the Eurasian lynx. *Journal of Animal Ecology*, **78**, 741–751.

Nilsen, E. B., Gaillard, J. M., Andersen, R. *et al.* (2009b). A slow life in hell or a fast life in heaven: demographic analyses of contrasting roe deer populations. *Journal of Animal Ecology*, **78**, 585–594.

Nolan, J. W., Russell, R. H. and Anderka, F. (1984). Transmitters for monitoring Aldrich snares set for grizzly bears. *Journal of Wildlife Management*, **48**, 942–945.

Nordstrom, C. A., Wilson, L. J., Iverson, S. J. and Tollit, D. J. (2008). Evaluating quantitative fatty acid signature analysis (QFASA) using harbor seals *Phoca vitulina richardsi* in captive feeding studies. *Marine Ecology Progress Series*, **360**, 245–263.

Norris, J. L. and Pollock, K. H. (1995). Capture-recapture model M_{bh}, Bivariate MLE of capture-recapture distribution and resulting model selection. *Journal of Environmental and Ecological Statistics*, **2**, 305–313.

Norris, J. L. and Pollock, K. H. (1996). Nonparametric maximum likelihood estimators for population size under two closed capture-recapture models with heterogeneity. *Biometrics,* **52**, 639–649.

North, M. P. and Reynolds, J. H. (1996). Microhabitat analysis using radiotelemetry locations and polytomous logistic regression. *Journal of Wildlife Management*, **60**, 639–653.

Northcott, T. H., and Slade, D. (1976). A live trapping technique for river otters. *Journal of Wildlife Management*, **40**, 163–164.

Norwegian Consensus-Platform for Alternatives (Norecopa). (2008). *A consensus document from all the participants*. Harmonisation of the Care and Use of Animals in Field Research, Gardermoen, Norway, 21–22 May 2008. Available at: www.norecopa.no/sider/tekst.asp?side=8 (accessed 26 March 2011).

Novak, M. (1981). The foot-snare and the leg-hold traps, a comparison. In J. A. Chapman and D. Pursley (eds). *Proceedings Worldwide Furbearer Conference*, pp. 1671–1685. International Association of Fish & Wildlife, Washington, DC.

Nowak, R. (2003). Wolf evolution and taxonomy. In L. Mech and L. Boitani (eds). *Wolves. Behavior, ecology and conservation,* pp. 239–258. Chicago University Press, Chicago, IL.

Nowell, K. and Jackson, P. (1996). *Wild cats: status survey and action plan*. IUCN, Gland, Switzerland.

Nusser, S. M., Clark, W. R., Otis, D. L. and Huang, L. (2008). Sampling considerations for disease surveillance in wildlife populations. *Journal of Wildlife Management,* 72, 52–60.

Nutter, F. B., Levine, J. F., Stoskopf, M. K., Gamble, H. R. and Dubey, J. P. (1998). Seroprevalence of *Toxoplasma gondii* and *Trichinella spiralis* in North Carolina black bears (*Ursus americanus*). *Journal of Parasitology*, **84**, 1048–1050.

Nyariki, D. M., Mwang'ombe, A. W. and Thompson, D. M. (2009). Land-use change and livestock production challenges in an integrated system: the Masai-mara ecosystem, Kenya. *Journal of Human Ecology*, **26**, 163–173.

Nyholm, E. (1959). Kärpästä ja lumikosta ja nïïden talvisista elinpïïreistä. *Suomen Riista*, **13**, 106–116.

Nyhus, P. J., Osofsky, S. A., Ferraro, P., Madden, F., and Fischer, H. (2005). Bearing the costs of human-wildlife conflict: the challenges of compensation schemes. In R. Woodroffe, S. Thirgood, and A. Rabinowitz (eds), pp. 107–121. *People and wildlife: conflict or co-existence*. Cambridge University Press Cambridge.

Ó Néill, L., De Jongh, A., Ozolinš, J., De Jong, T, and Rochford, J. (2007). Minimizing leg-hold trapping trauma for otters with mobile phone technology. *Journal of Wildlife Management*, 71, 2776–2780.

O'Brien, S. J. and Mayr, E. (1991). Bureaucratic mischief: recognizing endangered species and subspecies. *Science,* **251**, 1187–1188.

O'Connell, A. F., Talancy, N. W., Bailey, L. L., Sauer, J. R., Cook, R. and Gilbert, A. T. (2006). Estimating site occupancy and detection probability parameters for mammals in a coastal ecosystem. *Journal of Wildlife Management*, **70**, 1625–1633.

O'Keefe, J. and Dostrovsky, J. (1971). The hippocampus as a spatial map. Preliminary evidence from unit activity in the freely-moving rat. *Brain Research*, **34**, 171–175.

O'Keefe, J. and Nadel, L. (1978). *The hippocampus as a cognitive map*. Clarendon, Oxford.

O'Neill, R. V., DeAngelis, D. L., Waide, J. B. and Allen, T. F. H. (1986). *A hierarchical concept of ecosystems*. Princeton University Press, Princeton.

Oakleaf, J. K., Mack, C. M. and Murray, D. L. (2003). Effects of wolves on livestock calf survival and movements in central Idaho. *Journal of Wildlife Management*, 67, 299–306.

Oakleaf, J. K., Murray, D. L., Oakleaf, J. R. *et al.* (2006). Habitat Selection by Recolonizing Wolves in the Northern Rocky Mountains of the United States. *Journal of Wildlife Management*, **70**, 554–563.

Obbard, M. E., Howe, E. J. and Kyle, C. J. (2010). Empirical comparison of density estimators for large carnivores. *Journal of Applied Ecology*, 47, 76–84.

Odden, J., Linnell, J. D. C., Moa, P. F., Herfindal, I., Kvam, T. and Andersen, R. (2002). Lynx depredation on domestic sheep in Norway. *Journal of Wildlife Management*, **66**, 98–105.

Odden, J., Linnell, J. D. C. and Andersen, R. (2006). Diet of Eurasian lynx, *Lynx lynx*, in the boreal forest of southeastern Norway: the relative importance of livestock and hares at low roe deer density. *European Journal of Wildlife Research*, **52**, 237–244.

Odden, J., Herfindal, I., Linnell, J. D. C. and Andersen, R. (2008). Vulnerability of domestic sheep to lynx depredation in relation to roe deer density. *Journal of Wildlife Management*, 72, 276–282.

O'Donoghue, M., Boutin, S., Krebs, C. J. and Hofer, E. J. (1997). Numerical responses of coyotes and lynx to the snowshoe hare cycle. *Oikos*, **80**, 150–162.

O'Donoghue, M., Boutin, S., Krebs, C. J., Zuleta, G., Murray, D. L. and Hofer, E. J. (1998). Functional responses of coyotes and lynx to the snowshoe hare cycle. *Ecology*, **79**, 1193–1208.

Ogada, M. O., Woodroffe, R., Oguge, N. O. and Frank, L. G. (2003). Limiting depredation by African carnivores: the role of livestock husbandry. *Conservation Biology*, 17, 1521–1530.

Ogilvie, S. C. and Eason, C. T. (1998). Evaluation of iophenoxic acid and rhodamine B for marking feral ferrets (*Mustelo furo*). *New Zealand Journal of Zoology*, **25**, 105–108.

Ogutu, J., Bhola, N. and Reid, R. (2005). The effects of pastoralism and protection on the density and distribution of carnivores and their prey in the Mara ecosystem of Kenya. *Journal of Zoology*, **265**, 281–293.

Okarma, H. and Jędrzejewski, W. (1997). Live-trapping wolves with nets. *Wildlife Society Bulletin*, **25**, 78–82.

Okarma, H., Jędrzejewski, W., Schmidt, K., Śnieżko, S., Bunevich, A. N. and Jędrzejewska, B. (1998). Home ranges of wolves in Białowieża primeval forest, Poland, compared with other Eurasian populations. *Journal of Mammalogy*, **79**, 842–852.

Okoniewski, J. C. and Chambers, R. E. (1984). Coyote vocal response to an electronic siren and human howling. *Journal of Wildlife Management*, **48**, 217–222.

Oksanen, T., Power, M. E. and Oksanen, L. (1995). Ideal free habitat selection and consumer-resource dynamics. *American Naturalist*, **146**, 565–585.

Oli, M. K. (1993). A key for the identification of the hair of mammals of a snow leopard (*Panthera uncia*) habitat in Nepal. *Journal of Zoology*, **231**, 71–93.

Olsen, G. H., Linhart, S. B., Holmes, R. A., Dasch, G. J. and Male, C. B. (1986). Injuries to coyotes caught in padded and unpadded steel foothold traps. *Wildlife Society Bulletin*, **14**, 219–223.

Olsen H. H., Linscombe, R. G., Wright, V. L. and Holmes, R. A. (1988). Reducing injuries to terrestrial furbearers by using padded foothold traps. *Wildlife Society Bulletin*, **16**, 303–307.

Onderka, D. K., Skinner, D. L. and Todd, A. W. (1990). Injuries to coyotes and other species caught by four models of footholding devices. *Wildlife Society Bulletin*, **18**, 175–182.

Onorato, D., White, C., Zager, P. and Waits, L. P. (2006). Detection of predator presence at elk mortality sites using mtDNA analysis of hair and scat samples. *Wildlife Society Bulletin*, **34**, 815–820.

Otali, E. and Gilchrist, J. S. (2004). The effects of refuse feeding on body condition, reproduction, and survival of banded mongooses. *Journal of Mammalogy*, **85**, 491–497.

Otis, D. L., Burnham, K. P., White, G. C. and Anderson, D. L. (1978). Statistical inference from capture data on closed animal populations. *Wildlife Monographs* **62**.

Oyler-McCance, S. J. and Leberg, P. L. (2005). Conservation genetics in wildlife management. In C. E. Braun (ed.). *Techniques for wildlife investigations and management*, pp. 632–657. The Wildlife Society, Bethesda, MD.

Packard, J. M., Mech, L. D. and Seal, U. S. (1983). Social influences on reproduction in wolves. In L. Carbyn (ed.). *Wolves in Canada and Alaska: their status, biology and management*, pp. 78–85. Canadian Wildlife Service, Ottawa.

Paetkau, D., Calvert, W., Stirling, I. and Strobeck, C. (1995). Microsatellite analysis of population-structure in Canadian polar bears. *Molecular Ecology*, **4**, 347–354.

Page, L. K., Swihart, R. K. and Kazacos, K. R. (1999). Implications of raccoon latrines in the epizootiology of baylisascariasis. *Journal of Wildlife Diseases*, **35**, 474–480.

Paisley, S. and Garshelis, D. L. (2006). Activity patterns and time budgets of Andean bears (*Tremarctos ornatus*) in the Apolobamba Range of Bolivia. *Journal of Zoology*, **268**, 25–34.

Palme, R. (2005). Measuring fecal steroids guidelines for practical application. *Annals of the New York Academy of Sciences*, **1046**, 75–80.

Palmer, W. E., Wellendorf, S. D., Gillis, J. R. and Bromley, P. T. (2005). Effect of field borders and nest-predator reduction on abundance of northern bobwhites. *Wildlife Society Bulletin*, **33**, 1398–1405.

Palmeria, F. B. L., Crawshaw, P. G., Haddad, C. M., Ferraz, K. M. P. M. B. and Verdade, L. M. (2008). Cattle depredation by puma (*Puma concolor*) and jaguar (*Panthera onca*) in central-western Brazil. *Biological Conservation*, **141**, 118–125.

Palomares, F. (1994). Site fidelity and effects of body mass on home-range size of Egyptian mongooses. *Canadian Journal of Zoology*, 72, 465–469.

Palomares, F., Gaona, P., Ferreras, P. and Delibes, M. (1995). Positive effects on game species of top predators by controlling smaller predator populations: an example with lynx, mongooses and rabbits. *Conservation Biology*, **9**, 295–305.

Palomares, F., Delibes, M., Revilla, E., Calzada, J. and Fedriani, J. M. (2001). Spatial ecology of Iberian lynx and abundance of European rabbits in southwestern Spain. *Wildlife Monographs No.* **148**.

Pamperin, N. J., Follmann, E. H. and Person, B. T. (2008). Sea-ice use by arctic foxes in northern Alaska. *Polar Biology*, **31**, 1421–1426.

Pangle, K. L., Peacor, S. D. and Johannsson, O. E. (2007). Large nonlethal effects of an invasive invertebrate predator on zooplankton population growth rate. *Ecology*, **88**, 402–412.

Panzacchi, M., Linnell, J. D. C., Serrao, G. *et al.* (2008a). Evaluation of the importance of roe deer fawns in the spring-summer diet of red foxes in southeastern Norway. *Ecological Research*, **23**, 889–896.

Panzacchi, M., Linnell, J. D. C., Odden, J., Odden, M. and Andersen, R. (2008b). When a generalist becomes a specialist: patterns of red fox predation on roe deer fawns under contrasting conditions. *Canadian Journal of Zoology*, **86**, 116–126.

Parks, J. E. and Graham, J. K. (1992). Effects of cryopreservation procedures on sperm membranes. *Theriogenology*, **38**, 209–222.

Parks, E. K., Derocher, A. E. and Lunn, N. J. (2006). Seasonal and annual movement patterns of polar bears on the sea ice of Hudson Bay. *Canadian Journal of Zoology*, **84**, 1281–1294.

Pascal, M. and Delattre, P. (1981). Comparison de différentes méthods de détermination de l'âge individual chez le vision (*Mustela vison* Schreiber). *Canadian Journal of Zoology*, **59**, 202–211.

Pastalkova, E., Itskov, V., Amarasingham, A. and Buzsáki, G. (2008). Internally generated cell assembly sequences in the rat hippocampus. *Science*, **321**, 1322–1327.

Patterson, B. D., Kasiki, S. M., Selempo, E. and Kays, R. W. (2004). Livestock predation by lions (*Panthera leo*) and other carnivores on ranches neighboring Tsavo National Parks, Kenya. *Biological Conservation*, **119**, 507–516.

Pawlina, I. and Proulx, G. (1999). Factors affecting trap efficiency, a review. In G. Proulx (ed.). *Mammal trapping*, pp. 95–115. Alpha Wildlife Research and Management Ltd., Sherwood Park, Alberta.

Paxinos, E., McIntosh, C., Ralls, K. and Fleischer, R. (1997). A noninvasive method for distinguishing among canid species: amplification and enzyme restriction of DNA from dung. *Molecular Ecology*, **6**, 483–486.

Pearce, J. L. and Boyce, M. S. (2006). Modelling distribution and abundance with presence-only data. *Journal of Applied Ecology*, **43**, 405–412.

Pedersen, V. A., Linnell, J. D. C., Andersen, R., Andren, H., Linden, M. and Segerstrom, P. (1999). Winter lynx *Lynx lynx* predation on semi-domestic reindeer *Rangifer tarandus* in northern Sweden. *Wildlife Biology*, **5**, 203–211.

Pearson, O. P. and Enders, R. K. (1944). Duration of pregnancy in certain mustelids. *Journal of Experimental Zoology*, **95**, 21–35.

Pelican, K. M., Tunwattana, W., Wildt, D. E., Fazio, J. M., Mesa-Cruz, J. B. and Howard, J. G. (2007). Fecal steroid hormones remain stable for at least 2 days under field conditions in the Thailand clouded leopard. *Proceedings of the meeting on Carnivore reproduction*. The Wildlife Society Bethesda, MD.

Pellikka, J., Rita, H. and Linden, H. (2005). Monitoring wildlife richness—Finnish applications based on wildlife triangle censuses. *Annales Zoologici Fennici*, **42**, 123–134.

Pelton, M. R. and Marcum, L. C. (1975). The potential use of radioisotopes for determining densities of black bears and other carnivores. In R. L. Phillips and C. Jonkel (eds). *Proceedings 1975 predator symposium*, pp. 221–236. Montana Forest and Conservation Experiment Station, University of Montana, Missoula, MO.

Pennock, D. S. and Dimmick, W. W. (1997). Critique of the evolutionarily significant unit as a definition for "distinct population segments" under the U.S. Endangered Species Act. *Conservation Biology*, **11**, 611–619.

Pérez-Garnelo, S. S., Garde, J., Pintado, B. *et al.* (2004). Characteristics and in vitro fertilizing ability of giant panda (*Ailuropoda melanoleuca*) frozen-thawed epididymal spermatozoa obtained 4 hours post-mortem: a case report. *Zoo Biology*, **23**, 279–285.

Perrin, M. R. and D'Inzillo Carranza, I. (1999). Capture, immobilization and measurements of the spotted-necked otter in the Natal Drakensberg, South Africa. *South African Journal of Wildlife Research*, **29**, 52–53.

Person, D. K. and Hirth, D. H. (1991). Home range and habitat use of coyotes in a farm region of Vermont. *Journal of Wildlife Management*, **55**, 433–441.

Peters, R. (1978). Communication, cognitive mapping, and strategy in wolves and hominids. In R. L. Hall and H. S. Sharp (eds). *Wolf and man, evolution in parallel*, pp. 95–108. Academic Press, New York.

Peters, R. and Mech, L. D. (1975). Scent-marking in wolves. *American Scientist*, **63**, 628–637.

Peterson, A. T. and Robins, C. R. (2003). Using ecological-niche modeling to predict barred owl invasions with implications for spotted owl conservation. *Conservation Biology*, 17, 1161–1165.

Petit, E., and Valiere, N. (2006). Estimating population size with noninvasive capture-mark-recapture data. *Conservation Biology*, **20**, 1062–1073.

Petrides, G. A. (1950). The determination of sex and age ratios in fur animals. *American Midland Naturalist*, **43**, 355–382.

Pettorelli, N., Hilborn, A., Broekhuis, F. and Durant, S. M. (2009). Exploring habitat use by cheetahs using ecological niche factor analysis. *Journal of Zoology*, **277**, 141–148.

Phillips, R. L. (1996). Evaluation of 3 types of snares for capturing coyotes. *Wildlife Society Bulletin*, **24**, 107–110.

Phillips, S. J. and Dudik, M. (2008). Modeling of species distributions with Maxent, new extensions and a comprehensive evaluation. *Ecography*, **31**, 161–175.

Phillips, D. L. and Koch, P. L. (2002). Incorporating concentration dependence in stable isotope mixing models. *Oecologia*, **130**, 105–113.

Phillips, R. L. and Gruver, K. S. (1996). Performance of the Paws-I-Trip pan tension device on 3 types of traps. *Wildlife Society Bulletin*, **24**, 119–122.

Phillips, R. L., Gruver, K. S. and Williams, E. S. (1996). Leg injuries to coyotes captured in three types of foothold traps. *Wildlife Society Bulletin*, **24**, 260–263.

Phillips, S. J., Anderson, R. P. and Schapire, R. E. (2006). Maximum entropy modeling of species geographic distributions. *Ecological Modelling*, **190**, 231–259.

Pielou, E. C. (1975). *Ecological diversity*. Wiley, New York.

Pianka, E. R. (1974). Niche overlap and diffuse competition. *Proceedings of the National Academy of Sciences of the United States of America*, 71, 2141–2145.

Pikunov, D. G., Aramilev, V. V., Fomenko, V. V. *et al.* (2000). Endangered species: The decline of the Amur leopard in the Russian Far East. *Russian Conservation News*, **24**, 19.

Pilgrim, K. L., McKelvey, K. S., Riddle, A. E. and Schwartz, M. K. (2005). Felid sex identification based on noninvasive genetic samples. *Molecular Ecology Notes*, **5**, 60–61.

Pitt, J. A., Larivière, S. and Messier, F. (2006). Condition indices and bioelectrical impedance analysis to predict body condition of small carnivores. *Journal of Mammalogy*, **87**, 717–722.

Pitt, J. A., Lariviere, S. and Messier, F. O. (2008). Social organization and group formation of raccoons at the edge of their distribution. *Journal of Mammalogy*, **89**, 646–653.

Pitt, W. C., Box, P. W. and Knowlton, F. F. (2003). An individual-based model of canid populations, modelling territoriality and social structure. *Ecological Modelling*, **166**, 109–121.

Pledger, S. (2000). Unified maximum likelihood estimates for closed capture-recapture models using mixtures. *Biometrics*, **56**, 434–442.

Pledger, S., Pollock, K. H. and Norris, J. L. (2003). Open capture-recapture models with heterogeneity, 1. Cormack-Jolly-Seber model. *Biometrics*, **59**, 786–794.

Pledger, S., Pollock, K. H. and Norris, J. L. (2010). Open capture-recapture models with heterogeneity, 2. Jolly-Seber model. *Biometrics*, **66**, 883–890.

Poessel, P., Breck, S., Biggins, D., Livieri, T., Crooks, K. and Angeloni, L. (2011). Landscape features influence postrelease predation on endangered black-footed ferrets. *Journal of Mammalogy*, **92**, 732–741.

Polasky, S., Camm, J. D., Solow, A. R., Csuti, B., White, D. and Ding, R. G. (2000). Choosing reserve networks with incomplete species information. *Biological Conservation*, **94**, 1–10.

Polisar, J., Maxit, I., Scognamillo, D., Farrell, L., Sunquist, M. E. and Eisenberg, J. F. (2003). Jaguars, pumas, their prey base, and cattle ranching: ecological interpretations of a management problem. *Biological Conservation*, **109**, 297–310.

Pollock, K. H. (1982). A capture-recapture design robust to unequal probability of capture. *Journal of Wildlife Management*, **46**, 752–757.

Pollock, K. H., Winterstein, S. R. and Conroy, M. J. (1989b). Estimation and analysis of survival distributions for radio-tagged animals. *Biometrics*, **45**, 99–109.

Pollock, K. H., Winterstein, S. R., Bunck, C. M. and Curtis, P. D. (1989a). Survival analysis in telemetry studies, the staggered entry design. *Journal of Wildlife Management*, **53**, 7–15.

Pollock, K. H., Nichols, J. D., Brownie, C. and Hines, J. E. (1990). Statistical-inference for capture-recapture experiments. *Wildlife Monographs*, **107**, 1–97.

Pollock, K. H., Nichols, J. D., Simons, T. R., Farnsworth, G. R., Bailey, L. L. and Sauer, J. R. (2002). Large scale wildlife monitoring studies, statistical methods for design and analysis. *Environmetrics*, **13**, 1–15.

Pollock, K. H., Marsh, H., Bailey, L. L., Farnsworth, G. L., Simons, T. L., and Alldredge, M. W. (2004). Separating components of detection probability in abundance estimation: An overview with diverse examples. In W. L. Thompson (ed). Sampling Rare and Elusive Species: Concepts, Designs and Techniques for Estimating Population Parameters. Island Press, Washington D. C.

Porton, I. J. (2005). The ethics of wildlife contraception. In C. S. Asa and I. J. Porton (eds). *Wildlife contraception: issues, methods, and applications*, pp. 3–16. Johns Hopkins University Press, Baltimore, MD.

Post, D. M. (2002). Using stable isotopes to estimate trophic position: models, methods, and assumptions. *Ecology*, **83**, 703–718.

Pottinger, T. G. (1999). The impact of stress on animal reproductive activities. In P. H. M. Balm (ed.). *Stress physiology in animals*, pp. 130–177. CRC Press, Boca Raton, FL.

Potvin, F., Beton, L. and Patenaude, R. (2004). Field anesthesia of American martens using isoflurane. In D. J. Harrison, A. K. Fuller and G. Proulx (eds). *Martens and fishers (Martes) in human-altered environments, an international perspective*, pp. 265–273. Springer Science +Business Media, New York.

Powell, R. A. (1978). A comparison of fisher and weasel hunting behavior. *Carnivore*, **1**, 28–34.

Powell, R. A. (1979). Ecological energetics and foraging strategies of the fisher (*Martes pennanti*). *Journal of Animal Ecology*, **48**, 195–212.

Powell, R. A. (1987). Black bear home range overlap In North Carolina and the concept of home range applied to black bears. *International Conference on Bear Research and Management*, 7, 235–242.

Powell, R. A. (1989). Effects of resource productivity, patchiness and predictability on mating and dispersal strategies. In V. Standen and R. A. Foley (eds). *Comparative Socioecology of Mammals and Humans*, pp. 101–123. British Ecological Society Symposium, Blackwell, Oxford.

Powell, R. A. (1993a). Why do some forest carnivores exhibit intrasexual territoriality and what are the consequences for management? *Proceedings of the International Union Game Biologists*, **21**, 268–273.

Powell, R. A. (1993b). *The fisher: life history, ecology, and behavior*, 2nd edn. University of Minnesota Press, Minneapolis, MN.

Powell, R. A. (1994). Structure and spacing of *Martes* populations. In S. W. Buskirk, A. S. Harestad, M. G. Raphael and R. A. Powell (eds). *Martens, sables and fishers, biology and conservation*, pp. 101–121. Cornell University Press, Ithaca, NY.

Powell, R. A. (2000). Animal home ranges and territories and home range estimators. In L. Boitani and T. K. Fuller (eds). *Research techniques in animal ecology*, pp. 65–110. Columbia University Press, New York.

Powell, R. A. (2004). Home ranges, cognitive maps, habitat models and fitness landscapes for *Martes*. In D. J. Harrison, A. K. Fuller and G. Proulx (eds). *Marten and fishers (Martes) in human-altered environments, An international perspective*, pp. 135–146. Kluwer Academic Publishers, Norwell, MA.

Powell, R. A. (2005). Evaluating welfare of American black bears (*Ursus americanus*) captured in foot snares and in winter dens. *Journal of Mammalogy*, **86**, 1171–1177.

Powell, R. A. and Fried, J. J. (1992). Helping by juvenile pine voles (*Microtus pinetorum*), growth and survival of younger siblings, and the evolution of pine vole sociality. *Behavioral Ecology*, 4, 325–333.

Powell, R. A. and Mitchell, M. S. (1998). Topographical constraints and home range quality. *Ecography*, **21**, 337–341.

Powell, R. A. and Mitchell, M. S. (In press). What is a home range? *Journal of Mammalogy*.

Powell, R. A. and Proulx, G. (2003). Trapping and marking terrestrial mammals for research, integrating ethics, standards, techniques, and common sense. *Institute of Laboratory Animal Research Journal*, **44**, 259–276.

Powell, R. A., Zimmerman, J. W. and Seaman, D. E. (1997). *Ecology and behaviour of North American black bears, home ranges, habitat and social organization*. Chapman and Hall, London.

Powell, L. A., Conroy, M. J., Hines, J. E., Nichols, J. D. and Krementz, D. G. (2000). Simultaneous use of mark-recapture and radio telemetry to estimate survival, movement, and capture rates. *Journal of Wildlife Management*, **64**, 302–313.

Powell, R. A., Lewis, J. C., Slough, B. G. *et al.* (In press). Evaluating *Martes* translocations; Testing model predictions with field data. In K. W. J. Aubry, W. J. Zielinski, M. Raphael, G. Proulx and S. W. Buskirk (eds). *Biology and conservation of martens, sables, and fishers, A new synthesis*. Cornell University Press, Ithaca, New York.

Pradel, R. (1996). Utilization of capture-mark-recapture for the study of recruitment and population growth rate. *Biometrics*, **52**, 703–709.

Pradel, R., Hines, J. E., Lebreton, J. D. and Nichols, J. D. (1997). Capture-recapture survival models taking account of "transients." *Biometrics*, **53**, 60–72.

Preisser, E. L., Bolnick, D. I. and Benard, M. F. (2005). Scared to death? The effects of intimidation and consumption in predator-prey interactions. *Ecology*, **86**, 501–509.

Priede, I. G. and Swift, S. M. (eds) (1992). *Wildlife telemetry: remote monitoring and tracking of animals*. Ellis Horwood, West Sussex, UK.

Pritchard, J. K., Stephens, M. and Donnelly, P. (2000). Inference of population structure using multilocus genotype data. *Genetics*, **155**, 945–959.

Proctor, M. F., McLellan, B. N., Strobeck, C. and Barclay, R. M. R. (2005). Genetic analysis reveals demographic fragmentation of grizzly bears yielding vulnerably small populations. *Proceedings of the Royal Society B-Biological Sciences*, **272**, 2409–2416.

Proffitt, K. M., Grigg, J. L., Hamlin, K. L. and Garrott, R. A. (2009). Contrasting effects of wolves and human hunters on elk behavioral responses to predation risk. *Journal of Wildlife Management*, **73**, 345–356.

Programme d'éducation en sécurité et en conservation de la faune. (1988). *Piégeage et gestion des animaux à fourrure*. Programme d'éducation en sécurité et en conservation de la faune (PESCOF). Association Provinciale des Trappeurs Indépendants, et Direction de l'Éducation du Ministère du Loisir de la Chasse et de la Pêche, Quebec.

Proulx, G. (1991). *Humane trapping program—annual report 1990/91*. Alberta Research Council, Edmonton, Alberta.

Proulx, G. (1995). An innovative and selective live-trap for raccoons (*Procyon lotor*). *Small Carnivore Conservation*, **12**, 19–20.

Proulx, G. (1997). A preliminary evaluation of four types of traps to capture northern pocket gophers, *Thomomys talpoides*. *Canadian Field-Naturalist*, **111**, 640–643.

Proulx, G. (1999a). Review of current mammal trap technology in North America. In G. Proulx (ed.). *Mammal trapping*, pp. 1–46. Alpha Wildlife Publications, Sherwood Park, Alberta.

Proulx, G. (1999b). The Bionic, an effective marten trap. In G. Proulx (ed.). *Mammal trapping*, pp. 79–87. Alpha Wildlife Publications, Sherwood Park, Alberta.

Proulx, G. (2005). Long-tailed weasel, *Mustela frenata*, movements and diggings in alfalfa fields inhabited by northern pocket gophers, *Thomomys talpoides*. *Canadian Field-Naturalist*, **119**, 175–180.

Proulx, G. and Barrett, M. W. (1989). Animal welfare concerns and wildlife trapping, ethics, standards and commitments. *Transactions of the Western Section of the Wildlife Society*, **25**, 1–6.

Proulx, G. and Barrett, M. W. (1990). Assessment of power snares to effectively kill red fox. *Wildlife Society Bulletin*, **18**, 27–30.

Proulx, G. and Barrett, M. W. (1991a). Evaluation of the Bionic[R] trap to quickly kill mink (*Mustela vison*) in simulated natural environments. *Journal of Wildlife Diseases*, **27**, 276–280.

Proulx, G. and Barrett, M. W. (1991b). Ideological conflict between animal rightists and wildlife professionals over trapping wild furbearers. *Transactions of the North American Wildlife and Natural Resources Conference*, **56**, 387–399.

Proulx, G. and Barrett, M. W. (1993a). Evaluation of the Bionic[R] trap to quickly kill fisher (*Martes pennanti*) in simulated natural environments. *Journal of Wildlife Diseases*, **29**, 310–316.

Proulx, G. and Barrett, M. W. (1993b). Evaluation of mechanically improved Conibear 220 taps to quickly kill fisher (*Martes pennanti)* in simulated environments. *Journal of Wildlife Diseases*, **29**, 317–323.

Proulx, G, and Barrett, M. W. (1994). Ethical considerations in the selection of traps to harvest martens and fishers. In S. W. Buskirk, A. S. Harestad, M. G. Raphael and R. A. Powell (eds). *Martens, sables, and fishers, biology and conservation*, pp. 192–196. Cornell University Press, Ithaca, NY.

Proulx, G. and Drescher, R. K. (1994). Assessment of rotating-jaw traps to humanely kill raccoons (*Procyon lotor*). *Journal of Wildlife Diseases*, **30**, 335–339.

Proulx, G., Cook, S. R., and Barrett, M. W. (1989a). Assessment and preliminary development of the rotating-jaw Conibear 120 trap to effectively kill marten (*Martes americana*). *Canadian Journal of Zoology*, **67**, 1074–1079.

Proulx, G., Barrett, M. W. and Cook, S. R. (1989b). The C120 Magnum, an effective quick-kill trap for marten. *Wildlife Society Bulletin*, **17**, 294–298.

Proulx, G., Barrett, M. W. and Cook, S. R. (1990). The C120 Magnum with pan trigger, a humane trap for mink (*Mustela vison*). *Journal of Wildlife Diseases*, **26**, 511–517.

Proulx, G., Onderka, D. K., Kolenosky, A. J., Cole, P. J., Drescher, R. K. and Badry, M. J. (1993a). Injuries and behavior of raccoons (*Procyon lotor*) captured in the Soft Catch[TM] and the EGG[TM] traps in simulated natural environments. *Journal of Wildlife Diseases*, **29**, 447–452.

Proulx, G., Kolenosky, A. J., Badry, M. J., Cole, P. J. and Drescher, R. K. (1993b). Assessment of the Sauvageau 2001–8 trap to effectively kill arctic fox. *Wildlife Society Bulletin*, **21**, 132–135.

Proulx, G. and Barrett, M. W. (1993c). Field testing the C120 Magnum trap for mink. *Wildlife Society Bulletin*, **21**, 421–426.

Proulx, G., Pawlina, I. M., Onderka, D. K., Badry, M. J. and Seidel, K. (1994). Field evaluation of the No. 1–1/2 steel-jawed leghold and the Sauvageau 2001–8 traps to humanely capture arctic fox. *Wildlife Society Bulletin*, **22**, 179–183.

Proulx, G., Kolenosky, A. J., Cole, P. J. and Drescher, R. K. (1995). A humane killing trap for lynx (*Felis lynx*), the Conibear 330 with clamping bars. *Journal of Wildlife Diseases*, **31**, 57–61.

Provencher, P. (1969). *Le manuel pratique du trappeur Québécois*. Paul Provencher and Sports/famille, Montreal, Quebec.

Prugh, L. R. and Ritland, C. E. (2005). Molecular testing of observer identification of carnivore feces in the field. *Wildlife Society Bulletin*, **33**, 189–194.

Prugh, L. R., Ritland, C. E., Arthur, S. M. and Krebs, C. J. (2005). Monitoring coyote population dynamics by genotyping faeces. *Molecular Ecology*, **14**, 1585–1596.

Prugh, L., Arthur, S. and Ritland, C. (2008). Use of faecal genotyping to determine individual diet. *Wildlife Biology*, **14**, 318–330.

Pruss, S. D., Cool, N. L., Hudson, R. J. and Gaboury, A. R. (2002). Evaluation of a modified neck snare to live-capture coyotes. *Wildlife Society Bulletin*, **30**, 508–516.

Pulliam, H. R. (2000). On the relationship between niche and distribution. *Ecology Letters*, **3**, 349–361.

Purvis A., Mace G. M. and Gittleman J. L. (2004). Past and future carnivore extinctions: a phylogenetic perspective. In J. L. Gittleman., S. M. Funk, D. W. Macdonald and R. K. Wayne (eds). *Carnivore conservation,* pp. 11–34. Cambridge University Press, Cambridge.

Pusey, A. E. and Packer, C. (1994). Infanticide in lions: Consequences and counterstrategies. In S. Parmigiani and F. S. Von Saal (eds). *Infanticide and parental care*, pp. 277–299. Harwood Academic Press, Chur, Switzerland.

Putman, R. J. (1984). Facts from faeces. *Mammal Review,* **14**, 79–97.

Pyke, G. H. (1984). Optimal foraging theory, a critical review. *Annual Review of Ecology and Systematics*, **15**, 523–575.

Pyke, G. H., Pulliam, H. R. and Charnov, E. L. (1977). Optimal foraging, a selective review of theory and tests. *Quarterly Review of Biology*, **52**, 137–154.

Queyras, A. and Carosi, M. (2004). Noninvasive techniques for analysing hormonal indicators of stress. *Annali Istituto Superiore di Sanità*, **40**, 211–221.

Quinn, N. W. S. and Gardner, J. F. (1984). Relationships of age and sex to lynx pelt characteristics. *Journal of Wildlife Management*, **48**, 953–956.

Rabin, L.A. (2003). Maintaining behavioural diversity in captivity for conservation, natural behaviour management. *Animal Welfare*, **12**, 85–94.

Räikkönen, J., Vucetich, J. A., Peterson, R. O. and Nelson, M. P. (2009). Congenital bone deformities and the inbred wolves (*Canis lupus*) of Isle Royale. *Biological Conservation,* **142**, 1025–1031.

Ralls, K. and White, P. J. (1995). Predation on San Joaquin kit foxes by larger canids. *Journal of Mammalogy,* **76**, 723–729.

Ralls, K., Siniff, D. B., Williams, T. D. and Kuechle, V. B. (1989). An intraperitoneal radio transmitter for sea otters. *Marine Mammal Science*, **5**, 376–381.

Ramsay, M. A. and Stirling, I. (1986). Long-term effects of drugging and handling free-ranging polar bears. *Journal of Wildlife Management*, **50**, 619–626.

Ramsay, E. C., Sleeman, J. M. and Clyde, V. L. (1995). Immobilization of black bear (*Ursus americanus*) with orally administered carfentanil citrate. *Journal of Wildlife Diseases*, **31**, 391–393.

Randall, D. A., Marino, J., Haydon, D. T. *et al.* (2006). An integrated disease management strategy for the control of rabies in Ethiopian wolves. *Biological Conservation,* **131**, 151–162.

Raphael, M. G. (1994). Techniques for monitoring populations of fishers and American Martens. In S. W. Buskirk, A. S. Harestad, M. G. Raphael and R. A. Powell (eds). *Martens, Sables, and Fishers biology and conservation*, pp. 224–240. Cornell University Press, Ithaca, New York.

Rasmussen, G. S. A. (1999). Livestock predation by the painted hunting dog *Lycaon pictus* in a cattle ranching region of Zimbabwe: a case study. *Biological Conservation,* **88**, 133–139.

Rasmussen, R. V. and Borsting, C. F. (2000). Effects of variations in dietary protein levels on hair growth and pelt quality in mink (*Mustela vison*). *Canadian Journal of Animal Science*, **80**, 633–642.

Rasmussen, H. B., Wittemyer, G. and Douglas-Hamilton, I. (2005). Estimating age of immobilized elephants from teeth impressions using dental silicon. *African Journal of Ecology*, **43**, 215–219.

Ratnasingham, S., and P. D. N. Hebert. (2007). The Barcode of Life Data System (www. barcodinglife.org). *Molecular Ecology Notes*, 7, 355–364.

Ratnayeke, S., van Manen, F. T. and Padmalal, U. K. G. K. (2007). Home ranges and habitat use of sloth bears *Melursus ursinus inornatus* in Wasgomuwa National Park, Sri Lanka. *Wildlife Biology*, **13**, 272–284.

Rausch, R. L. (1953). On the status of some arctic mammals. *Arctic*, **6**, 91–148.

Rausch, R. L. (1963). Geographic variation in size in North American brown bears, *Ursus arctos* L., as indicated by condylobasal length. *Canadian Journal of Zoology*, **41**, 33–45.

Rausch, R. L. (1967). Some aspects of the population ecology of wolves, Alaska. *American Zoologist*, 7, 253–265.

Rausch, R. L. (1969). Morphogenesis and age-related structure of permanent canine teeth in the brown bear, *Ursus arctos* L., in arctic Alaska. *Zeitshrift fur Morphologie der Tiere*, **66**, 167–188.

Ray, J. C. (1997). Comparative ecology of two African forest mongooses, *Herpestes naso* and *Atilax paludinosus*. *African Journal of Ecology*, **35**, 237–253.

Read, M. R. (2002). Long acting neuroleptic drugs. In D. Heard (ed.). *Zoological restraint and anesthesia*. International Veterinary Information Service (www.ivis.org), Ithaca, New York.

Read, M. R., Caulkett, N. A., Symington, A. and Shury, T. K. (2001). Treatment of hypoxemia during xylazine-tiletamine-zolazepam immobilization of wapiti. *Canadian Veterinary Journal*, **42**, 661–664.

Reddy, S. and Davalos, L. M. (2003). Geographical sampling bias and its implications for conservation priorities in africa. *Journal of Biogeography*, **30**, 1719–1727.

Reed, J. M., Murphy, D. D. and Brussard, P. F. (1998). Efficacy of population viability analysis. *Wildlife Society Bulletin*, **26**, 244–251.

Reed, M., Mills, L. S., Dunning Jr., J. B. *et al.* (2002). Emerging issues in population viability analysis. *Conservation Biology*, **16**, 1–7.

Reed, D. H., O'Grady, J. J., Brook, B. W. *et al.* (2003). Estimates of minimum viable population sizes for vertebrates and factors influencing those estimates. *Biological Conservation*, **113**, 23–34.

Rees, P.A. (2001). Is there a legal obligation to reintroduce animal species into their former habitats? *Oryx*, **35**, 216–223.

Rempel, R. S., Rodgers, A. R. and Abraham, K. F. (1995). Performance of a GPS animal location system under boreal forest canopy. *Journal of Wildlife Management*, **59**, 543–551.

Renfree, M. B. and Calaby, J. M. (1981). Background to delayed implantation and embryonic diapause. *Journal of Reproduction and Fertility*, **Suppl. 29**, 1–9.

Reynolds, J. C. and Aebisher, N. J. (1991). Comparison and quantification of carnivore diet by faecal analysis: a critique, with recommendations, based on the study of the red fox *Vulpes vulpes*. *Mammal Review*, **21**, 97–122.

Reynolds, R. D. and Laundrè, J. W. (1990). Time intervals for estimating pronghorn and coyote home ranges and daily movements. *Journal of Wildlife Management*, **54**, 316–322.

Reynolds-Hogland, M. J. and Mitchell, M. J. (2007a). Effects of roads on habitat quality for bears in the Southern Appalachians, a long-term study. *Journal of Mammalogy*, **88**, 1050–1061.

Reynolds-Hogland, M. J. and Mitchell, M. S. (2007b). Three axes of ecological studies, Matching process and time in landscape ecology. In J. A. Bissonette and I. Storch (eds). *Temporal Dimensions of Landscape Ecology, Wildlife Responses to Variable Resources*, pp. 174–194. Springer, New York.

Rhodes, C. J., Atkinson, R. P. D., Anderson, R. M. and Macdonald, D. W. (1998). Rabies in Zimbabwe: reservoir dogs and the implications for disease control. *Philosophical Transactions of The Royal Society B*, **353**, 999–1010.

Rice, M. B., Ballard, W. B., Fish, E. B., Wester, D. B. and Holdermann, D. (2007). Predicting private landowner support toward recolonizing black bears in the trans-Pecos region of Texas. *Human Dimensions of Wildlife*, **12**, 405–415.

Riddle, A. E., Pilgrim, K. L., Mills, L. S., McKelvey, K. S. and Ruggiero, L. F. (2003). Identification of mustelids using mitochondrial DNA and noninvasive sampling. *Conservation Genetics*, **4**, 241–243.

Rigg, R. (2001). *Livestock guarding dogs: their current use world wide.* URL: http://www.canids.org/occasionalpapers/:IUCN/SSC Canid Specialist Group Occasional Paper No.1 [online].

Riley, S. P. D. (2006). Spatial ecology of bobcats and gray foxes in urban and rural zones of a national park. *Journal of Wildlife Management*, **70**, 1425–1435.

Riley, S. P. D., Sauvajot, R. M., Fuller, T. K. *et al.* (2003). Effects of urbanization and habitat fragmentation on bobcats and coyotes in Southern California. *Conservation Biology*, **17**, 566–576.

Riley, S. P.D., Pollinger, J. P., Sauvajot, R. M. *et al.* (2006). A southern California freeway is a physical and social barrier to gene flow in carnivores. *Molecular Ecology*, **15**, 1733–1741.

Riley, S. P. D., Bromley, C., Poppenga, R. H., Uzal, F. A., Whited, L. and Sauvajot, R. M. (2007). Anticoagulant exposure and notoedric mange in bobcats and mountain lions in urban southern California. *Journal of Wildlife Management*, **71**, 1874–1884.

Robbins, C. T., Schwartz, C. C. and Felicetti, L. A. (2004). Nutritional ecology of ursids: a review of newer methods and management implications. *Ursus*, **15**, 161–171.

Robertson, B. A. and Hutto, R. L. (2006). A framework for understanding ecological traps and an evaluation of existing evidence. *Ecology*, **87**, 1075–1085.

Robinson, S. and Milner-Gulland, E. J. (2003). Political change and factors limiting numbers of wild and domestic ungulates in Kazakhstan. *Human Ecology*, **31**, 87–110.

Robinson, H. S., Wielgus, R. B., Cooley, H. S. and Cooley, S. W. (2008). Sink populations in carnivore management: cougar demography and immigration in a hunted population. *Ecological Applications*, **18**, 1028–1037.

Robinson, S. J., Waits, L. P. and Martin, I. D. (2009). Estimating abundance of American black bears using DNA-based capture-mark-recapture models. *Ursus*, **20**, 1–11.

Roelke-Parker, M. E., Munson, L., Packer, C. *et al.* (1996). A canine distemper virus epidemic in Serengeti lions (*Panthera leo*). *Nature*, **379**, 441–445.

Roelofs, J. B., Van Eerdenburg, F. J. C. M., Soede, N. M. and Kemp, B. (2005). Pedometer readings for estrous detection and as predictor for time of ovulation in dairy cattle. *Theriogenology*, **64**, 1690–1703.

Roemer, G. W. and Donlan, C. J. (2005). Biology, policy and law in endangered species conservation, II, a case history in adaptive management of the island fox on Santa Catalina Island, California. *Endangered Species Update*, **22**, 144–156.

Rogers, L. L. (1977). *Social relationships, movements, and population dynamics of black bears in northeastern Minnesota.* PhD thesis, University of Minnesota. Minneapolis, MN.

Rogers, L. L. (1978). Interpretation of cementum annuli in first premolars of bears. *Proeedings Eastern Workshop on Black Bear Research and Management*, 4, 102–112.

Rogers, L. L. (1987). Effects of food supply and kinship on social behavior, movements, and population growth of black bears in northeastern Minnesota. *Wildlife Monographs*, **97**, 1–72.

Rogers, L. M., Hounsome, T. D. and Cheeseman, C. L. (2002). An evaluation of passive integrated transponders (PITs) as a means of permanently marking badgers (*Meles meles*). *Mammal Review*, **32**, 63–65.

Rolley, R.E. (1987). Bobcat. In M. Novak, J. A. Baker, M. E. Obbard and B. Malloch (eds). *Wild furbearer management and conservation in North America*, pp. 671–681. Ministry of Natural Resources, Toronto, Ontario.

Roloff, G. J., Millspaugh, J. J, Gitzen, R. A. and Brundige, G. C. (2001). Validation tests of a spatially explicit habitat effectiveness model for Rocky Mountain elk. *Journal of Wildlife Management*, **65**, 899–914.

Romero, L. M. (2004). Physiological stress in ecology: lessons from biomedical research. *Trends in Ecology and Evolution*, **19**, 249–255.

Rondinini, C. and Boitani, L. (2002). Habitat use by beech martens in a fragmented landscape. *Ecography*, **25**, 257–264.

Rondinini, C. and Boitani, L. (2007). Systematic conservation planning and the cost of tackling conservation conflicts with large carnivores in Italy. *Conservation Biology*, **21**, 1455–1462.

Rondinini, C., Stuart, S. and Boitani, L. (2005). Habitat suitability models and the shortfall in conservation planning for african vertebrates. *Conservation Biology*, **19**, 1488–1497.

Rondinini, C., Ercoli, V. and Boitani, L. (2006a). Habitat use and preference by polecats (*Mustela putorius*) in a Mediterranean agricultural landscape. *Journal of Zoology*, **269**, 213–219.

Rondinini, C., Wilson, K. A., Boitani, L., Grantham, H. and Possingham, H. P. (2006b). Tradeoffs of different types of species occurrence data for use in systematic conservation planning. *Ecology Letters*, **9**, 1136–1145.

Rood, J. P. (1980). Mating relationships and breeding suppression in the dwarf mongoose. *Animal Behaviour*, **28**, 143–150.

Rood, J. P. (1986). Ecology and social evolution in the mongooses. In D. I. Rubenstein and R. W. Wrangham (eds). *Ecological aspects of social evolution*, pp. 131–152. Princeton University Press, Princeton, NJ.

Rood, J. P. and Nellis, D. W. (1980). Freeze marking mongooses. *Journal of Wildlife Management*, **44**, 500–502.

Rook, A. J., Dumont, B., Isselstein, J. *et al.* (2004). Matching type of livestock to desired biodiversity outcomes in pastures -a review. *Biological Conservation*, **119**, 137–150.

Roon, D. A., Thomas, M. E., Kendall, K. C. and Waits, L. P. (2005). Evaluating mixed samples as a source of error in noninvasive genetic studies using microsatellites. *Molecular Ecology*, **14**, 195–201.

Rooney, N. J., Gaines, S. A., Bradshaw, J. W. S. and Penman, S. (2007). Validation of a method for assessing the ability of trainee specialist search dogs. *Applied Animal Behaviour Science*, **103**, 90–104.

Ropert-Coudert, Y. and Wilson, R.P. (2005). Trends and perspectives in animal-attached remote sensing. *Frontiers in Ecology and Environment*, **3**, 437–444.

Rosatte, R. C. (1987). Striped, spotted, hooded, and hog-nosed skunk. In M. Novak, J. A. Baker, M. E. Obbard and B. Malloch (eds). *Wild furbearer management and conservation in North America*, pp. 599–613. Ministry of Natural Resources, and Ontario Trappers Association, Toronto, Ontario.

Rosatte, R. C. (1988). Rabies in Canada—history, epidemiology and control. *The Canadian Veterinary Journal*, **29**, 362–365.

Rosatte, R. C., Power, M. J., MacInnes, C. D. and Campbell, J. B. (1992). Trap-vaccinate-release and oral vaccination for rabies control in urban skunks, raccoons and foxes. *Journal of Wildlife Diseases*, **28**, 562–571.

Rosenzweig, M. L. (1966). Community structure in sympatric carnivore. *Journal of Mammalogy*, 47, 602–612.

Rosenzweig, M. L. (1981). A theory of habitat selection. *Ecology*, **62**, 327–335.

Rota, C. T., Fletcher, J., Dorazio, R. M. and Betts, M. G. (2009). Occupancy estimation and the closure assumption. *Journal of Applied Ecology*, **46**, 1173–1181.

Roth, J. D. (2002). Temporal variability in arctic fox diet as reflected in stable-carbon isotopes; the importance of sea ice. *Oecologia*, **133**, 70–77.

Roth, J. D., Marshall, J. D., Murray, D. L., Nickerson, D. M. and Steury, T. D. (2007). Geographical gradients in diet affect population dynamics of Canada lynx. *Ecology*, **88**, 2736–2743.

Roth, J. D., Murray, D. L. and Steury, T. D. (2008). Spatial dynamics of sympatric canids, modeling the impact of coyotes on red wolf recovery. *Ecological Modelling*, **214**, 391–403.

Rowcliffe, J. M., Field, J., Turvey, S. T. and Carbone, C. (2008). Estimating animal density using camera traps without the need for individual recognition. *Journal of Applied Ecology*, **45**, 1228–1236.

Rowe-Rowe, D. T. and Green, B. (1981). Steel-jawed traps for live capture of black-backed jackals. *South African Journal of Wildlife Research*, **11**, 63–65.

Royle, J. A. (2004). Modeling abundance index data from anuran calling surveys. *Conservation Biology*, **18**, 1378–1385.

Royle, J. A. (2006). Site occupancy models with heterogeneous detection probabilities. *Biometrics*, **62**, 97–102.

Royle, J. A. and Dorazio, R. M. (2008). *Hierarchical modeling and inference in ecology*. Academic Press, New York.

Royle, J. A. and Gardner, B. (2011). Hierarchical spatial capture-recapture models for estimating density from trap arrays. In A. F. O'Connell, J. D. Nichols, and K. U. Karanth (eds). *Camera-traps in animal ecology, methods and analyses*, pp. 163–190. Springer, New York.

Royle, J. A. and Kery, M. (2007). A Bayesian state-space formulation of dynamic occupancy models. *Ecology*, **88**, 1813–1823.

Royle, J. A. and Link, W. A. (2005). A general class of multinomial mixture models for anuran calling survey data. *Ecology*, **86**, 2505–2512.

Royle, J. A. and Link, W. A. (2006). Generalized site occupancy models allowing for false positive and false negative errors. *Ecology*, **87**, 835–841.

Royle, J. A. and Nichols, J. D. (2003). Estimating abundance from repeated presence absence data or point counts. *Ecology*, **84**, 777–790.

Royle, J. A. and Young, K. V. (2008). A hierarchical model for spatial capture-recapture data. *Ecology*, **89**, 2281–2289.

Royle, A. J., Nichols, J. D., Karanth, K. U. and Gopalaswamy, A. M. (2009a). A hierarchical model for estimating density in camera-trap studies. *Journal of Applied Ecology*, **46**, 118–127.

Royle, J. A., Karanth, K. U., Gopalaswamy, A. M. and Kumar, N. S. (2009b). Bayesian inference in camera trapping studies for a class of spatial capture-recapture models. *Ecology*, **90**, 3233–3244.

Royle, J. A., Stanley, T. R. and Lukas, P. M. (2008). Statistical modelling and inference from carnivore survey data. In R. A. Long, P. MacKay, W. J. Zielinski and J. C. Ray (eds). *Noninvasive survey methods for carnivores*, pp. 293–312. Island Press, Washington DC.

Roze, U. (2009). *The North American Porcupine*, 2nd edn. Cornell University Press, Ithaca, New York.

Rudran, R. (1996). General marking techniques. In D. E. Wilson, F. R. Cole, J. D. Nichols, R. Rudran and M. S. Foster (eds). *Measuring and monitoring biological diversity. standard methods for mammals*, pp. 299–304. Smithsonian Institution Press, Washington, DC.

Rueness, E. K., Asmyhr, M. G., Sillero-Zubiri, C. *et al.* (2011). The cryptic African wolf: *Canis aureus lupaster* is not a golden jackal and is not endemic to Egypt. *PLoS ONE*, **6**(1), e16385.

Ruette, S., Stahl, P. and Alaberet, M. (2003). Applying distance-sampling methods to spotlight counts of red foxes. *Journal of Applied Ecology*, **40**, 32–43

Rühe, F., Ksinsik, M. and Kiffner, C. (2008). Conversion factors in carnivore scat analysis: sources of bias. *Wildlife Biology*, **14**, 500–506.

Russell, W. M. S. and Burch, R. L. (1959). *The principles of humane experimental technique*. Methuen, London.

Rust, C. C. (1968). Procedure for live trapping weasels. *Journal of Mammalogy*, **49**, 318–319.

Ruth, T. K., Smith, D. W., Haroldson, M. A. *et al.* (2003). Large-carnivore response to recreational big-game hunting along the Yellowstone National Park and Absaroka-Beartooth Wilderness boundary. *Wildlife Society Bulletin*, **31**, 1150–1161.

Ruth, T. K., Buotte, P. C. and Quigley, H. B. (2010). Comparing ground telemetry and Global Positioning System methods to determine cougar kill rates. *Journal of Wildlife Management*, 74, 1122–1133.

Rutz, C. and Hays, G. C. (2009). New frontiers in biologging science. *Biology Letters*, **5**, 289–292.

Ryder, O. A. (1986). Species conservation and systematics: the dilemma of subspecies. *Trends in Ecology and Evolution*, **1**, 9–10.

Ryser, A., Schol, M., Zwahlen, M., Oetliker, M., Ryser-Degiorgis, M.-P. and Breitenmoser, U. (2005). A remote-controlled teleinjection system for the low-stress capture of large mammals. *Wildlife Society Bulletin*, **33**, 721–730.

Saarenmaa, H., Stone, N. D, Folse, L. J. *et al.* (1988). An artificial intelligence modelling approach to simulating animal/habitat interactions. *Ecological Modelling*, **44**, 125–141.

Sabean, B. and Mills, J. (1994). *Raccoon—6" x 6" Body-Gripping Trap Study*. Nova Scotia Department of Natural Resources, Unpublished report.

Sacks, B. N. (2005). Reproduction and body condition of California coyotes (*Canis latrans*). *Journal of Mammalogy*, **86**, 1036–1041.

Sacks, B. N., Blejwas, K. M. and Jaeger, M. M. (1999). Relative vulnerability of coyotes to removal methods on a northern California ranch. *Journal of Wildlife Management*, **63**, 939–949.

Sager-Fradkin, K. A., Jenkins, K. J., Happe, P. J., Beecham, J. J., Wright, R. G. and Hoffman, R. A. (2008). Space and habitat use by black bears in the Elwha Valley prior to dam removal. *Northwest Science*, **82**, 164–178.

Sahr, D. P. and Knowlton, F. F. (2000). Evaluation of tranquilizer trap devices (TTDs) for foothold traps used to capture gray wolves. *Wildlife Society Bulletin*, **28**, 597–605.

Samuel, M. D. and Fuller, M. R. (1994). Wildlife radiotelemetry. In T.A. Bookout (ed.). *Research and management techniques for wildlife and habitats*, 5th edn, pp. 370–418. The Wildlife Society, Bethesda, MD.

Samuel, M. D., Pierce, D. J. and Garton, E. O. (1985). Identifying areas of concentrated use within the home range. *Journal of Animal Ecology*, **54**, 711–719.

Sand, H., Zimmermann, B., Wabakken, P., Andren, H. and Pedersen, H. C. (2005). Using GPS technology and GIS cluster analyses to estimate kill rates in wolf-ungulate ecosystems. *Wildlife Society Bulletin*, **33**, 914–925.

Sand, H., Liberg, O., Aronson, A., Forslund, P. *et al.* (2010). *Den Skandinaviska Vargen en sammanställning av kunskapsläget 1998–2010 från det skandinaviska vargforskningsprojektet SKANDULV, Grimsö forskningsstation, SLU*. Rapport till Direktoratet for Naturforvaltning, Trondheim, Norway.

Sandell, M. (1989). The mating tactics and spacing patterns of solitary carnivores. In J. L. Gittleman (ed.). *Carnivore behavior, ecology, and evolution*, pp. 164–182. Cornell University Press, Ithaca, NY.

Sandell, M. (1990). The evolution of seasonal delayed implantation. *Quarterly Review of Biology*, **65**, 23–42.

Sanderson, E., Forrest, J., Loucks, C. *et al.* (2006). *Setting priorities for the conservation and recovery of wild tigers 2005–2015. A technical report*. Wildlife Conservation Society, World Wildlife Fund, Smithsonian, and National Fish and Wildlife Foundation–Save the Tiger Fund. New York and Washington DC.

Sanderson, G. C. (1961). Techniques for determining age of raccoons. Ill. *Natural History Survey Biological Notes*, **45**, 1–16.

Sanderson, G. C. and Nalbandov, A. V. (1973). The reproductive cycle of the raccoon in Illinois. *Bulletin of Illinois Natural History Survey*, **31**, 29–85.

Sands, J. and Creel, S. (2004). Social dominance, aggression and fecal glucocorticoid levels in a wild population of wolves, *Canis lupus*. *Animal Behavior*, **67**, 387–396.

Sangay, T. and Vernes, K. (2008). Human-wildlife conflicts in the Kingdom of Bhutan: patterns of livestock predation by large mammalian carnivores. *Biological Conservation*, **141**, 1272–1282.

Sanson G., Brown, J. L. and Farstad, W. (2005). Noninvasive faecal steroid monitoring of ovarian and adrenal activity in farmed blue fox (*Alopex lagopus*) females during late pregnancy, parturition and lactation onset. *Animal Reprodcution Science*, **87**, 309–319.

Santos-Reis, M., Santos, M. J., Lorenço, S., Marques, J. T., Pereira, I. and Pinto, B. (2004). Relationships between stone martens, genets, and cork oak woodlands in Portugal. In D. J. Harrison, A. K. Fuller and G. Proulx (eds). *Martens and fishers (Martes) in human-altered environments, an international perspective*, pp. 147–172. Springer Science+Business Media, New York.

Sargeant, G. A., Johnson, D. H. and Berg, W. E. (1998). Interpreting carnivore scent station surveys. *Journal of Wildlife Management*, **62**, 1235–1245.

Sargeant, G. A, Sovada, M. A., Slivinski, C. C. and Johnson, D. H. (2005). Markov chain Monte Carlo estimation of species distribution: a case study of the swift fox in western Kansas. *Journal of Wildlife Management*, **69**, 483–497.

Sargolini, F., Fyhn, M. and Hafting, T. (2006). Conjunctive representation of position, direction, and velocity in entorhinal cortex. *Science*, **312**, 758–762.

Sarmento, P., Cruz, J., Eira, C. and Fonseca, C. (2009). Evaluation of camera trapping for estimating red fox abundance. *Journal of Wildlife Management*, **73**. 1207–1212.

Sarrazin, F. and Barbault, R. (1996). Reintroduction, challenges and lessons for basic ecology. *Trends in Ecology and Evolution*, **11**, 474–478.

Sasaki, H. and Ono, Y. (1994). Habitat use and selection of the Siberian weasel *Mustela sibirica coreana* during the non-mating season. *Journal of the Mammalogical Society of Japan*, **19**, 21–32.

Sastre, N., Francino, O., Lampreave, G. *et al.* (2009). Sex identification of wolf (*Canis lupus*) using noninvasive samples. *Conservation Genetics*, **10**, 555–558.

Sato, M., Tsubota, T., Komatsu, T. *et al.* (2001). Changes in sex steroids, gonadotropins, prolactin and inhibin in pregnant and nonpregnant Japanese black bears (*Ursus thibetanus japonicus*). *Biology of Reproduction*, **65**, 1006–1013.

Sauer, J. R. and Knutson, M. G. (2008). Objectives and metrics for wildlife monitoring. *Journal of Wildlife Management*, **72**, 1663–1664.

Saunders, J. K. (1964). Physical characteristics of the Newfoundland lynx. *Journal of Mammalogy*, **45**, 36–47.

Savarie, P. J., Johns, B. E. and Gaddis, S. E. (1992). A review of chemical and particle marking agents used for studying vertebrate pests. In J. E. Borrecco and R. E. Marsh (eds). *Proceedings 15th vertebrate pest conference*, pp. 252–257. University of California, Davis, CA.

Savolainen, V., R. S. Cowan, A. P. Vogler, G. K. Roderick, and R. Lane. (2005). Towards writing the encyclopaedia of life: an introduction to DNA barcoding. *Philosophical Transactions of the Royal Society B-Biological Sciences*, **360**, 1805–1811.

Sawyer, H., Nielson, R. M., Lindzey, F. and McDonald, L. L. (2006). Winter habitat selection of mule deer before and during development of a natural gas field. *Journal of Wildlife Management*, **70**, 396–403.

Sawyer, H., Kauffman, M. J., Nielson, R. M. and Horne, J. S. (2009). Identifying and prioritizing ungulate migration routes for landscape-level conservation. *Ecological Applications*, **19**, 2016–2025.

Schaller, G. B. (1967). *The deer and the tiger: aA study of wildlife in India*. University of Chicago Press, Chicago, IL.

Schaller, G. B. (1972). *The Serengeti Lion*. University of Chicago Press, Chicago, IL.

Scheaffer, R. L., Mendenhall III, W and Ott, L. (1996). *Elementary survey sampling*. (Fifth edn). Duxbury Press, Belmont, CA.

Scheepers, J. L. and Venzke, K. A. E. (1995). Attempts to reintroduce African wild dogs *Lycaon pictus* into Etosha National Park, Namibia. *South African Journal of Wildlife Research*, **25**, 138–140.

Scheffer, V. B. (1950). Growth layers on the teeth of pinnipedia as an indication of age. *Science*, **112**, 309–311.

Schemnitz, S. D. (2005). Capturing and handling wild animals. In C. E. Braun (ed.). *Techniques for wildlife investigations and management*, pp. 239–285. The Wildlife Society, Bethesda, MD.

Schiess-Meier, M., Ramsauer, S., Gabababapelo, T. and König, B. (2007). Livestock predation—insights from problem animal control registers in Botswana. *Journal of Wildlife Management*, **71**, 1267–1274.

Schillaci, M. A., Jones-Engel, L., Heidrich, J.E., Miller, G.P. and Froehlich, J.W. (2001). Field methodology for lateral cranial radiography of nonhuman primates. *American Journal of Physiological Anthropology*, **116**, 278–284.

Schipper, J. (2007). Camera-trap avoidance by kinkajous *Potos flavus*: rethinking the "non-invasive" paradigm. *Small Carnivore Conservation*, **36**, 38–41.

Schipper, J., Chanson, J. S., Chiozza *et al.* (2008). The status of the world's land and marine mammals: diversity, threat, and knowledge. *Science*, **322**, 225–230.

Schluter, D. and McPhail, J. (1992). Ecological character displacement and speciation in sticklebacks. *American Naturalist*, **140**, 85–108.

Schmidt, K., and Kowalczyk, R. (2006). Using scent-marking stations to collect hair samples to monitor Eurasian lynx populations. *Wildlife Society Bulletin*, **34**, 462–466.

Schmidt, A. M., Nadal, L. A., Schmidt, M. J. and Beamur, N. B. (1979). Serum concentrations of oestradiol and progesterone during the normal oestrus cycle and early pregnancy in the lion (*Panthera leo*). *Journal of Reproduction and Fertility*, **57**, 267–272.

Schmidt, A. M., Hess, D. L., Schmidt, M. J., Smith, R. C. and Lewis, C. R. (1988). Serum concentrations of oestradiol and progesterone and sexual behaviour during the normal oestrous cycle in the leopard (*Panthera pardus*). *Journal of Reproduction and Fertility*, **82**, 43–49.

Schmidt, A. M., Hess, D. L., Schmidt, M. J. and Lewis, C. R. (1993). Serum concentrations of oestradiol and progesterone and frequency of sexual behaviour during the normal oestrous cycle in the snow leopard (*Panthera uncia*). *Journal of Reproduction and Fertility*, **98**, 91–95.

Schoener, T. W. (1971). Theory of feeding strategies. *Annual review of ecology and systematics*, **2**, 369–404.

Schoener, T. W. (1982). The controversy over interspecific competition. *American Scientist*, **70**, 27–39.

Schroeder, M. T. (1987). Blood chemistry, hematology, and condition evaluation of black bears in northcoastal California. *International Conference Bear Research and Management*, 7, 333–349.

Schwartz, C. C. and Arthur, S. M. (1999). Radiotracking large wilderness mammals: Integration of GPS and Argos Technology. *Ursus*, **11**, 261–273.

Schwartz, M. and Monfort, S. (2008). Genetic and endocrine tools for carnivore surveys. In R. A. Long, P. MacKay, W. J. Zielinski and J. C. Ray (eds). *Noninvasive survey methods for carnivores*, pp. 238–264. Island Press, Washington DC.

Schwartz, C. C., Swenson, J. E., and Miller, S. D. (2003). Large carnivores, moose, and humans: a changing paradigm of predator management in the 21st century. *Alces*, **39**, 41–63.

Schwartz, C. C., Haroldson, M. A. and White, G. C. (2010). Hazards affecting grizzly bear survival in the Greater Yellowstone Ecosystem. *Journal of Wildlife Management*, **74**, 654–667.

Schwarz, C. J. and Arnason, A. N. (1996). A general methodology for the analysis of capture-recapture experiments in open populations. *Biometrics*, **52**, 860–873.

Schwarz, C. J. and Stobo, W. T. (1997). Estimating temporary emigration using the robust design. *Biometrics*, **53**, 178–194.

Schwarz, C. J., Schweigert, J. F. and Arnason, A. N. (1993). Estimating migration rates using tag-recovery data. *Biometrics*, **49**, 177–193.

Schwarzenberger, F. (2007). The many uses of noninvasive faecal steroid monitoring in zoo and wildlife species. *International Zoo Yearbook*, **41**, 52–74.

Schwarzenberger, F., Möstl, E., Palme, R. and Bamberg, E. (1996). Faecal steroid analysis for non-invasive monitoring of reproductive status in farm, wild and zoo animals. *Animal Reproduction Science*, **42**, 515–526.

Schwarzenberger, F., Fredriksson, G., Schaller, K. and Kolter, L. (2004). Fecal steroid analysis for monitoring reproduction in the sun bear (*Helarctos malayanus*). *Theriogenology*, **62**, 1677–1692.

Schwerdtner, K. and Gruber, B. (2007). A conceptual framework for damage compensation schemes. *Biological Conservation*, **134**, 354–360.

Scott, J. M., Heglund, P., Morrison, M. L. and Raven, P. H. (2002). Predicting species occurrences: issues of accuracy and scale. Island Press, Washington DC.

Seal, U. S., Plotka, E. D., Packard, J. M. and Mech, L. D. (1979). Endocrine correlates of reproduction in the wolf. I. Serum progesterone, estradiol, and LH during the estrous cycle. *Biology of Reproduction*, **21**, 1057–1066.

Seal, U. S., Plotka, E. D., Smith, J. D., Wright, F. H., Reindl, N. J., Taylor, R. S. and Seal, M. F. (1985). Immunoreactive luteinizing hormone, estradiol, progesterone, testosterone and androstenedione levels during the breeding season and anestrus in Siberian tigers (*Panthera tigris altaica*). *Biology of Reproduction*, **32**, 361–368.

Seaman, D. E. (1993). *Home range and male reproductive optimization in black bears*. PhD thesis. North Carolina State University, Raleigh, NC.

Seaman, D. E. and Powell, R. A. (1990). Identifying patterns and intensity of home range use. *International Conference on Bear Research and Management*, **8**, 243–249.

Seaman, D. E. and Powell, R. A. (1996). Accuracy of kernel estimators for animal home range analysis. *Ecology*, 77, 2075–2085.

Seber, G. A. F. (1965). A note on the multiple recapture census. *Biometrika*, **52**, 249–259.

Seber, G. A. F. (1982). *The estimation of animal abundance and related parameters*, 2nd edn. Macmillan Publishing Company, New York.

Seddon, P. J. (1999). Persistence without intervention, assessing success in wildlife re-introductions. *Trends in Ecology and Evolution*, **14**, 503-ff.

Seddon, P. J., VanHeezik, Y. and Maloney, R. F. (1999). Short- and medium-term evaluation of foothold trap injuries in two species of fox in Saudi Arabia. In G. Proulx (ed.). *Mammal trapping*, pp. 67–78. Alpha Wildlife Publications, Sherwood Park, Alberta.

Selye, H. (1951). The general adaptation syndrome. *Annual Reviews in Medicine*, **2**, 327–342.

Sepulveda, M. A., Bartheld, J. L., Monsalve, R., Gomez, V. and Medina-Vogel, G. (2007). Habitat use and spatial behaviour of the endangered Southern river otter (*Lontra provocax*) in riparian habitats of Chile: Conservation implications. *Biological Conservation*, **140**, 329–338.

Sequin, E.S., Jaeger, M.M., Brussard, P.F. and Barrett, R. H. (2003). Wariness of coyotes to camera traps relative to social status and territory boundaries. *Canadian Journal of Zoology*, **81**, 2015–2025.

Serfass, T. L., Brooks, R. P., Swimley, T. J., Rymon, L. M. and Hayden, A. H. (1996). Considerations for capturing, handling, and translocating river otters. *Wildlife Society Bulletin*, **24**, 21–24.

Servheen, C., Thier, T. T., Jonkel, C. J. and Beaty, D. (1981). An ear-mounted transmitter for bears. *Wildlife Society Bulletin*, **9**, 56–57.

Servheen, C., Herrero, S. and Peyton, B. (1999). *Bears: status survey and conservation action plan*. IUCN Publications, Gland, Switzerland.

Shaffer, M. L. (1981). Minimum population sizes for species conservation. *BioScience*, **31**, 131–134.

Shaw, C. N., Wilson, P. J. and White, B. N. (2003). A reliable molecular method of gender determination for mammals. *Journal of Mammalogy*, **84**, 123–128.

Sheffield, S. R., Sullivan, J. P. and Hill, E. F. (2005). Identifying and handling contaminant-related wildlife mortality/morbidity incidents in the field. In C. E. Braun (ed.). *Techniques for wildlife investigations and management*, pp. 213–238. The Wildlife Society, Bethesda, MD.

Sheldon, W. G. (1949). A trapping and tagging technique for wild foxes. *Journal of Wildlife Management*, **13**, 309–311.

Shideler, S. E., Munro, C. J., Johl, H. K., Taylor, H. W., and Lasley, B. L. (1995). Urine and fecal sample collection on filter paper for ovarian hormone evaluations. *American Journal of Primatology*, **37**, 305–315.

Shille, V. M., Haggerty, M. A., Shackleton, C. and Lasley, B. L. (1990). Metabolites of estradiol in serum, bile, intestine and feces of the domestic cat (*Felis catus*). *Theriogenology*, **34**, 779–794.

Shirley, M. G., Linscombe, R. G. and Sevin, L. R. (1983). A live trapping and handling technique for river otter. *Proceedings Annual Conference of the Southeastern Association of Fish and Wildlife Agencies*, **37**, 182–189.

Shivik, J. A. (2002). Odor-adsorptive clothing, environmental factors, and search-dog ability. *Wildlife Society Bulletin*, **30**, 721–727.

Shivik, J. A. (2006). Tools for the edge: what's new for conserving carnivores. *BioScience*, **56**, 253–259.

Shivik, J. A., Treves, A. and Callahan, P. (2003). Nonlethal techniques for managing predation: primary and secondary repellents. *Conservation Biology*, **17**, 1531–1537.

Shivik, J. A., Martin, D. J., Pipas, M. J., Turnam, J. and DeLiberto, T. J. (2005). Initial comparison, jaws, cables, and cage-traps to capture coyotes. *Wildlife Society Bulletin*, **33**, 1375–1383.

Short, R. V. and King, J. M. (1964). The design of a crossbow and dart for the immobilization of wild animals. *The Veterinary Record*, **76**, 628–630.

Sidorovich, V. E., Tikhomirova, L. L. and Jędrzejewska, B. (2003). Wolf *Canis lupus* numbers, diet and damage to livestock in relation to hunting and ungulate abundance in northeastern Belarus during 1990–2000. *Wildlife Biology*, **9**, 103–112.

Sih, A., Hanser S. F. and McHugh, K. A. (2009). Social network theory, new insights and issues for behavioral ecologists. *Behavioral Ecology and Sociobiology*, **63**, 975–988.

Sillero-Zubiri, C. and Laurenson, K. (2001). Interactions between carnivores and local communities: Conflict or co-existence? In J. Gittleman, S. Funk, D. W. Macdonald and R. K. Wayne (eds). *Carnivore conservation*, pp. 282–312. Cambridge University Press, Cambridge.

Sillero-Zubiri, C., Gottelli, D. and Macdonald, D. W. (1996). Male philopatry, extra-pack copulations and inbreeding avoidance in Ethiopian wolves (*Canis simensis*). *Behavioural Ecology and Sociobiology*, **38**, 331–340.

Sillero-Zubiri, C., Hoffmann, M. and Macdonald, D. W. (2004). *Canids: Foxes, wolves, jackals and dogs: status survey and conservation action plans*. IUCN Publications, Gland, Switzerland.

Sillero-Zubiri, C., Sukumar, R. and Treves, A. (2006). Living with wildlife: the roots of conflict and the solutions. In D. W. Macdonald and K. Service (eds). *Key topics in conservation biology*, pp. 253–270. Blackwell Publishing, Oxford.

Silva, R. C., Costa, G. M. J., Andrade, L. M. and França, L. R. (2010). Testis stereology, seminiferous epithelium cycle length, and daily sperm production in the ocelot (*Leopardus pardalis*). *Theriogenology*, **73**, 157–167.

Silver, H. and Silver, W. T. (1969). Growth and behavior of the coyote-like canid of northern New England with observations on canid hybrids. *Wildlife Monographs*, **17**.

Silver, S. C., Ostro, L. E. T., Marsh, L. K. *et al.* (2004). The use of camera traps for estimating jaguar *Panthera onca* abundance and density using capture/recapture analysis. *Oryx*, **38**, 148–154.

Silverman, B. W. (1990). *Density estimation for statistics and data analysis*. Chapman and Hall, London.

Silvy, N. J., Lopez, R. R. and Peterson, M. K. (2005). Wildlife marking techniques. In C. E. Braun (ed.). *Research and management techniques for wildlife and habitats*, pp. 339–376. The Wildlife Society, Bethesda, MD.

Simcharoen, S., Barlow, A. C. D., Simcharoen, A. and Smith, J. L. D. (2008). Home range size and daytime habitat selection of leopards in Huai Kha Khaeng Wildlife Sanctuary, Thailand. *Biological Conservation*, **141**, 2242–2250.

Simón, M. A., Cadenas, R., Gil-Sánchez, J. M. *et al.* (2009). Conservation of free-ranging Iberian lynx (*Lynx pardinus*) populations in Andalusia. In A. Vargas, Ch. Breitenmoser and U. Breitenmoser U. (eds). *Iberian lynx ex situ conservation: a multidisciplinary approach*, pp. 42–55. Fundación Biodiversidad, Madrid.

Sinclair, A. R. E., Fryxell, J. and Caughley, G. (eds) (2005). *Wildlife ecology and management*, 2nd edn. Blackwell Science, London.

Sinha, A. A., Conaway, C. H., and Kenyon, K. W. (1966). Reproduction in the female sea otter. *Journal of Wildlife Management*, **39**, 121–130.

Skalski, J. R. and Robson, D. S. (1992). *Techniques for wildlife investigations.* Academic Press, San Diego.

Skinner, D. L. and Todd, A. W. (1990). Evaluating efficiency of footholding devices for coyotes capture. *Wildlife Society Bulletin*, **18**, 166–175.

Slate, D., Rupprecht, C. E., Rooney, J. A., Donovan, D., Lein, D. H. and Chipman, R. B. (2005). Status of oral rabies vaccination in wild carnivores in the United States. *Virus Research*, **111**, 68–76.

Slater, P. J. B. and Lester, N. P. (1982). Minimising errors in splitting behaviour into bouts. *Behaviour*, **79**, 153–161.

Slough, B. G. (1996). Estimating lynx population age ratio with pelt-length data. *Wildlife Society Bulletin*, **24**, 495–499.

Slovic, P. (1987). Perceptions of risk. *Science*, **236**, 280–285.

Smith D. W. and Bangs E. E. (2009). Reintroduction of wolves to Yellowstone National Park: history, values and ecosystem restoration. In M. W. Hayward and M. J. Somers (eds). *Reintroduction of top-order predators*, pp. 99–125. Wiley-Blackwell Publishing, Oxford.

Smith, W. P., Borden, D. L. and Endres, K. M. (1994). Scent-station visits as an index to abundance of raccoons, an experimental manipulation. *Journal of Mammalogy*, **75**, 637–647.

Smith, M. E., Linnell, J. D. C., Odden, J. and Swenson, J. E. (2000a). Methods for reducing livestock losses to predators. A: Livestock guardian animals. *Acta Agriculturæ Scandinavica*, **50**, 279–290.

Smith, M. E., Linnell, J. D. C., Odden, J. and Swenson, J. E. (2000b). Methods for reducing livestock losses to predators: B. Aversive conditioning, deterrents and repellents. *Acta Agriculturæ Scandinavica*, **50**, 304–315.

Smith, D. A., Ralls, K., Hurt, A. *et al.* (2003). Detection and accuracy rates of dogs trained to find scats of San Joaquin kit foxes (*Vulpes macrotis mutica*). *Animal Conservation*, **6**, 339–346.

Smith, D. R., Brown, J. A. and Nancy, C. H. L. (2004). Application of adaptive sampling to biological populations. In W. L. Thompson (ed.). *Sampling rare or elusive species: concepts, designs, and techniques for estimating population parameters*, pp. 77–122. Island Press, Washington DC.

Smith, D. A., Ralls, K., Cypher, B. L. and Maldonado, J. E. (2005). Assessment of scat-detection dog surveys to determine kit fox distribution. *Wildlife Society Bulletin*, **33**, 897–904.

Smith, D. A., Ralls, K., Cypher, B. L. *et al.* (2006). Relative abundance of endangered San Joaquin kit foxes (*Vulpes macrotis mutica*) based on scat-detection dog surveys. *Southwestern Naturalist*, **51**, 210–219.

Smith, J. B., Jenks, J. A. and Klaver, R. W. (2007). Evaluating detection probabilities for American marten in the Black Hills, South Dakota. *Journal of Wildlife Management*, **71**, 2412–2416.

Smouse, P. E., Focardi, S., Moorcroft, P. R., Kie, J. G., Forester, J. D. and Morales, J. M. (2010). Stochastic modeling of animal movement. *Philosophical Transactions of The Royal Society B*, **365**, 2201–2211.

Smuts, G. L., Anderson, J. L. and Austin, J. C. (1978). Age determination of the African lion *Panthera leo*. *Journal of Zoology*, **185**, 115–146.

Smuts, G. L., Whyte, I. J. and Dearlove, T. W. (1977). A mass capture technique for lions. *African Wildlife Journal*, **15**, 81–87.

Snyder, N. F., Derrickson, S. R., Beissinger, S. R. *et al.* (1996). Limitations of captive breeding in endangered species recovery. *Conservation Biology*, **10**, 338–348.

Sober, E. (2008). *Evidence and evolution.* Cambridge University Press, Cambridge.

Soberon, J. (2007). Grinnellian and Eltonian niches and geographic distributions of species. *Ecology Letters*, **10**, 121–131.

Socha, R. and Zemek, R. (2003). Wing morph-related differences in the walking pattern and dispersal in a flightless bug, *Pyrrhocoris apterus* (Heteroptera). *Oikos*, **100**, 35–42.

Soisalo, M. K. and Cavalcanti, S. M. C. (2006). Estimating the density of a jaguar population in the Brazilian Pantanal using camera-traps and capture-recapture sampling in combination with GPS radio-telemetry. *Biological Conservation*, **129**, 487–496.

Solstad, T., Boccara, C. N., Kropff, E., Moser, M.-B. and Moser, E. I. (2008). Representation of geometric borders in the entorhinal cortex. *Science*, **322**, 1865–1868.

Sommer, C. (1996). Ecotoxicology and developmental stability as an in situ monitor of adaptation. *Ambio*, **25**, 374–376.

Soorae, P. S. and Stanley Price, M. R. (1997). *Successful re-introductions of large carnivores-what are the secrets?* Paper presented at the 11th International Conference on Bear Research and Management, Graz, Austria.

Sork, V. L. and Waits, L. P. (2010). Contributions of landscape genetics: approaches, insights, and future potential. *Molecular Ecology*, **19**, 3489–3495.

Soto-Azat, C., Boher, F., Fabry, M., Pascual, P. and Medina-Vogel, G. (2008). Surgical implantation of intra-abdominal radiotransmitters in marine otters (*Lontra felina*) in central Chile. *Journal of Wildlife Diseases*, **44**, 979–982.

Soulsbury, C. D., Baker, P. J., Iossa, G. and Harris, S. (2008). Fitness costs of dispersal in red foxes (*Vulpes vulpes*). *Behavioral Ecology and Sociobiology*, **62**, 1289–1298.

South, A. (1999). Extrapolating from individual movement behaviour to population spacing patterns in a ranging mammal. *Ecological Modelling*, **117**, 343–360.

Spady, T. J., Lindburg, D. G. and Durrant, B. S. (2007). Evolution of seasonality in bears. *Mammal Review*, **37**, 21–53.

Spalton, J. A. and Al Hikmani, H. M. (2006). The leopard in the Arabian peninsula—distribution and subspecies status. *Cat News Special Issue*, **1**, 4–8.

Spaulding, R., Krausman, P. R. and Ballard, W. B. (2000). Observer bias and analysis of gray wolf diets from scats. *Wildlife Society Bulletin*, **28**, 947–950.

Spear, L. P. (2000). The adolescent brain and age-related behavioral manifestations. *Neuroscience and Biobehavioral Reviews*, **24**, 417–63.

Spence, C. E., Kenyon, J. E., Smith, D. R., Hayes, R. D. and Baer, A. M. (1999). Surgical sterilization of free-ranging wolves. *Canadian Veterinary Journal*, **40**, 118–121.

Spencer, S. R., Cameron, G. N. and Swihart, R. K. (1990). Operationally defining home range, temporal dependence exhibited by hispid cotton rats. *Ecology*, **71**, 1817–1822.

Spencer, W. D. (In press). Home ranges and the value of spatial information. *Journal of Mammalogy*.

Spiegelhalter D., Thomas A., and Best, N. G. (2009). *WinBUGS User Manual.* MCR Biostatistics Unit, Cambridge.

Spong, G. and Creel, S. (2001). Deriving dispersal distances from genetic data. *Proceedings of the Royal Society B*, **268**, 2571–2574.

Sponheimer, M., Robinson, T, Ayliffe, L. *et al.* (2003). Nitrogen isotopes in mammalian herbivores: hair δ15N values from a controlled feeding study. *International Journal of Osteoarchaeology*, **13**, 80–87.

Stahl, P., Vandel, J. M., Ruette, S., Coat, L., Coat, Y., and Balestra, L. (2002). Factors affecting lynx predation on sheep in the French Jura. *Journal of Applied Ecology*, **39**, 204–216.

Stallknecht, D. E. (2007). Impediments to wildlife disease surveillance, research, and diagnostics. *Wildlife and Emerging Zoonotic Diseases: The Biology, Circumstances and Consequences of Cross-Species Transmission*, **315**, 445–461.

Stander, P. E. (1997). Field age determination of leopards by tooth wear. *African Journal of Ecology*, **35**, 156–161.

Stander, P. E. and Morkel, P. vdB. (1991). Field immobilization of lions using disassociative anaesthetics in combination with sedatives. *African Journal of Ecology*, **29**, 137–148.

Stander, P., Ghau, X., Tsisaba, D. and Txoma, X. (1996). A new method of darting, stepping back in time. *African Journal of Ecology*, **34**, 48–53.

Steinetz, B. G., Brown, J. L., Roth, T. L. and Czekala, N. (2005). Relaxin concentrations in serum and urine of endangered species: correlations with physiologic events and use as a marker of pregnancy. *Annals of the New York Academy of Sciences*, **1041**, 367–378.

Steinetz, B. G., Goldsmith, L. T. and Leist, G. (1987). Plasma relaxin levels in pregnant and lactating dogs. *Biology of Reproduction*, **37**, 719–725.

Steneck, R. S. (2005). An ecological context for the role of large carnivores in conserving biodiversity. In J. C. Ray, K. H. Redford, R. S. Steneck and J. Berger J. (eds). *Large carnivores and the conservation of biodiversity*, pp. 9–33. Island Press, Washington DC.

Stenson, G. B. (1988). Oestrus and the vaginal smear cycle of the river otter, *Lutra canadensis*. *Journal of Reproduction and Fertility*, **83**, 605–610.

Stephens, D. W. and Krebs, J. R. (1987). *Foraging theory*. Princeton University Press, Princeton, NJ.

Stephens, P. A., Zaumyslova, O. Y., Miquelle, D. G., Myslenkov, A. I. and Hayward, G. D. (2006). Estimating population density from indirect sign, track counts and the Formozov-Malyshev-Pereleshin formula. *Animal Conservation*, **9**, 339–348.

Stephenson, A. B. (1977). Age determination and morphological variation of Ontario otters. *Canadian Journal of Zoology*, **55**, 1577–1583.

Stephenson, T. R., Testa, J. W., Adams, G. P., Sasser, R. G., Schwartz, C. C. and Hundertmark, K. J. (1995). Diagnosis of pregnancy and twinning in moose by ultrasonography and serum assay. *Alces*, **31**, 167–172.

Sterling, B., Conley, W. and Conley, M. R. (1983). Simulations of demographic compensation in coyote populations. *Journal of Wildlife Management*, **47**, 1177–1181.

Sterner, R. T., Meltzer, M. I., Shwiff, S. A. and Slate, D. (2009). Tactics and economics of wildlife oral rabies vaccination, Canada and the United States. *Emerging Infectious Diseases*, **15**, 1176–1184.

Stetz, J. B., Kendall, K. C. and Servheen, C. (2010). Evaluation of bear rub surveys to monitor grizzly bear population trends. *Journal of Wildlife Management*, **74**, 860–870.

Stevens, S. S., and Serfass, T. L. (2008). Visitation patterns and behavior of Nearctic river otters (*Lontra canadensis*) at latrines. *Northeastern Naturalist*, **15**, 1–12.

Stewart, P. D. and Macdonald, D. W. (1997). Age, sex, and condition as predictors of moult and the efficacy of a novel fur-clip technique for individual marking of the European badger (*Meles meles*). *Journal of Zoology*, **241**, 543–550.

Stewart-Oaten, A., Murdoch, W. W. and Parker, K. R. (1986). Environmental impact assessment, "pseudoreplication" in time? *Ecology*, **67**, 929–940.

Stillman, T. A., Goss-Custard, J. D. and Alexander, J. (2000). Predator search pattern and the strength of interference through prey depression. *Behavioral Ecology*, **11**, 597–605.

Stirling, I. (1989). *Polar bears*. Fitzhenry and Whiteside, Toronto, Ontario.

Stirling, I., Jonkel, D., Smith, P., Robertson, R. and Cross, D. (1977). The ecology of the polar bear (*Ursus maritimus*) along the western coast of Hudson Bay. *Canadian Wildlife Service Occasional Papers*, **33**, 1–64.

Stöhr, K. and Meslin, F. M. (1996). Progress and setbacks in the oral immunisation of foxes against rabies in Europe. *Veterinary Record*, **139**, 32–35.

Stoms, D. M. (1992). Effects of habitat map generalization in biodiversity assessment. *Photogrammetric Engineering and Remote Sensing*, **58**, 1587–1591.

Stoner, D. C., Rieth, W. R., Wolfe, M. L. and Neville A. (2008). Long-distance dispersal of a female cougar in a basin and range landscape. *Journal of Wildlife Management*, **72**, 933–939.

Storfer, A., M. A. Murphy, J. S. Evans, C. S. Goldberg, S. Robinson, S. F. Spear, R. Dezzani, E. Delmelle, L. Vierling, and L. P. Waits. (2007). Putting the 'landscape' in landscape genetics. *Heredity*, **98**, 128–142.

Stoskopf, M. K., Beck, K., Fazio, B. *et al.* (2005). Integrating science and management, implementing the recovery of the Red Wolf. *Wildlife Society Bulletin*, **33**, 1145–1152.

Strand, O., Skogland, T. and Kvam, A. T. (1995). Placental scars and estimation of litter size: an experimental test in the arctic fox. *Journal of Mammalogy*, **76**, 1220–1225.

Strathearn, S. M., Lotimer, J. S., Kolenosky, G. B. and Lintack, W. M. (1984). An expanding break-away radio collar for black bear. *Journal of Wildlife Management*, **48**, 939–942.

Stuart-Hill, G., Grossman, D. (1993). Parks, profits and professionalism, lion return to Pilanesberg. *African Wildlife*, **47**, 267–280.

Stump, W. Q., and Anthony, R. G. (1983). Use of eye lens protein for estimating age in *Microtus pennsylvanicus*. *Journal of Mammalogy*, **64**, 697–700.

Sundell, J., Norrdahl, K., Korpimaki, E. and Hanski, I. (2000). Functional response of the least weasel, *Mustela nivalis nivalis*. *Oikos*, **90**, 501–508.

Sundell, J., Kojola, I. and Hanski, I. (2006). A new GPS-GSM-based method to study behavior of brown bears. *Wildlife Society Bulletin*, **34**, 446–450.

Sundqvist, C., Amador, A. G. and Bartke, A. (1989). Reproduction and fertility in the mink (*Mustela vison*). *Journal of Reproduction and Fertility*, **85**, 413–441.

Sundqvist, A. K., Ellegren, H. and Vilà, C. (2008). Wolf or dog? Genetic identification of predators from saliva collected around bite wounds on prey. *Conservation Genetics*, **9**, 1275–1279.

Sutor, A. (2008). Dispersal of the alien raccoon dog *Nyctereutes procyonoides* in Southern Brandenburg, Germany. *European Journal of Wildlife Research*, **54**, 321–326.

Svartberg, K. (2002). Shyness-boldness predicts performance in working dogs. *Applied Animal Behaviour Science*, **79**, 157–174.

Swann, D. E., Hass, C. C., Dalton, D. C. and Wolf, S. A. (2004). Infrared-triggered cameras for detecting wildlife: an evaluation and review. *Wildlife Society Bulletin*, **32**, 357–365.

Swenson, J. E. and Andrén, H. (2005). A tale of two countries: large carnivore depredation and compensation schemes in Sweden and Norway. In R. Woodroffe, S. Thirgood and A. Rabinowitz (eds). *People and wildlife: conflict or coexistence?*, pp. 323–339. Cambridge University Press, Cambridge.

Swenson, J. E., Adamič, M., Huber, D. and Stokke, S. (2007). Brown bear body mass and growth in northern and southern Europe. *Oecologia*, **153**, 37–47.

Swenson, J. E., Wabakken, P., Sandegren, F. *et al.* (1995). The near extinction and recovery of brown bears in Scandinavia in relation to the bear management policies of Norway and Sweden. *Wildlife Biology*, **1**, 11–25.

Swenson, J. E., Støen, O.-G., Zedrosser, A. *et al.* (2010). *Bjørnens status och økologi i Skandinavia (Status and Ecology of the Bear in Scandinavia)* 2010–3 fra Det skandinaviske bjørneprojektet. Skandinaviska Björnprojektet, Orsa.

Swihart, R. K., Slade, N. A. and Bergstrom, B. J. (1988). Relating body size to the rate of home range use in mammals. *Ecology*, **69**, 393–399.

Swiss Academy of Medical Sciences and Swiss Academy of Sciences. (2005). *Ethical principles and guidelines for scientific experiments on animals*. SAMS, Petersplatz 13, CH-4051, Bern, and SAS, Bärenplatz 2, CH-3011 Bern, Switzerland.

Taber, R. D. and Cowan, I. McT. (1969). Capturing and marking wild animals. In R. H. Giles, Jr., (ed.). *Wildlife management techniques*, pp. 277–317. The Wildlife Society, Washington DC.

Tallmon, D. A., Koyuk, A., Luikart, G. and Beaumont, M. A. (2008). ONeSAMP: a program to estimate effective population size using approximate Bayesian computation. *Molecular Ecology Resources*, **8**, 299–301.

Tambling, C. J., Cameron, E. Z., Du Toit, J. T. and Getz, W. M. (2010). Methods for locating African lion kills using Global Positioning System movement data. *Journal of Wildlife Management*, **74**, 549–556.

Tanaka, H. (2005). Seasonal and daily activity patterns of Japanese badgers (*Meles meles anakuma*) in Western Honshu, Japan. *Mammal Study*, **30**, 11–17.

Taylor, L. H. and Latham, S. M. (2001). Risk factors for human disease emergence. *Philosophical Transactions of the Royal Society of London, B*, **356**, 983–989.

Teerink, B. J. (2003). *Hair of West European mammals: atlas and identification key*. Cambridge University Press, Cambridge.

Theuerkauf, J. and Jędrzejewski, W. (2002). Accuracy of radiotelemetry to estimate wolf activity and locations. *Journal of Wildlife Management*, **66**, 859–864.

Theuerkauf, J., Jędrzejewski, W., Schmidt, K. *et al.* (2003). Daily patterns and duration of wolf activity in the Białowieża forest, Poland. *Journal of Mammalogy*, **84**, 243–253.

Thiemann, G. W., Iverson, S. J. and Stirling, I. (2006). Seasonal, sexual and anatomical variability in the adipose tissue of polar bears (*Ursus maritimus*). *Journal of Zoology*, **269**, 65–76.

Thomas, J. W. and Pletscher, D. H. (2002). The "lynx affair"—professional credibility on the line. *Wildlife Society Bulletin*, **30**, 1281–1286.

Thompson, S. K. (1990). Adaptive cluster sampling. *Journal of American Statistical Association*, **85**, 1050–1059.

Thompson, S. K. (2002). *Sampling*, 2nd edn. John Wiley and Sons, New York.

Thompson, J. L. (2007). *Abundance and density of fisher on managed timberlands in north coastal California*. MS thesis. Humboldt State University, Arcata, CA.

Thompson, C. M. and Gese, E. M. (2007). Food webs and intraguild predation: community interactions of a native mesocarnivore. *Ecology*, **88**, 334–346.

Thompson, S. K. and Seber, G. A. F. (1996). *Adaptive sampling*. John Wiley and Sons, New York.

Thompson, M. E. and Wrangham, R. W. (2008). Diet and reproductive function in wild female chimpanzees (*Pan troglodytes schweinfurthii*) at Kibale National Park, Uganda. *American Journal of Physical Anthropology*, **135**, 171–181.

Thompson, W. L. (2004). *Sampling rare or elusive species, concepts, designs and techniques for estimating population parameters*. Island Press, Washington DC.

Thompson, W. L., White, G. C. and Gowan, C. (1998). *Monitoring vertebrate populations*. Academic Press, San Diego, CA.

Thompson, D. J., Fecske, D. M., Jenks, J. A. and Jardin, A. R. (2009). Food habits of recolonizing cougars in the Dakotas: prey obtained from prairie and agricultural habitats. *American Midland Naturalist*, **161**, 69–75.

Thorn, M., Scott, D. M., Green, M., Bateman, P. W and Cameron, E. Z. (2009). Estimating brown hyaena occupancy using baited camera traps. South African *Journal of Wildlife Research*, **39**, 1–10.

Thorne, E. T. and Williams, E. S. (1988). Disease and endangered species: the black-footed ferret as a recent example. *Conservation Biology*, **2**, 66–74.

Thurber, J. M., Peterson, R. O., Drummer, T. D. and Thomasma, T. A. (1994). Gray wolf response to refuge boundaries and roads in Alaska. *Wildlife Society Bulletin*, **22**, 61–68.

Tieszen, L. L., Boutton, T. W., Tesdahl, K. G. and Slade, N. A. (1983). Fractionation and turnover of stable carbon isotopes in animal tissues: implications for $\delta\ ^{13}C$ analysis of diet. *Oecologia*, **57**, 32–37.

Tilman, D., May, R. M., Lehman, C. L. and Nowak, M. A. (2002). Habitat destruction and the extinction debt. *Nature* **371**, 65–66.

Timm, S. F., Munson, L., Summers, B. A. *et al.* (2009). A suspected canine distemper epidemic as the cause of a catastrophic decline in Santa Catalina Island foxes (*Urocyon littoralis catalinae*). *Journal of Wildlife Diseases*, **45**, 333–343.

Tinker, M. T., Doak, D. F., Estes, J. A., Hatfield, B. B., Staedler, M. M. and Bodkin, J. L. (2006). Incorporating diverse data and realistic complexity into demographic estimation procedures for sea otters. *Ecological Applications*, **16**, 2293–2312.

Toal, R. L., Walker, M. A. and Henry, G. A. (1986). A comparison of real-time ultrasound, palpation, and radiography in pregnancy detection and litter size determination in the bitch. *Veterinary Radiology*, **27**, 102–108.

Tobler, M. W., Carrillo-Percastegui, S. E., Pitman, R. L., Mares, R. and Powell, G. (2008). An evaluation of camera traps for inventorying large- and medium-sized terrestrial rain-forest mammals. *Animal Conservation*, **11**, 169–178.

Tomkiewicz, S. M., Fuller, M. R., Kie, J. G. and Bates, K. K. (2010). Global Positioning System (GPS) and associated technologies in animal behaviour and ecological research. *Philosophical Transactions of The Royal Society B*, **365**, 2163–2176.

Totton, S. C., Tinline, R. R., Rosatte, R. C. and Bigler, L. L. (2002). Contact rates of raccoons (*Procyon lotor*) at a communal feeding site in rural eastern Ontario. *Journal of Wildlife Diseases*, **38**, 313–319.

Touma, C. and Palme, R. (2005). Measuring fecal glucocorticoid metabolites in mammals and birds: the importance of validation. *Annals of the New York Academy of Sciences*, **1046**, 54–74.

Trap Effectiveness Research Team. (1995). *Annual report. Trap effectiveness project. April 1994–March 1995*. Alberta Environmental Centre, Vegreville, Alberta.

Tredick, C. A., and Vaughan, M. R. (2009). DNA-based population demographics of black bears in coastal North Carolina and Virginia. *Journal of Wildlife Management*, **73**, 1031–1039.

Treves, A. (2009). Hunting for large carnivore conservation. *Journal of Applied Ecology*, **46**, 1350–1356.

Treves, A. and Karanth, K. U. (2003). Human-carnivore conflict and perspectives on carnivore management worldwide. *Conservation Biology*, 17, 1491–1499.

Treves, A. and Naughton-Treves, L. (2005). Evaluating lethal control in the management of human-wildlife conflict. In R. Woodroffe, S. Thirgood and A. Rabinowitz (eds). *People and wildlife: conflict or coexistence?*, pp. 86–106. Cambridge University Press, Cambridge.

Treves, A., Naughton-Treves, L., Harper, E. K. *et al.* (2004). Predicting carnivore-human conflict: a spatial model derived from 25 years of data on wolf predation on livestock. *Conservation Biology*, **18**, 114–125.

Tullar, B. F., Jr. (1984). Evaluation of a padded foothold trap for capturing foxes and raccoons. *New York Fish and Game Journal*, **31**, 97–103.

Tumanov, I. L. (1998). Reproductive characteristics of captive European brown bears and growth rates of their cubs in Russia. *Ursus*, **10**, 63–65.

Turchin, P. (1996). Fractal analyses of animal movement, a critique. *Ecology*, 77, 2086–2090.

Turchin, P. (1998). *Quantitative analysis of movement: measuring and modeling population redistribution in animals and plants*. Sinauer, Sunderland, MA.

Turek, F. W. and Campbell, C. S. (1979). Photoperiodic regulation of neuroendocrine-gonadal activity. *Biology of Reproduction*, **20**, 32–50.

Turner, M. G., Wu, Y., Wallace, L. L., Romme, W. H. and Brenkert, A. (1994). Simulating winter interactions among ungulates, vegetation, and fire in northern Yellowstone Park. *Ecological Applications*, **4**, 472–496.

Tuyttens, F. A. M., Delahay, R. J., Macdonald, D. W., Cheeseman, C. L., Long, B. and Donnelly, C. A. (2000). Spatial perturbation caused by a badger (*Meles meles*) culling operation, implications for the function of territoriality and the control of bovine tuberculosis (*Mycobacterium bovis*). *Journal of Animal Ecology*, **69**, 815–828.

Tuyttens, F. A. M., Macdonald, D. W. and Roddam, A. W. (2002). Effects of radio-collars on European badgers (*Meles meles*). *Journal of Zoology*, **257**, 37–42.

Tveraa, T., Fauchald, P., Henaug, C. and Yoccoz, N. G. (2003). An examination of a compensatory relationship between food limitation and predation in semi-domestic reindeer. *Oecologia*, **137**, 370–376.

Tyre, A. J., Possingham, H. P. and Lindenmayer, D. B. (2001). Inferring process from pattern: can territory occupancy provide information about life history parameters?. *Ecological Applications*, **11**, 1722–1737.

Tyre, A. J., Tenhumberg, B., Field, S. A., Niejalke, D., Parris, K. and Possingham, H. P. (2003). Improving precision and reducing bias in biological surveys: Estimating false-negative error rates. *Ecological Applications*, **13**, 1790–1801.

US Fish and Wildlife Service—Alaska Region (2008). *Marine mammals management—polar bears*. http://alaska.fws.gov/fisheries/mmm/polarbear/pbmain.htm (August 2011).

Ulizio, T. J., Squires, J. R., Pletscher, D. H., Schwartz, M. K., Claar, J. J. and Ruggiero, L. F. (2006). The efficacy of obtaining genetic-based identifications from putative wolverine snow tracks. *Wildlife Society Bulletin*, **34**, 1326–1332.

Uphyrkina, O., Johnson, W. E., Quigley, H. B. *et al.* (2001). Phylogenetics, genome diversity and origin of modern leopard, *Panthera pardus*. *Molecular Ecology*, **10**, 2617–2633.

Urbano, F., Cagnacci, F., Calenge, C., Dettki, H., Cameron, A. and Neteler, M. (2010). Wildlife tracking data management: a new vision. *Philosophical Transactions of The Royal Society B*, **365**, 2177–2185.

Vajta, G. and Kuwayama, M. (2006). Improving cryopreservation systems. *Theriogenology*, **65**, 236–244.

Valdespino, C., Asa, C. S. and Bauman, J. E. (2002). Estrous cycles, copulation, and pregnancy in the fennec fox (*Vulpes zerda*). *Journal of Mammalogy*, **83**, 99–109.

Valeix, M., Loveridge, A. J., Chamaill-Jammes, S. *et al.* (2009). Behavioral adjustments of African herbivores to predation risk by lions: Spatiotemporal variations influence habitat use. *Ecology*, **90**, 23–30.

Valentini, A., Pompanon, F. and Taberlet, P. (2008). DNA barcoding for ecologists. *Trends in Ecology and Evolution*, **24**, 110–117.

Vali, U., Einarsson, A., Waits, L. and Ellergren, H. (2008). To what extent do microsatellite markers reflect genome-wide genetic diversity in natural populations? *Molecular Ecology*, **17**, 3808–3817.

Valkenburg, P., Tobey, R. W. and Kirk, D. (1999). Velocity of tranquilizer darts and capture mortality of caribou calves. *Wildlife Society Bulletin*, **27**, 894–896.

Valtonen, M., Rajakoski, E. J. and Mäkelä, J. I. (1977). Reproductive features in the female raccoon dog. *Journal of Reproduction and Fertility*, **51**, 517-NP.

Van Ballenberghe, V. (1984). Injuries to wolves sustained during live-capture. *Journal of Wildlife Management*, **48**, 1425–1429.

van Beest, F. M., Mysterud, A., Loe, L. E. and Milner, J. M. (2010). Forage quantity, quality and depletion as scale-dependent mechanisms driving habitat selection of a large browsing herbivore. *Journal of Animal Ecology*, **79**, 910–922.

Van Bommel, L., de Vaate, M. D. B., de Boer, W. F. and de Iongh, H. H. (2007). Factors affecting livestock predation by lions in Cameroon. *African Journal of Ecology*, **45**, 490–498.

Van Der Merwe, N. J. (1953). The jackal. *Fauna Flora (Pretoria)*, **4**, 3–82.

Van Etten, K. W., Wilson, K. R. and Crabtree, R. L. (2007). Habitat use of red foxes in Yellowstone National Park based on snow tracking and telemetry. *Journal of Mammalogy*, **88**, 1498–1507.

Van Heerden, J. and Kuhn, F. (1985). Reproduction in captive hunting dogs *Lycaon pictus*. *South African Journal of Wildlife Research*, **15**, 80–84.

Van Horn, R. C., McElhinny, T. L. and Holekamp, K. E. (2003). Age estimation and dispersal in the spotted hyaena (*Crocuta crocuta*). *Journal of Mammalogy*, **84**, 1019–1030.

Van Horne, B. (1983). Density as a misleading indicator of habitat quality. *Journal of Wildlife Management*, 47, 893–901.

Van Moorter, B., Visscher, D., Benhamou, S., Börger, L., Boyce M. S. and Gaillard, J-M. (2009). Memory keeps you at home, a mechanistic model for home range emergence. *Oikos*, **118**, 641–652.

Van Tienhoven, A. (1983). *Reproductive physiology of vertebrates*. Comstock Publishing Associates, Ithaca, NY.

Van Valkenburgh, B. (1988). Incidence of tooth breakage among large, predatory mammals. *American Naturalist*, **131**, 291–300.

Van Winkle, W. (1975). Comparison of several probabilistic home-range models. *Journal of Wildlife Management*, **39**, 118–123.

Vanak, A. T. and Gompper, M. E. (2007). Effectiveness of noninvasive techniques for surveying activity and habitat use of the Indian fox *Vulpes bengalensis* in southern India. *Wildlife Biology*, **13**, 219–224.

Vandermeer, J. (1981). *Elementary mathematical ecology*. John Wiley and Sons, New York.

Vangen, K. M., Persson, J., Landa, A., Andersen, R. and Segerström, P. (2001). Characteristics of dispersal in wolverines. *Canadian Journal of Zoology*, **79**, 1641–1649.

Vashon, J. H., Vaughan, M. R., Vashon, A. D., Martin, D. D. and Echols, K. N. (2003). An expandable radiocollar for black bear cubs. *Wildlife Society Bulletin*, **31**, 380–386.

Vashon, J. H., Meehan, A. L., Organ, J. F. *et al.* (2008). Spatial ecology of a Canada lynx population in northern Maine. *Journal of Wildlife Management*, **72**, 1479–1487.

Velloso, A. L., Wasser, S. K., Monfort, S. L. and Diety, J. M. (1998). Longitudinal fecal steroid excretion in maned wolves (*Chrysocyon brachyurus*). *General and Comparative Endocrinology*, **112**, 96–107.

Vercillo, F., Lucentini, L., Mucci, N., Ragni, B., Randi, E. and Panara, F. (2004). A simple and rapid PCR-RFLP method to distinguishing *Martes martes* and *Martes foina*. *Conservation Genetics*, **5**, 869–871.

Verts, B. J. (1960). A device for anesthesizing skunks. *Journal of Wildlife Management*, **24**, 335–336.

Vial, F., Cleaveland, S., Rasmussen, G. and Haydon, D. T. (2006). Development of vaccination strategies for the management of rabies in African wild dogs. *Biological Conservation*, **131**, 180–192.

Vickers, T. W., Munson, L. and Garcelon, D. K. (2007). *Unusually high prevalence of ceruminous gland carcinomas in Channel Island foxes on Santa Catalina Island, California*. Abstract presented at the 2007 Wildlife Disease Association Meeting, Estes Park, CO.

Vickery, S. S. and Mason, G. J. (2003). Behavioural persistence and its implication for reintroduction success. *Ursus*, **14**, 35–46.

Voigt, G. L. (2000). *Hematology Techniques and Concept: fore veterinary technicians*. Iowa State University Press, Ames, Iowa.

Vucetich, J. A. and Peterson, R. O. (2004) The influence of prey consumption and demographic stochasticity on population growth rate of Isle Royale wolves *Canis lupus*. *Oikos*, **107**, 309–320.

Vucetich, J. A. and Peterson, R. O. (2008). *Ecological Studies of Wolves on Isle Royal—Annual Report 2007–2008*. Michigan Technical University, MI.

Vucetich, J. A., Peterson, R. O. and Schaefer, C. L. (2002). The effect of prey and predator densities on wolf predation. *Ecology,* **83**, 3003–3013.

Wabakken, P., Sand, H., Liberg, O. and Bjarvall, A. (2001). The recovery, distribution, and population dynamics of wolves on the Scandinavian peninsula, 1978–1998. *Canadian Journal of Zoology,* **79**, 710–725.

Wade-Smith, J., Richmond, M. E., Mead, R. A., and Taylor, H. (1980). Hormonal and gestational evidence for delayed implantation in the striped skunk, *Mephitis mephitis*. *General and Comparative Endocrinology,* **42**, 509–515.

Waits, L. P. (2004). Using genetic sampling to detect and estimate abundance of rare wildlife species. In W. L. Thompson (ed.). *Sampling rare or elusive species*, pp. 211–228. Island Press, Washington DC.

Waits, J. L. and Leberg, P. L. (2000). Biases associated with population estimation using molecular tagging. *Animal Conservation*, **3**, 191–199.

Waits, L. P. and Paetkau, D. (2005). Noninvasive genetic sampling tools for wildlife biologists: A review of applications and recommendations for accurate data collection. *Journal of Wildlife Management*, **69**, 1419–1433.

Waits, L. P., Luikart, G. and Taberlet, P. (2001). Estimating the probability of identity among genotypes in natural populations, cautions and guidelines. *Molecular Ecology,* **10**, 249–256.

Waits, L. P., Buckley-Beason, V. A., Johnson, W. E., Onorato, D. and McCarthy, T. (2007). A select panel of polymorphic microsatellite loci for individual identification of snow leopards (*Panthera uncia*). *Molecular Ecology Notes*, 7, 311–314.

Walker, M. L. and Herndon, J. G. (2008). Menopause in nonhuman primates? *Biology of Reproduction*, 79, 398–406.

Wallace, M. P. (2000). Retaining natural behaviour in captivity for re-introduction programmes. In L. M. Gosling and W. J. Sutherland (eds). *Behaviour and conservation*, pp. 300–313. Cambridge University Press, Cambridge.

Wallace, R. L. (2003). Social influences on conservation, lessons from U.S. recovery programs for marine mammals. *Conservation Biology*, 17, 104–115.

Waller, J. and Servheen, C. (2005). Effects of transportation infrastructure on grizzly bears in northwestern Montana. *Journal of Wildlife Management,* **69**, 985–1000.

Walston, J., Robinson, J. G., Bennett, E. L. *et al.* (2010). Bringing the Tiger Back from the Brink—The Six Percent Solution. *PLoS Biol,* **8**(9): e1000485.

Walters, J. R., Doerr P. D. and Carter, J. H., III. (1988). The cooperative breeding system of the red-cockaded woodpecker. *Ethology*, **78**, 275–305.

Walters, J. R., Copeyon, C. K. and Carter, J. H., III. (1992). Test of the ecological basis of cooperative breeding in red-cockaded woodpecker. *Auk*, **109**, 90–97.

Walton, L. R., Cluff, H. D., Paquet, P. C. and Ramsay, M. A. (2001a). Performance of 2 models of satellite collars for wolves. *Wildlife Society Bulletin*, **29**, 180–186.

Walton, L. R., Cluff, H. D., Paquet, P. C. and Ramsay, M. A. (2001b). Movement patterns of barren-ground wolves in the central Canadian Arctic. *Journal of Mammalogy*, **82**, 867–876.

Wambuguh, O. (2007). Interactions between humans and wildlife: landowner experiences regarding wildlife damage, ownership and benefits in Laikipia District, Kenya. *Conservation and Society,* **5**, 408–428.

Wang, S. W. and Macdonald, D. W. (2006). Livestock predation by carnivores in Jigme Singye Wangchuck National Park, Bhutan. *Biological Conservation,* **129**, 558–565.

Waples, R. S. (1991). Pacific salmon, *Oncorhynchus* spp., and the definition of "species" under the Endangered Species Act. *Marine Fisheries Review,* **53**, 11–22.

Warburton, B. (1982). Evaluation of seven trap models as humane and catch-efficient possum trap. *New Zealand Journal of Zoology,* **9**, 409–418.

Warburton, B., Poutu, N., Peters, D. and Waddington, P. (2008). Traps for killing stoats (*Mustela erminea*), improving welfare performance. *Animal Welfare,* **17**, 111–116.

Warren, J. J. and Mysterud, I. (2001). Mortality of lambs in free-ranging domestic sheep (*Ovis aries*) in northern Norway. *Journal of Zoology,* **254**, 195–202.

Washburn, B. E. and Millspaugh, J. J. (2002). Effects of simulated environmental conditions on glucocorticoid metabolite measurements in white-tailed deer feces. *General and Comparative Endocrinology,* **127**, 217–222.

Wasser, S. K., Risler, L., and Steiner, R. A. (1988). Excreted steroids in primate feces over the menstrual cycle and pregnancy. *Biology of Reproduction,* **39**, 862–872.

Wasser, S. K., Thomas, R., Nair, P. P. *et al.* (1993). Effects of dietary fibre on faecal steroid measurements in baboons (*Papio cynocephalus cynocephalus*). *Journal of Reproduction and Fertility,* **97**, 569–574.

Wasser, S. K., Davenport, B., Ramage, E. R. *et al.* (2004). Scat detection dogs in wildlife research and management: application to grizzly and black bears in the Yellowhead Ecosystem, Alberta, Canada. *Canadian Journal of Zoology,* **82**, 475–492.

Watson, A., Moss, R., Parr, R., Mountford, M. D. and Rothery, P. (1994). Kin landownership, differential aggression between kin and non-kin, and population fluctuations in red grouse. *Journal of Animal Ecology,* **63**, 39–50.

Watson, P. J. and Thornhill, R. (1994). Fluctuating asymmetry and sexual selection. *Trends in Ecology and Evolution,* **9**, 21–25.

Way, G. W., Ortega, I. M., Auger, P. J. and Strauss, E. G. (2002). Box-trapping eastern coyotes in southeastern Massachusetts. *Wildlife Society Bulletin,* **30**, 695–702.

Wayne, R. K., Modi, W. S. and O'Brien, S. J. (1986). Morphological variability and asymmetry in the cheetah (*Acinonyx jubatus*), a genetically uniform species. *Evolution,* **40**, 78–85.

Weaver, J. L. (1993). Refining the equation for interpreting prey occurrence in gray wolf scat. *Journal of Wildlife Management,* **57**, 534–538.

Weaver, J. L., Wood, P., Paetkau, D. and Laack, L. L. (2005). Use of scented hair snares to detect ocelots. *Wildlife Society Bulletin,* **33**, 1384–1391.

Webb, N., Hebblewhite, M. and Merrill, E. H. (2008). Statistical methods for identifying wolf kill sites in a multiple prey system using GPS collar locations. *Journal of Wildlife Management,* **72**, 798–807.

Weckwerth, R. P. and Hawley, V. D. (1962). Marten food habits and population fluctuations in Montana. *Journal of Wildlife Management,* **26**, 55–74.

Weese, J. S. and Jack, D. C. (2008). Needlestick injuries in veterinary medicine. *Canadian Veterinary Journal,* **49**, 780–784.

Wegge, P., Pokheral, C. P. and Jnawali, S. R. (2004). Effects of trapping effort and trap shyness on estimates of tiger abundance from camera trap studies. *Animal Conservation,* **7**, 251–256.

Wei, K., Zhang, Z. H., Zhang, W. P. *et al.* (2008). PCR-CTPP: a rapid and reliable genotyping technique based on ZFX/ZFY alleles for sex identification of tiger (*Panthera tigris*) and four other endangered felids. *Conservation Genetics*, **9**, 225–228.

Weir, R. D. and Corbould, F. B. (2007). Factors affecting diurnal activity of fishers in north-central British Columbia. *Journal of Mammalogy*, **88**, 1508–1514.

Weir, B. J. and Rowlands, I. W. (1973). Reproductive strategies of mammals. *Annual Review of Ecology and Systematic*, **4**, 139–163.

Wemmer, C. and Sunquist, M. E. (1988). Felid reintroductions, economic and energetic considerations. In H. Freeman (ed.). *Proceedings of the 5th International Snow Leopard Symposium*, pp. 193–205. International Snow Leopard Trust and The Wildlife Institute India, Delhi, India.

Wen, Z. (2009). *Superpopulation capture-recapture models augmented with genetic data.* PhD Thesis, Operations Research Graduate Program and Department of Statistics, North Carolina State University, Raleigh, NC.

Wen, Z., Pollock, K. H., Nichols, J. D. and Waser, P. (2010). Augmenting superpopulation, capture–recapture models with population assignment data. *Biometrics Online*, DOI: 10.1111/j.1541-0420.2010.01522.x

West N. H. and Van Vliet, B. N. (1986). Factors influencing the onset and maintenance of bradycardia in mink. *Physiological Zoology*, **59**, 451–463.

West, G., Heard, D. and Caulkett, N. (eds). (2007). *Zoo Animal and Wildlife Immobilization and Anesthesia.* Blackwell Publishing Professional, Ames, IO.

White, G. C. (2000). Population viability analysis: data requirements and essential analyses. In L. Boitani and T. K. Fuller (eds). *Research techniques in animal ecology: controversies and consequences*, pp. 288–331. Columbia University Press, New York.

White, G. C. and Burnham, K. P. (1999). Program MARK, survival estimation from populations of marked animals. *Bird Study*, **46**, S120–S139.

White, G. C. and Garrott, R. A. (1990). *Analysis of wildlife radio-tracking data.* Academic Press, New York.

White, G. C. and Shenk, T. M. (2001). Population estimation with radio marked animals. In J. J. Millspaugh and J. M. Marzluff (eds). *Design and analysis of radio telemetry studies*, pp. 329–350. Academic Press, San Diego, CA.

White, P. J., Kreeger, T. J., Seal, U. S. and Tester, J.R. (1991). Pathological response of red foxes to capture in box traps. *Journal of Wildlife Management*, **55**, 75–80.

White, P. C. L., Harris, S. and Smith, G. C. (1995). Fox contact behaviour and rabies spread: a model for the estimation of contact probabilities between urban foxes at different population densities and its implications for rabies control in Britain. *Journal of Applied Ecology*, **32**, 693–706.

Whitney, L. F. and Underwood, A. B. (1952). *The raccoon.* Practical Science Publishing Co., Orange, Connecticut, 177 pp.

Whittier, C. A., Cranfield, M. R. and Stoskopf, M. K. (2010). Real-time PCR Detection of *Campylobacter spp.* in free-ranging Mountain Gorillas (*Gorilla beringei beringei*). *Journal of Wildlife Diseases*, **46**, 791–802.

Whittington, J., St Clair, C. C. and Mercer, G. (2005). Spatial responses of wolves to roads and trails in mountain valleys. *Ecological Applications*, **15**, 543–553.

Wikelski, M. and Cooke, S. J. (2006). Conservation physiology. *Trends in Ecology and Evolution*, **21**, 38–46.

Wildt, D. E., Meltzer, D., Chakraborty, P. K. and Bush, M. (1984). Adrenal-testicular-pituitary relationships in the cheetah subjected to anesthesia/electroejaculation. *Biology of Reproduction*, **30**, 665–672.

Wildt, D. E., Howard, J. G., Hall, L. L. and Bush, M. (1986). Reproductive physiology of the clouded leopard: I. Electroejaculates contain high proportions of pleiomorphic spermatozoa throughout the year. *Biology of Reproduction*, **34**, 937–947.

Williams, B. K. (1997). Logic and science in wildlife biology. *Journal of Wildlife Management*, **61**, 1007–1015.

Williams, E. S. and Barker, I. K. (2001). *Infectious diseases of wild mammals*. Wiley-Blackwell, Boston, MS.

Williams, C. L. and Johnston, J. J. (2004). Using genetic analyses to identify predators. *Sheep and Goat Research Journal*, **19**, 85–88.

Williams, W. F., Osman, A. M., Shehata, S. H. M. and Gross, T. S. (1986). Pedometer detection of prostaglandin F2alpha-induced luteolysis and estrus in the Egyptian buffalo. *Animal Reproduction Science*, **11**, 237–241.

Williams, E. S., Thorne, E. T., Appel, M. J. G. and Belitsky, D. W. (1988). Canine distemper in black-footed ferrets (*Mustela nigripes*) from Wyoming. *Journal of Wildlife Diseases*, **24**, 385–398.

Williams, C. K., Ericsson, G. and Heberlein, T. A. (2002a). A quantitative summary of attitudes toward wolves and their reintroduction (1972–2000). *Wildlife Society Bulletin*, **30**, 575–584.

Williams, B. K., Nichols, J. D. and Conroy, M. J. (2002b). *Analysis and management of animal populations, modeling, estimation, and decision making*. Academic Press, San Diego, CA.

Williams, C. L., Blejwas, K. M., Johnston, J. J. and Jaeger, M. M. (2003). A coyote in sheep's clothing: predator identification from saliva. *Wildlife Society Bulletin*, **31**, 926–932.

Wilson, D. E. and Reeder, D. M. (eds). (2005). *Mammal species of the world. A taxonomic and geographic reference*, 3rd edn. Johns Hopkins University Press, Baltimore, MD.

Wilson, K. A., Westphal, M. I., Possingham, H. P. and Elith, J. (2005). Sensitivity of conservation planning to different approaches to using predicted species distribution data. *Biological Conservation*, **122**, 99–112.

Wilson, K. R., and Anderson, D. R. (1985). Evaluation of two density estimators of small mammal population size. *Journal of Mammalogy*, **66**, 13–21.

Wimsatt, W. A. (1963). Delayed implantation in the Ursidae with particular reference to the black bear (*Ursus americanus* Pallus). In A. C. Enders (ed.). *Delayed implantation*. Chicago, University of Chicago Press.

Wimsatt, W. A. (1975). Some comparative aspects of implantation. *Biology of Reproduction*, **12**, 1–40.

Windberg, L. A. (1995). Demography of a high density coyote population. *Canadian Journal of Zoology*, **73**, 942–954.

Windberg, L. A. (1996). Coyote responses to visual and olfactory stimuli related to familiarity with an area. *Canadian Journal of Zoology*, **74**, 2248–2254.

Windberg, L. A. and Knowlton, F. F. (1988). Management implications of coyote spacing patterns in southern Texas. *Journal of Wildlife Management*, **52**, 632–640.

Wingfield, J. C. (2005). The concept of allostasis: coping with a capricious environment. *Journal of Mammalogy*, **86**, 248–254.

Wingfield, J. C., Hegner, R. E., Dufty, J. A. M. and Ball, G. F. (1990). The "challenge hypothesis": theoretical implications for patterns of testosterone secretion, mating systems, and breeding strategies. *American Naturalist*, **136**, 829–846.

Wingfield, J. C., Hunt, K., Breuner, C. *et al.* (1997). Environmental stress, field endocrinology, and conservation biology. In J. R. Clemmons and R. Buchholz (eds). *Behavioral Approaches to Conservation in the Wild*, pp. 95–131. Cambridge University Press, Cambridge.

Winnie, J. and Creel, S. (2007). Sex-specific behavioural responses of elk to spatial and temporal variation in the threat of wolf predation. *Animal Behaviour*, **73**, 215–225.

Winstanley, R. K., Buttermer, W. A. and Saunders, G. (1999). Fat deposition and seasonal variation in body composition of red foxes (*Vulpes vulpes*) in Australia. *Canadian Journal of Zoology*, 77, 406–412.

Wintle, B. A., Kavanagh, R. P., McCarthy, M. A. and Burgman, M. A. (2005). Estimating and dealing with detectability in occupancy surveys for forest owls and arboreal marsupials. *Journal of Wildlife Management*, **69**, 905–917.

Wisely, S., Santymire, R. M., Livieri, T. M., Mueting S. A. and Howard, J. (2008). Genotypic and phenotypic consequences of reintroduction history in the black-footed ferret (*Mustela nigripes*). *Conservation Genetics*, **9**, 389–399.

With, K. A. (1994). Using fractal analysis to assess how species perceive landscape structure. landscape. *Ecology*, **9**, 25–36.

With, K. A. and Crist, T. O. (1996). Translating across scales, simulating species distributions as the aggregate response of individuals to heterogeneity. *Ecological Modelling*, **93**, 125–137.

Withey, J. C., Bloxton, T. D. and Marzluff, J. M. (2001). Effects of tagging and location error in wildlife radiotelemetry studies. In J. J. Millspaugh and J. M. Marzluff (eds). *Radio tracking and animal populations*, pp. 43–75. Academic Press, San Diego, CA.

Witt, H. V. (1980). The diet of the red fox. In E. Zimen (ed.). *The red fox Biogeographica* **18**, pp. 65–69. W. Junk B.V Publishers, Hague, Netherlands.

Wobeser, G. A. (1992). Traumatic degenerative and developmental lesions in wolves and coyotes from Saskatchewan. *Journal of Wildlife Diseases*, **28**, 268–275.

Wobeser, G. A. (1994). *Investigation and management of disease in wild animals*. Plenum Press, New York.

Wobeser, G. A. (1996). Forensic (medico-legal) necropsy of wildlife. *Journal of Wildlife Diseases*, **32**, 240–249.

Wobeser, G. A. (2007). *Disease in wild animals: investigation and management*. Springer Verlag, Berlin.

Wolfe, G. J. (1974). *Siren-elicited howling response as a coyote census technique*. M.S. thesis, Colorado State University, Fort Collins, CO.

Wolfe, T. R. and Bernstone, T. (2004). Intranasal drug delivery, an alternative to intravenous administration in selected emergency cases. *Journal of Emergency Nursing*, **30**, 141–147.

Wolff, J. O. (1989). Social behavior. In G. L. Kirkland and J. N. Layne (eds). *Advances in the study of Peromyscus (Rodentia)*, pp. 271–291. Texas Tech University Press, Lubbock, TX.

Wolff, J. O. (1993). Why are female small mammals territorial? *Oikos*, **68**, 364–369.

Wolff, J. O. and Van Horn, T. (2003) Vigilance and foraging patterns of American elk during the rut in habitats with and without predators. *Canadian Journal of Zoology*, **81**, 266–271.

Wong, D., Wild, M. A., Walburger, M. A. *et al.* (2009). Primary pneumonic plague contracted from a mountain lion carcass. *Clinical Infectious Diseases*, **49**, 33–38.

Wong, S. T., Servheen, C. W. and Ambu, L. (2004). Home range, movement and activity patterns, and bedding sites of Malayan sun bears *Helarctos malayanus* in the rainforest of Borneo. *Biological Conservation*, **119**, 169–181.

Woodford, M. H. and Rossiter, P. B. (1994). Disease risks associated with wildlife translocation projects. In P. J. S. Olney, G. M. Mace and A. T. C. Feistner (eds). *Creative conservation, interactive management of wild and captive animals*, pp. 178–198. Chapman and Hall, London.

Woodroffe, R. (2001). Strategies for carnivore conservation: lessons from contemporary extinctions. In J. L. Gittleman, S. M. Funk, D. W. Macdonald and R. K. Wayne (eds). *Carnivore conservation.* Cambridge University Press, Cambridge.

Woodroffe, R. and Frank, L. G. (2005). Lethal control of African lions (*Panthera leo*): local and regional population impacts. *Animal Conservation*, **8**, 91–98.

Woodroffe, R. and Ginsberg, J. R. (1998). Edge effects and the extinction of populations inside protected areas. *Science*, **280**, 2126–2128.

Woodroffe, R., Lindsey, P., Romanach, S., Stein, A. and Ranah, S. M. K. (2005a). Livestock predation by endangered African wild dogs (*Lycaon pictus*). in northern Kenya. *Biological Conservation*, **124**, 225–234.

Woodroffe, R., Bourne, F. J., Cox, D. R. *et al.* (2005b). Welfare of badgers (*Meles meles*) subject to culling, patterns of trap-related injury. *Animal Welfare*, **14**, 11–17.

Woodroffe, R., Donnelly, C. A., Cox, D. R. *et al.* (2006). Effects of culling on badger *Meles meles* spatial organization: implications for the control of bovine tuberculosis. *Journal of Applied Ecology*, **43**, 1–10.

Woodroffe, R., Frank, L. G., Lindsey, P. A., Ranah, S. M. K. O. and Romañach, S. (2007). Livestock husbandry as a tool for carnivore conservation in Africa's community rangelands: a case-control study. *Biodiversity and Conservation*, **16**, 1245–1260.

Woods, J. G., Paetkau, D., Lewis, D., MacLellan, B. N., Proctor, M. and Strobeck, C. (1999). Genetic tagging free-ranging black and brown bears. *Wildlife Society Bulletin*, **27**, 616–627.

Woodward, K. N. (2005). Veterinary pharmacovigilance. Part 4. Adverse reactions in humans to veterinary medicinal products. *Journal of Veterinary Pharmacology and Therapeutics*, **28**, 185–201.

Woolf, A. and Nielsen, C. (2002). *The Bobcat in Illinois.* Southern Illinois University, Carbondale, IL. (Mimeograph).

Woolfenden, G. E. and Fitzpatrick, J. W. (1984). *The Florida scrub jay. Demography of a cooperative-breeding bird.* Princeton University Press, Princeton.

Worton, B. J. (1989). Kernel methods for estimating the utilization distribution in home-range studies. *Ecology*, **70**, 164–168.

Wright, S. (1931). Evolution in Mendelian populations. *Genetics*, **16**, 97–159.

Wright, P. L. (1963). Variations in reproductive cycles in North American mustelids. In A. C. Enders (ed.). pp. 77–97. *Delayed implantation.* University of Chicago Press. Chicago.

Wright, P. L. (1966). Observations on the reproductive cycles of the American badger. *Symposium of the Zoological Society of London,* **15**, 27–45.

Wright, B. E. (2010) Use of chi-square tests to analyze scat-derived diet composition data. *Marine Mammal Science,* **26**, 395–401.

Wright, P. L. and Rausch, R. (1955). Reproduction in the wolverine (*Gulo gulo*). *Journal of Mammalogy,* **36**, 346–355.

Wright, J. A., Barker, R. J., Schofield, M. R., Frantz, A. C., Byrom, A. E. and Gleeson, D. M. (2009). Incorporating genotype uncertainty into mark-recapture-type models for estimating abundance using DNA samples. *Biometrics,* **65**, 833–840.

Wultsch, C. (2008). Molecular scatology—optimizing noninvasive genetic monitoring of jaguars (*Panthera onca*) and other elusive feline species in the tropics. *Wild Felid Monitor,* **2**, 21.

Yamaguchi, N., Dugdale, H. L. and Macdonald, D. W. (2006). Female receptivity, embryonic diapause, and superfetation in the European badger (*Meles meles*): Implications for the reproductive tactics of males and females. *Quarterly Review of Biology,* **82**, 33–48.

Yoccoz, N. G., Nichols, J. D. and Boulinier, T. (2001). Monitoring of biological diversity in space and time. *Trends in Ecology and Evolution,* **16**, 446–453.

Yoshizaki, J. (2007). *Use of natural tags in closed population capture-recapture studies, modeling misidentification.* PhD Thesis, Biomathematics Graduate Program and Department of Zoology, North Carolina State University, Raleigh, NC.

Yoshizaki, J., Brownie, C., Pollock, K. H. and Link, W. A. (2011). Modeling misidentification errors that results from use of genetic tags in capture-recapture studies. *Environmental and Ecological Statistics,* **18**, 27–55.

Yoshizaki, J., Pollock, K. H, Brownie, C., Webster, R. A. (2009). Modeling misidentification errors in capture-recapture studies using photographic identification of evolving marks. *Ecology,* **90**, 3–9.

Young, S. P. and Goldman, E. A. (1946). *The puma, mysterious American cat.* Dover Publications, New York.

Young, W. G. and Marty, T. M. (1986). Wear and microwear on the teeth of a moose *Alces alces* population in Manitoba Canada. *Canadian Journal of Zoology,* **64**, 2467–2479.

Young, J. K. and Shivik, J. A. (2006). What carnivore biologists can learn from bugs, birds, and beavers: a review of spatial theories. *Canadian Journal of Zoology,* **84**, 1703–1711.

Young, K. M., Walker, S. L., Lanthier, C., Waddell, W. T., Monfort, S. L. and Brown, J. L. (2004). Noninvasive monitoring of adrenalcortical activity in carnivores by fecal glucocorticoid analyses. *General and Comparative Endocrinology,* **137**, 148–165.

Young, A. J., Carlson, A. A., Monfort, S. L., Russell, A. F., Bennett, A. C. and Clutton-Brock, T. (2006). Stress and the suppression of subordinate reproduction in cooperatively breeding meerkats. *Proceedings of the National Academy of Sciences,* **103**, 12005–12010.

Young, A. J., Clutton-Brock, T. and Monfort, S. L. (2008). The causes of physiological suppression among female meerkats: a role for subordinate restraint due to the threat of infanticide? *Hormones and Behavior,* **53**, 131–139.

Zabel, A. and Holm-Müller, K. (2008). Conservation performance payments for carnivore conservation in Sweden. *Conservation Biology,* **22**, 247–251.

Zabel, J. and Taggart, S. J. (1989). Shift in red fox, *Vulpes vulpes*, mating system associated with El Niño in the Bering Sea. *Animal Behaviour*, **38**, 830–838.

Zhang, H., Li, D., Wang, C., and Hull, V. (2009). Delayed implantation in giant pandas: the first comprehensive empirical evidence. *Reproduction*, **138**, 979–986.

Zielinski, W. J. (2000). Weasels and martens—carnivores in northern latitudes. *Ecological Studies*, **141**, 95–118.

Zielinski, W. J. and Kucera, T. E. (1995). *American marten, fisher, lynx, and wolverine: survey methods for their detection*. General Technical Report PSW-GTR-157. Pacific Southwest Research Station, Forest Service, US Department of Agriculture. Albany, CA.

Zielinski, W. J., Schlexer, F. V., Pilgrim, K. L. and Schwartz, M. K. (2006). The efficacy of wire and glue hair snares in identifying mesocarnivores. *Wildlife Society Bulletin*, **34**, 1152–1161.

Zimmerman, J. W. (1992). *A habitat suitability model for black bears in the southern Appalachian region, evaluated for location error*. PhD Thesis. North Carolina State University, Raleigh, NC.

Zimmermann, B., Wabakken, P., Sand, H., Pedersen, H. C. and Liberg, O. (2007) Wolf movement patterns: a key to estimation of kill rate? *Journal of Wildlife Management*, **71**, 1177–1182.

Zoellick, B. W. and Smith, N. S. (1986). Capturing desert lit foxes at dens with box traps. *Wildlife Society Bulletin*, **14**, 284–286.

Zoellick, B. W. and Smith, N. W. (1992). Size and spatial organization of home ranges of kit foxes in Arizona. *Journal of Mammalogy*, **73**, 83–88.

Zweifel-Schielly, B. and Suter, W. (2007). Performance of GPS telemetry collars for red deer *Cervus elaphus* in rugged alpine terrain under controlled and free-living conditions. *Wildlife Biology*, **13**, 299–312.

Index

Note: Page numbers in *italic* refer to Figures and Tables.

The manufacturer's authorised representative in the EU for product safety is Oxford University Press España S.A. of El Parque Empresarial San Fernando de Henares, Avenida de Castilla, 2 - 28830 Madrid (www.oup.es/en or product.safety@oup.com). OUP España S.A. also acts as importer into Spain of products made by the manufacturer.
Printed and bound by CPI Group (UK) Ltd, Croydon, CR0 4YY

11/07/2026

02164418-0001